Musimathics

Musimathics
The Mathematical Foundations of Music

Volume 2

Gareth Loy

Foreword by John Chowning

The MIT Press
Cambridge, Massachusetts
London, England

First MIT Press paperback edition, 2011
© 2007 Gareth Loy

MIT Press books may be purchased at special quantity discounts for business or sales promotional use. For information, please e-mail <special_sales@mitpress.mit.edu.

This book was set in Times Roman by MPS Limited, a Macmillan Company
Printed and bound in the United States of America.

Library of Congress Cataloging-in-Publication Data

Loy, D. Gareth.
Musimathics : the mathematical foundations of music / Gareth Loy.
    p. cm.
Includes bibliographical references and indexes.
ISBN 978-0-262-12285-6 (v. 2: hc.: alk. paper)—978-0-262-12282-5 (v. 1:hc.: alk. paper)
ISBN 978-0-262-51656-3 (v. 2: pb.: alk. paper)—978-0-262-51655-6 (v. 1:pb.: alk. paper)
1. Music in mathematics education. 2. Mathematics—Study and teaching. 3. Music theory—Mathematics.
4. Music—Acoustics and physics. 5. Composition (Music). I. Title.

QA19.M87L69   2007
781.2—dc22

                                                                              2005051090

10 9 8 7 6 5 4

This book is dedicated to the memory of
Maxine Flora Reitz Loy
Harold Amos Loy
Thomas Harold Loy

Contents

Foreword

During the 1960s and 1970s the Artificial Intelligence Laboratory at Stanford University was a multidisciplinary facility populated by enormously gifted and dedicated workers trained in the sciences, engineering sciences, social sciences, and music. The musicians were part of the Center for Computer Research in Music and Acoustics (CCRMA) and shared the use of the A. I. Lab's computer. While the scientists had no professional interest in music, their scientific and technical knowledge was of critical importance to the musicians in learning to make effective use of the ever-evolving hardware and software. We had been seduced by Max Mathews's now famous statement, "There are no theoretical limitations to the performance of the computer as a source of musical sounds, in contrast to the performance of ordinary instruments."[1] Music was to enter an entirely new domain.

An implicit understanding within this community of individual users went something like this: Any question asked would be cheerfully and completely answered, to any level of detail . . . *once!* It was assumed that the questioner would then make the effort to follow up and fully comprehend the answer and all its implications before returning with another question. For the musicians, disciplined and well educated but most having little mathematical or scientific training and none in the use of computers, this was an opportune and accommodating intellectual environment.

Gareth Loy was one of us who in the mid-1970s, as a graduate student at CCRMA, made full use of the opportunities presented to him in this extraordinary environment. He assumed responsibility for writing the software for the Samson Box, by far the most powerful and complex digital synthesizer/processor of the day, named after its primary designer, Peter Samson.[2] Because Loy was trained as a composer, composition was his ultimate purpose, and on completing the software he composed *Nekyia,* a beautiful and powerful composition in four channels that fully exploited the capabilities of the Samson Box. As an integral part of this community, Loy paid back many times over all that he had learned, by conceiving the system with maximal generality such that it could be used for research projects in psychoacoustics as well as for hundreds of compositions by a host of composers having diverse compositional strategies. These accomplishments, both musical and technical, further revealed to Loy the profundity of his capabilities and led him to pursue his interest in this shared territory of music and mathematics, ultimately to the benefit of all of us.

The two volumes of *Musimathics* are a kind of instantiation of the process of learning that had such a powerful facilitating effect on the work at CCRMA in those years: explanations presented with wit and in great detail but here logically ordered and ever available, not just *once!* This second volume of *Musimathics* is comprehensive. Loy focuses on the digital domain, from elemental binary numbers through digital signal processing and synthesis to such heady topics as Gabor and acoustical quanta, all in terms of their mathematical underpinning and all clearly explained with elegant and illuminating graphics. Reflecting his intellectual journey—the questions, the answers, the study, and most important, the motivation: music!—but now with the wisdom from years of teaching and study, Loy is an extraordinarily gifted guide. Excellent texts inspire, and this one certainly does.

John Chowning

Preface

This second volume of *Musimathics* continues the story of music engineering begun in volume 1. It takes a deeper cut into the mathematics of music and sound, including

- Digital audio, sampling, binary numbers
- Complex numbers and how they simplify representation of musical signals
- Fourier transform, convolution, and filtering
- Resonance, the wave equation, and the behavior of acoustical systems
- Sound synthesis
- The short-time Fourier transform and the wavelet transform

The material in volume 1 was all preparatory to the subjects introduced in this volume, although this volume can certainly be read independently. Cross-references to volume 1 occur wherever there is an antecedent concept required in this volume. Additional mathematical orientation is provided as necessary.

Musimathics takes an uncommon approach to presenting mathematics. It cultivates the reader's common sense. I believe that enlightened common sense and inference are the whole of mathematics and that inference itself flows from enlightened common sense. The cure for any lack of mathematical preparation on the reader's part is simply to focus on what makes the most sense, and the rest will follow. This is my personal experience and a major premise of this book.

Inference without common sense leads nowhere. But to the naive reader, this is exactly where treatises on mathematics seem to lead. The problem is how mathematics is presented in print. If authors had to state explicitly all the assumptions that underlie an argument, even trivial mathematical assertions would be too long-winded to print. Instead, the commonsense foundations of mathematical arguments are assumed so that the focus can be placed on the interesting and possibly surprising inferences that are being reported. This means that *most of the common sense has been removed on purpose,* rendering this splendid and remarkable subject off-limits to the mathematically unprepared reader. The cure for this is to "rehydrate" the common sense back into it. By seeing the rationale alongside the mathematics, the reader can gain a deeper insight into both music and mathematics.

Musimathics provides two aids to the reader. The first is common sense about music and sound; the second is patience about the process of inference. *Musimathics* provides complete derivations of important concepts together with explanations of the steps. Breathtaking vistas can be opened up by starting from humble assumptions and climbing the ladder of inference. But it's easy to fall off the ladder if the reader misses a step. I know what that fall feels like, so I've labored to make the climb as secure and straightforward as possible.

The Web site http://www.Musimathics.com contains additional source material, animations, figures, and sources for other program examples in this book. Also, try saying "Musimathics" to your favorite Web browser and see what happens.

Acknowledgments

I am grateful for the loving support I have received from my wife, Lisa, and my family and friends over the decade it has taken to write *Musimathics*. Thanks to all, including Bernard Mont-Raynaud, Mark Dolson, Dana Massie, and Charles Seagrave for reviewing chapters of this volume. Thanks to Linda Graham and Barbara Cook Loy for inspiration and support.

I am continually grateful to all whose scholarship and insight have fed into the rich stream of knowledge that this book can at best sample and summarize. The enormous list of these individuals begins with the bibliography of this book and extends recursively through all the influences they cite. If there is anything to praise in this work, it is because it reflects these antecedents; if there is fault, it is mine alone.

In closing, let me express my heartfelt thanks to the *Musimathics* team at the MIT Press: Doug Sery and Valerie Geary (acquisitions editor and assistant), Deborah Cantor-Adams (production editor) and Alice Cheyer (freelance copyeditor), Sharon Deacon Warne (designer), Janet Rossi (production coordinator), Mary Reilly (graphics coordinator), and Patrick Ciano (cover designer). I am especially grateful to Doug Sery, whose clear vision and steady hand helped guide me from initial contact through completed project. His belief in the value of this effort has sustained me and helped make publication possible.

Gareth Loy
Corte Madera, California, October, 2006

Musimathics

1   Digital Signals and Sampling

The gods confound the man who first found out how to distinguish the hours! Confound him, too, who in this place set up a sundial, to cut and hack my days so wretchedly into small portions!
—Plautus

Digital audio has fundamentally changed the way music is made, distributed, and shared. It is now so pervasive that most of the music we hear is digitally stored and processed. A good deal of it is also created digitally.

The public has benefited enormously from the technological advances of digital audio, but at a price. Legal efforts to limit and regulate music copying have come about largely because of digital audio's ability to make perfect copies of recordings. Digital audio has become the proverbial lion in the pathway that our society must address in order to restore balance between artistic, commercial, and social aims. To fully understand its promise and pitfalls, we all—musicians, audio engineers, and listeners alike—must come to grips with this technology.

This chapter introduces the digital representation of signals, which (in the words of Plautus) consists of cutting and hacking time into very small portions. The focus is on intuitive understanding. The mathematical underpinnings of sampling theory are delayed until section 4.8.

1.1   Measuring the Ephemeral

The dimensions of a wooden board remain relatively static over time, but the dimensions of sound waves are transitory, changing from moment to moment. How are we to measure something so ephemeral? To begin with, let's consider the ocean's tides, which have the explanatory advantage over sound waves of changing slowly and being visible.

In order to record wave motion, we could construct a float attached to a pier in the ocean (figure 1.1). An angled bar attached to the float is pushed up and down by the waves. The rollers that connect the bar to the pier constrain it to travel only vertically. A pen mounted on the end of the bar leaves a mark on a piece of paper wrapped around a rotating drum. The apparatus monitors wave motion continuously—at every instant. The track mark created by the wave fluctuations is a *continuous function* of time.

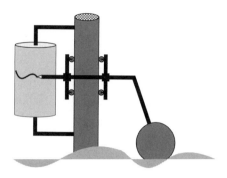

Figure 1.1
Float system for measuring waves.

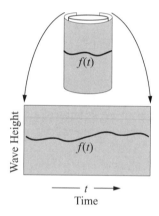

Figure 1.2
Unwrapping the paper from the drum.

Suppose we stop the drum after it has rotated once around, and unwrap the paper to examine the mark left on it by the pen (figure 1.2). Because of the paper's position on the drum, the x-axis represents the passage of time and the y-axis represents the fluctuating height of the waves.

If we let $f(t)$ represent the track mark, we can determine the height of the waves at any time t by evaluating $f(t)$ for the particular value of t that we wish to examine. For example, suppose it took 1 minute for the drum to revolve once, and the width of the paper is 1 meter. Then the height of the wave that occurred 30 seconds after we began is analogous to the height of the mark at 0.5 meter.

The function $f(t)$ is analogous in two senses: it analogizes time to place (on the x-axis), and wave height to place (on the y-axis). So $f(t)$ is an *analog function* of time. If $w(t)$ is the actual wave height, we can relate it to $f(t)$ by writing

$$f(t) \propto w(t),$$

where $\propto$ means "is proportional to." Analogies are very useful but can sometimes be misleading. For example, whereas real time flows inexorably forward, the paper analogy of time (represented by the variable t on the x-axis) is not similarly restricted. We can select any position along the x-axis, but we cannot return to the corresponding moment in real time. So the analogy is not perfect. We must remain alert to the limitations of analogies lest they confuse our thinking.

1.1.1   Sampling

An entirely different approach to measuring ocean waves is to sample wave height at periodic time intervals. Suppose once a day precisely at noon we go to a lagoon by the sea and look at a tide-measuring pole in the water showing the median height of the waves (figure 1.3). Numbered marks on the pole indicate the waves' height. We identify the mark that seems to be nearest to wave height, and record the measurements sequentially in a log book. After a month, we may end up with a list of measurements like those in figure 1.4.

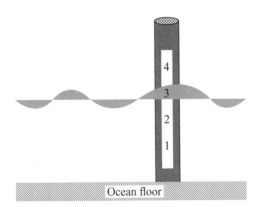

Figure 1.3
Sampling wave height.

January Tide Log	
Date	Height
Jan. 1	3
Jan. 2	4
Jan. 3	3
⋮	⋮
Jan. 29	2
Jan. 30	3
Jan. 31	1

Figure 1.4
Tide log.

Whereas the float recorder measures wave height continuously, sampling measures it discontinuously, "every so often." Taken together, the samples represent a *discrete function* of time. If the discrete function g consists of the samples in the tide log in figure 1.4, then $g(n)$ indexes the nth tide sample record, where n is an integer. For example, if $n = 1$, then by figure 1.4, $g(n) = 3$. Similarly, $g(2) = 4$, $g(3) = 3$, and so on.

Though they have similar mathematical notation, continuous functions like $f(t)$ and discrete functions like $g(n)$ are very different. Whereas a continuous function can be evaluated at any real-valued index, there is nothing in between the integer-valued indexes of a discrete function. Nothing at all. Whatever the waves were doing while we weren't measuring is lost forever. The only way to tell for sure if a function is discrete or continuous is to check the type of its argument: if it is an integer, the function is discrete; otherwise it is continuous. Throughout this book, if I don't state otherwise, assume that the index of a function is a real value, that is, taken from continuous measurements.

1.1.2 Sampling Rate

If a constant time interval T elapses between observations, the process is called *periodic sampling* with sampling period T. The expression nT, where n is an integer, corresponds to the moment when sample n was taken. The rate at which samples are taken, called the *sampling rate* or *sampling frequency* f_s, is the reciprocal of the sampling period:

$$f_s = \frac{1}{T}.$$ *Sampling Frequency* (1.1)

1.1.3 Capturing Frequency Information

Periodic sampling not only captures amplitude information about the changing height of the waves, it also captures frequency information. Increasing the sampling rate increases the highest frequency that can be recorded.

For example, suppose we wish to study the ebb and flow of the ocean's tides, the vertical rise and fall of the sea level surface that is caused primarily by the change in gravitational attraction of the moon and, to a lesser extent, the sun. Sampling the tide once a month is not frequent enough to get meaningful data because the tides have a period of about 24 hours, 50 minutes. If we want to capture useful tidal information, we are *undersampling*.

We are still undersampling if we sample once a day because the time between a low tide and subsequent high tide is somewhat more than 6 hours. If we sample at least four times a day, we start getting reasonable data about the flow of tides. Sampling every hour provides a better view but requires making more measurements (and storing more information). If we sample every minute, we are probably *oversampling* because the tide changes less than about 1 centimeter per minute at its fastest rate. If we sample every second, we are now recording the individual waves as they splash past the measuring pole.

Thus, increasing the sampling rate increases the highest frequency of fluctuation that can be tabulated.

1.1.4 Improving the Process

There are a couple of problems with the tide sampling method just outlined. When we look at the tide-measuring pole, a jumble of waves obscures our observation of the slow-changing average tide. Let's remove these extraneous waves so that they don't interfere with the measurements.

Lowpass Filtering We can create a sampling system to record only slow-changing tide fluctuations. Mount a hollow tube vertically in the ocean floor. Its bottom rests on the sea floor and is closed; its top rises above the highest tide and is open. Sea water can flow in and out of the tube through a small-diameter pipe that is attached to its side well below sea level (figure 1.5). The narrow pipe restricts the rate at which water flows into and out of the tube, preventing rapid changes in water level. Since the small pipe prevents rapid fluctuation of water level—and since frequency is proportional to rate of fluctuation—the small pipe blocks high-frequency wave energy from entering the tube. The small-diameter pipe acts as a *lowpass filter*, meaning it only allows low frequencies to enter the tube.

Sample-and-Hold Although the lowpass filter slows down the water's rate of fluctuation inside the tube, the water level is still constantly changing, albeit at a slower rate. If we could make the level inside the tube as stationary as possible during measurements, accuracy would improve. One last refinement takes care of this: install a hinged lid that can seal the small pipe shut on command. The system in figure 1.6 now functions as a *sample-and-hold*.

Ordinarily, the lid is left open so that water can flow in and out of the tube as before. When we wish to measure the tide, we close the lid and wait a while for any turbulence to die down so that the water level becomes constant. Now we can measure the water's height with confidence.

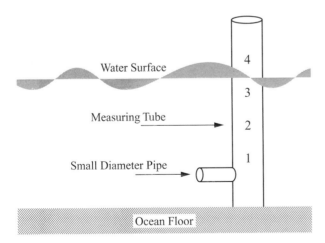

Figure 1.5
Lowpass filtering.

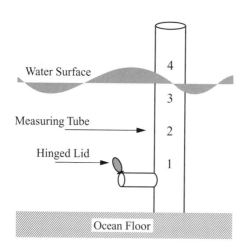

Water Surface

Measuring Tube

Hinged Lid

Ocean Floor

4

3

2

1

Figure 1.6
Sample-and-hold.

The sample-and-hold gets its name from its two states: sampling and holding. The hold state occurs when the valve is closed, while the tube holds the water level steady to be measured. During the time the valve is open, it is continuously sampling the height of the water level. The characteristic response of the sample-and-hold is shown in figure 1.7. The time required for the valve to close—to go from sampling to holding—is called the *aperture time* T_{ap}. The time required to go from holding back to sampling—from the moment the valve is opened again until the water level inside the tube is the same as the tide level outside—is called the *acquisition time* T_{acq}. The switching transients shown in the figure are the shudders sent through the system when the valve is suddenly opened or closed. If there is any leakage in the system when the valve is closed, the hold value will droop, as shown in the figure. A well-designed sample-and-hold will have little droop and mild transients.

A control function labeled Hold in figure 1.7 indicates the beginning of the aperture time and the beginning of the acquisition time. When the Hold function is low, the sample-and-hold is continuously sampling, and when it is high, the sample-and-hold is holding. Measuring the level and converting its height to a discrete value can commence anytime after the aperture time ends and before the acquisition time begins. Conversion is signaled by the Convert control function, as shown in the figure. It goes high in the middle of the hold period to trigger sample capture. The *conversion time* T_{conv} is the minimum sample time.

Figure 1.8 shows the operation of the sample-and-hold through several samples. The Hold and Convert signals are shown below the input and output functions. When the Hold function goes high, the sample-and-hold starts to hold its current output value. We see the output function of the sample-and-hold flatten and start to droop a little during the time Hold is high (marked with vertical lines). In the middle of the hold time, after the aperture time has elapsed, the Convert signal goes high. During that time (marked with boxes) the magnitude of the output function is converted. When

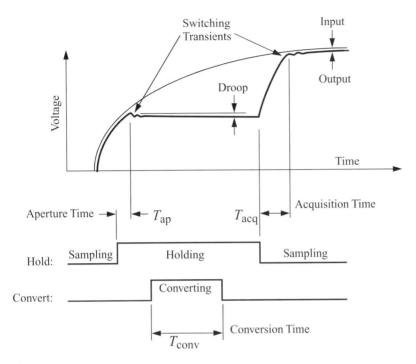

Figure 1.7
Sample-and-hold response characteristics. Adapted from Ramsay (1996).

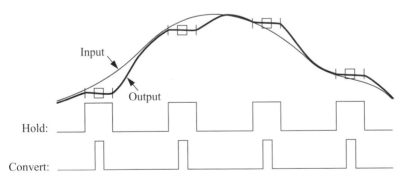

Figure 1.8
Sample-and-hold operation.

Hold goes low again, the sample-and-hold starts to acquire the input function again, catching up with it before Hold is reasserted.

Slew Rate   Notice in figure 1.8 that the output is relatively slow to catch up with the input after the Hold signal releases. In terms of the tide-measuring tube, the water level in the measuring tube requires some time to return to the level of the surrounding water. This is because the small pipe restricts the *slew rate* of water flow in and out of the pipe. The maximum slew rate is an indication of how steep a slope the output function can achieve; hence it is a measure of the maximum rate of change of the output signal. Slew rate is the ratio of *rise* to *rise time*. The *rise* usually covers the distance from 10–90 percent of maximum excursion; the *rise time* is the time required to cover that distance.

If r_0 is the 10 percent level and r is the 90 percent level, then the rise is $\Delta r = r - r_0$, and the slew rate S is

$$S = \frac{\Delta r}{t_r},$$ *Slew Rate* (1.2)

where t_r is the rise time.

For the tide-measuring system, we might measure slew rate in millimeters per second. For an electronic circuit, we might measure millivolts per microsecond.

We can see how good a sample-and-hold is at tracking the input signal by making a ratio of input slew rate S_i to output slew rate S_o. If

$$\frac{S_i}{S_o} > 1,$$

the input function can change faster than the output function, and the system is *slew-rate-limited*. Lowpass filters are slew-rate-limited by definition: they are designed to retard the rate of change of the input signal. Other systems, such as high-fidelity amplifiers, are designed to have extremely high slew rate ratios so as to retain utmost fidelity to the input signal.

Real Time   The sample-and-hold's maximum sustainable rate of operation depends on the minimum time required for the conversion operation. The aperture time T_{ap}, conversion time T_{conv}, and acquisition time T_{acq} are fixed by the design of the sample-and-hold. Since these times must be sequential, the minimum sampling time of the sample-and-hold is $T_{min} = T_{ap} + T_{conv} + T_{acq}$. The sampling period T must be greater than or equal to T_{min} to operate reliably,

$$T_{min} \leq T.$$ *Timely Sampling Criterion* (1.3)

The maximum sustainable sampling rate is

$$f_{max} = \frac{1}{T_{min}} \text{ samples per second.}$$ *Fastest Possible Sampling Rate* (1.4)

Conversely, if at least R samples per second are required, then we must have $T_{min} \leq 1/R$.

An operation takes place in *real time* only if it can perform its task in a timely manner with respect to the larger dynamical system of which it is a part. So long as the sample period T is longer than the shortest sample-and-hold time T_{min}, the sample-and-hold can run in real time. But if the "timely manner" criterion is violated, the sample-and-hold cannot complete its operations before the next sample must begin. Its operation becomes unstable and ceases to function in real time.

Real-time operation is crucial to all forms of music. Consider what happens when a performer sight-reading a sheet of music too fast can't keep up with the beat: at best, errors creep into the performance and at worst the music stops. The music also stops at least momentarily if a music-playing device can't retrieve data from its memory in time. Music is a time-based art form that depends upon a subtle interplay of psychological anticipation and expectation, and if this expectation is not met, music vanishes. For this reason, music performance and all forms of audio recording and playback are referred to as *hard-deadline real-time* tasks, because they suffer critical failure if real-time constraints are violated.[1]

Some forms of hard-deadline real-time failure are more egregious than others. Performers who skip a beat simply pick up where they left off; although there is interruption, there is little loss of information. But a premium is placed on the reliability of data recorders: if a measured system can't be monitored in a timely manner, any missed sampling opportunities are irrecoverably lost. Interruption of data playback may or may not be a catastrophe, but interruption of recording is an irredeemable catastrophe because any data missed are lost irretrievably. "Time waits for no man." The best one can do is try to record again, if appropriate.

Real-time operations don't necessarily have to be fast, but they do have to be fast enough to keep up with the dynamical system that they are monitoring. For example, the tide-sampling system discussed earlier only requires sampling operations on the order of six to eight times a day for reasonable real-time operation. Sampling for audio and music must be done much more rapidly, on the order of 50 thousand times a second.

1.2 Analog-to-Digital Conversion

In general, tide sampling involves these elements:

- *Lowpass filtering*   Removing unneeded high-frequency energy.
- *Synchronization*   Waiting for the right moment to sample.
- *Sample-and-hold*   Stabilizing wave height to facilitate measurement.
- *Discretization*   Observing where the tide line hits the measuring pole.
- *Quantization*   Rounding to the nearest mark on the pole.
- *Recording*   Storing the observations in a time-ordered sequence.

Together these elements are referred to as *analog-to-digital conversion* (ADC). Audio sampling uses the same basic process as tide sampling, but the sampling rate is much faster. In order to capture

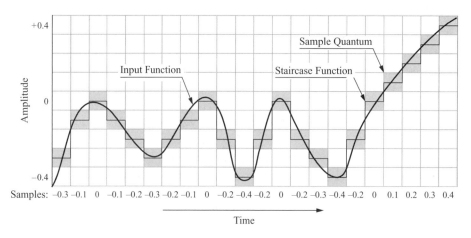

Figure 1.9
Quantization of a signal in time and amplitude.

the band of frequencies of interest to human listeners, digital audio recorders must sample inputs on the order of once every 20 microseconds, achieving sampling rates on the order of 50,000 samples per second.

Figure 1.9 shows how these elements are combined. The smoothly curved line represents the continuous analog input signal after lowpass filtering. The staircase function represents the output of the sample-and-hold. The columns represent the sample periods, and the rows represent the sample magnitudes. The individual squares of the grid in figure 1.9, called *quanta*, demarcate the boundaries of time and amplitude for each possible measurement. The shaded boxes show which quanta correspond to the input function.

A sample is really just an indication of which quantum the input function passed through. The only information that remains after the sampling is finished is the list of quanta that the input function visited. In the end, we have no other information about the sampled function than this.

In figure 1.9 the quanta are relatively large and provide a very low-resolution picture of the trajectory of the input function. The resolution can be improved by reducing the area occupied by the individual quanta, which means either increasing measurement precision (if possible) or increasing the sampling rate (if possible) or both. In terms of figure 1.9, doubling the precision would halve the height of each row; doubling the sampling rate would halve the width of each column. Doing either would increase the amount of information to be stored or transmitted: doubling the precision doubles the number of quanta per unit of magnitude, and doubling the sampling rate doubles the number of samples per unit of time. Reducing the area of the quantum increases the fidelity of the sampling process at the cost of increasing the information storage and transmission requirements. Determining an appropriate quantum size for audio recording depends upon the sensitivity of the ear and the available storage or transmission technology (see section 1.11).

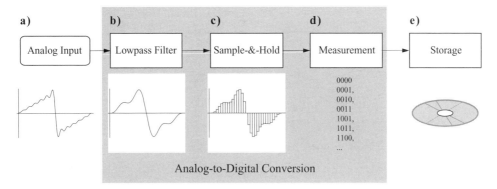

Figure 1.10
Analog-to-digital conversion.

The ADC itself imposes limits on sampling rate and sampling precision. Equation (1.4) shows that the maximum sampling rate depends upon the design of the sample-and-hold and conversion logic. There are also practical limits on the precision of the ADC: electrical circuits have a certain amount of inherent noise, which limits their useful sensitivity. Precision much in excess of the noise floor is of little use because what would be measured would be mostly noise.

The analog-to-digital conversion process for audio is summarized in figure 1.10. The input (a) may come from any analog source, such as a microphone signal. The lowpass filter (b) removes unwanted high-frequency energy. Synchronized by a clock, the sample-and-hold (c) captures the current instantaneous value of the filtered input signal for measurement (d). The measurements are converted and stored sequentially on a digital recording medium (e) such as computer memory or disk storage.

1.3   Aliasing

Sampling discards all information about what happens in between samples. This raises an important issue about capturing frequency information digitally. To understand the issue, suppose we turn a bicycle upside down so that its wheels can freely rotate. Now paint one spoke red, and set the wheel spinning counterclockwise at a rate of 1 Hz. Then turn out the lights and set a strobe light to blink once per second. The first time it blinks, the wheel is at 0 radians (horizontal and pointing to the right). If the wheel is traveling at exactly 1 Hz and the strobe blinks exactly at 1-second intervals, the position of the red spoke will appear to be stationary. (For this thought experiment, ignore the effect of friction, which would otherwise slow down the wheel.) One second later the red spoke has made a full circle, so we see it at 2π radians, which is equivalent to 0 radians. One second later it has made another full circle and is now at 4π radians, which is also equivalent to 0 radians,

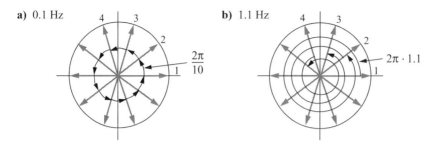

Figure 1.11
Ten snapshots of a 0.1 Hz phasor.

and in general at time $t = 0, 1, 2, \ldots$ seconds, the position of the wheel will appear to be at $2\pi t \equiv 0$ radians, that is, it will appear to be stationary.

Now spin the wheel at 2 Hz counterclockwise with the strobe still blinking at 1-second intervals. The sequence of angles when the strobe flashes will be $0\pi, \ 4\pi, \ 8\pi, \ldots$, and the red spoke still looks to be stationary because at time $t = 0, 1, 2, \ldots$ seconds, $4\pi t \equiv 0$ radians.

Now imagine spinning the wheel at 0.1 Hz counterclockwise with the strobe still set to blink at 1-second intervals. The angular velocity is $2\pi/10$ radians per second, so the red spoke takes 10 seconds to rotate once. We'd see the sequence of images shown in figure 1.11a as the wheel rotates.

Now, if the wheel rotates at 1.1 Hz, the red spoke goes through 1.1 revolutions per second, (figure 1.11b). Interestingly, when the strobe flashes at 1-second intervals, we see exactly the same sequence of points for 1.1 Hz as we saw for 0.1 Hz.

This is not just an optical illusion. *There is no way for us to tell the difference* without turning on a light, allowing us to see the continuous motion of the wheel. (We won't be able to tell the difference only if the strobe is really instantaneous. If the aperture time of the strobe light is long, the red spoke will appear blurrier at faster radian velocities.) An uninformed viewer would be unable to tell what the "real" frequency of the wheel is. In fact, from the viewer's perspective, the correct answer could be any of 0.1 Hz, 1.1 Hz, 2.1 Hz, or in general, $n + 0.1$ Hz for $n = 0, 1, 2, \ldots$.

Think about it. The expression $n + 0.1$ evaluated for every $n = 0, 1, 2, \ldots$ is an infinite series. So there is an infinite number of possible frequencies to choose from: 0.1 Hz, 1.1 Hz, 2.1 Hz, and so on. If the only information about the frequency of the rotating wheel is what we can observe while the strobe lamp is blinking, then we can't know which of the infinitely many choices is the correct frequency. It certainly is reasonable to assume that the wheel is actually rotating only at one frequency, but if we can watch only with the strobe light blinking, then all other frequencies related by $n + 0.1$ Hz for $n = 0, 1, 2, \ldots$ are equivalent as far as our observations are concerned. Since one choice is just as good as another, the set of all these frequencies are *aliases* of each other. The phenomenon we are observing is called *aliasing,* or *foldover.*

Now have a motor gradually accelerate the bicycle wheel as it turns counterclockwise while the strobe light continues to blink at a constant rate of once per second. As the angular velocity goes from 0 Hz to 0.5 Hz, we see the distance between successive strobe light images of the red spoke increasing. We can convince ourselves that it is still turning counterclockwise, picking up speed. When the speed reaches exactly 0.5 Hz, a new phenomenon arises, and we begin to see the red spoke alternating right and left.

As the counterclockwise speed continues to increase, it seems as if the red spoke gradually starts to turn clockwise. If we turn on a steady light source, we are easily satisfied that this is not true, but when the light is off, the strange effect takes over again. As the counterclockwise speed of the wheel continues to increase, it seems that the clockwise rotation of the red spoke is slowing down, even though we know the wheel is turning counterclockwise at ever-higher speed. When the speed of the wheel reaches exactly 1 Hz, it appears (as we have previously observed) that the red spoke has stopped moving. As the speed continues to increase above 1 Hz, it again appears that the red spoke starts turning counterclockwise with increasing speed. If we continued to watch, this pattern would repeat itself over and over as the wheel gains speed. The same effect occurs if the wheel spins clockwise at an increasing rate. The only difference is that the apparent motion of the red spoke is reversed. Thus, the effect is inverted for negative frequencies (clockwise rotation). The classic example of this is to watch wagon wheels turning in old Western movies. As the wagon picks up speed, the wheels appear to speed up, slow down, and change direction in contradiction to the movement of the wagon.

We've seen that for a strobe interval of 1 second, the apparent frequency matched the actual frequency of the wheel's rotation only in the frequency range of 0 Hz to 0.5 Hz (counterclockwise rotation) and in the negative frequency range 0 Hz to –0.5 Hz (clockwise rotation). If the sampling rate of the strobe light is increased to 10 blinks per second, we start experiencing aliasing when the wheel rotates faster than 5 Hz or –5 Hz. Generally, aliasing occurs at frequencies greater than or equal to $\pm 1/2$ the sampling rate. Conversely, the valid frequencies—the ones where the apparent frequency matches the actual frequency—are all less than $\pm 1/2$ the sampling rate. Because of the way sampling works, all frequencies beyond $\pm 1/2$ of the sampling rate are said to *alias* back into this range. The range of valid frequencies—that is, all frequencies in the range $\pm 1/2$ of the sampling rate—is called the *baseband*. The frequencies in the baseband are called the *principal frequencies* to distinguish them from the aliased frequencies.

1.3.1 Nyquist Sampling Theorem

In general, for sampling frequency f_s, all frequencies f within the range $-f_s/2 < f < f_s/2$ Hz are not aliased and all frequencies outside this range are aliased. The aliasing effect occurs because frequencies outside this range are *indistinguishable* from frequencies within this range because of the way sampling affects observation.

Aliasing is a consequence of the *Nyquist sampling theorem,* named after Harry Nyquist, who worked out the mathematics of sampling theory in the 1930s.[2] To honor his work, the proper term for the frequency aliasing limit of $\pm f_s/2$ is the Nyquist barrier, Nyquist limit, or *Nyquist frequency* (Nyquist 1928). The Nyquist frequency is the barrier where aliasing begins.

Let us design a formula that expresses the relation we have observed between actual frequency, apparent frequency, and sampling rate. In the example of the bicycle wheel, let us define f as the actual frequency of the spinning wheel and define f_a as the apparent frequency, the rate at which we observe the wheel to be spinning.

First, we observed that when actual frequency is within the baseband, the actual frequency and apparent frequency are equal, and $f = f_a$. But when f lies outside the baseband, f_a increases and decreases in a nonuniform way in response to uniform changes in f because of aliasing. We must find a way to relate f and f_a when f lies outside the baseband that doesn't invalidate the simple case when f lies inside it.

As the experiment with the bicycle wheel showed, actual frequency f can range over $\pm\infty$, while apparent frequency remains captured in the baseband, so that $-f_s/2 < f_a < f_s/2$. To construct a formula for apparent frequency f_a that models what we've observed, we could offset actual frequency f like this:

$$f_a = f + kf_s,$$

but the value of k would have to change depending on the value of f to keep f_a within the baseband. We can express this requirement by placing conditions on the formula. The general form of a conditional equation is $a = b|_P$ which means a equals b only if P is true. In our case, the condition P is that apparent frequency f_a must remain in the baseband regardless of f, in other words, P is true if and only if $-f_s/2 < f_a < f_s/2$. Combining these ideas, we can write

$$f_a = f + k(f)f_s\big|_{-f_s/2 < f_a < f_s/2},$$

where $k(f)$ is an integer function of f that we must invent to ensure the condition $-f_s/2 < f_a < f_s/2$ remains true regardless of the value of f or f_s.

All we need now is to define function $k(f)$. Let's look at an example. If f lies within the baseband, then $k = 0$ because the apparent frequency equals the actual frequency, and the equation above reduces to $f_a = f$. If f exceeds the Nyquist frequency ($f \geq f_s/2$) then k must compensate by becoming -1 so that $f_a = f - f_s$, causing f_a to remain in the baseband at the aliased frequency. If $f \leq -f_s/2$, then k becomes $+1$ to ensure the condition remains true. In general, as f goes above integer multiples of the Nyquist frequency, k follows the sequence $k = -1, -2, -3, \ldots$. Similar reasoning holds for negative frequencies. Let us define $k(x)$ as follows:

$$k(x) = -\left\lfloor x + \frac{1}{2} \right\rfloor. \tag{1.5}$$

The minus sign reminds us that as the frequency goes up, k must go down to compensate, and vice versa.[3] Figure 1.14a shows a plot of this function over the range $-2, 2$. With this definition for k, we can write the formula for aliased frequency as follows:

$$f_a = f + k\left(\frac{f}{f_s}\right)f_s. \qquad\qquad \textit{Apparent Frequency} \tag{1.6}$$

Consider a sampling frequency $f_s = 100$ Hz and a frequency $f = 60$ Hz. Because $f > f_s/2$, aliasing will occur. By equation (1.5), we have $k = -\lfloor (60/100) + (1/2) \rfloor = -1$, and by equation (1.6), the apparent frequency is therefore $f_a = 60 - (1 \cdot 100) = -40$ Hz. The fact that the result is a negative frequency agrees with the observation that the bicycle wheel (which in this example was actually turning counterclockwise) appeared to be turning clockwise (corresponding to negative frequencies) when the actual frequency f was in the range $f_s/2 < f < f_s$. (For more discussion of positive and negative frequencies, see section 2.6.7.)

Figure 1.12 illustrates aliasing. The vertical lines indicate sample boundaries. Figure 1.12a shows a cosine wave at 1/8 of the sampling rate. The wave is *oversampled* in the sense that we could

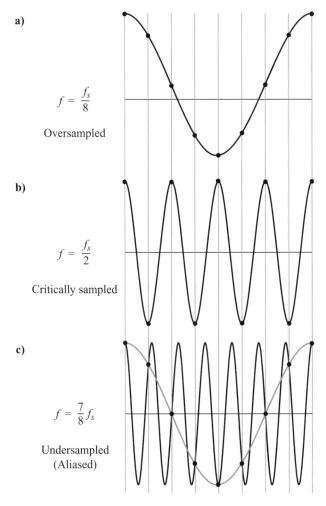

a)

$f = \dfrac{f_s}{8}$

Oversampled

b)

$f = \dfrac{f_s}{2}$

Critically sampled

c)

$f = \dfrac{7}{8} f_s$

Undersampled
(Aliased)

Figure 1.12
Aliasing.

a)

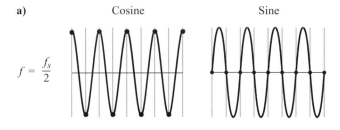

$$f = \frac{f_s}{2}$$

b)

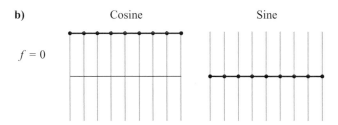

$$f = 0$$

Figure 1.13
Cosine and sine at the Nyquist frequency and at zero.

sample less often and still capture its characteristic features. Figure 1.12b shows a cosine wave at 1/2 the sampling rate. It is *critically sampled* because there are just enough samples to capture it. Figure 1.12c shows a cosine wave at 7/8 of the sampling rate aliased to 1/8 of the rate. It is *under-sampled* because there aren't enough samples to accurately represent it.

Notice in figure 1.13a that a cosine wave with frequency equal to the Nyquist frequency and zero phase offset has positive energy. This is because the waveform is sampled when its peaks line up with the sample boundaries. But a sine wave at the Nyquist frequency with zero phase offset has no energy because the zero crossings line up with the sample boundaries. Similarly, in figure 1.13b, at 0 Hz the cosine "wave" is maximal (a constant 1) and the sine "wave" is minimal (a constant 0).

If we were to compute the apparent frequency f_a for all values of f in the range $\pm 4f_s$ we'd see the function of apparent frequency shown in figure 1.14b. This figure reveals all the features of the bicycle wheel experiment: how the red spoke speeds up, blinks left/right, then switches to a negative frequency, slows to zero, then starts all over again. We see that 0 Hz and the Nyquist frequency act like a pair of reflectors, trapping the apparent frequency in a kind of hall of mirrors, a range of frequencies from which it can never escape.

1.3.2 Consequences of Aliasing

Aliasing has a very dramatic and usually very bad effect on digital audio recording. This is best illustrated with an example. Suppose an ADC is running at a sample rate of $f_s = 10$ kHz, and we are recording a violin tone with a fundamental frequency of 750 Hz. The Nyquist frequency is

a)

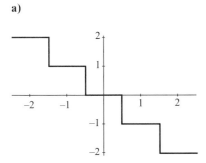

b)

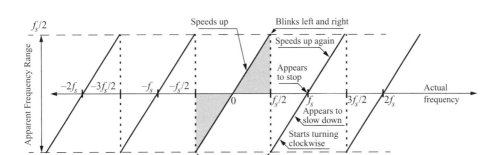

Figure 1.14
Aliasing and the Nyquist frequency barrier.

5 kHz. The violin's harmonics are 750, 1500, 2250, 3000, 3750, 4500, 5250 6000, 6750, 7500, 8250, 9000, 9750 Hz as shown in figure 1.15a. But all frequencies above 5000 Hz are aliased according to equation (1.6). For example, the apparent frequency of the spectral component at 5250 Hz is calculated as $k = -\lfloor(5{,}250/10{,}000) + 0.5\rfloor = -1$, and therefore

$$f_a = 5{,}250 + (-1)10{,}000 = -4{,}750 \text{ Hz.}$$

But there's a problem: −4750 Hz is not in the harmonic series of the violin tone. In fact, none of the other aliased components are in the violin tone's harmonic sequence. The result is that inharmonic components are added to the violin tone by aliasing, distorting the tone. Let us see how this comes about. Here are the harmonics of the violin tone (f) and their aliased frequencies (f_a) according to equation 1.6, assuming the Nyquist frequency is 5000 Hz:

f:	750	1500	2250	3000	3750	4500	5250	6000	6750	7500	8250	9000	9750
f_a:	750	1500	2250	3000	3750	4500	−4750	−4000	−3250	−2500	−1750	−1000	−250

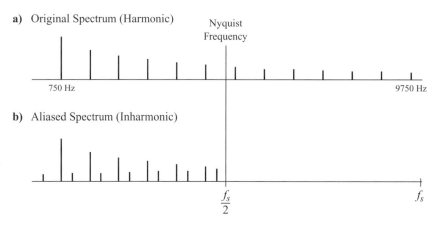

a) Original Spectrum (Harmonic)

750 Hz

9750 Hz

b) Aliased Spectrum (Inharmonic)

Figure 1.15
Aliased spectrum of harmonic tone.

The ear does not distinguish negative frequency tones from positive ones (see volume 1, chapter 6). For circular motion, the eye can distinguish negative and positive frequencies because the direction of rotation is different. But for wave motion, the only difference is whether the wave rises then falls, or falls then rises (see volume 1, chapter 5), which the ear ignores. So a harmonic at –4750 Hz sounds exactly the same to the ear as a harmonic at 4750 Hz. Thus, in order to determine the spectrum that the ear will actually hear, we must neglect the sign of the aliased harmonic frequencies. That means the ear will hear these frequencies:

f_a:	750	1500	2250	3000	3750	4500	4750	4000	3250	2500	1750	1000	250

Rearranging them into ascending order:

f_a:	250	750	1000	1500	1750	2250	2500	3000	3250	3750	4000	4500	4750

This sequence is shown in figure 1.15b. Clearly, this set of frequencies is no longer a harmonic spectrum because the components are not integer multiples of the fundamental. Figure 1.15 shows why aliasing is sometimes called foldover. If you printed figure 1.15a on a piece of paper and folded it over at the Nyquist frequency, you would obtain the order of components in figure 1.15b. Aliasing has reflected all components over 5000 Hz back into the baseband at nonharmonic frequencies, introducing a large amount of harmonic distortion into the violin recording.

The ADC can't discriminate and reject aliased components any better than we could when trying to identify the frequency of the rotating wheel under a strobe light. What is the solution to this problem?

Fortunately, the lowpass filter is a technology that allows us to discriminate against unwanted frequencies above a certain limit. We must lowpass-filter any signal we wish to record to remove all positive and negative frequencies outside the Nyquist frequency limit. Otherwise, components outside this range will alias, and corrupt the recording. Lowpass filters designed for this purpose are called *anti-aliasing filters*.

To summarize, it is crucial that digitized signals receive proper anti-aliasing filtering before conversion. Aliased components are impossible to remove once they are converted because they are all folded into the baseband wherever they happen to fall. Only signals that are properly band-limited so all components are less than the Nyquist rate will be safe from aliasing. Non-band-limited waveforms, such as the so-called geometric waveforms (square, triangle, impulse, sawtooth, see chapter 9) or any signal with energy above the Nyquist rate, will *always* create aliasing.

Aliasing rears its ugly head most commonly in two specific applications: audio recording and sound synthesis. For audio recording, so long as the input signal receives proper anti-aliasing filtering, all should be well.

1.3.3   Sampling Rate and Radian Velocity

Let's return to the bicycle wheel experiment. When the bicycle wheel is spinning at a rate equal to the sampling rate, it is turning exactly one whole revolution between samples (2π radians per second), which is why the red spoke appears stationary. When the bicycle wheel is spinning with a frequency equal to the Nyquist rate, it is turning exactly 1/2 of a rotation between samples, and its angular velocity is therefore π radians per second.

So there is an equivalence between the sampling rate and angular velocity: frequency f is to the Nyquist frequency $f_s/2$ as the corresponding angular velocity θ is to π. In other words,

$$\frac{f}{f_s/2} = \frac{\theta}{\pi},$$

where θ is the angular velocity of the radial motion. In particular, the Nyquist frequency barrier $\pm f_s/2$ Hz corresponds to $\pm\pi$ radians per second. We can restate the Nyquist sampling theorem, substituting angular velocity for frequency.

Only those angular velocities θ that satisfy the inequality $-\pi < \theta < \pi$ are not aliased; all other angular velocities are aliased.

Relating these terms allows us to study aliasing via the relation $\theta : \pi$ instead of having to use the relation $f : f_s/2$. That way we can investigate frequency measurements without having to specify the rate f_s at which the signal was sampled, simplifying things enormously. This is sometimes called *normalized frequency*. In this way of thinking, π is equivalent to the Nyquist frequency barrier $f_s/2$, and θ varies between $\pm\pi$ in the same way that frequency f varies between the Nyquist frequency barriers $\pm f_s/2$.

Here are some useful equivalencies between frequency and phase angle. For any arbitrary sampling rate f_s, the frequency f corresponding to phase angle θ is as follows.

Phase Angle	Corresponding Frequency	Phase Angle	Corresponding Frequency
$\theta = \pi$	$f = \dfrac{f_s}{2}$	$\theta = -\pi$	$f = -\dfrac{f_s}{2}$
$\theta = \dfrac{\pi}{2}$	$f = \dfrac{f_s}{4}$	$\theta = -\dfrac{\pi}{2}$	$f = -\dfrac{f_s}{4}$
$\theta = \dfrac{\pi}{4}$	$f = \dfrac{f_s}{8}$	$\theta = -\dfrac{\pi}{4}$	$f = -\dfrac{f_s}{8}$

1.4 Digital-to-Analog Conversion

If we think of a sample sequence as a "dehydrated" version of the original analog signal, we "rehydrate" it to recover the original analog waveform using *digital-to-analog conversion* (DAC) (figure 1.16). It looks rather like analog-to-digital conversion done backwards.

Samples are retrieved from storage in the same order and at the same rate that they were recorded (a). The voltage corresponding to the magnitude of each sample is reconstructed (b). A sample-and-hold circuit stabilizes and sustains each reconstructed voltage for the duration of the sample period. The result is a staircase function (c). A lowpass filter smooths away the sharp edges of the stair steps, restoring the wave to its former band-limited shape (see figure 1.10b).

The sample reconstruction process can be likened to combining weights of various heft on a weighing scale to achieve a particular weight. For instance, if we have weights 1, 2, 4, and 8 g and need 5 g, we would combine the 4-g and 1-g weights (figure 1.17). Since the weight holders in that

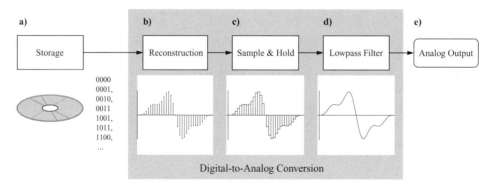

Figure 1.16
Digital-to-analog conversion.

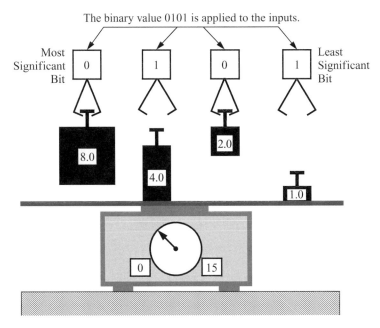

Figure 1.17
Sample reconstruction by combining weights.

figure have but two positions, open and closed, it is convenient to use binary digits which have but two values, 0 and 1, to represent their state. The set of 1s and 0s ordered by weight can be used to encode weight combinations in the available magnitude range from 0 to 15 g in 1-g increments. The presence of a weight on the scale is indicated by a 1 in the corresponding bit position. A further discussion of binary numbers is presented in section 1.5.

The equivalent electrical circuit (figure 1.18) has four resistors (-⌇⌇⌇-), each connected to a power source on one side and a switch on the other. The switches determine whether the voltage across a particular resistor adds to the output. The magnitude of a particular sample can be reconstructed by closing the appropriate combination of switches.

The switches require a small amount of time to settle when a new sample value is first applied. It is virtually impossible to arrange for them to change at exactly the same time because of inevitable manufacturing variations and because each switch is responsible for a different amount of voltage. This nonuniform switching time can cause the output voltage to swing wildly for a brief time while the switches settle. Such unpredictable and unwanted swings in the output voltage are called *glitches*.

The voltage reconstruction circuit is followed by a sample-and-hold to get rid of the glitches. The sample-and-hold is timed to capture the reconstructed voltage only in the middle of the sample period, after the glitches have died away and before the next sample time arrives. The result is a staircase function. The sharp discontinuities at the edges of each stair step contain high-frequency energy that must be removed in order to reconstruct the analog waveform.

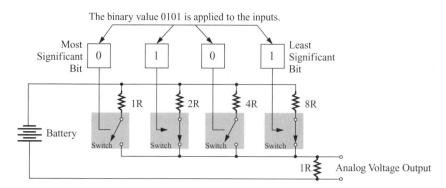

Figure 1.18
Sample reconstruction electrical circuit.

The lowpass filter, also called the *reconstruction filter,* smooths the abrupt transitions in the staircase function, reconstructing the analog waveform to its band-limited form (see figure 1.10b). The reconstruction filter must eliminate all frequencies above the Nyquist frequency by attenuating them to a level that is less than the noise floor in order to prevent aliasing. Ideally, it must not change the amplitude of frequencies that are less than the Nyquist frequency and within the band of interest (the baseband).

Aliasing also crops up in digital-to-analog conversion. Sound synthesis in particular presents a tricky problem. It is quite easy to synthesize non-band-limited signals (see chapter 9), but the analog reconstruction filter in the DAC won't eliminate aliased components represented in the digital signal fed to it. That's too late in the process. Any aliased components in the digital waveform must be removed by a digital anti-aliasing filter operating in advance of the DAC.

The output of the digital-to-analog converter is not the full-bandwidth original input signal (see figure 1.10a). At best, sampling can recover frequencies only up to the Nyquist frequency limit because of aliasing (see section 1.3). Though some high-frequency information may be lost, it is lost by design: if a recording system is well designed, f_s will be above any frequency of interest to listeners, typically in the range of 44,100 to 50,000 Hz.

1.5   Binary Numbers

Computers use binary arithmetic because circuits that have only two states are more stable than ones that have, say, ten. This stability allows transistors—the basic elements of such circuits—to be made very small without being less reliable.

1.5.1   Small Is Beautiful

Smaller electrical components use less current and do not require as much time to charge and discharge, so they can run faster. As they use less current, they dissipate less heat, requiring less effort to cool them. As circuit elements shrink in size, more components can fit in the same area, allowing increased complexity without increased physical size; alternatively, the same number of components can be made to fit in a smaller space. Since smaller chips cost less to fabricate, the compelling advantages of smallness explain why the electronics industry has worked so hard to reduce the size of integrated circuits.

In 1965, only four years after the introduction of the integrated circuit, Gordon Moore (then director of research and development at a large semiconductor manufacturer) famously observed that the electronics industry was realizing an exponential growth in the number of transistors per integrated circuit, and he predicted that this trend would continue indefinitely (G. E. Moore 1965). Dubbed *Moore's law* by the press, it has held remarkably true over the last 40 years, with circuit density doubling approximately every 18 months. Digital technology is one of the few industries that has seen both a dramatic increase in functionality and an equally dramatic reduction in cost. This has helped drive digital technology into many facets of modern life, including digital audio.

1.5.2 Binary Number System

A single binary digit, called a *bit,* can have the value 0 or 1. Bits are grouped using *place value,* as in the decimal system. With decimal integers, the rightmost position is the ones place, representing values from 0 to 9; the second position is the tens place, representing values from 10 to 19; then the hundreds place, thousands place, and so on. The magnitude of each place in the decimal system corresponds to a weighting of the place value by 10 raised to the power of the index of the place. For example, the decimal value 123 can be represented as the polynomial $(1 \times 10^2) + (2 \times 10^1) + (3 \times 10^0)$. The binary place system uses the same approach but raises each place value to a power of 2. So, for example, the binary value 1111011 can be represented as the polynomial

$$(1 \times 2^6) + (1 \times 2^5) + (1 \times 2^4) + (1 \times 2^3) + (0 \times 2^2) + (1 \times 2^1) + (1 \times 2^0)$$

$$= 64 + 32 + 16 + 8 + 0 + 2 + 1$$

$$= 123.$$

Fractional values can be represented in the decimal system using negative exponents of 10. So, for example, 0.75 can be written in polynomial form as $(7 \times 10^{-1}) + (5 \times 10^{-2})$. Fractional binary place values use negative exponents of 2. The decimal value 0.75 can be written as the polynomial $(1 \times 2^{-1}) + (1 \times 2^{-2}) = 0.5 + 0.25 = 0.75$.

The weighting system in figure 1.17 can be used to think about representing magnitudes with binary numbers. That figure shows a set of weights where each subsequent weight is twice as heavy as the one before. There are four such weights, thus $2^4 = 16$ combinations. If a weight is present in a measurement, we mark a 1 in that place; otherwise we mark a 0. If the weights were 1, 2, 4, and 8 grams, we could measure the weights as shown in table 1.1.

Another way to show this sequence is circularly (figure 1.19). This representation shows that adding 1 to any value creates the next value in sequence, and when we get to the maximum, 1111,

Table 1.1
Binary Counting Sequence

Weight	Value		Weight	Value
0000	0		1000	8
0001	1		1001	9
0010	2		1010	10
0011	3		1011	11
0100	4		1100	12
0101	5		1101	13
0110	6		1110	14
0111	7		1111	15

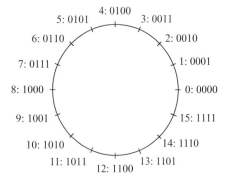

Decimal: Binary

Figure 1.19
Binary counting.

adding 1 returns us to 0. Numbers outside the binary range 0000 to 1111 wrap back into this range, like an odometer on a car. A number x represented with N bits of precision will have the binary value $((x))_{2^N}$, which means x taken modulo 2^N (see appendix section A.5). For example, if we have four bits of precision, we set $N = 4$. If the number we want to represent is $x = 17$, which is outside the range that four binary bits can represent, the odometer effect will cause it to be represented as $((17))_{2^4} = 1$. This can be seen in figure 1.19: advancing through 17 values counterclockwise around the circle from 0000 causes us to overflow the numeric precision of the counting system; the result wraps around to 0001. Numerical overflow and wrap-around can have devastating consequences for digital audio (see section 1.10.4).

For sound, the values to be weighted are the samples output from an analog-to-digital converter. At each sample time, we must find a combination of weights that matches the output of the sample-and-hold to the tolerance of the available precision, that is, the change in weight introduced by the smallest weight, finer than which we cannot measure. For a digital-to-analog converter, we use the weighting system in the other direction: we apply the weights as indicated by the sample's binary value to reconstruct the sample magnitude to the tolerance of the available precision.

1.5.3 Two's Complement Number System

The binary system described in section 1.5.2 represents unsigned magnitude, that is, positive values in the range 0 to $2^N - 1$, where N is the binary precision. But because sound fluctuates above and below standard atmospheric pressure, we must adapt the binary number system to represent signed values. One solution is to split the available binary precision so that half the numeric range represents positive amplitudes and the other half represents negative amplitudes.

The *two's complement* system is the most popular way to represent signed integer quantities because it uses the same rules as unsigned binary arithmetic for addition and subtraction. It is used universally in modern computers. Figure 1.20a shows the two's complement system using four bits of precision.

Notice that the binary counting pattern is the same as in figure 1.19. But the sign of numbers in the lower half of the circle is taken to be negative. The most significant bit of all positive numbers is 0 and the most significant bit of all negative numbers is 1, so the most significant bit can be interpreted as the *sign bit*.

Whereas the unsigned binary counting sequence shown in figure 1.19 has a discontinuity between the representation for 15 and 0, the two's complement system has a discontinuity between the most positive (7) and the most negative (–8) values.

Negative and positive values can be converted into each other by inverting the bits and adding 1. If a bit was 0, it becomes 1, and vice versa. For example, to find the negative two's complement value corresponding to 2, find its binary equivalent (0010), invert the bits (1101), and add 1 (=1110). This

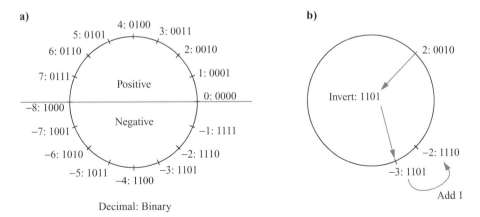

Decimal: Binary

Figure 1.20
Two's complement representation of signed numbers.

pattern is shown in figure 1.20b. Here's another example. To find the positive value corresponding to –3, find its binary equivalent (1101), invert the bits (0010), and add 1 (= 0011).

The rules for binary addition are the same as for decimal addition but using only 0 and 1. This means $1 + 1 = 0$ and carry the 1. For this example, the precision is 4 bits, so any result that overflows is discarded.

$$-3 + 4 = 1 \qquad 2 + (-3) = -1 \qquad -2 + (-3) = -5$$

$$\begin{array}{r} 1101 \\ +0100 \\ \hline 0001 \end{array} \qquad \begin{array}{r} 0010 \\ +1101 \\ \hline 1111 \end{array} \qquad \begin{array}{r} 1110 \\ +1101 \\ \hline 1011 \end{array}$$

There is an asymmetry in the two's complement system. Every negative value has a corresponding positive value except the most negative value (–8, in this case). To prove this, we look for its corresponding positive value. From figure 1.20a, the two's complement binary equivalent of –8 is 1000; inverting, we have 0111; adding 1 gives 1000 = –8. So the most negative value is its own inverse. Apply the same reasoning to zero and see what happens.

1.5.4   Lining Up the Bits

How can we use binary numbers to measure the instantaneous amplitude of an audio waveform? In particular, how do we map the two's complement binary magnitudes to the range of voltages coming from the microphone?

To keep it simple, suppose we want to encode the range of ±3 volts with three bits, giving $2^3 = 8$ binary values to work with, 4 positive and 4 negative. Using the two's complement number system, we associate the most negative binary value (100) with –3 V and the most positive (011) with +3 V. Unfortunately, we end up without a weight corresponding to the magnitude of 0 V, which dangles unrepresented between 0 and –1 (figure 1.21).

We'd prefer to have a value for 0 to represent silence. Let's try again, this time shifting the scale to the left until binary 000 lines up with 0 V (figure 1.22). Now binary 000 equals 0 V, but –3 V does not line up with any binary value, and +3 V exceeds the largest positive binary number. Let's try one more time, adjusting the scale of the voltages by squeezing its range to fit (figure 1.23). Now we have one binary value—the most negative—that is left dangling. But if we never use it, the rest of the scale is aligned and balanced. So we agree never to use the most negative value (100).

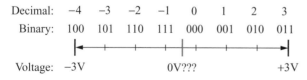

Figure 1.21
Binary encoding, first attempt.

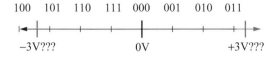

Figure 1.22
Binary encoding, second attempt.

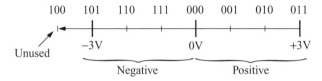

Figure 1.23
Binary encoding, third attempt.

Table 1.2
Hexadecimal Notation

Decimal	0	1	2	3	4	5	6	7
Hexadecimal	0	1	2	3	4	5	6	7
Binary	0000	0001	0010	0011	0100	0101	0110	0111
Decimal	8	9	10	11	12	13	14	15
Hexadecimal	8	9	A	B	C	D	E	F
Binary	1000	1001	1010	1011	1100	1101	1110	1111

1.5.5 Hexadecimal Notation

Digital audio applications typically require between 16 and 24 bits of precision for good-quality audio. In 16-bit arithmetic as used by compact discs, positive amplitudes range from 0 to $2^{16-1} - 1 = 32{,}767$, and negative amplitudes range from $-32{,}767$ to 0. (The most negative value for 16 bits is $-32{,}768$, which we've agreed not to use.) The decimal value 32,767 is a rather unwieldy number in binary: 0111111111111111.

Binary values can be made more compact by representing them in a higher base. The most common choice is *hexadecimal* (base 16). Since 16 is a power of 2, binary numbers fit evenly inside hexadecimal digits. Since hexadecimal notation requires counting from 0 to 15, hexadecimal digit values are represented by the numbers 0 to 9 and six additional letters: A, B, C, D, E, and F (table 1.2). For example, the decimal value 4096 is 1000000000000 in binary, but 1000 in hex. The decimal value $-32{,}767$ is written in hex as 8001, and the decimal value 32,767 is written in hex as 7FFF.

1.6   Synchronization

Sampling is a kind of selection process that arises when an observer momentarily regards an aspect of an object as a consequence of a triggering event. Ideally, observer and object are at least nominally independent so that the one can perceive the other objectively, without affecting it so much as to spoil the measurement.

Depending upon the intention of the experimenter, the trigger may be derived from some feature of the observed object, or it may be something independent, such as the expiration of a timer. Control over the triggering circumstances is *synchronization*. In the example of measuring the tides, the synchronizing trigger was a periodic timer (the position of the sun at noon). Periodic sampling is *synchronous* with a time-based trigger and *asynchronous* with the measured system. (Of course, the sun is a factor in the flow of the tides, so the trigger and observed system are not entirely independent.) If instead we wanted to observe the height of wave crests, we would sample synchronously with arrival of wave crests (and asynchronously with time).

Synchronization is fundamentally about how a nominally independent system (the observer) is made temporally dependent on another system (the object).

If the observed object provides both the conditions for the trigger and the state to be sampled, then the operation is synchronous; if the trigger and state arise from separate systems, the operation is asynchronous.

We may need to sample just once or repeatedly depending upon the intention. To determine the winner of a foot race, we only need a single static sample of the runners' positions when the first runner crosses the finish line. But we must sample the tide repeatedly to capture its frequency behavior. For all engineering projects, intention dictates system design: sampling requires an observer; hence sampling is subject to all the complications that sentient beings impose.

Sampling must be strictly periodic in order to accurately capture frequency information. What if the sampling rate varies from sample to sample for some reason? Inconstancy in the sampling period results in a form of distortion called *phase jitter*. Because this is a form of distortion involving time, and because time determines frequency and phase information, it follows that phase jitter distorts the phases and frequencies of the recorded signal. A more precise description of phase jitter is given in section 9.2.8.

1.7   Discretization

Something is *discrete* if it is indivisible. It is *indivisible* if it loses its identity when divided. A person is indivisible; a tree is indivisible. The members of a symphony orchestra are discrete because they can only be divided into subgroups that preserve the identity of the members. If there are three oboists, half the oboe section can't be sent to a rehearsal. The natural numbers are discrete because they are indivisible: each has a minimum size limit of one unit. (This is why fractions were invented, to express parts of whole numbers as ratios of whole numbers.)

A length is *continuous* in the sense that no matter how we divide it, it retains its identity, its "lengthness." Area and volume are continuous for the same reason. Real numbers were invented to represent continuities. Whereas there is no whole number that lies between two adjacent whole numbers, there is at least one other real number between any two real numbers.

Identity is the criterion of what is discrete and what is continuous, because a loss of identity occurs when scaling a discrete system but not when scaling a continuous system.

Discretization is fixing a point on a continuous function so that it becomes individually discernible. Discretization and quantization are not the same. If I locate two points on a line, I discretize the line at those points. If I want to measure the distance between the points, I must quantize the distance. Thus, discretization is the precursor step to quantization.

1.8 Precision and Accuracy

Precision and accuracy are not the same thing. *Precision* is generally the amount of information we have about a measurement; *accuracy* relates to whether the measurement is true. A measurement is *true* if there is *fidelity* between the value being observed and the resulting measurement. Inaccuracies introduce distortion into the measuring process.

Increasing the precision of a measuring system increases the amount of information it provides. *Information* is a quantity that relates to the number of facts necessary to convey a measurement. For example, if we know that middle C is 261 Hz, we have less information than if we know it is 261.626 Hz.

1.9 Quantization

Whenever we make a measurement, we must decide how precise we wish it to be. Measurement with too little precision may provide unusably crude results, but too much precision can burden calculations without providing much benefit. Useful measurement requires *quantization,* reducing the true magnitude of a measured object to the available precision and discarding the difference between the true value and its quantized approximation.

1.9.1 Linear Quantization

Since all millimeter marks on a meter stick span a uniform distance, it is linearly quantized. We can define *linear quantization* as

$$Q(x) = \frac{r(kx)}{k}, \qquad -1 \le x \le 1, \hspace{3cm} \textit{Linear Quantization} \quad (1.7)$$

where $Q(x)$ is an integer function of a real variable. The function r is the *rounding operator.* If the fractional part of a real number is less than 0.5, rounding reduces the value to the nearest lower integer; a fractional part greater than or equal to 0.5 is promoted to the next higher integer. Rounding is defined as

$$r(a) = \begin{cases} a - \lfloor a \rfloor < 0.5, & \lfloor a \rfloor, \\ a - \lfloor a \rfloor \geq 0.5, & \lceil a \rceil, \end{cases}$$

where the operator $\lfloor a \rfloor$ is called the *floor function,* that is, the largest integer less than a, and $\lceil a \rceil$ is the *ceiling function,* the smallest integer greater than a.

For binary quantization, if we set $k = 2^{N-1}$, where N is the number of bits, then equation (1.7) can be written in terms of binary quantization, Q_b:

$$Q_b(x) = \frac{r(2^{N-1}x)}{2^{N-1}}, \qquad -1 \leq x \leq 1. \tag{1.8}$$

Equation (1.8) partitions the signed unit interval $[-1, 1]$ into N steps. Figure 1.24 shows a binary quantization function with $N = 8$. The quantization function partitions the continuous input value x into a discrete set of N ranges. Each range corresponds to an integer output level $Q_b(x)$, also called a *reproduction level*. Rounding causes each x to be mapped by the function to its nearest reproduction level. In this example, values of x in the range $-1/4$ to nearly $1/4$ map to the reproduction level value of 0.

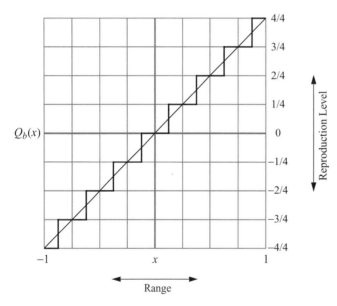

Figure 1.24
Quantization into $N = 8$ steps.

We can also perform *dequantization,* $Q_b^{-1}(x)$, the inverse operation of quantization, which reproduces the input value x from the value encoded by the quantizer. The value that the dequantizer reproduces is the centroid of the reproduction level.[4] For example, suppose we set the input to $x = 0.74$. Then using the quantizer shown in figure 1.24, $Q_b(x) = 3/4$. Dequantizing, we have $Q_b^{-1}(3/4) = [5/8, 7/8]$, and the centroid of the range $[5/8, 7/8]$ is 0.75, which is *not* 0.74 . The quantizer/dequantizer has discarded the difference between the true input value and the nearest quantization step.

The difference between the output value (0.75) and the input value (0.74) is called the *quantization error*. It is the difference between the input value of the quantizer and the corresponding output value of the dequantizer.

What is the worst quantization error possible? Suppose we have a 100 m tape that is marked in meters. The most precise measurement we can make reliably with this tape is to choose the meter mark that is nearest to the true length of the measured object, which can be at most 1/2 meter away. Therefore, in general, the maximum quantization error is one half of the quantum. The more precise the measurement system, the smaller will be both the quantum and the maximum quantization error.

A measurement system is characterized by its range, its origin, and the span of its quantum. For example, the range of the piano keyboard is 88 semitones, its origin is A0, and its span is A0 to C8. The equation for the decibel establishes a range (from the threshold of hearing to the limit of hearing), an origin (the reference intensity I), and the span of a quantum (the decibel) (see volume 1, equation (4.40)).

1.9.2   Beat Quantization

The quantum need not be of uniform size. The frequency span of a semitone increases as one goes higher up the piano keyboard. Similarly, the intensity span of a decibel grows as one goes higher up the decibel scale. Such measuring systems have *nonlinear quantization*. Rhythmic quantization in music can also be nonlinear. An audio example of nonlinear-quantizing codecs is given in section 1.12.1.

When taking musical dictation, a transcribing musician must accurately write down the pitches and rhythmic durations of a performing musician. Even if the performer follows the beat of a metronome, there will still be small discrepancies in the performance that must be ignored in order to determine the underlying rhythm of the music. That is, the transcriber must quantize any misplaced beats in the act of reducing the performance to notation.

The same transcription problem is faced by MIDI sequencer software that is designed to receive musical performance information from a MIDI keyboard or other MIDI controller.[5] By quantizing notes to the nearest beat, MIDI sequencers allow a poorly timed performance to be justified to the beat, and many musicians, the author included, have used this technology to clean up the beat timing of a ragged performance.

The simplest approach just rounds or truncates each performed note to the beginning of the nearest beat. But the size of the beat quantum must be picked carefully. For instance, if the beat quantum is the quarter note, two consecutive eighth notes played within a quarter note's duration will be quantized to start at the beginning of the same beat, wrecking the music (figure 1.25). Similarly,

Figure 1.25
Quarter note quantization wrecks eighth notes.

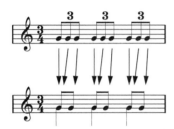

Figure 1.26
Eighth note quantization wrecks triplet eighths.

if the beat quantum is the eighth note, and the music contains triplet eighths, their identity as triplets will be destroyed (figure 1.26).

This is reminiscent of the discussion in section 1.7 about the *loss of identity* that occurs when scaling a discrete system. Even though music is a time-based art form and time is continuous, *musical time is discrete*. To convey a particular rhythm requires that beats of various sizes be held in strict proportion, and any scaling or quantization that compromises this proportionality destroys the music.

But that can't be the whole story. Continuous time obviously plays a role in musical expression. Performers speed up (accelerando), slow down (ritardando), hold notes (fermata), perform notes briefly (staccato), and take other liberties with the beat. Jazz performers lag behind the beat in order to give a relaxed, "cool" sound; rhythmic phrasing in Mozart's music sometimes has a mincing quality; Beethoven's music has a kind of swagger. We are incredibly attuned to the subtleties of rhythmic timing, and we glean an amazing amount of information from it. Temporal nuance lends a signature to the music that allows us to identify not just a particular style, not just a particular era, composer, or musician, but even how the performer was feeling that day. We may favor or disfavor musicians depending upon whether their interpretation of rhythmic minutiae moves us emotionally.[6]

Subdivisions of a phrase, measure, or beat that show a characteristic pattern of temporal distortion are called *grooves*. Grooves are regular but nonlinear quantizations of musical time. For example, consider swing rhythm. Figure 1.27a shows the way eighth notes in a measure of 4/4 time are interpreted using even quantization. The written duration of each eighth note is identical to the performed duration, as shown by the proportions of the boxes below the notes. Figure 1.27b shows

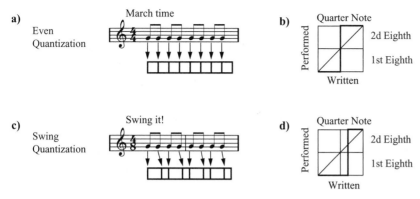

Figure 1.27
Straight time and swing time.

the quantization graph for this groove. Written time flows left to right, performed time flows bottom to top. The quantization function assigns the written beat duration evenly to both performed eighth notes.

Figure 1.27c shows how the swing groove works. The music notation still shows straight eighths, but the written durations and the performed durations are different: the second eighth of each beat is always delayed. Ordinarily, the proportion between two swing eighths is 2/3 + 1/3, as shown by the proportion of the boxes below the notes. If swing rhythm were written explicitly, it would look like this: ♩♪. This notation is a lot more tedious than writing straight eighths with an accompanying note specifying a beat quantization pattern to skew them during performance.

Figure 1.27d shows the quantization graph for this groove. Again, written time flows left to right, performed time flows bottom to top. Here, the quantization function assigns 2/3 of the written beat duration to the first performed eighth and the remaining 1/3 to the second performed eighth. The result is a rhythm akin to that produced by a child skipping.

1.10 Noise and Distortion

All systems for transducing, recording, and transmitting audio are plagued to a certain extent with loss of fidelity in the form of noise and distortion.

1.10.1 Noise

Noise is an unwanted signal added to one being measured. If an input signal to a system is $x(t)$ and the output of the system is $f(t)$, we can describe the noise contributed by the system as

$$f(t) \propto x(t) + n(t)$$

where $n(t)$ is the noise signal that we must contend with. Noise is *additive* because the resulting measurement is proportional to the sum of the input signal and the noise.

Noise comes in many sorts. It may be unrelated to the input signal, or it may be the result of some horrid interaction between the input signal and the recording equipment. Noise may be broadband, containing energy at all frequencies in equal proportion. In analog electrical systems such as microphones and amplifiers, noise arises naturally as a consequence of the random thermal motion of matter. Narrowband noise such as hum from power lines can also sometimes be a problem.[7]

Noise in digital systems comes mostly from artifacts of quantization. As mentioned in section 1.9.1, quantization discards the difference between the true input x and the nearest quantization step Q, resulting in a quantization error of $\varepsilon = x - Q$. The quantization error signal is $\varepsilon(t) = x(t) - Q(t)$ measured at each sample time t.

The shape of the error signal $\varepsilon(t)$ depends serendipitously upon precisely how the input signal happens to land among the quantum boundaries at each sample. Suppose $x(t)$ is a white noise signal, that is, one that is completely random and unpredictable from moment to moment. Then at any given sampling time t, $x(t)$ is equally likely to lie anywhere within the quantum that it falls into. Since $x(t)$ is random and uncorrelated to the quantization function $Q(t)$, $\varepsilon(t)$ is also random and uncorrelated to $x(t)$. This allows us to think of $\varepsilon(t)$ simply as an uncorrelated white noise signal added to the recording by the digitization process. This is in fact what $\varepsilon(t)$ commonly sounds like in practice. The quantization error signal is usually small in comparison to the magnitude of the recorded signal because it is bounded by one half quantum whereas the recorded signal can range over the entire available precision.

This simplistic analysis breaks down somewhat if $x(t)$ is a musical signal. Since music is not utterly random, it will show some similarity (correlation) with the quantization function $Q(t)$. In this case, $\varepsilon(t)$ will also show some correlation with $x(t)$. The more correlated $x(t)$ is with $Q(t)$, the more $\varepsilon(t)$ ends up resembling a distorted version of $x(t)$. The ear finds it easier to ignore $\varepsilon(t)$ if it is a colorless (white) noise, but $\varepsilon(t)$ can be a magnet to the ear if it changes sound color in a way that is related to the input signal. Techniques to work around this problem are discussed in section 1.13.4.

1.10.2   Signal-to-Noise Ratio

We can characterize the amount of noise in a system by relating the amplitude of its largest useful signal A_{max} to the amplitude of the noise signal A_ε. The ratio, the *signal-to-noise ratio* (SNR), is defined as

$$\text{SNR} = 20 \log_{10} \frac{A_{max}}{A_\varepsilon} \text{ dB.} \qquad\qquad \textit{Signal-to-Noise Ratio} \quad (1.9)$$

See volume 1, equation (5.31).

A_{max} and A_ε are usually measured in terms of their RMS amplitude level (see volume 1, section 5.5.1).

Since quantization error is the main source of digital noise, what is the quantization SNR for typical digital audio systems? Suppose the full-scale range of the ADC is $\pm A$. A signal that uses all available precision will have a positive amplitude of $+A$ and a negative amplitude of $-A$, for a total range of $2A$. The maximum quantization error will be one half the quantum size, or $2A/2^N$, where N is the number of binary bits in a sample. So the question is, What is the SNR corresponding to these values? To begin with, let $A_{max} = 2A$, and let $A_\varepsilon = 2A/2^N$. Then

$$\frac{A_{max}}{A_\varepsilon} = \frac{2A}{2A/2^N} = 2^N.$$

Plugging this value into the equation for SNR yields

$$\text{SNR} = 20 \log_{10} 2^N \cong 6N \text{ dB.} \qquad \textit{Quantization Error Signal-to-Noise Ratio} \quad (1.10)$$

Equation (1.10) says that each bit of sample precision provides a 6 dB increase of SNR. So, for example, standard 16-bit compact disc audio quality has an SNR equal to $6 \cdot 16 = 96$ dB. That's pretty good, actually. However, this only means that the *loudest possible signal* will be 96 dB above the quantization error noise. Weaker signals will have correspondingly worse SNR. Moreover, as the signal becomes small with respect to the size of the quantum, the reconstructed waveforms increasingly resemble square waves. In the extreme, when the signal is so small that only one bit is changing, quantization will cause the signal to be reconstructed as a square wave regardless of its original shape. Thus, at low amplitudes, digital encoding creates harmonic distortion as well as noise. These objectionable artifacts of digital audio technology have spurred industry to introduce better conversion systems.

Input signals $x(t)$ that are weaker than the error signal $\varepsilon(t)$ will be masked by it (see volume 1, section 6.6). Because the amplitude of the error signal A_ε of most digital systems for most types of audio signals is relatively constant, A_ε represents a good rule of thumb for the minimum intensity of recordable signals. The strength of the maximum useful signal A_{max} is typically a fixed constant for most types of recording systems. Signals louder than A_{max} will be subject to distortion. So the useful *dynamic range* that a system can process is limited by A_{max} at the top and A_ε on the bottom. On the whole, it's important for recording engineers to adjust the input sensitivity of the recorder to keep the signals being recorded near A_{max} so that the softest sounds will hopefully be substantially louder than $\varepsilon(t)$. Of course, the engineer must simultaneously be sure never to exceed A_{max}. Digital systems are particularly bad at handling inputs in excess of A_{max} for reasons that are described in section 1.10.3.

Like cobwebs and dust bunnies, noise is a ubiquitous fact of life. Though it's best to keep noise out to begin with, that's not always an option, as is the case with historical recordings. A wax cylinder recording of the composer Johannes Brahms playing the piano was made at the dawn of recording technology, but the recording has become so degraded over time as to be practically unrecognizable as music. Some exciting noise-reduction techniques have recently become available that have made it possible, as it were, to reach into the jaws of entropy and snatch the information back out (see section 10.4.1).

1.10.3   Distortion

Distortion, like a fun house mirror, disarranges the proportions of the signal being recorded. Many kinds of systems including audio recorders can be thought of as mirrors that reflect their input to

a) Linear System

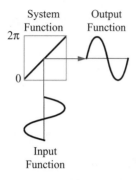

b) Nonlinear System

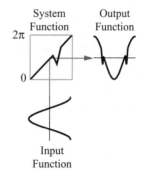

c) Clipping

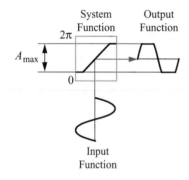

d) Frequency Doubling

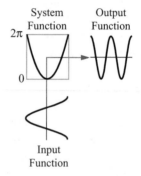

Figure 1.28
Linear and nonlinear system functions.

their output. If the system is linear, the mirror is flat and the input is reflected by the system to the output undistorted. Figure 1.28a shows a linear system. In fact, a system is linear because the mirror—properly, the *system function*—is a straight line. The mirror is not flat if the system is nonlinear. The system function in figure 1.28b has a discontinuity in its middle and the center of the output signal is distorted by it.

Nonlinear distortion called *clipping* (figure 1.28c) arises when the input signal exceeds A_{max}, the upper limit of the system's dynamic range. The system is unable to reproduce the input signal when its magnitude exceeds the system's precision limit. If we wish to capture the input signal undistorted, we must keep its magnitude in the linear range, less than A_{max}. Of course, distortion is not always a bad thing. People like to see themselves in fun house mirrors. Distortion can be artistic, such as the "fuzz box" effects applied by rock-and-roll guitarists to give themselves an edgy sound. The effect of these devices is essentially the same as the one in figure 1.28c.

Clipping can be used to show an important property of linear systems. A linear system may scale the signal to be bigger or smaller, but the characteristic shape of the signal will not change. The system in figure 1.28c is nonlinear because doubling the input will not double the output since the limit A_{max} has already been exceeded and the output is clipped. Suppose x_1 and x_2 are input signals, y_1 and y_2 are output signals, and a and b are scaling terms. If a system behaves such that $ax_1 \propto y_1$ and $bx_2 \propto y_2$, then it is linear if it preserves $ax_1 + bx_2 \propto y_1 + y_2$ for all possible values of a and b. In other words, if input x_1 produces output y_1, and input x_2 produces output y_2, then any weighted combination of x_1 and x_2 will result in a proportional combination of outputs y_1 and y_2 if the system is linear.

In all cases, the effect of distortion is to modify the spectrum of the input signal. A strikingly simple example is shown in figure 1.28d. The system function is a parabola, and it exactly doubles the frequency of the input signal without otherwise changing its shape: one period of the input waveform results in two periods squeezed into the same time interval on the output. There is an interesting artistic application for this type of distortion (see section 9.4.11).

Distortion appears to be a result of a function (the system function) that modifies another function (the input function). This suggests that we can model distortion mathematically as a system function S that modifies the input function w:

$$f(t) \propto S(w(t)). \hspace{4cm} \textit{Distortion} \ (1.11)$$

If S is a linear function, such as the simple gain scaling function $S(x) = ax$, then there is no distortion. However, if S is nonlinear, such as $S(x) = x^2$, then $f(t)$ will be distorted.

When distortion and noise are considered together, the result is the distorted input signal plus the noise:

$$f(t) \propto S(w(t)) + n(t).$$

1.10.4 Overflow, Wrap-around, and Clipping

In the two's complement system, the number that is 1 greater than the most positive number is the most negative number. With 3 bits of precision, the maximum positive value is +3, or 011 binary, because the most significant bit is taken as the sign bit. In binary arithmetic, $011 + 1 = 100$, but in two's complement notation, binary 100 is –4, not +4. Similarly, it follows that 1 less than the most negative value is the most positive value.

If the magnitude of a sample ever exceeds the available precision, there will be a sudden full-scale discontinuity in the recording because the value wraps around (figure 1.29a). If even just one sample wraps around, listeners may hear a click. If one sample or a few samples wrap around every period, a distorted buzzing will be heard. If many samples wrap around, sound quality quickly degrades into a dreadful garbled crashing noise.

One way to fix wrap-around error is to use saturating arithmetic to prevent the limit of precision from being exceeded. If the magnitude overflows the most positive value, saturating arithmetic forces its result to remain at the positive maximum; similarly, if the magnitude overflows the most

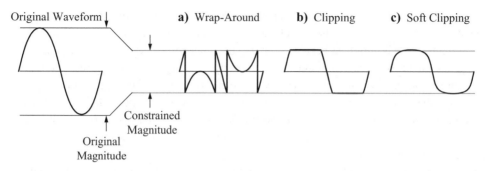

Figure 1.29
Managing overflow.

negative value, the result is forced to remain at the most negative value (figure 1.29b). *Saturation* is defined as

$$\sigma(x) = \begin{cases} A_{max}, & x \geq A_{max}, \\ x, & A_{max} > x > -A_{max}, \\ -A_{max}, & x \leq -A_{max}. \end{cases} \qquad \textit{Saturation} \quad (1.12)$$

Saturation is equivalent to clipping. Though saturation is an improvement over wrap-around, it can still create very harsh distortion.

A further refinement to handling overflow is *soft clipping,* which is really a kind of dynamic range limiting: as the signal strays closer and closer to the limits of precision, it is progressively attenuated so it comes in for a soft landing, so to speak. The harmonic distortion is less severe because the signal doesn't suddenly bump into a hard clipping boundary (see figure 1.29c). This is still a form of distortion, although its effect on the recording is much milder than wrap-around error. The only way to completely avoid clipping is to keep signal levels within the available range.

1.11   Information Density of Digital Audio

Take another look at figure 1.9. After sampling, the only information that remains about the input signal is the list of quanta that it visited. Except for the value of the signal at the discrete moments when it is sampled, all other information, such as what may have happened between measurements, is discarded. The true magnitude of the measured phenomenon is discarded by quantization, and we preserve only the quantized measurement.

The dimensions of the quantum are therefore a measure of the uncertainty in the measuring process: the bigger the area of the quantum, the less certain we can be about the true value of the input signal. Certainty is thus proportional to precision and sampling rate because as these parameters increase, the area of the quantum decreases. Increasing measurement precision narrows the span of the quantum and reduces uncertainty. But increasing precision increases the amount of information that each measurement carries, making the data less tractable to store and manipulate. Increasing the sampling rate narrows the size of the sampling period and also reduces uncertainty. But increasing the number of samples likewise increases the amount of information, making the data set larger and calculations less tractable.[8]

Sampling systems must strike a balance between too much uncertainty and too much information. To strike the right balance between information and uncertainty for audio recording requires that we understand how the limits of hearing interact with sampling theory. The ear is capable of discerning a dynamic range of about 120 dB. At 6 dB per bit, a minimum of 20 bits of precision is required to adequately capture this dynamic range. Fresh young ears can hear frequencies up to about 20 kHz. To remove all frequencies above the Nyquist barrier, practical anti-aliasing filters require a cutoff frequency of about 40 percent of the sampling rate. So, to achieve a passband of 20 kHz requires a sampling frequency of about 50 kHz.

From these calculations, we see that the compact disc standard of 16 bits at 44.1 kHz does not cover the whole range of human hearing. Nonetheless, since it is the least common denominator of current audio standards, let us use the compact disc to develop some ballpark figures for the amount of information required to store digital audio.

1.11.1   Storage Requirements of Digital Audio

One second's worth of 16-bit audio at 44.1 kHz requires $16 \cdot 44{,}100 = 705.6$ kilobits/s. For stereo, we must double this to 1411.2 kilobits/s. If we were to store 1 second of CD-quality audio on a computer disk, how much memory would be required? Computer disks are conventionally measured in megabytes (MB). A *byte* is 8 bits of data, and so a *megabyte* is ... well, there is more than one definition for this basic unit of storage. It is defined either as $2^{20} = 1{,}048{,}576$ bytes or as $10^6 = 1{,}000{,}000$ bytes. The first definition derives from the common computer practice of counting bytes in powers of 2 because computers universally use binary arithmetic. By this reasoning, a *kilobyte* of memory is 1024 bytes, so a megabyte would be $1024^2 = 2^{20}$ bytes. Common usage overwhelmingly favors this definition of a megabyte. On the other hand, world standards bodies have legislated in favor of the second definition. An advantage of the second definition is that it is consistent with the SI prefix *mega,* meaning 1 million. Since this book attempts to be consistent with SI units, I'll use the 10^6 definition.[9]

So, 1 second of CD-quality stereo audio at 1411.2 kilobits/s equals 1.4112 megabits/s, or 0.1764 SI megabytes/s. That works out to 5.67 seconds of stereo audio per SI megabyte. So an hour of 16-bit 44.1 kHz stereo audio on a computer disk would require about 340 SI megabytes. Standard CDs store about 74 minutes of music, so their audio information storage capacity must be on the order of 783 SI MB.[10]

1.11.2   Computational Requirements

If we wish to generate music in real time, synthesizing each sample of audio as we go along, how fast a computer will we require? How many operations per sample can today's computers support?

Of course, the answer depends in part upon how complex a calculation is required to make interesting-sounding music.

How much real time do we need in order to calculate a single audio sample at CD-quality rates? With a sample frequency $f_s = 44.1$ kHz, the sampling period T equals $1/44,100 = 0.226 \times 10^{-4}$ s (about 23 microseconds). Since we have to calculate two samples for stereo in that amount of time, we actually have half of that duration for each sample. Thus we have about 11.3 μs to calculate each sample. Suppose it takes 1000 machine operations to calculate one sample. That means the computer must execute each instruction in the time of 0.114×10^{-7} s in order to operate in real time at audio sample rates. The computer would have to execute on the order of 88 million operations per second to keep up with this rate.

Fortunately, today's computers are able to perform well beyond this level, meaning that a reasonable amount of digital audio processing can be calculated in real time on most modern computers. This is a fairly recent development. In 1990, F. Richard Moore figured that the fastest computers of that day would require "about two or so orders of magnitude improvement" in speed "to allow much general-purpose computer music processing to be accomplished in real time" (Moore 1990, note, p. 54). It is really only since about the year 2000 that the requisite computational bandwidth has become affordable in inexpensive personal computer systems. Up to that time specialized digital hardware was required for high-quality audio signal processing tasks. Modern affordable general-purpose computers are increasingly capable of performing interesting real-time audio signal processing without specialized hardware. This is really good news, in my opinion.[11]

1.12   Codecs

A *codec* (coder/decoder) is any device that encodes and decodes a signal. The analog-to-digital and digital-to-analog converter are a codec pair.

Pulse Code Modulation (PCM) is the name of the codec technology discussed so far in this chapter. The term PCM was coined by engineers to describe the steps of sampling, quantization, binary encoding, and transmission of binary data via a carrier signal. It's called PCM because to transmit digital data, a transmission medium such as light, electricity, or microwaves was switched on and off (modulated) in pulses of varying length according to a binary code.

There are many kinds of codecs besides PCM. Standard PCM can be very high quality, but it requires a great deal of data to store and bandwidth to transmit. Two alternative codec technologies emphasize data compression. μ-Law quantization for telecommunications and MP3 for Internet audio both focus on reducing the amount of data required to store or transmit audio signals.

1.12.1   Nonlinear-Quantizing Codecs

For speech, intelligibility is most important. Speech signals have a wide dynamic range, but only intermittently use their full range, spending the majority of their time at relatively low amplitudes.

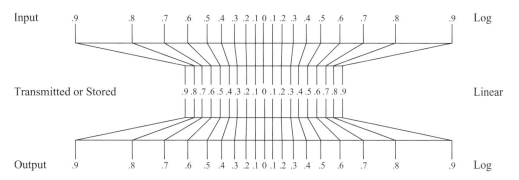

Figure 1.30
Companding codec.

Unfortunately, linear quantizers have the worst SNR at low amplitude and a relatively limited dynamic range. Therefore, linear quantizers are suboptimal for telephony.

We can increase the SNR for speech at low amplitudes by warping the quantization scale so that quanta near zero are small. This lowers the quantization error for speech signals that are mostly low-amplitude. The quantum size (and the attendant quantization error) grows with increasing amplitude (figure 1.30). Since quantization error increases with amplitude, the SNR is constant at all dynamic levels.

The μ-Law codec used in telephony in the United States and Japan, and the A-Law codec used in Europe, perform nonlinear quantization. μ-Law encoding compresses the wide dynamic range of hearing into a mere 8 bits per sample. The dynamic range is expanded again during decoding. A compander is a system that compresses and then expands a signal, and so μ-Law and A-Law are *companding codecs*.

To reduce the data bandwidth further, the sampling rate f_s is lowered to 8 kHz. Using the 0.4 scaling coefficient for practical anti-aliasing filters, this results in an upper frequency limit of 3.2 kHz. This is okay for speech because the highest vocal formant frequency is less than this limit.

The combination of companding and lowered sampling rate provides a useful and very economical codec for voice communications. It provides adequate intelligibility for speech at low data rates.

1.12.2 Lossy Quantization—MP3

MP3 audio format is widely used to transmit music on the Internet and to store music on portable audio players. MP3 is short for its official name: ISO-MPEG Audio Layer-3, set forth in IS 11172-3 and IS 13818-3.

MP3 is a member of the class of lossy codecs. Here's why: Regular linear digital audio (such as for compact discs) encodes a verbatim copy of the sampled acoustical waveform. All the information that was captured in the recording process is stored and reproduced regardless of whether the ear can detect the information or whether the information is redundant.

The MP3 codec achieves high audio data compression without significant loss of quality by being smarter about these two points. Clearly, if no one can perceive certain information, there's no need to store or transmit it. MP3 includes a psychoacoustic model of hearing that allows it to avoid encoding information that we can't hear (see volume 1, section 6.9.4, for a discussion of the psychoacoustic model of critical bands). The information that is lost because of this step is unrecoverable because it is not encoded at all. Therefore, the psychoacoustically driven encoding step is lossy. Because it is based on psychoacoustics, MP3 is sometimes classified as a *psychoacoustic codec*.

MP3 also includes a lossless data compression step that comes after the psychoacoustic encoding. This step squeezes out all redundancy in the encoded signal. To understand what I mean by *redundancy*, suppose we have a text message that reads "AAAAABBBBCCCDDE". If we sent the message verbatim, it would require transmission of 15 bytes of data (one per character). Instead, we institute a rule at the sending and receiving ends that says in effect, "Send only the first instance of a new letter followed by a count of its immediate repetitions." By this rule, we might transmit something like "A5B4C3D2E1". We've reduced the transmission to 10 bytes for a savings of 33 percent by eliminating redundancy in the original message. At the decoding end, we reverse the rule and are able to reconstruct a verbatim copy of the original message, demonstrating that removing redundancy is lossless. This is a useful technique if there is a high degree of redundancy in the signal, as is the case with most audio signals.

Because there is a good deal of musical information that we can't hear because of psychoacoustical masking effects, and because there is quite a lot of redundancy in most music, MP3 can generally achieve quite dramatic data compression without substantially degrading audio quality. This format has been a boon to the distribution of music over relatively slow channels such as the Internet. The combination of the Internet, MP3, and personal computer hardware and software has created an industry-shaking worldwide revolution in the distribution of music. A fuller discussion of MP3 is presented in chapters 3 and 10.

1.13   Further Refinements

The lowpass filters required for A/D and D/A conversion present converter designers with some serious challenges. For a 44.1 kHz sampling rate, the Nyquist sampling theorem allows signals up to exactly 22.05 kHz to pass without aliasing and requires all signals above exactly 22.05 kHz to be attenuated by at least 96 dB so that they are under the level of the quantization noise (assuming 16 bits of precision). The ideal lowpass filter would behave like a brick wall to frequencies above the Nyquist limit and would have no effect on frequencies below the limit. Unfortunately, realizable lowpass filters are not able to switch from passing to rejecting frequencies so quickly.

1.13.1   Realizable Anti-Aliasing Filters

Practical lowpass filters can only gradually switch from passing low frequencies to attenuating high frequencies. The *transition band* is the span of frequencies required to switch from passing to

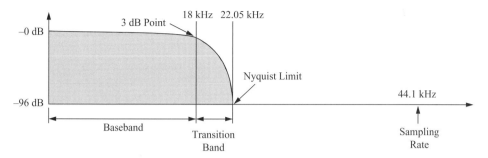

Figure 1.31
Idealized response of an anti-aliasing filter.

attenuating. The frequency where the filter begins attenuating is the *cutoff frequency*. The cutoff frequency is that frequency where the signal strength is attenuated by 3 dB from its maximum. For this reason, the cutoff frequency is also sometimes called the −3 dB point. For a 44.1 kHz sampling rate, a practical lowpass filter's transition band will be about 4.5 kHz wide, so the −3 dB point is about $22.05 - 4.5 \cong 18$ kHz, which is dangerously close to humanly hearable frequencies we'd like to preserve (see figure 1.31). We'd like the lowpass filter to have a very abrupt transition band so we don't lose any frequencies of interest in the baseband. But the narrower we make the transition band, generally the worse the lowpass filter performs. Problems can include amplitude, phase and harmonic distortion artifacts introduced by the filter. Figure 1.31 shows an idealized view of the response of an anti-aliasing filter. Depending upon its design, the transition band may be wider or narrower and may include amplitude ripples in the response in exchange for a sharper transition.

How can we reduce the artifacts of anti-aliasing filters?

1.13.2   Solution 1: Increase the Sampling Rate

Rather than squeezing the transition band of the lowpass filter and degrading the audio quality, a relatively simple fix is to increase the sampling rate. For example, with a sampling rate of 88.2 kHz, the anti-aliasing filter only needs to eliminate frequencies above 44.1 kHz. If we set the −3 dB point of the filter to 22.05 kHz (comfortably above the highest frequency we can hear), the transition band can stretch over an ample 22.05 kHz to reach the −96 dB point at the Nyquist frequency.

If both recording and playback take place at this higher rate, we're done. If the playback sampling rate is to remain at 44.1 kHz, we must add a sampling rate reduction step, called a *decimation filter*. Since this filter is digital, it is uncomplicated to design and does not impose any untoward effects on the signal, so the net effect is positive.

Increasing the sampling rate above the minimum required by the Nyquist sampling theorem is *oversampling*. The two-times oversampling example just given can be extended arbitrarily to N-times oversampling. Not only does this continue to relax the anti-aliasing filter design requirements but it also reduces the quantization noise without increasing the sampling precision. How is that possible?

a) Standard 44.1 kHz Sampling Frequency

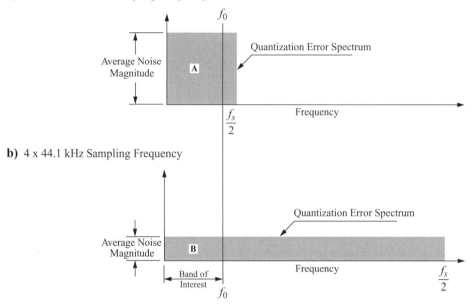

Figure 1.32
Oversampling reduces audible quantization noise.

The Nyquist sampling theorem implies that *all* frequencies lie within the baseband. Therefore, *all* the quantization noise in such a signal also appears in the baseband. Assume that the quantization noise is spread uniformly over the baseband (it is white noise). If we increase f_s, we spread the white noise out over a wider band. We are spreading out a constant amount of white noise into a broader frequency band; therefore the average noise in any part of the band is reduced. In particular, the noise in the band of hearable frequencies goes down as the sampling rate goes up. Let's call f_0 the highest frequency of interest. Let's say $f_0 = 20\text{kHz}$. Then we can define the *oversampling ratio* (OSR) as $f_s/(2f_0)$. It can be shown that oversampling reduces the in-band quantization noise by the square root of the OSR. This can be demonstrated intuitively. Figure 1.32a shows the quantization noise spectrum of a standard ADC converter. Figure1.32b shows the effects on quantization noise of oversampling. Since the Nyquist frequency limit $f_s/2$ is much higher, the quantization noise is spread over a wider range. Although the total quantization noise is unchanged, the quantization noise that lies in the band of interest is greatly reduced. We see this by comparing the areas of rectangles A and B in the figure.

The combination of a gentler transition band for the anti-aliasing filter and reduced quantization noise makes the technique of oversampling a useful improvement. The decimation filter used to reduce the sampling rate back to the desired lower rate actually helps increase sample resolution: we are creating N extra samples at the higher rate for each one sample at the final rate. The decimation filter will basically average those extra samples together, increasing the resolution. For example, if we sample at $16f_s$ and then decimate by a 16:1 ratio, we end up with $\log_2 16 = 4$ extra bits of dynamic range. Of course, the price to be paid for this extra resolution is to record at a much higher sampling rate.

1.13.3   Solution 2: Sigma-Delta Modulation

Delta modulation means quantizing the *change* in the signal from sample to sample. This is accomplished by subtracting the current sample from the previous sample and encoding the difference. This allows a radical simplification of the digitization process: instead of needing at least 16 bits to indicate absolute sample value, we need just one bit running at $16f_s$. At each sampling time, we simply observe whether the signal has gone up or down and output a corresponding 1 or 0. We lowpass-filter this 1-bit signal to reconstruct the original analog input. (While this represents an improvement in circuit design, it does not affect the amount of information that must be encoded.)

A refinement of this process, *sigma-delta conversion,* has an additional benefit: it lowpass-filters the signal and highpass-filters the quantization noise so that the spectrum of the quantization noise is no longer uniform. Quantization noise density is highpass-filtered out of the band of interest into inaudible higher frequencies. (Precisely how this works is outside the scope of this book.)

Sigma-delta conversion combines the advantages of oversampling, noise shaping, and decimation filtering. It is inexpensive and easy to manufacture because the analog anti-aliasing filter requirements are less demanding, and the sample-and hold circuit for a 1-bit converter is easy to design. Sigma-delta converters can be more linear than standard converters, and the background noise level is independent of the input signal level. They can be easily fabricated in large integrated circuits with relatively simple analog design criteria, and they are in widespread use today.

1.13.4   Solution 3: Dithering

Even where the input signal is strong, a great deal of important audio information—ambience, warmth, stereo sound field, reverberation, and the final decay of instruments—lies in the lowest dynamic ranges and therefore is captured in the least significant bits of the quantized audio signal. Quantization error is introduced during analog-to-digital conversion and also during certain mastering operations whenever sample precision must be reduced. Quantization error can mask and distort this sensitive and important low-amplitude information. Dithering provides a way to reduce the impact of quantization error.

In section 1.10.1 we saw that if the input signal $x(t)$ fed into a quantizer is completely random, then the quantization error signal $\varepsilon(t)$ is also completely random and sounds like white noise hiss. But if $x(t)$ is a musical signal, $\varepsilon(t)$ is more correlated to $x(t)$. In the worst case, where $\varepsilon(t)$ is highly correlated to $x(t)$, the quantization error signal resembles a horribly distorted version of $x(t)$. A constant white noise signal is relatively easy for the ear to forget about, rather as we stop noticing that the sky is blue, but noise that noticeably changes color over time is much harder to ignore. This coloration can sound dreadfully bad, especially for low-amplitude signals where $\varepsilon(t)$ is not far below $x(t)$ in amplitude.

Since $\varepsilon(t)$ sounds worse the more correlated it is to $x(t)$, an ingenious way to improve things is to add to $x(t)$ a very low-amplitude noise signal $u(t)$. This *dither signal* decorrelates the lowest-amplitude components of $x(t)$ from the quantization function $Q(t)$ and makes the quantization error signal $\varepsilon(t)$ sound like white noise again. The type of noise added is critical: it must decorrelate weak components of $x(t)$ without adding much noise on its own account. There are many types of dither signals that have been found to be useful, for example, least significant bit triangular probability density function dither (LSB TPDF).

The disadvantage of dithering is that some noise is added to the signal, although the reduction in quantization error noise far outweighs this liability. The dithering noise itself can be reduced by oversampling and sigma-delta modulation (see section 1.13.3).

Dithering the ADC Input   Quantization noise is inherent in analog-to-digital conversion because the input $x(t)$ is continuous but the output of the ADC is of finite precision. Adding dithering to $x(t)$ prior to conversion prevents the quantization error signal from being correlated to low-level signals in $x(t)$. The dithering noise in this case must be analog because it must be added to $x(t)$ prior to conversion.

Employing high oversampling (in the range of $64f_s$ up to $512f_s$) allows us to spread the added noise from dithering over a wider frequency range, as shown in figure 1.32b.

Dithering during Mastering   Mastering is the final step in preparing a recording for mass replication. Final alterations include gain scaling, filtering, and reverberation. In the digital domain these are arithmetic operations such as multiplication that result in an increased sample precision. For example, if we have a sample value of 0.875, and scale its amplitude by 1/2 (that is, reduce it by 6 dB), the result is 0.4375. Notice that the multiplication converted three digits of precision to four. If we can't preserve all four digits of this result, then we must truncate it back to three, but then we are throwing away whatever information resided in the least significant digit.

Many signal-processing operations result in an increase of precision, so these calculations are typically done using extended precision. For example, 16-bit audio samples are typically processed using 24- or 32-bit binary arithmetic. In this way, little or no low-level information is lost during intermediate calculations because it is carried along in the least significant bits of the result.

However, when it comes time to go back to 16-bit samples, the excess precision must be truncated, which introduces quantization error. But dithering can be used to decorrelate the error signal prior to truncation. It is important to note that once final dithering has been performed, any additional operations on the signal will undo the benefits of dithering. This is because any such additional operations will likely require another quantization/dithering step. Dithering multiple times will build up the noise and degrade the signal quality. Thus, dithering should be the very last step performed during mastering.

1.14   Cultural Impact of Digital Audio

Digital audio allows music to be copied without introducing any additional noise. Audio editing has benefited enormously from this. To modify an analog recording without destroying it requires that it be copied, and each new generation introduces additional noise and distortion. Noiseless digital copying has dramatically improved the ability of artists to edit, process, and archive their work. Nondestructive editing systems make copies of the audio to be edited, allowing changes to be seamlessly undone if necessary.

Of course, unlimited verbatim copying is a double-edged sword. Once a recording is published, anyone possessing the requisite equipment (which can be obtained for relatively little money) can copy it perfectly as many times as they like. It is possible for fans to "love an artist to death" by widely sharing the artist's work without paying a royalty. This is not a new problem, actually. Mozart carefully guarded his musical scores to prevent piracy, handing out parts to musicians at rehearsals and performances, and gathering them up again at the end.

The problem of digital audio copying has led some music industry groups to seek limits on copying and sharing music recordings. Tactics include adoption of mandated copy protection systems, digital watermarking, legislation to outlaw circumvention of antipiracy measures (DMCA 1998), and even lawsuits against their own customers. However, though art may be regarded by some as a commodity, it is actually an integral part of our shared culture, and transmission of culture from place to place and across the generations is accomplished by sharing, which involves copying, borrowing, and imitation. Richard Dawkins coined the term *meme* to mean "a unit of cultural transmission, or a unit of *imitation*." As examples, he cited "tunes, ideas, catch-phrases, clothes fashions, ways of making pots or of building arches."[12]

The fair use doctrine is a social compact by which our society attempts to balance the need of artists to earn money from their art against the culture's need for freedom to copy and imitate its cultural memes. This doctrine is currently experiencing great strains because of digital audio. The health of our musical culture hangs in the balance. We will all be impoverished if the commonwealth of music is shackled by forces whose only concern is the commercial value of copyright. But we will be similarly impoverished if artists cannot achieve success because they have no viable means of earning a livelihood.

There seems to be a kind of conservation law that something is gained and something is lost with every technological innovation. When we get what we want, our world is changed by it. All technology has this kind of good-news/bad-news nature: who would guess that the same technology that allowed Daedalus to escape from the prison of King Minos would lead his son Icarus to fall into the sea and perish? What is required for a good outcome in the debate over fair use of digital audio is simply that the participants not make Icarus's prideful mistake.

Summary

A continuous function of time such as the pressure of air on a microphone diaphragm can be recorded as an analog function of time (on a phonograph or analog magnetic tape) or as a discontinuous sampled function of time. The result of sampling is a time-ordered set of discrete measurements. Digital audio recording employs an analog-to-digital converter (ADC) to sample the continuous input audio signal.

Precision characterizes how much information a measurement yields; accuracy has to do with the fidelity between the value being observed and the resulting measurement. Inaccuracies and imprecision introduce distortion into the measuring process.

Discretization is the isolation of a point on a continuous function. Quantization establishes the nearest measuring mark to that point. Sampling quantizes the value being measured at a discrete point in time.

The analog-to-digital conversion process begins with an anti-aliasing filter that removes energy content from the signal above the Nyquist barrier. It allows us to specify whatever sampling rate we desire by preventing aliasing of components above the Nyquist barrier. The lowpass-filtered analog signal is then fed to a sample-and-hold system that stabilizes the signal during sample capture. The sample-and-hold system's value is then measured and converted into a binary representation. The binary samples are then stored on a medium, such as a writable compact disc.

The digital-to-analog conversion process starts with fetching the binary samples one per sample period from the medium. The binary values are converted to an analog voltage and sent to a sample-and-hold system to convert them into a staircase function. The staircase function is then lowpass-filtered with a reconstruction filter that is equivalent to the filter used for anti-aliasing. The result is an analog signal equivalent to the anti-aliased input signal.

For a given sampling frequency, all frequencies greater than 1/2 of that rate are aliased. The aliasing effect occurs because actual frequencies outside this range are indistinguishable from frequencies within this range because of the way sampling affects our observations. The Nyquist theorem states that we must sample at twice the rate of the highest frequency we wish to represent. Given the difficulties of designing good anti-aliasing filters, a transition band or guard band near the Nyquist rate progressively attenuates components to prevent aliasing.

There is an equivalence for sampled waveforms between sampling rate and angular velocity. We can say that frequency f is to the Nyquist barrier $f_s/2$ as angular velocity θ is to π. Therefore, when plotting a spectrogram, we can simply plot frequency as radian velocity between $-\pi$ and π and leave sampling rate out of the picture entirely. This allows us to graphically interpret frequency information on a uniform scale without reference to the underlying sampling rate. The magnitude of spectral components can be represented in two dimensions by graphing positive and negative amplitude on the y-axis and using the x-axis to show phase angle between $-\pi$ and π instead of frequency.

Linear-quantizing digitizers can be quite data-intensive for audio. To reduce bandwidth, a logarithmic quantization is used for telephony; MP3 exploits psychoacoustics to remove unhearable components. Audio codecs use oversampling, 1-bit sigma-delta modulation, and noise shaping to achieve highest quality.

2 Musical Signals

Music is fashioned wholly in the likeness of numbers. Whatever is delightful in song is brought about by number. Sounds pass quickly away, but numbers, which are obscured by the corporeal element in sounds and movements, remain.
—"Scholia Enchiriadis"

It has been written that the shortest and best way between two truths of the real domain often passes through the imaginary one.
—Jacques Hadamard

Mathematicians aren't above imagining new kinds of numbers when circumstances warrant. The natural numbers—whole numbers greater than zero—are probably as old as civilization. But when the natural numbers failed to solve equations such as $c = a - b$ for all possible natural numbers a and b, mathematicians invented negative numbers. The result was the birth of the integers. Rational numbers were developed when integers failed to solve equations such as $c = a \div b$ for all possible integers a and b. When a careful look at irrational numbers such as π showed the limitations of rational numbers, mathematicians invented real numbers. Yet there are straightforward mathematical situations that can't be handled by real numbers either.

2.1 Why Imaginary Numbers?

Table 2.1 shows a sequence of simple equations requiring increasingly advanced number systems for their solution, starting with natural numbers and progressing through solutions requiring real numbers. But what about that last equation? How can it be solved?

Table 2.1
Some Simple Equations and Their Requirements

Equation	Solution	Requires
$2x = 4$	$x = 2$	Natural integers
$2x + 4 = 0$	$x = -2$	Signed integers
$4x = 2$	$x = 1/2$	Rational numbers
$x^2 = 2$	$x = \sqrt{2},\ x = -\sqrt{2}$	Real numbers
$x^2 + 2 = 0$	$x = ?$	?

First, rearrange it a bit:

$$x^2 = -2. \tag{2.1}$$

Solving for x means taking the square root of both sides. This would require a number on the right-hand side that, when squared, results in a negative number. But squaring is multiplying a number by itself, and the result is always positive. So there can be no solution to (2.1) under the standard rules of mathematics. Although (2.1) is simple to write, its solution poses a contradiction to familiar mathematical understanding.

What if we attempted to "quarantine" the minus sign, to factor it away from the 2? For example, we can rewrite equation (2.1) as $x^2 = 2(-1)$. In algebra we frequently let a variable or an expression stand for an unknown quantity. Let's put the unknown and troublesome aspect of (2.1) into this expression: $i^2 = (-1)$. Using this definition, we can rewrite (2.1) as $x^2 = 2 \cdot i^2$. Solving for x, we obtain

$$x = \sqrt{2} \cdot i. \tag{2.2}$$

We managed to banish the minus sign by embedding it in i, but we're hardly any better off because we don't know how to interpret i, which is still unknown. All we know is that it would have to be a number that, when multiplied by itself, equals -1. But we also know that *there is no such number* under the rules of mathematics as we currently understand them.

We have a choice: stick with the rules we have (which we'd very much like to do) or fiddle around with the rules (which might lead to chaos, so we'd rather not). But having observed that the current rules do not cover all outcomes, we can't just ignore this problem.

So, consider inventing a new kind of number that, when multiplied by itself, produces a negative result. We'd have to modify the rules of mathematics carefully to allow such a number without falsifying anything we already know to be true.

To pick up where we left off, let

$$i^2 = -1. \qquad\qquad\qquad\qquad\qquad \textit{Imaginary Rule} \quad (2.3)$$

Now let's assert that the square root of i^2 is

$$i = \sqrt{-1}. \qquad\qquad\qquad\qquad\qquad \textit{Imaginary Number} \quad (2.4)$$

Without a doubt, equations (2.3) and (2.4) stand conventional mathematics on its ear.[1] But these turn out to be just what we need. Start with equation (2.2): $x = \sqrt{2} \cdot i$. Square it to prove that it leads back to equation (2.1).

$x^2 = (\sqrt{2} \cdot i)^2$ Square both sides.

$\quad = (\sqrt{2})^2 \cdot i^2$ Square the terms separately because $(ab)^2 = a^2 \cdot b^2$.

$\quad = (\sqrt{2})^2(-1)$ Substitute -1 for i^2 by equation (2.3).

$\quad = 2(-1)$

$\quad = -2.$

Using (2.3), the Imaginary Rule, we have found a solution to $x^2 = -2$. But what sort of number is created in equation (2.4) to facilitate this solution? It is certainly not a real number because no real number when multiplied by itself produces a negative result. Mathematicians have named $\sqrt{-1}$ the *imaginary number* (although it is worth pointing out that in fact all numbers are imaginary insofar as they are all free creations of the human mind). The usefulness of the imaginary number is actually quite real; its use leads to some particularly beautiful insights about music and sound.

2.2 Operating with Imaginary Numbers

Let's take a moment to summarize. In order to solve all the equations in table 2.1, we had to create a new kind of number—the imaginary number—which produces a negative result when squared. So far, there's one number in this entire class of numbers, $i = \sqrt{-1}$.

This is unsettling because we have to relate this new kind of number to all other kinds. Clearly, if we have some numbers that produce only positive results when squared and others that produce negative results, we must figure out how to tell them apart and keep them distinct. Some things we already know about imaginary numbers can help.

For instance, whenever the imaginary number is multiplied by a real number, the result is also an imaginary number. For example, if $x = \sqrt{2} \cdot i$, then x is also an imaginary number. We can prove this by squaring it: if the result is negative, x is imaginary, and if the result is positive, x must be a real number. This was already demonstrated, but here it is again:

$$x^2 = (\sqrt{2}i)^2$$
$$= 2(-1) \tag{2.5}$$
$$= -2.$$

So, we know several facts about multiplying imaginary numbers.

The product of an imaginary number and a real number is an imaginary number.

Squaring an imaginary number turns it back into a real number.

If the square of a number is negative, then it is imaginary.

But can we combine imaginary numbers with other mathematical operations, like addition? We can't add real numbers and imaginary numbers any more than we can directly add apples and oranges because their differing characteristics require that we treat them distinctly. However, we could add apples and oranges together if we kept them distinct. Imagine the following conversation:

"Hi. How many apples do you have in your bag?"
   "Five."
   "And how many oranges do you have?"

"Four."

"Here, take these three apples and these three oranges; now how many do you have?"

"I have eight apples and seven oranges."

This conversation can be notated using a for apples and o for oranges: $(5a + 4o) + (3a + 3o) = 8a + 7o$. Everything is fine so long as we preserve the distinction between apples and oranges and operate on them separately. We could even formalize this rule by defining a new class of object named "fruit" and declaring that "fruit consists of a certain number of apples *plus* a certain number of oranges." For example, we could have the following quantities of fruit: $(5a + 4o)$ or $(3a + 3o)$ or, if we had none, $(0a + 0o)$ or, if I had no apples and owed you an orange, $(0a - 1o)$.

We can combine real numbers and imaginary numbers the same way. Just as we defined the term *fruit* to mean the combination of apples and oranges, mathematicians have adopted the term *complex numbers* to mean the combination of a regular number and an imaginary number. I think this name is unfortunate because it suggests these numbers are complicated. In fact, complex numbers are no more complicated than regular numbers.

2.3   Complex Numbers

In the preceding example, I represented a sum of apples and oranges using a notation that allowed combining them but keeping them distinct. We can take a similar approach to constructing complex numbers.

A complex number is the sum of a real and an imaginary number.
The imaginary part of the sum is distinguished by i.

Table 2.2 shows some examples of complex numbers. The third column shows how these numbers are sometimes abbreviated in practice.

2.3.1   Operations on Complex Numbers

We need a way to isolate parts of complex numbers so we can take them apart. We do so with two functions, Re and Im. If $z = a + bi$, then

$a = \text{Re}\{z\}$.   Evaluates to the real part of z.

$b = \text{Im}\{z\}$.   Evaluates to the imaginary part of z as a real number.

Table 2.2
Some Complex Numbers

Complex Number	Description	Short Equivalent
$0 + 0i$	Zero real part and zero imaginary part	0
$0 + bi$	Zero real part and imaginary part with value b	bi
$0 + 1i$	Zero real part and imaginary part of 1	i
$a + (-b)i$	Real part a and imaginary part $-b$	$a - bi$

For example,

$$\mathrm{Re}\{a + (-b)i\} = a,$$

$$\mathrm{Im}\{a + (-b)i\} = -b.$$

Notice that the result returned by $\mathrm{Im}\{\ \}$ is the imaginary part *converted back into a real number.* That is, $\mathrm{Im}\{a + bi\} = b$, not bi.

Remembering always to keep the real and imaginary parts of a complex number distinct, we can make up rules for how complex numbers behave under standard mathematical operations. First, let $u = a + bi$ and $v = c + di$, where a and c are the real parts and bi and di are the imaginary parts of u and v, respectively. We now can define the following operations.

Complex Equality   For two complex numbers to be equal, their real parts must match *and* their imaginary parts must match. That is, if $u = v$, then $a = c$, and $b = d$.

Complex Addition   If we add two complex numbers, we must add the real parts and the imaginary parts separately:

$$u + v = (a + bi) + (c + di)$$

$$= (a + c) + (bi + di) \qquad\qquad\qquad \textit{Complex Addition} \ (2.6)$$

$$= (a + c) + (b + d)i.$$

Complex Multiplication   If we multiply two complex numbers, we follow the usual procedures for multiplication, remembering that a real number times an imaginary number yields an imaginary product, and that an imaginary number squared yields a negative product:

$$u \cdot v = (a + bi)(c + di)$$

$$= ac + bdi^2 + adi + bci \qquad\qquad \textit{Complex Multiplication} \ (2.7)$$

$$= ac - bd + bci + adi$$

$$= (ac - bd) + (bc + ad)i.$$

This rather complicated arithmetic result will become a lot clearer with a graphical technique (see section 2.3.4).

Complex Negation   Negating inverts the sign of both the real and the imaginary parts of the complex number u:

$$-u = -(a + bi) = -a - bi. \qquad\qquad\qquad \textit{Complex Negation} \ (2.8)$$

Complex Conjugation   Since complex numbers provide two sign values to operate on, it would be convenient to be able to change the sign of the real part or the sign of the imaginary part independently.

The* conjugate *of a complex number is the negation of its imaginary part.

I indicate the conjugation operation by putting a bar over the quantity to be conjugated:

$$\bar{z} = \overline{a + bi} = a - bi,$$ *Complex Conjugate* (2.9)

which reads, "The complex conjugate of $a + bi$ is $a - bi$."

When we multiply a complex number $(a - bi)$ and its conjugate, the imaginary component drops out, and the result is real:

$$(a - bi)(a + bi) = a^2 + abi - abi - b^2 i^2$$

$$= a^2 + b^2.$$

A complex number multiplied by its complex conjugate is a real value.

Complex Division   Dividing a complex number by a real number is easy; divide the imaginary and real parts separately. For example, $(6 + 8i)/2 = 3 + 4i$. But what if the denominator is not real?

Sometimes it's easier to work around a problem than face it head-on. What if we could make the imaginary component drop out from the denominator? Then the problem would revert to trivial division by a real value. We know we can convert a complex number into a real number by multiplying it by its conjugate. But whatever we do to the denominator we must also do to the numerator to keep the ratio balanced. For example, find the dividend of

$$\frac{2 + 2i}{3 - 2i}.$$

Multiplying the numerator and denominator by the complex conjugate of the denominator will make the denominator real and allow us to divide the complex numerator by a real denominator:

$$\frac{(2 + 2i)(3 + 2i)}{(3 - 2i)(3 + 2i)} = \frac{6 + 10i + 4i^2}{9 - 4i^2} = \frac{2 + 10i}{13} = \frac{2}{13} + \frac{10}{13}i.$$

Consider $a + bi$ divided by $c + di$:

$$\frac{(a + bi)(c - di)}{(c + di)(c - di)} = \frac{(ac + bd)}{c^2 + d^2} + \frac{(bc - ad)i}{c^2 + d^2}.$$ *Complex Division* (2.10)

Fortunately, we won't be needing this rather complicated equation because there is a much simpler method for complex division (see section 2.3.6). Meanwhile, the following rule may be helpful.

To divide complex numbers, multiply the numerator and denominator by the conjugate of the denominator, then reduce.

2.3.2   Graphical Representation of Complex Numbers

If carrying around two numbers in order to represent one complex number seems difficult, remember we do this with rational numbers, too. Like rational numbers, complex numbers combine two numbers to create a new kind of quantity.

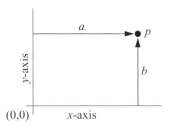

Figure 2.1
A point p on the Cartesian plane.

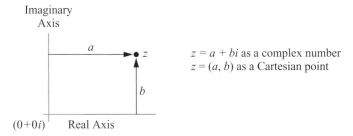

Figure 2.2
A complex number z on the complex plane.

Number pairs also combine to create a new quantity in plane geometry: a *point* in the Cartesian plane is defined as a pair of numbers, one for the x-axis, and one for the y-axis. For example (figure 2.1), a point p on the plane can be defined by a combination of horizontal and vertical values (a, b).

If we add Cartesian point p to another point q, we must add the x and y values separately. That is, if $p = (a, b)$, and $q = (c, d)$, then

$$p + q = [(a + c), (b + d)].$$

This is like operating on the two halves of a complex number separately. In fact, this suggests a way to interpret complex numbers graphically.

If we associate real numbers with the x-axis and imaginary numbers with the y-axis, then we could think of a complex number as forming a point on the *complex plane*.[2] For example, we could associate the complex number $z = a + bi$ with the point $z = (a, b)$ in the Cartesian plane (figure 2.2). This would allow us to apply geometry to understand complex numbers.

Let's look at some examples. Numbers like 1 and 3.14 are pure real numbers (complex numbers with a 0 imaginary part) lying on the real axis, whereas numbers like $1i$ and $3.14i$ are pure imaginary (with a 0 real part) lying on the imaginary axis. All other numbers $z = a + bi$, such that $a \neq 0$ and $b \neq 0$, are complex numbers in the complex plane. Note that since $i = 1i$, i is represented graphically as 1 on the imaginary axis (figure 2.3). The figure shows the position of some constants on the complex plane and the complex number z, its negation $-z$, its conjugate $\bar{z}$, and its negated conjugate $-\bar{z}$.

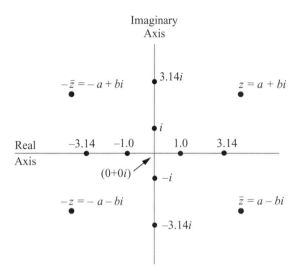

Figure 2.3
Numbers on the complex plane.

2.3.3   Trigonometric Representation of Complex Numbers

Complex numbers really begin to pay off when we view them through trigonometry. They provide a compact, powerful representation for sinusoids that we will come to depend upon. Then the complicated algebraic rules of complex math become unnecessary.

Figure 2.4 shows a triangle in the complex plane. The location of the complex point z can be found in two ways. First, using geometry, if we define the lengths of the sides of the triangle as a and b, then the point is $z = a + bi$.

Equivalently, we could find z by determining the magnitude (length) of the hypotenuse r and its angle θ above the horizontal plane. If $r \neq 0$, and θ is the angle of the hypotenuse relative to the positive real axis, then by trigonometry,

$$a = r \cdot \cos\theta,$$

$$bi = r \cdot i\sin\theta.$$

(See appendix section A.2 for an introduction to trigonometry.)

Substitute these trigonometric definitions for a and bi back into the geometric definition for z:

$$z = a + bi$$

$$= (r \cdot \cos\theta) + (r \cdot i\sin\theta).$$

Factoring out the common term r,

$$z = r(\cos\theta + i\sin\theta). \qquad \textit{Trigonometric Form of a Complex Number} \quad (2.11)$$

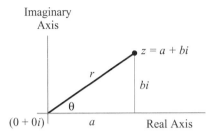

Figure 2.4
A complex number as a magnitude and an angle.

Equation (2.11) provides a way to find a complex number just by knowing its distance r from the origin of the complex plane and its angle θ. More important, (2.11) suggests that we can treat complex numbers as vectors. A *vector* is simply the combination of a magnitude and a direction. Equation (2.11) identifies a complex number as a magnitude (r) and a direction (θ). Furthermore, using (2.11), we can view any complex number equally as the sum of two orthogonal vectors (lying at a 90° angle from one another), the first on the real axis with a magnitude of a and the second on the imaginary axis with a magnitude of b.

The larger significance of equation (2.11) is that, by relating trigonometric functions to the construction of complex numbers, we now have a bridge between these two realms.

The Angle and the Magnitude In figure 2.4 the variable θ is called the *angle* of z.[3] The variable r, showing the distance from the point z to the origin, is called the *absolute value* or the *magnitude* of z. The latter definition does not conflict with the absolute value of a real number, which is similarly the distance from a point to the origin; we're just expanding the definition to cover points other than those on the real line. The absolute value of a complex number z is written $r = |z|$, just as with integers and real numbers. For example, $|-3| = 3$, and $|-3i| = 3i$. But what about complex numbers that do not lie on the real or imaginary axis?

Suppose we have determined the location of a point in the complex plane by its magnitude r and angle θ, but now we wish to rediscover its Cartesian coordinates a and b. Trigonometry again comes to the rescue because this is the same as finding the length of the sides of a right triangle from the length of its hypotenuse and its angle. The length of the side along the real axis is $a = r\cos\theta$, and the length of the side along the imaginary axis is $b = r\sin\theta$. So now we can convert back and forth between the complex and Cartesian coordinate systems.

Pythagoras Revisited Another useful relation between the complex number $z = a + bi$ and the magnitude of its hypotenuse r involves z and its conjugate $\bar{z}$. The relation is $r^2 = z\bar{z}$. Here's how to see it. Define a right triangle anchored at the origin of the Cartesian plane with sides a and b, and hypotenuse r (figure 2.5). Then by the Pythagorean theorem, we can write

$$r^2 = a^2 + b^2 .$$

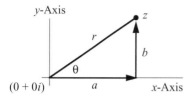

Figure 2.5
Triangle in the Cartesian plane.

Now add the term $abi - abi$ to the middle of the right-hand side. Clearly we can do this because adding and then subtracting the same value from an equation does not alter the equality. So we have

$$r^2 = a^2 + abi - abi + b^2 . \qquad (2.12)$$

Focus on the b^2 term for a moment. Because of the nature of i,

$$b^2 = -b^2 \cdot i^2 = bi(-bi). \qquad (2.13)$$

We can do this because by equation (2.3), $i^2 = -1$. Repeating equation (2.12) but substituting the definition for b^2 from (2.13), we get

$$r^2 = a^2 + abi - abi + bi(-bi).$$

Factoring,

$$r^2 = (a + bi)(a - bi) \qquad (2.14)$$
$$= z\bar{z},$$

by the definition of the complex conjugate.[4] Now if $r^2 = z\bar{z}$, then it must be that the hypotenuse

$$r = \sqrt{z\bar{z}} = \sqrt{a^2 + b^2} .$$

This will come in handy later.

2.3.4   Multiplication Interpreted Trigonometrically

Equation (2.11) showed that we can find a complex number z if we know its angle θ and its distance from the complex origin r:

$$z = a + bi = r(\cos\theta + i\sin\theta). \qquad \textit{Complex Number Interpreted Trigonometrically} \quad (2.15)$$

What would happen if we multiplied the trigonometric form of two complex numbers? If $u = r(\cos\theta + i\sin\theta)$, and $v = s(\cos\phi + i\sin\phi)$, their product is

$$uv = r(\cos\theta + i\sin\theta) \cdot s(\cos\phi + i\sin\phi).$$

We can simplify this if we let $a = \cos\theta$, $b = \sin\theta$, $c = \cos\phi$, and $d = \sin\phi$. Substituting these definitions into the equation yields

$$uv = r(a + ib) \cdot s(c + id).$$

We can then expand it as follows:

$$uv = rs(a + ib)(c + id) = rs(ac + aid + ibc + bdi^2).$$

Remembering that $i^2 = -1$, we have

$$uv = rs(ac + aid + ibc - bd).$$

Grouping terms into complex number format,

$$uv = rs[(ac - bd) + i(bc + ad)].$$

Now if we substitute back the original terms for a, b, c, and d, we end up with

$$uv = rs[(\cos\theta\cos\phi - \sin\theta\sin\phi) + i(\sin\theta\cos\phi + \cos\theta\sin\phi)] \tag{2.16}$$

Where is all this leading? I believe there's an unwritten rule in mathematics that equations must get longer before they can get shorter. The good news is, we've reached the point where this one starts getting shorter. But first, there are two tools we need. A modest application of trigonometry (see appendix section A.2) demonstrates that

$$\cos\theta\cos\phi - \sin\theta\sin\phi = \cos(\theta + \phi),$$

and

$$\sin\theta\cos\phi + \cos\theta\sin\phi = \sin(\theta + \phi).$$

These trigonometric identities allow us to reduce the size of equation (2.16) substantially. Substituting these identities back into (2.16), we get:

$$uv = rs[\cos(\theta + \phi) + i\sin(\theta + \phi)]. \qquad \textit{Complex Multiplication Interpreted} \atop \textit{Trigonometrically} \tag{2.17}$$

Equation 2.17 tells a much simpler story about the product of uv than (2.16) does. The product of uv is scaled by the product of the magnitudes r and s, and the angle of uv is the sum of the angles θ and ϕ.

The product of two complex numbers is the product of their magnitudes and the sum of their angles.

This is much simpler and more intuitive than the algebraic definition given in equation (2.7).

Interpreting i Geometrically   How can we represent i itself as a complex number in trigonometric form? What are its angle and magnitude? By definition, i corresponds to the value $+1$ on the imaginary axis (figure 2.6). So we can represent i as having magnitude $r = 1$ and angle $\theta = 90°$. With

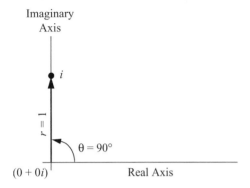

Figure 2.6
Geometric value of i.

this information, and recalling equation (2.15), we can write the trigonometric form of i:

$$i = 1(\cos 90° + i \sin 90°)$$

$$= 1\left(\cos\frac{\pi}{2} + i\sin\frac{\pi}{2}\right). \tag{2.18}$$

Now, $\cos(\pi/2) = 0$ and $\sin(\pi/2) = 1$. Substituting, we have

$$i = 1\left(\cos\frac{\pi}{2} + i\sin\frac{\pi}{2}\right)$$

$$= 0 + i.$$

We have demonstrated that the complex value of i is $(0 + i)$, ending up right back where we started. But now we also know its trigonometric expression and have an idea of how to visualize it graphically.

Multiplying by i   What happens if we multiply i times a complex number z? Recalling equations (2.17) and (2.18),

$$z \cdot i = r[\cos(\theta + 90°) + i\sin(\theta + 90°)]. \tag{2.19}$$

We just rotate z counterclockwise by 90°. Suppose we set $z = 1 + 0i$, which makes it a positive vector lying on the real axis of magnitude 1. Its trigonometric form is $z = 1(\cos 0 + i \sin 0)$ because this is also a positive vector lying on the real axis of magnitude 1. Equation (2.19) says that if we multiply z times i, we end up with

$$z = [\cos(0 + 90°) + i\sin(0 + 90°)],$$

which is a positive vector lying on the imaginary axis of magnitude 1. In other words, we've just rotated z by 90° counterclockwise, leaving its length the same.

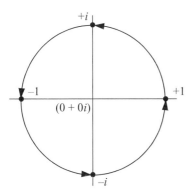

Figure 2.7
Successive multiplications of 1 by i.

Multiplying a number by i rotates it 90° counterclockwise; its magnitude remains the same.

Note that this rule works for complex numbers, real numbers, and imaginary numbers alike. For example, if we start with 1, and multiply it by i, we rotate it by 90° counterclockwise, obtaining $1.0i$. Similarly, multiplying $i \cdot i$ rotates i by 90° counterclockwise, and gives -1 (because $i^2 = -1$). Multiplying $-1 \cdot i$ yields $-i$, and multiplying $-i \cdot i$ gives 1 again. This is shown graphically in figure 2.7.

2.3.5 Squaring a Complex Number

Using equation (2.17), we know that z^2 will have an absolute value of r^2 and an argument of $\theta + \theta = 2\theta$:

$$z^2 = r^2(\cos 2\theta + i\sin 2\theta).$$

To square a complex number, square the magnitude and double the angle.

2.3.6 Complex Division with Trigonometry

If multiplying two complex numbers means multiplying the magnitudes and adding the angles, it follows that division means dividing the magnitudes and subtracting the angles. The ratio of two complex numbers u and v can be expressed as

$$\frac{u}{v} = \frac{r(\cos\theta + i\sin\theta)}{s(\cos\phi + i\sin\phi)} = \frac{r}{s}[\cos(\theta - \phi) + i\sin(\theta - \phi)].$$

The angle of the argument will be rotated in the clockwise direction. What if the denominator is i? In that case,

$$\frac{u}{i} = \frac{r[\cos\theta + i\sin\theta]}{1[\cos 90° + i\sin 90°]} = r[\cos(\theta - 90°) + i\sin(\theta - 90°)].$$

To divide a number by i, rotate it clockwise 90°; its magnitude remains unchanged.

2.3.7 Tricks with i

Remember these useful tricks, they will come in handy soon:

• Multiplying by $-i$ is the same as dividing by i. That is, $1/i = -i$. To see this, rotate a complex vector of unit length clockwise by $90°$ (figure 2.7).

• $(1/i)(1/i) = (-i)(-i) = -1$. To see this, rotate a complex vector of unit length clockwise by $180°$ (figure 2.7).

2.4 de Moivre's Theorem

So far, we have been laying the groundwork for our trip up Complex Mountain, buying supplies and trekking to the edge of the foothills. In the following sections we begin to see some lofty sights. The first is de Moivre's theorem.

The square of a complex number z is

$$z^2 = r^2(\cos 2\theta + i \sin 2\theta).$$

Multiplying again by z, we have $z^2 \cdot z = r^3(\cos 3\theta + i \sin 3\theta)$, and in general,

$$z^n = r^n(\cos n\theta + i \sin n\theta), \qquad n = 0, 1, 2, \ldots. \tag{2.20}$$

Consider the set of all complex numbers whose magnitude is 1. They are all a unit distance from complex zero, $0 + 0i$, the origin of the complex plane, which means they form a circle around complex zero (figure 2.7). The points are complex numbers on the unit circle because they are all a unit distance from complex zero.

If we view complex numbers as vectors, complex multiplication and division are nothing more than rotating these vectors around complex zero and scaling their magnitudes.

What if the magnitudes of the two complex numbers being multiplied are both exactly unity? Since we are multiplying unities, we'd expect that all we do is spin the vector around while the magnitude remains the same, and this is indeed what happens. For example, in equation (2.20), let $r = 1$, and observe that no matter what value we assign to n, z^n will always have a magnitude of 1, placing it always on the unit circle.

Though perhaps it is not obvious at first, we can easily rewrite the term $\cos n\theta + i \sin n\theta$ from (2.20) as $(\cos \theta + i \sin \theta)^n$. Clearly, $(\cos \theta + i \sin \theta)^n$ is a complex number with unity magnitude raised to a power. Let $z^n = (\cos \theta + i \sin \theta)^n$.

But we also know that any complex number z^n can be written as $z^n = r^n(\cos n\theta + i \sin n\theta)$ for some value of r. And, in this case, $r = 1$ because we are looking only at points on the unit circle with unity magnitude:

$$z^n = r^n(\cos \theta + i \sin \theta)^n = r^n(\cos n\theta + i \sin n\theta)$$

$$= 1^n(\cos n\theta + i \sin n\theta),$$

and since $1^n = 1$, we have shown that

$$\cos n\theta + i \sin n\theta = (\cos \theta + i \sin \theta)^n.$$

de Moivre's[5] Theorem: *Raising a complex number with unity magnitude to a power n is equivalent to multiplying the angle of the complex number by n.*

For example, if we raise a complex number with unity magnitude to increasing powers of n, we set the number to spin counterclockwise around the unit circle. The greater the value of θ, the greater the angular distance covered by each increase in n.

If the magnitude of the vector r is not unity, then in general,

$$r^n(\cos n\theta + i \sin n\theta) = [r(\cos \theta + i \sin \theta)]^n. \qquad\qquad \textit{de Moivre's Theorem} \quad (2.21)$$

This provides a simple formula for calculating powers of complex numbers. If $z = r(\cos \theta + i \sin \theta)$, then

$$z^n = r^n(\cos n\theta + i \sin n\theta).$$

2.4.1 Taylor Series for Sine and Cosine

Brook Taylor[6] knew that some series had been shown to be equivalent to trigonometric functions. For example, he knew that

$$\sin x = x - \frac{x^3}{3!} + \frac{x^5}{5!} - \frac{x^7}{7!} + \cdots \qquad\qquad (2.22)$$

and

$$\cos x = 1 - \frac{x^2}{2!} + \frac{x^4}{4!} - \frac{x^6}{6!} + \cdots, \qquad\qquad (2.23)$$

where x is a real number in radians. The ellipses in these equations mean these series must extend to infinity before they are equal to the expressions on the left side.[7] The ! operator is the *factorial* operator in mathematics. For example, $3! = 3 \cdot 2 \cdot 1$. In general, $n! = n(n-1)(n-2) \cdots (1)$.

Note the symmetry of these equations. Both begin with positive terms, then alternate signs, $+, -, +, \ldots$. The series for $\sin x$ is defined using only odd numbers, and the series for $\cos x$ is defined using only even numbers.

Many such series were developed by Brook Taylor and others because they wanted to find quick ways to approximate the numeric value of trigonometric relations. Each subsequent term in equations (2.22) and (2.23) is much smaller than the preceding one, so these series converge quickly to their target values. Hence, we can compute the sine or cosine of an angle to any degree of precision desired simply by summing more and more terms of these equations. The more terms that are summed, the more precise the result. Summing an infinite number of terms produces the value exactly. Although with modern computers this application of the Taylor series is no longer a pressing concern, (2.22) and (2.23) can be combined in a kind of Chinese puzzle that reveals a most startling result.

2.4.2   Value of e

What is e? A dictionary will indicate that it is the base of the natural logarithms, whose symbol honors the great mathematician Leonhard Euler.[8] The first few decimal places of e offer very little additional insight into its nature. They are 2.718281828459045235. . . . Like π, e is an irrational constant that is useful in a variety of mathematical contexts. We can take a kind of "black box" approach to it, that is, use it without thinking too much about what it is.

Then why did I even bring up the subject of e? Well, it turns out that the Taylor expansion of e,

$$e = \sum_{n=0}^{\infty} \frac{1}{n!} = \frac{1}{0!} + \frac{1}{1!} + \frac{1}{2!} + \cdots, \qquad \textit{Taylor Expansion of } e \quad (2.24)$$

links up with the sine and cosine series in (2.22) and (2.23) in a way that has great practical bearing. Consider this series for e raised to a power x:

$$e^x = 1 + \sum_{n=1}^{\infty} \frac{x^n}{n!} = 1 + \frac{x}{1!} + \frac{x^2}{2!} + \frac{x^3}{3!} + \frac{x^4}{4!} + \frac{x^5}{5!} + \cdots. \qquad (2.25)$$

As with the sine and cosine series, the series in (2.25) provides a way to compute an approximate value to any desired precision of e to any power x.

2.5   Euler's Formula

If we substitute the complex number z into equations (2.22), (2.23), and (2.25), we have

$$\sin z = z - \frac{z^3}{3!} + \frac{z^5}{5!} - \frac{z^7}{7!} + \cdots, \qquad (2.26)$$

$$\cos z = 1 - \frac{z^2}{2!} + \frac{z^4}{4!} - \frac{z^6}{6!} + \cdots, \qquad (2.27)$$

and

$$e^z = 1 + \frac{z}{1!} + \frac{z^2}{2!} + \frac{z^3}{3!} + \frac{z^4}{4!} + \frac{z^5}{5!} + \cdots. \qquad (2.28)$$

The similarities in the patterns of (2.26), (2.27), and (2.28) are striking. It seems that the series for e^z is made up of the sine and cosine series interleaved together in some way . . . except that the sine and cosine series alternate plus and minus signs whereas the series for e^z only sums positive terms. If we could find a way to relate these three series, we'd have a path toward linking e to the cosine and sine functions.

It looks like the expansion of e^z somehow combines the terms for the expansions of $\sin z$ and $\cos z$. What if we sum the sine and cosine series just to see what they look like together?

$$\cos z + \sin z = +1 + z - \frac{z^2}{2!} - \frac{z^3}{3!} + \frac{z^4}{4!} + \frac{z^5}{5!} - \frac{z^6}{6!} - \frac{z^7}{7!}. \tag{2.29}$$

Repeating pattern of signs

Notice that the signs in (2.29) show a repeating pattern: $+,+,-,-,+,+,-,-$. This feels familiar. Looking back at figure 2.7, recall that the powers of i have the same periodicity of signs:

$$i^0 = 1 \qquad +$$
$$i^1 = i \qquad +$$
$$i^2 = -1 \qquad -$$
$$i^3 = -i \qquad -$$
$$i^4 = 1 \qquad +$$
$$\vdots \qquad \vdots$$

So the series for e is identical to the series for $\cos z + \sin z$ except that the terms of the $\cos z + \sin z$ series switch signs with the same periodicity as successive powers of i. If we modify the equation for e^z to be e^{iz}, the effect on the right-hand side of equation (2.28) would be as follows:

$$e^{iz} = 1 + \frac{iz}{1!} + \frac{i^2 z^2}{2!} + \frac{i^3 z^3}{3!} + \frac{i^4 z^4}{4!} + \frac{i^5 z^5}{5!} + \frac{i^6 z^6}{6!} + \frac{i^7 z^7}{7!} + \cdots$$

$$= 1 + \frac{iz}{1!} - \frac{z^2}{2!} - \frac{iz^3}{3!} + \frac{z^4}{4!} + \frac{iz^5}{5!} - \frac{z^6}{6!} - \frac{iz^7}{7!} + \cdots. \tag{2.30}$$

We can see the destination hidden in equation (2.30). Notice that the even exponents correspond to the series for $\cos z$, and the odd exponents correspond to the series for $\sin z$. This might be clearer if we rearrange (2.30) to group the even exponents and then group the odd exponents, which all contain i:

$$e^{iz} = 1 - \frac{z^2}{2!} + \frac{z^4}{4!} - \frac{z^6}{6!} + \cdots + i\left(\frac{z}{1!} - \frac{z^3}{3!} + \frac{z^5}{5!} - \frac{z^7}{7!} + \cdots\right). \tag{2.31}$$

Since the left group of terms equals $\cos z$, and the right group of terms equals $\sin z$, we have shown that

$$e^{iz} = \cos z + i \sin z. \qquad\qquad\qquad \textit{Euler's Formula} \quad (2.32)$$

This famous result is known today as *Euler's formula* or *Euler's equation.* It links the hyperbolic functions involving e to trigonometric functions involving π.

2.5.1 But Where Is π?

If equation (2.32) indeed links e and π, where is π?

First, let's add a new variable a on both sides of equation (2.32):

$$e^{i(z+a)} = \cos(z+a) + i\sin(z+a). \tag{2.33}$$

Second, recall from trigonometry that because there are 2π radians in a circle, $\cos z = \cos(z+2\pi)$ and $\sin z = \sin(z+2\pi)$. In general, the same is true for any integer multiple of 2π. This means that the sine and cosine functions are *periodic:*

$$\cos z = \cos(z+n2\pi),$$

$$\sin z = \sin(z+n2\pi),$$

where n is any integer. As shown in figure 2.8, this makes intuitive sense if we remember that adding $n2\pi$ to some angle z returns us to the same spot on the circle each time, so long as n is an integer.

Now, if we let $a = 2\pi$ in equation (2.33), we have

$$e^{i(z+2\pi)} = \cos(z+2\pi) + i\sin(z+2\pi).$$

But, as we've just seen, this is identical to

$$e^{i(z+0)} = \cos(z+0) + i\sin(z+0),$$

and therefore we've shown that

$$e^{i(z+2\pi)} = e^{iz}. \tag{2.34}$$

As promised, we've introduced π to Euler's formula. But that's not all we get for the effort. It follows that e^{iz} is periodic with period 2π just as the sine and cosine functions are:

$$e^{i(z+n2\pi)} = e^{iz}, \tag{2.35}$$

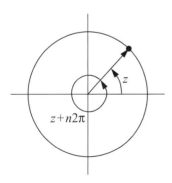

Figure 2.8
Scaling an angle by $n2\pi$.

where n is any integer. Perhaps this is not so surprising when we remember that we have related e^{iz} to functions of circles, and circles are . . . well, circular.

2.5.2 The Most Beautiful Formula

To see more of this interesting view, let's climb a little higher. Because $\cos 2\pi n = 1$, and $\sin 2\pi n = 0$, for integer values of n, it follows that

$$\begin{aligned} e^{i2\pi n} &= \cos(2\pi n) + i \sin(2\pi n) \\ &= 1 + 0i, \end{aligned}$$
(2.36)

revealing that $e^{i2\pi n}$ equals 1 *regardless* of the value of n (so long as n is an integer). This is a pretty startling result, actually. We are within striking distance of a truly breathtaking vista. Suppose we simplify equation (2.36) by setting $n = 1/2$ (relaxing the requirement that n be an integer). Substituting, we have

$$e^{i2\pi(1/2)} = e^{i\pi} = \cos \pi + i \sin \pi,$$

and since we know that $\cos \pi = -1$ and $\sin \pi = 0$, we have

$$\begin{aligned} e^{i\pi} &= -1 + 0i \\ &= -1. \end{aligned}$$
(2.37)

Finally, if we rearrange equation (2.37) slightly, we get

$$e^{i\pi} + 1 = 0. \qquad\qquad \textit{Euler's Identity} \quad (2.38)$$

This equation brings together five of the most important numerical values in mathematics, e, i, π, 1, and 0, in one simple, elegant relation. Equation (2.38) has been described as the most beautiful formula in mathematics. It's like seeing the entire panoply of the planets together with a crescent moon at sunset. By analogy, this equation can also be seen to integrate the four main branches of mathematics: 0 and 1 from arithmetic, π from geometry, i from algebra, and e from analysis.

Equation (2.38) is the cornerstone of a major mathematical edifice that represents musical signals, among other things, in a crisp and penetrating way. This goes to show, as Sir D'Arcy Wentworth Thompson (1917) wrote, "The perfection of mathematical beauty is such . . . that whatsoever is most beautiful and regular is also found to be most useful and excellent."

2.5.3 What Is e^i by Itself?

We got such a fine result by introducing π into Euler's identity that it seems a shame to remove it, but to find out the value of e^i by itself, we go back to the Taylor series definition of e raised to a power, given in equation (2.25):

$$e^x = 1 + \frac{x}{1!} + \frac{x^2}{2!} + \frac{x^3}{3!} + \frac{x^4}{4!} + \frac{x^5}{5!} + \cdots.$$

We want to find the solution to this equation for $x = 0 + 1i$. Substituting, we get

$$e^{0+1i} = 1 + \frac{i}{1!} + \frac{i^2}{2!} + \frac{i^3}{3!} + \frac{i^4}{4!} + \frac{i^5}{5!} + \cdots$$

$$= 1 + i - \frac{1}{2!} - \frac{i}{3!} + \frac{1}{4!} + \frac{i}{5!} - \cdots.$$

Grouping even and odd terms yields

$$e^i = 1 - \frac{1}{2!} + \frac{1}{4!} - \frac{1}{6!} + \cdots + i\left(1 - \frac{1}{3!} + \frac{1}{5!} - \frac{1}{7!} + \cdots\right) \tag{2.39}$$

$$= \cos 1 + i \sin 1,$$

which says that e^i is a complex number with a real part equal to cos 1 and an imaginary part equal to sin 1. Remembering that the arguments to the sine and cosine functions are in units of radians, we can write

$$e^i = 1(\cos 1 + i \sin 1), \tag{2.40}$$

which allows us further to say that equation (2.40) represents a vector in the complex plane of unit length and angle of 1 radian. The Cartesian values are approximately $x = \cos 1 \cong 0.543$ and $y = \sin 1 \cong 0.841$.

2.6 Phasors

Consider the complex number $z = x + iy$. If we let $x = \cos\theta$ and $y = \sin\theta$, then

$$z = \cos\theta + i\sin\theta,$$

and we know by Euler's formula that

$$\cos\theta + i\sin\theta = e^{i\theta}. \tag{2.41}$$

Now, if we set $\theta = 1$, we've simplified back to equation (2.40), where we established that by itself e^i can be thought of as a vector from the origin of the complex plane with length 1 and angle of 1 radian. But let's leave θ in the equation and assume θ represents a real number.

Suppose θ gradually decreases from 1 to 0. At first, when $\theta = 1$, the value will be $e^{i1} = \cos 1 + i \sin 1$, just as in equation (2.41). But as θ decreases, $e^{i\theta}$ rotates clockwise (figure 2.9). When θ gets to 0, we have

$$e^{i0} = \cos 0 + i \sin 0$$

$$= 1 + i0,$$

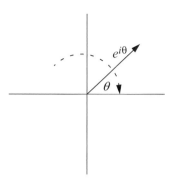

Figure 2.9
Unit vector as θ decreases toward zero.

that is, the vector $e^{i\theta}$ will be lying along the positive real axis:

Let's check out some other interesting values for θ. We've already seen by equation (2.38) that when $\theta = \pi$,

$$
\begin{aligned}
e^{i\pi} &= \cos \pi + i \sin \pi \\
&= -1 + 0i,
\end{aligned}
\tag{2.42}
$$

which means the vector $e^{i\theta}$ will be lying along the negative real axis:

Two other values of θ are noteworthy. If we set $\theta = \pi/2$:

$$
\begin{aligned}
e^{i(\pi/2)} &= \cos \frac{\pi}{2} + i \sin \frac{\pi}{2} \\
&= 0 + 1i,
\end{aligned}
\tag{2.43}
$$

which means the vector $e^{i\theta}$ will be lying along the positive imaginary axis. And if we set $\theta = \pi 3/2$,

$$
\begin{aligned}
e^{i(\pi 3/2)} &= \cos \frac{3\pi}{2} + i \sin \frac{3\pi}{2}, \\
&= 0 - 1i,
\end{aligned}
\tag{2.44}
$$

which means the vector $e^{i\theta}$ will be lying along the negative imaginary axis.

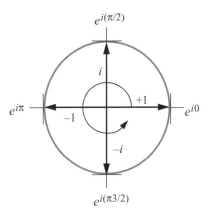

Figure 2.10
Complex unit circle.

Finally, if $\theta = 2\pi$,

$$e^{i2\pi} = \cos 2\pi + i \sin 2\pi$$
$$= 1 + 0i,$$

(2.45)

which means the vector $e^{i\theta}$ has gone back to lying along the positive real axis, making one complete rotation.

Thus, as θ goes from 0 to 2π, the vector $e^{i\theta}$ *spins once counterclockwise on its axis* in the complex plane. This is shown in figure 2.10. One complete rotation is called a *period*. The different positions the vector reaches on the unit circle during a period are referred to as its *phases*. Since the angle θ controls the phase, it is called the *phase angle*.

To summarize, e^i is a unit vector, that is, a vector of length 1 and angle of 1 radian. For $e^{i\theta}$, as θ goes from 0 to 2π, the unit vector visits every point on the complex unit circle, including $+1$, i, -1, and $-i$. As θ increases past 2π, $e^{i\theta}$ will just continue to spin, returning to $+1$ whenever θ is an integer multiple of 2π.

Notice how much more compact it is to write $e^{i\theta}$ rather than $\cos\theta + i\sin\theta$. This provides tremendous economy of expression for discussing wave motion later.

There's only one thing missing: a way to make the vector other than unit length. If we scaled $e^{i\theta}$ by a real variable r, we could change the length of the vector as well:

$$z = re^{i\theta}. \qquad\qquad \textit{Phasor} \quad (2.46)$$

As r changes, the vector's length changes, and as θ changes, the vector's direction changes. Equation (2.46) is convenient *polar representation* of any complex variable. It is easier to write $z = re^{i\theta}$ than to write $z = r(\cos\theta + i\sin\theta)$, and we get the intuitive advantage of visualizing z as a vector spinning around the origin of the complex plane. Polar representation of complex variables is so powerful that it has its own name: equation (2.46) is called the *phasor*.

The two variables r and θ in (2.46) allow us to identify uniquely any point on the complex plane. Imagine a line of length r called the *radial coordinate* with its base anchored at the origin of the complex plane. The counterclockwise angle of this line above the real axis is given by θ, called the *angular coordinate,* or polar angle. Together, r and θ are called the *polar coordinates*. They are related to Cartesian coordinates by

$$x = r\cos\theta, \quad \text{and} \quad y = r\sin\theta,$$

where r is the radial distance from the origin, and θ is the angle from the real axis traveling counterclockwise. In terms of x and y, by the Pythagorean theorem,

$$r = \sqrt{x^2 + y^2},$$

$$\theta = \tan^{-1}\frac{y}{x}.$$

2.6.1 Circular Motion

In the previous section, we scaled Euler's formula by a real variable r,

$$re^{i\theta} = r(\cos\theta + i\sin\theta),$$

defined as a phasor of magnitude r. Figure 2.9 showed that as θ decreases, the phasor spins *clockwise,* and figure 2.10 showed that as θ increases, the phasor spins *counterclockwise.*

2.6.2 The Upside-Down Bicycle

Imagine a phasor as a spoke on a bicycle wheel that has been painted red. It spins one way or another depending on whether θ is increasing (becoming more positive) or decreasing (becoming more negative). The variable r can be thought of as the length of the spoke and hence the size of the wheel.[9] One spoke on the other wheel is also painted red so we can track them both easily. Now spin one wheel counterclockwise, corresponding to θ growing more positive, and spin the other clockwise, corresponding to θ growing more negative.

As the wheels spin, how many revolutions do the wheels make per second? Suppose we observe that the red spokes on both wheels rotate once every second, or 1 Hz. We can relate the speed of rotation to the rate at which the angle of the red spoke on each wheel changes. Since the spokes are rotating at 1 Hz, the angle of each red spoke travels through 2π radians per second. The faster they spin, the higher the frequency and the greater the angular velocity.

Now close your eyes and quickly blink them open and shut, noting the phase angles of the red spokes on the two wheels. If you blink quickly enough, the spokes seem not to be moving. You have observed the *instantaneous phase angle* of the spokes.

Now stop the bicycle wheels and spin one, and then the other, clockwise. Observe the instantaneous phase angles again. Since one wheel started after the other, their instantaneous phase angles will not be equal even if they rotate with the same frequency: one wheel will lead the other by some amount. This is their *phase difference.*

If the wheels travel at the same exact frequency, the phase difference will be a constant. If, as is more likely, one wheel travels faster than the other, the phase difference will change gradually as the wheels turn, and one red spoke will overtake the other and then pass it. One wheel is said to be *precessing* the other. The *rate of precession* is the time it takes for the phase difference between them to return to the initial phase difference.

Again, spin one wheel clockwise and the other counterclockwise so that they are traveling at the same rate of speed in opposite rotation. Even if they travel at exactly the same speed, the angular velocities of the two wheels are not equal because the radian velocity of the wheel turning counterclockwise is 2π radians per second, whereas the radian velocity of the wheel turning clockwise is -2π radians per second. Thus, if the wheel turning counterclockwise with positive radial velocity has a frequency of 1 Hz, the one turning clockwise with negative radial velocity must have a frequency of –1 Hz. *Positive frequencies* correspond to counterclockwise rotation and *negative frequencies* correspond to clockwise rotation.

2.6.3   Positive and Negative Frequencies

How can we express positive and negative radian velocity mathematically? We can understand positive frequencies with the phasor

$$re^{i\theta} = r(\cos\theta + i\sin\theta), \tag{2.47}$$

such that as θ increases, the phasor spins counterclockwise, corresponding to positive frequencies.

Can we represent negative frequencies simply by inverting the sign of θ? That is, what about $e^{i(-\theta)} = e^{-i\theta}$? It should work, but let's check. Going back to Euler's formula, we see that if $e^{i\theta} = \cos\theta + i\sin\theta$, then

$$e^{-i\theta} = (\cos -\theta) + (i\sin -\theta) \tag{2.48}$$

$$= \cos\theta - i\sin\theta$$

because $\cos -\theta = \cos\theta$, and $\sin -\theta = -\sin\theta$.

By (2.48), as θ begins to increase from 0, the real part, $\cos\theta$, will go from 1 toward 0 (shrinking along the x-axis toward 0) while the imaginary part, $-i\sin\theta$, will go from 0 toward $-i$ (growing along the negative y-axis). When $\theta = 0$, the negative phasor $e^{-i\theta}$ starts off lying along the real axis line just like its positive cousin $e^{i\theta}$. But as θ begins to increase from 0, the negative phasor drops *down and to the left,* beginning a clockwise rotation, just as we wanted.

Thus, we can use the *negative-frequency phasor* $e^{-i\theta}$ to represent negative, clockwise-turning frequencies and the *positive-frequency phasor* $e^{i\theta}$ to represent positive, counterclockwise-turning frequencies. This is summarized in figure 2.11.

2.6.4   Complex Harmonic Motion

We know that simple harmonic motion is the projection of circular motion onto one-dimensional displacement (see volume 1, section 5.1, especially volume 1, figure 5.7). And sinusoidal motion

Positive-Frequency Phasor:

$$e^{i\theta} = \cos\theta + i\sin\theta$$

Negative-Frequency Phasor:

$$e^{-i\theta} = \cos\theta - i\sin\theta$$

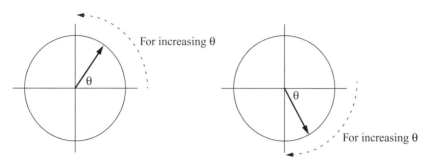

For increasing θ

For increasing θ

Figure 2.11
Positive- and negative-frequency phasors.

Figure 2.12
Projecting sine and cosine from the same circular motion.

is the projection of simple harmonic motion through time (see volume 1, figure 5.9). The root motion governing both is circular motion.

Figure 2.12 shows two spotlights at right angles to each other projecting the shadow of a cone mounted on a turntable onto two screens. As the turntable turns, the harmonic motion of the shadows projected on the screens will show a phase difference of 90°, precisely the phase difference between sine and cosine. If we think of the cone on the turntable as a phasor on the complex plane, then the shadows describe the motion of the projected sine and cosine harmonic motions. The only difference between sine and cosine is the angle of projection.

If we look at figure 2.12 from directly overhead, as shown in figure 2.13, we see that the phasor $e^{i\theta}$ does indeed embody both the cosine and sine relations simultaneously. In this graph the circle

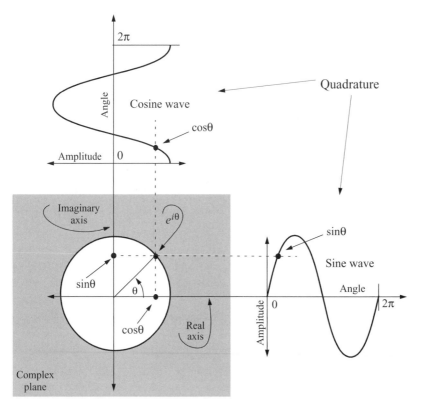

Figure 2.13
Complex phasor projected on the real axis and the imaginary axis.

is plotted in the complex plane, but the projected sine and cosine waves are plotted as real values of amplitude against angle θ as it moves in time from 0 to 2π.

Project along the real axis in figure 2.13 as θ goes from 0 to 2π. When $\theta = 0$, the phasor lies to the right of complex zero along the real axis, and it swings around counterclockwise as θ increases. We see that the phasor $e^{i\theta}$ makes a full circle counterclockwise, beginning and ending at $e^{i0} = 1 + 0i$. As the phasor turns, we see by inspection that the projected point $\sin\theta$ describes a sine wave because it begins at amplitude 0, gradually increases in amplitude to 1, works its way back to 0, then to -1, and finally returns to 0, just as a sine wave does. Similarly, the projected point $\cos\theta$ describes a cosine wave because it begins at amplitude 1, decreases in amplitude to -1, then works its way back to 1.

When signals are lock-stepped at a 90° phase difference like the sine and cosine projections of circular motion, the signals are said to be in *quadrature*. Although quadrature has a number of meanings in mathematics, in this context it means a 90° phase relation between two periodic quantities varying with the same period.

2.6.5 Sinusoids

Looking at figure 2.13, we see that if we project vertically across the real axis, the phasor $e^{i\theta}$ generates $\cos\theta$ as θ increases, and if we project horizontally across the imaginary axis, the phasor generates $\sin\theta$ as θ increases. Recalling Euler's formula, figure 2.13 shows that a phasor is indeed the sum of a cosine and a sine in quadrature because they both emerge as projections of $e^{i\theta} = \cos\theta + i\sin\theta$.

But why restrict ourselves to projecting just along the two dimensions of the complex plane? We can swivel the projector around to any arbitrary angle and create a host of different but related wave functions. Swiveling the light to $45°$, we produce a wave that is halfway between a sine and cosine wave (figure 2.14). The formula for this wave is

$$\cos(\theta + 45°) + i\sin(\theta + 45°).$$

As the projector is swiveled around an entire circle, we observe all possible combinations of sine and cosine wave. A slight modification of Euler's formula allows us to represent this process of projecting from different angles:

$$e^{i(\theta+\phi)} = \cos(\theta+\phi) + i\sin(\theta+\phi), \hspace{2cm} \textit{Sinusoids} \quad (2.49)$$

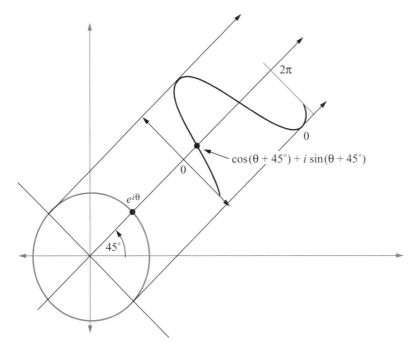

Figure 2.14
Phasor projected at 45°.

where ϕ is the *phase angle* of the projector. The family of curves defined by (2.49) are the *sinusoids*. By allowing the projected angle to vary, we allow all projections through a phasor from all possible angles.

2.6.6   Mixing Sine and Cosine to Create Sinusoids

Another way to create a sinusoid of any phase does not require complex arithmetic: we add a sine and a cosine together, varying the strengths of each to get the desired phase offset. For example, $a \cos\theta + b \sin\theta$, with $a = 1$, $b = 0$, reduces to $\cos\theta$; and with $a = 0$, $b = 1$, to $\sin\theta$; and with $a = b = 1$, to an equal mixture of the two. What does that look like? Plot sine and cosine waves of equal amplitude and sum them point by point (figure 2.15). Note that this figure has the same shape as figure 2.14 but an amplitude of $\pm\sqrt{2}$.

We have shown experimentally that $\cos\theta + \sin\theta = \sqrt{2}\sin(\theta + 45°)$. Using trigonometry, we can generalize this to show that

$$\sin(\theta + \phi) = a \cos\theta + b \sin\theta \tag{2.50}$$

2.6.7   Positive and Negative Frequencies and Amplitudes

Recall from section 2.6.1 that frequencies are positive or negative depending upon the direction of their circular motion. As shown in figure 2.11, a negative-frequency phasor $e^{-i\theta}$ turns clockwise, and a positive-frequency phasor $e^{i\theta}$ turns counterclockwise. Both these phasors have positive magnitudes. What is the behavior of negative-amplitude phasors $-e^{i\theta}$ and $-e^{-i\theta}$?

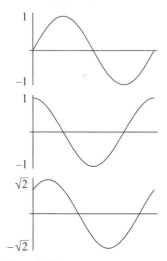

Figure 2.15
Sum of cosine and sine.

A negative-amplitude, positive-frequency phasor $-e^{i\theta}$ still generates a positive frequency because it still turns counterclockwise as θ increases. However, it has *negative length,* which means that it points in the opposite direction to a corresponding positive-length phasor. Consider the phasor $e^{i0} = 1 + 0i$, which is a unit vector lying on the real axis, anchored at complex zero and pointing to the right along the real axis:

If we negate it, we have $-e^{i0} = -1 + 0i$, which points to the left along the real axis:

Recall from equation (2.42) that $e^{i\pi} = -1 + 0i$, so $-e^{i0} = e^{i\pi}$. Think of it this way: if we start out with e^{i0}, a real unit vector pointing to the right, there are two ways we can make it point in the opposite direction: we can rotate it around complex zero (clockwise or counterclockwise) by a half circle, $e^{i\pi}$, or we can negate it, $-e^{i0}$.

Similarly, if a negative-frequency phasor $e^{-i\theta}$ is negated, it becomes $-e^{-i\theta}$. It still has a negative frequency because it turns clockwise as θ increases.

If we want to reverse the direction of a negative-frequency phasor, we have the same two choices as with the positive-frequency phasor: we can either rotate the phasor a half-circle around complex zero or flip its direction by negation.

These concepts are shown graphically in figure 2.16. A positive-frequency phasor with negative amplitude is identical to that same positive-frequency phasor rotated forward or backward by π radians (180°). In other words, $-e^{i\theta} = e^{i\theta \pm \pi}$. Similarly, $-e^{-i\theta} = e^{-i\theta \pm \pi}$.

Another way to view this is to look directly at the sine and cosine of θ when we add π to it (figure 2.17). We can see by inspection for sine and cosine that adding π to any angle is the same as negating it.

2.6.8 Phasors and Sinusoids

What would happen if we took a positive-frequency phasor $e^{i\theta} = \cos\theta + i\sin\theta$ spinning counterclockwise and a negative-frequency phasor $e^{-i\theta} = (\cos -\theta) + (i\sin -\theta)$ spinning clockwise and added them together?

When a positive-frequency phasor and a negative-frequency phasor are tied to the same angular displacement, they are in *conjugate symmetry,* so

$$e^{i\theta} + e^{-i\theta} = (\cos\theta) + (i\sin\theta) + (\cos -\theta) + (i\sin -\theta)$$

$$= \cos\theta + i\sin\theta + \cos\theta - i\sin\theta \qquad (2.51)$$

$$= 2\cos\theta.$$

Positive Complex Frequency

  with positive amplitude:
  $e^{i\theta} = \cos\theta + i\sin\theta$

  with negative amplitude:
  $-e^{i\theta} = -(\cos\theta + i\sin\theta)$

Negative Complex Frequency

  with positive amplitude:
  $e^{-i\theta} = \cos\theta - i\sin\theta$

  with negative amplitude:
  $-e^{-i\theta} = -(\cos\theta - i\sin\theta)$

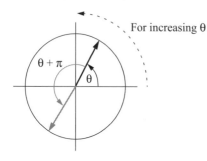
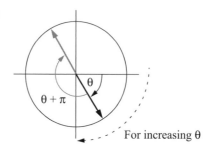

Figure 2.16
Positive and negative frequencies and amplitudes.

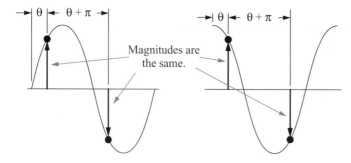

Figure 2.17
Adding π to an angle θ.

The result takes a bit of explaining, but it will prove to be crucial information. Rearranging (2.51) slightly, we get

$$\cos\theta = \frac{e^{i\theta} + e^{-i\theta}}{2} = \frac{e^{i\theta}}{2} + \frac{e^{-i\theta}}{2}.$$

Real Cosine as Sum of Two Phasors
in Conjugate Symmetry   (2.52)

That's a real cosine on the left side of equation (2.52). We've already seen a couple of other ways to create cosine waves, but this formula is the root of all other explanations. When summed, the real parts of phasors in conjugate symmetry add constructively and the imaginary parts cancel (figure 2.18). The sum of phasors in conjugate symmetry is always on the real number line. As θ varies, the length of the sum vector varies as $\cos\theta$.

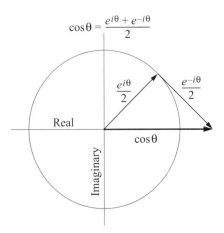

$$\cos\theta = \frac{e^{i\theta} + e^{-i\theta}}{2}$$

Figure 2.18
Sum of phasors in conjugate symmetry.

A real cosine consists of the vector sum of two half-amplitude phasors of conjugate symmetry.

Now, what if we subtract two phasors in conjugate symmetry?

$$e^{i\theta} + e^{-i\theta} = (\cos\theta) + (i\sin\theta) - (\cos-\theta) + (i\sin-\theta)$$

$$= \cos\theta + i\sin\theta - (\cos\theta + i\sin\theta) \tag{2.53}$$

$$= 2i\sin\theta.$$

Rearranging (2.53) slightly, we get

$$\sin\theta = \frac{e^{i\theta} - e^{-i\theta}}{2i}. \qquad \textit{Real Sine as Difference of Two Phasors in Conjugate Symmetry} \tag{2.54}$$

To get i out of the denominator, since $1/i = -i$, we can flip the sign of i:[10]

$$\sin\theta = -i\frac{e^{i\theta} - e^{-i\theta}}{2}. \tag{2.55}$$

This result might seem counterintuitive. How can a real sine wave be made of entirely imaginary components? When subtracted, the imaginary parts of phasors in conjugate symmetry add constructively, whereas the real parts cancel (figure 2.19). If we then multiply this imaginary difference by $-i$, we rotate it 90° clockwise to the real number line to obtain a real sinusoid. As θ varies, this difference varies as $\sin\theta$.

A real sine consists of the vector difference of two half-amplitude phasors of conjugate symmetry and amplitude.

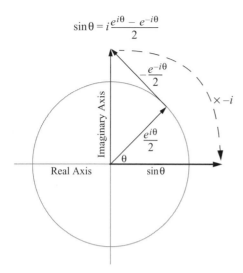

Figure 2.19
Difference of phasors in conjugate symmetry.

Equations (2.52) and (2.54) are really just variations on Euler's formula. They are the foundation of a great deal of important modern music technology. They are so important, in fact, that I present another visualization.

2.6.9 Cosine Machine

Figure 2.20 is a visual aid for equation (2.52) that I call the *cosine machine*. It forms the vector sum of two phasors mechanically. It has a motor with an arm attached to its rotor. At the end of the first arm is another arm of equal length connected to the first with a bearing. The end of the second arm slides from side to side in a slot. Figure 2.21 shows the cosine machine's stages of movement. Figure 2.22 shows the operation of the cosine machine with a pen attached, producing a cosine wave.

The length of each rotating arm is 1/2, so when the bars lie flat along the slot, they add up to 1. The arm attached to the motor turns counterclockwise (θ), and the outer arm turns clockwise ($-\theta$). The total side-to-side excursion of the arms is 2, and this motion outlines cosine movement as θ goes from 0 to 2π.

2.6.10 Sine Machine

Figure 2.23 shows an interpretation of equation (2.55) that I call the *sine machine*. This machine forms the vector difference of two phasors. We can split equation (2.55) into two phasors as follows:

$$\sin\theta = -i\frac{e^{i\theta} - e^{-i\theta}}{2} = i\frac{e^{-i\theta}}{2} - i\frac{e^{i\theta}}{2}.$$

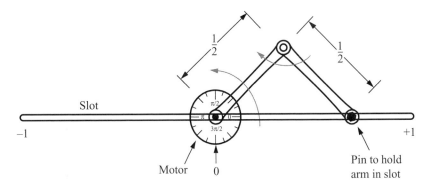

Figure 2.20
Cosine machine.

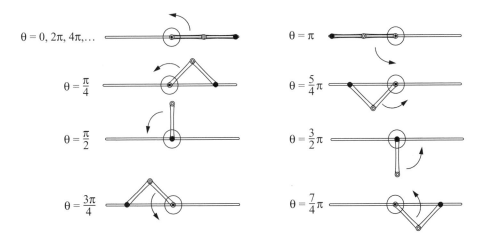

Figure 2.21
Stages of movement of the cosine machine.

The black arm in that figure spins clockwise and is associated with the negative-frequency phasor $e^{-i\theta}/2$; the white arm spins counterclockwise, matching the positive-frequency phasor $e^{i\theta}/2$. The black arm is connected at one end to a fixed bearing mounted behind the center of the slot that the black arm swivels around. The motor housing is attached to the other end of the black arm. The motor shaft is attached to the white arm. The other end of the white arm is attached to the slot via a pin to hold it in the slot. A pen is attached to the white arm in such a way that it always points straight down.

When the motor starts, both arms are directly above the center of the slot, corresponding to $\theta = 0$. As θ increases, the white arm moves clockwise, and the black arm moves counterclockwise. The vector difference generates a sine wave as θ goes from 0 to 2π.

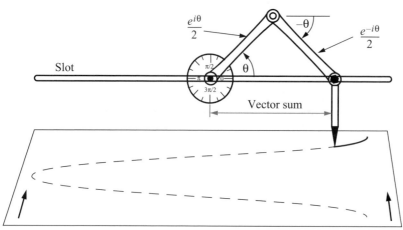

Figure 2.22
Rotating armature generating a cosine wave.

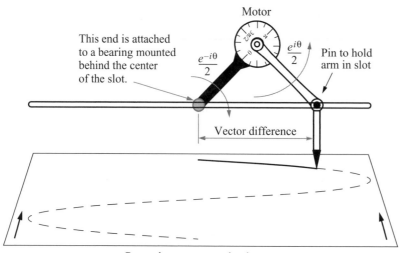

Figure 2.23
Sine machine.

Note that the cosine machine (figure 2.22) could have been used to create a sine wave by lifting the pen until $\theta = 90°$, then dropping it, introducing a phase delay. We could do exactly the same thing with the sine machine (figure 2.23) to generate a cosine wave. But the figures illustrate how to generate sine and cosine waves directly and in phase. For $0 \leq \theta \leq 2\pi$, the cosine machine (figure 2.22) generates a cosine wave, and the sine machine (figure 2.23) generates a sine wave.

2.6.11 Energy of a Phasor

Kinetic energy is proportional to the square of velocity (see volume 1, equation (4.28)). In terms of real waveforms, energy is the square of amplitude. But what corresponds to the energy of a phasor?

If the magnitude of the phasor $z = re^{i\theta}$ is r, then by equation (2.14), $r^2 = z\bar{z}$, and

$$r = \sqrt{z\bar{z}} = \sqrt{a^2 + b^2}.$$

Thus, if we associate the magnitude r of the phasor with the amplitude of a wave, r^2 is its energy.

2.6.12 Even and Odd Functions

Notice that the shape of the cosine wave is symmetrical around $x = 0$. The cosine function has the same value for both positive and negative x indexes. That is, $\cos x = \cos -x$, as shown in figure 2.24. Because of this, the cosine function is an *even function*. In general, a function f is even if $f(x) = f(-x)$.

The shape of the sine wave is antisymmetrical around $x = 0$. If we negate the sine function, we have $-f(x)$, shown as a bold curve in figure 2.25. For any x, the positive and negative functions are equal, and $\sin x = -\sin -x$. Because of this, the sine function is an *odd function*. In general, a function f is odd if $f(-x) = -f(x)$. Table 2.3 summarizes these observations.

With the exception of the zero function $f(x) = 0$, all functions are either even or odd, or a mixture of the two:

$$f(x) = f_e(x) + f_o(x). \tag{2.56}$$

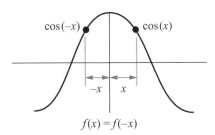

Figure 2.24
Cosine as an even function.

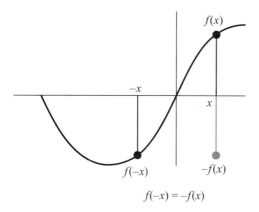

$$f(-x) = -f(x)$$

Figure 2.25
Sine as an odd function.

Table 2.3
Properties of Even and Odd Functions

Function	Definition	Property	Example
Even	$f(x) = f(-x)$	Symmetrical around 0	Cosine
Odd	$f(-x) = -f(x)$	Antisymmetrical around 0	Sine

Given the definitions for even and odd functions in table 2.3 and equation (2.56), it follows that:

$$f(-x) = f_e(-x) + f_o(-x) \qquad (2.57)$$
$$= f_e(x) - f_o(x).$$

Equation (2.56) says that a function of a positive index x is the sum of its even and odd parts. Equation (2.57) extends this slightly to say that a function of a negative index x is equal to the difference of its even and odd parts. If we add (2.56) and (2.57),

$$f(x) + f(-x) = f_e(x) + f_o(x) + f_e(x) - f_o(x)$$
$$= 2f_e(x),$$

and rearrange to solve for $f_e(x)$, we see that

$$f_e(x) = \frac{1}{2}[f(x) + f(-x)]. \qquad (2.58)$$

Equation (2.58) shows how to extract the even portion of any function $f(x)$: compute 1/2 times the value $f(x) + f(-x)$, and the result will be the even portion of the function.

If we subtract equation (2.57) from equation (2.56),

$$f(x) - f(-x) = f_e(x) + f_o(x) - f_e(x) + f_o(x)$$
$$= 2f_o(x)$$

and rearrange to solve for $f_o(x)$, we see that

$$f_o(x) = \frac{1}{2}[f(x) - f(-x)].$$ (2.59)

Equation (2.59) shows how to extract the odd portion of any function $f(x)$: compute 1/2 times the value $f(x) - f(-x)$, and the result will be the odd portion of the function.

Now, how can we be sure that equations (2.58) and (2.59) fully represent the whole of the function $f(x)$? Well, if we add them together, we should get back the original function, which we do:

$$f_e(x) + f_o(x) = \frac{1}{2}[f(x) + f(-x)] + \frac{1}{2}[f(x) - f(-x)]$$

$$= f(x).$$

We have proved that equations (2.58) and (2.59) fully represent the function $f(x)$, unrestricted in any way.

All functions can be broken down into even and odd parts using equations (2.58) and (2.59).

In particular, if we define $f(\theta) = e^{i\theta}$, we immediately demonstrate, as in equation (2.52), that

$$f_e(\theta) = \frac{1}{2}(e^{i\theta} + e^{-i\theta}),$$

and also demonstrate, as in equation (2.54), that

$$f_o(\theta) = \frac{1}{2}(e^{i\theta} - e^{-i\theta}).$$

Understanding even and odd functions will come in very handy in chapter 3.

2.6.13 Making Phasors Spin in Time

In figure 2.13 we saw that the phasor $e^{i\theta}$ makes one complete period as θ goes from 0 to 2π. If we wish to specify that this period should occur over a particular duration T, we can let $\theta = 2\pi t/T$, where t is time. For example, if we wish the phasor to complete one period in 1 second, we set $T = 1$. Then, as the real-valued time parameter t goes from 0 to 1 second, the phasor $e^{i2\pi t/T}$ goes through 2π radians, one full rotation.

Making a phasor spin at a particular rate puts it into the temporal realm. To signify this, let's say that the phasor with time t in its exponent is the *complex sinusoid*.

What if we want the complex sinusoid to go through two periods in 1 second? The most convenient approach would be to introduce a frequency variable f, so that $\theta = 2\pi f t/T$. Now the complex sinusoid is defined as

$$e^{i\theta} = e^{i2\pi f t/T}.$$ (2.60)

If we set $f = 2$, then the phasor $e^{i2\pi ft/T}$ will make two cycles in 1 second. If we set $f = 440$, it will make 440 cycles per second. Defining $\theta = 2\pi ft/T$ causes θ to express *angular velocity,* the rate at which the phasor spins.

It is common to simplify equation (2.60) by defining $\omega = 2\pi f$, so that the phasor becomes $e^{i\omega t/T}$. Since most often $T = 1$ (because we're mostly measuring frequency in Hz, which is cycles per second), we can simplify a bit further by leaving out T. The time-based phasor is then simply $e^{i\omega t}$. If we want amplitude to be other than unity, we can add an amplitude term A to scale the phasor's magnitude, so that we have the following canonical representation for the complex sinusoid:

$$A e^{i\omega t} = A e^{i2\pi ft}. \qquad\qquad\qquad\qquad \textit{Complex Sinusoid} \quad (2.61)$$

This powerful but economical representation of the complex sinusoid is used throughout the rest of the book.

2.7   Graphing Complex Signals

Suppose a bright light were mounted on the tip of an airplane's propeller as it flies past at night. The forward circular motion of the light would create a helix as it cuts through the air. To represent the complex sinusoid graphically (figure 2.26a), we use a 3-D representation where the y-axis is the imaginary number line, the x-axis (shown sloping *up* to the right) is the real number line, and the z-axis (sloping *down* to the right) is time. Figure 2.26b shows a complex sinusoid as a helix. We can project along the real axis to see just the sine wave component (shown on the "wall" behind the helix), or project along the imaginary axis to see just the cosine wave motion (shown on the "floor" below the helix).

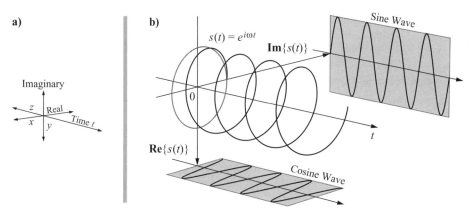

Figure 2.26
Projection of a complex signal.

If the equation for the helix is $s(t) = e^{i\omega t}$, then the cosine projection is just the real part, denoted $\text{Re}\{s(t)\}$, and the sine projection is just the imaginary part, denoted $\text{Im}\{s(t)\}$. The value of the helix when $t = 0$ in figure 2.26 is

$$s(0) = e^{i\omega 0} = \cos 0 + i \sin 0,$$

which in Cartesian coordinates corresponds to the 3-D point

$$(x, y, z) = (\text{Re}\{s(t)\}, \text{Im}(\{s(t)\}, t))$$

$$= (\cos 0, \sin 0, 0)$$

$$= (1, 0, 0).$$

The point on the helix for any other time can be similarly determined by plugging in the appropriate values of θ and t. The helix in figure 2.26 is spiraling counterclockwise, indicating positive frequency.

2.8 Spectra of Complex Sampled Signals

In section 1.3.3 we established an equivalence for sampled waveforms between sampling rate and angular velocity, saying that frequency f is to the Nyquist barrier $R/2$ as angular velocity θ is to π. Therefore, when plotting a spectrogram, we can simply plot frequency as radian velocity between $-\pi$ and π, and leave sampling rate out of it entirely. This is an advantage when comparing spectra that were sampled at different rates.

In section 2.6.7 we saw that there are positive and negative frequencies and amplitudes. As was shown in figure 2.11, the negative phasor $e^{-i\theta}$ turns clockwise, producing negative frequencies, and the positive phasor $e^{i\theta}$ turns counterclockwise, producing positive frequencies. Negating the sign of a phasor is the same as giving it a $180°$ phase shift. So, for some radian velocity θ, there are four possible phasors:

$e^{i\theta}$	Positive frequency, positive amplitude
$-e^{i\theta}$	Positive frequency, negative amplitude
$e^{-i\theta}$	Negative frequency, positive amplitude
$-e^{-i\theta}$	Negative frequency, negative amplitude

If we want to show the spectrum of a complex signal, we must find a way to represent each of these phasors distinctly. Figure 2.27 shows a complex spectrum with one of each kind of phasor. The positive-amplitude phasors are shown with bold arrows, and the corresponding negative-amplitude phasors are shown shaded. Positive and negative amplitude is graphed on the y-axis. The x-axis shows angular velocity between $-\pi$ and π instead of frequency.

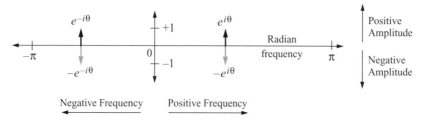

Figure 2.27
Complex spectrum of the four kinds of phasors.

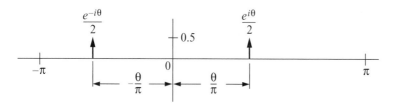

Figure 2.28
Complex spectrum of a real cosine wave.

2.8.1 Complex Spectrum of a Real Sampled Cosine Wave

In equation (2.52) we found that $\cos\theta = (e^{i\theta} + e^{-i\theta})/2$, which says that a real cosine equals the sum of two half-amplitude phasors of opposite frequency. Now that we have a way of representing complex spectral components mathematically, we can diagram spectrograms of these relations.

The complex spectrum of equation (2.52) is shown in figure 2.28. The Nyquist barrier is shown as $\pm\pi$, so the frequency θ corresponds to $\pm\theta/\pi$. The distance of the two arrows from zero along the x-axis represents the frequency of the two phasors: the left one has negative frequency, the right one has positive frequency. Both arrows point up because the signs of both phasors are positive, and they each have a magnitude of 0.5.

2.8.2 Complex Spectrum of a Real Sampled Sine Wave

The complex spectrum of the sine wave

$$\sin\theta = -i\frac{e^{i\theta} - e^{-i\theta}}{2} = i\frac{e^{-i\theta}}{2} - i\frac{e^{i\theta}}{2}$$

$$= i\left(\frac{e^{-i\theta}}{2} - \frac{e^{i\theta}}{2}\right)$$

is shown in figure 2.29. Both components are imaginary. This graphical representation does not allow us to show real and imaginary components together. Later in this chapter, I develop a 3-D representation of complex spectra that does. We see that real sine wave $\sin\theta$ is made of two imaginary

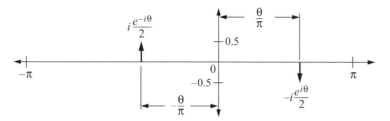

Figure 2.29
Complex spectrum of a real sine wave.

half-amplitude phasors, one with positive frequency and negative amplitude, the other with negative frequency and positive amplitude.

2.9 Multiplying Phasors

In section 2.3.4 we observed that to multiply two complex numbers, we multiply their magnitudes and add their angles. Similarly, to multiply two phasors, we multiply their magnitudes and add their angles. For complex sinusoids (that is, phasors containing time t in the exponent), we must multiply their magnitudes and add their angles *at every point in time*.

What happens if we multiply the time-based phasor $e^{i\omega t}$ by itself (thereby squaring it)? Recall that unless a scaling term is added the magnitude of a phasor is unity (that is, $|e^{i\omega t}| = 1$), so multiplying $e^{i\omega t}$ by itself won't change its magnitude. But its frequency will double because we sum the angles at each point in time, doubling the rotational velocity: $(\omega + \omega = 2\omega)$. Figure 2.30 shows the rotation of phasors through time as a helix. Figures 2.30a and 2.30b are identical phasors, so multiplying them is effectively a squaring operation. Figure 2.30c shows the product signal, which spins twice as fast but has the same amplitude. Multiplying phasors to raise their frequency is *modulation*.

If we multiply a positive-frequency phasor $e^{i\omega t}$ by a phasor of equal but negative frequency $e^{-i\omega t}$, the magnitude of the product will be unity, but the frequency will be 0 Hz, because $\omega - \omega = 0$. The result is a signal that has the value of complex unity $(1 + 0i)$ at all points (figure 2.31). Multiplying phasors to lower their frequency is *demodulation*.

We can change the frequency of a phasor by an arbitrary amount. Say we have two signals, $s_1 = A_1 e^{i2\pi f_1 t}$ and $s_2 = A_2 e^{i2\pi f_2 t}$. Their product is a signal with magnitude $A_1 \cdot A_2$ and frequency $f_1 + f_2$. For instance, if $f_1 = 4$ Hz and $f_2 = -3$ Hz, the product will be a phasor at 1 Hz with magnitude $A_1 \cdot A_2$ (figure 2.32).

Modulation and demodulation are used, for example, to convert between audio-frequency signals and radio-frequency signals. For instance, suppose a radio receiver detects a signal $f_1 = 1$ MHz. If the receiver has an internal oscillator tuned to $f_2 = 0.999$ mHz and multiplies these two signals together, the result would be an audio-frequency tone of 1 kHz.

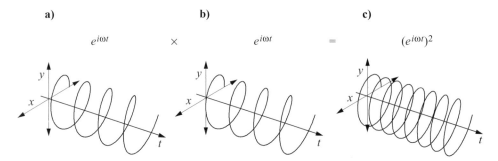

Figure 2.30
Squaring a phasor.

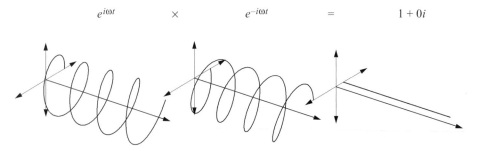

Figure 2.31
Multiplying identical positive- and negative-frequency phasors.

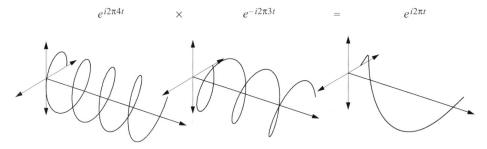

Figure 2.32
Changing the frequency of a phasor.

By far the most important musical application of modulation and demodulation is frequency detection. The Fourier transform operates somewhat like a radio receiver to tune in and register the frequencies present in a signal (see chapter 3).

Recall equation (2.52), which shows that a real cosine waveform is the sum of two half-amplitude phasors with opposite frequencies:

$$s(t) = \cos \omega t = \frac{e^{i\omega t}}{2} + \frac{e^{-i\omega t}}{2}.$$

If we set $\omega = 4\pi$, we could plot the spectrum of $s(t)$ as shown in figure 2.33.

Now let's define a complex waveform containing a single phasor:

$$m(t) = e^{i\phi t}.$$

If we set $\phi = -4\pi$, we could plot the spectrum of $m(t)$ as shown in figure 2.34.

What happens if we multiply the real signal $s(t)$ shown in figure 2.33 containing two phasors and the complex signal $m(t)$ shown in figure 2.34 containing just one phasor? The spectrum of the product of the waveforms,

$$m(t) \cdot s(t),$$

is shown in figure 2.35.

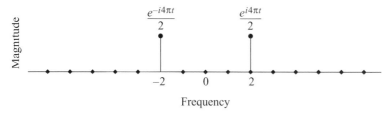

Figure 2.33
Spectrum of a real cosine signal.

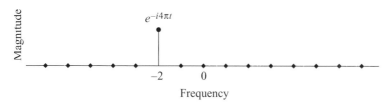

Figure 2.34
Spectrum of a phasor at -4π.

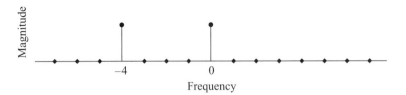

Figure 2.35
Spectrum of product of $m(t)$ and $s(t)$.

We can interpret figure 2.35 to say that all components of the real cosine signal $s(t)$ are shifted in frequency by the frequency of $m(t)$. In general, multiplying a signal by a phasor of frequency f adds frequency f to the frequencies of all components of the signal. All components of the signal are shifted by the same amount, no matter how complicated the signal is. (If the resulting spectrum is not conjugate symmetric, the resulting waveform is complex.)

2.10   Graphing Complex Spectra

Just as we need three dimensions to represent complex sinusoids in the time domain (as in figure 2.26), so we need three dimensions to represent complex spectra. For complex spectra, the z-axis represents frequency, and the y-axis and x-axis represent the imaginary and real number lines, respectively. The frequency of a sinusoid is represented by its position along the frequency axis (z-axis), but its magnitude is represented by a vector whose length is the sum of its real and imaginary parts.

For example, recall equation (2.52), which shows that a real cosine wave is the sum of two half-amplitude phasors with opposite frequencies. If in equation (2.52) we set $\theta = 2\pi ft$ then we have:

$$\cos 2\pi ft = \frac{e^{i2\pi ft}}{2} + \frac{e^{-i2\pi ft}}{2}.$$

Figure 2.36a shows the real cosine wave in the complex time domain, and figure 2.36b shows the complex spectrum of the real cosine. Each of the bold arrows in the spectrogram corresponds to one of the phasors in the cosine equation. The position of these arrows along the frequency axis corresponds to the frequency of the positive phasor f and the frequency of the negative phasor $-f$. The length of each arrow corresponds to the magnitude of each phasor (in this case, both have magnitudes of $1/2$). The orientation around the x-axis is determined by the vector sum of the real and imaginary components of the phasors. Since the magnitudes of both phasors are real (that is, the imaginary part of their magnitudes is zero), they lie parallel to the real axis. Since the amplitudes of both phasors are positive, they are on the positive side of the real axis.

Here's another example. If in equation (2.55) we set $\theta = 2\pi ft$ then we have:

$$\sin 2\pi ft = i\left(\frac{e^{-i2\pi ft}}{2} - \frac{e^{i2\pi ft}}{2}\right).$$

a) Complex Time Domain   **b)** Complex Frequency Domain
                                                  (Complex Spectrogram)

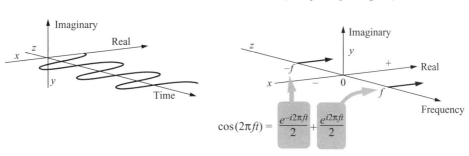

$$\cos(2\pi ft) = \frac{e^{-i2\pi ft}}{2} + \frac{e^{i2\pi ft}}{2}$$

Figure 2.36
Cosine wave in the complex time and frequency domains.

a)   **b)**

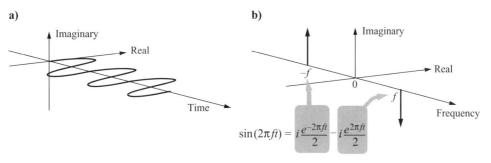

$$\sin(2\pi ft) = i\frac{e^{-2\pi ft}}{2} - i\frac{e^{2\pi ft}}{2}$$

Figure 2.37
Sine wave in the complex time and frequency domains.

This can be graphed as in figure 2.37. Both components are imaginary, so they are parallel to the imaginary axis. The positive-frequency component has a negative imaginary magnitude, so it points down.

Some useful terminology: components of a spectrum that are parallel to the the real axis are said to be *in phase,* and those that are parallel to the imaginary axis are said to be in *quadrature phase.* By this definition, the cosine wave's components are purely in phase, and the sine wave's components are purely in quadrature phase.

2.10.1 Graphical Proof of Euler's Formula

We can use 3-D representation of spectra to demonstrate a graphical proof of Euler's formula, equation (2.32),

$$e^{iz} = \cos z + i\sin z.$$

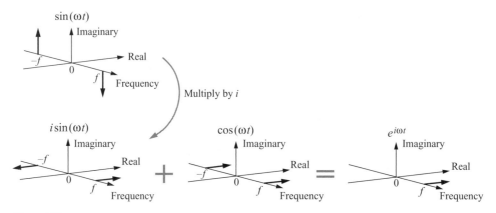

Figure 2.38
Geometric demonstration of Euler's function.

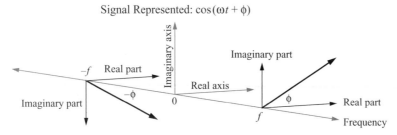

Figure 2.39
Real signal with a phase offset.

First, let $z = \omega t$ so $e^{i\omega t} = \cos \omega t + i \sin \omega t$. Let's start by graphing $\sin \omega t$, as shown at the top left of figure 2.38. Next, we must multiply this by i, which has the effect of rotating both phasors counterclockwise by 90°, as shown at the lower left of that figure. Last, we must add $\cos \omega t$ to the rotated phasors. Note that the negative-frequency components cancel, leaving only the positive component, e^{iz}.

What is the complex spectrum of a real signal with a phase offset? Consider the spectrum of the real signal $\cos(\omega t + \phi)$ (figure 2.39). Adding a phase offset to a real signal rotates its components in the complex spectral domain around the frequency axis. Notice that adding a phase offset to the cosine wave rotates the positive frequency phasor counterclockwise and the negative frequency phasor clockwise.

2.10.2 Frequency Components of Real Signals

Here are some useful conclusions to draw from the preceding examples:

• The magnitudes of the components of real signals are always balanced between negative and positive frequencies. For example, figures 2.36 and 2.37 show the positive- and negative-frequency components balanced in magnitude. If $|X(f)|$ represents the magnitude of a signal at frequency f, then for real signals we always have $|X(f)| = |X(-f)|$. This is true for real sines, real cosines, and in general, real sinusoids with any phase angle.

- The positive- and negative-frequency components of real signals that are in phase (that lie along the real axis) always have even symmetry around 0 Hz, that is, their phasors point in the same direction. For example, in figure 2.36, the components are real and in phase, and have even symmetry around 0 Hz (that is, both point in the same direction). These signals are invariably real cosine signals.

- The positive- and negative-frequency components of real signals that are in quadrature phase (that lie along the imaginary axis) always have odd symmetry around 0 Hz, that is, their phasors point in opposite directions. For example, in figure 2.37, the components are imaginary and in quadrature phase, and have odd symmetry around 0 Hz (that is, they point in opposite directions). These signals are invariably real sine signals.

For real signals, these rules always obtain. Therefore, the negative- and positive-frequency components of any real signal will, when summed, always cancel any imaginary magnitude, resulting in a signal that lies entirely along the real axis. This is what equations (2.52) and (2.55) show.

Complex signals have no such restrictions. When the negative- and positive-frequency components of a complex signal are combined, the result does not necessarily cancel the imaginary part.

If the real part of a signal is an even function, and its imaginary part is an odd function, its spectrum is said to be **Hermitian.**

The spectrum of every real signal is Hermitian.

The symmetry of a Hermitian spectrum allows us to discard all negative-frequency spectral information of a real signal because it is redundant with the positive-frequency information. We can regenerate it later if necessary because we know by the preceding rules exactly what the negative-frequency components will be with respect to their positive-frequency counterparts. This is why spectral plots of real signals are typically displayed only for positive frequencies: we can easily infer the negative frequencies by reflecting the positive frequencies around 0 Hz. In contrast, the spectrum of a complex signal must be explicitly specified over both positive and negative frequencies because the positive-frequency and negative-frequency components of a complex signal are independent.

2.11 Analytic Signals

A function is said to be *analytic* if it has no negative frequencies.[11] Analytic signals provide a convenient spectral representation of real signals because they remove the redundant negative-frequency information in such a way that it can be restored if needed.

2.11.1   Hilbert Transform

The method of creating an analytic signal from a real signal is based on the Hilbert transform.[12] We have already seen an example of the process in figure 2.38, where a single positive-frequency phasor is created from a real cosine plus a real sine multiplied by i. The Hilbert transform can be used to create signals in quadrature.

The Hilbert transform of a signal $x(t)$ is another signal $y(t)$ whose frequency components are all phase-shifted by 90° ($-\pi/2$ radians):

$$y(t) = H\{x(t)\}, \hspace{4cm} \textit{Hilbert Transform} \quad (2.62)$$

where $H\{\ \}$ is the Hilbert transform.

Using the Hilbert Transform to Create an Analytic Signal   We can interpret the processing in figure 2.38 using the Hilbert transform as follows. For some real frequency ω,

$$x(t) = A\cos\omega t. \hspace{2cm} \text{Begin with a real input signal.}$$

$$y(t) = \mathcal{H}\{x(t)\} \hspace{2cm} \text{Apply Hilbert transform (phase shift by 90°) to create } y(t).$$

$$= A\cos\left(\omega t - \frac{\pi}{2}\right)$$

$$= A\sin\omega t.$$

$$z(t) = x(t) + iy(t) \hspace{2cm} \text{Multiply } y(t) \text{ by } i \text{ and combine with input signal to create}$$

$$\hspace{5.5cm} \text{analytic signal } z(t).$$

$$= A\cos\omega t + i(A\sin\omega t)$$

$$= Ae^{i\omega t}.$$

The result, $z(t)$, is an analytic signal because the resulting phasor $Ae^{i\omega t}$ is a single positive amplitude phasor representing a single positive frequency component that has no complementary negative-frequency component.

Using the Hilbert Transform for Arbitrary Signals   The preceding method converts individual components of real signals to analytic form, but real-world signals tend to have many components in combination. We must generalize the procedure to all sinusoids in order to apply the Hilbert transform to more complicated signals.

Start by recalling that multiplying a phasor by $e^{i\pi/2} = i$ causes it to rotate 90° counterclockwise, and multiplying by $e^{-i\pi/2} = -i$ causes it to rotate 90° clockwise. We can create the quadrature signal $y(t)$ for any real signal $x(t)$ as follows:

- Rotate the positive-frequency components of $x(t)$ clockwise 90° by multiplying them by $-i$.

- Rotate the negative-frequency components counterclockwise 90° by multiplying them by i.

This procedure is shown graphically for real cosine and sine signals in figure 2.40. Notice that putting a cosine signal through the Hilbert transform produces a sine signal, and putting a sine signal through

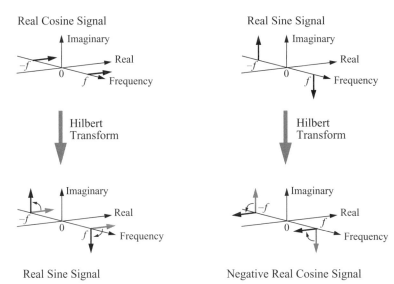

Real Cosine Signal

Real Sine Signal

Hilbert Transform

Hilbert Transform

Real Sine Signal

Negative Real Cosine Signal

Figure 2.40
Hilbert transform.

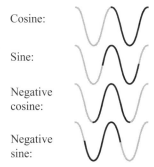

Cosine:

Sine:

Negative cosine:

Negative sine:

Figure 2.41
Quadrature phases of a sinusoid.

the Hilbert transform produces a negative cosine signal. A negative cosine signal passed through the Hilbert transform will produce a negative sine signal, and one more transformation will return it to the original cosine signal. We can also observe this by delaying sinusoids successively by 90° in the time domain (figure 2.41). The Hilbert transform is sometimes called a *quadrature filter* because it generates these four cyclic transformations of sinusoids (see section 3.9).

2.11.2 Creating an Analytic Signal

Applying the Hilbert transform is the first step in creating an analytic signal. To complete the operation, we multiply the Hilbert transform output $y(t)$ by i to rotate it an additional 90° counterclockwise. When

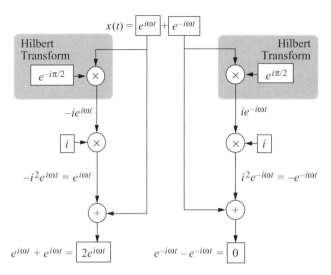

Figure 2.42
Analytic signal generator.

we add this signal to the original signal $x(t)$, the negative frequencies cancel, leaving just the positive frequencies. This process is pictured in figure 2.42 with a cosine wave input (shown in complex form).

Summing up, the analytic signal $x_a(t)$ of real signal $x(t)$ is

$$x_a(t) = x(t) + i\mathcal{H}\{x(t)\}$$

$$= x(t) + i\hat{x}(t).$$

Analytic Signal (2.63)

Figure 2.43 shows the process graphically for sines and cosines, although in fact this procedure will convert any real sinusoid into an analytic signal. Note that the amplitude of the analytic signal is doubled with respect to its original real signal. (The component at 0 Hz, the DC component, remains unchanged.) That's because this process effectively wraps all negative frequencies onto their corresponding positive frequencies with a phase inversion, thereby doubling their amplitude. So this is an energy-conserving process. Because it is energy-conserving, the Hilbert-transformed signal can be regenerated into the original real input signal $x(t)$.

2.11.3 Applications of the Hilbert Transform

The Hilbert transform has many musically relevant applications that are presented in later chapters. In the meantime, here are two straightforward applications.

Envelope Follower   We can use the Hilbert transform to extract the time-varying amplitude envelope from a musical tone. Suppose the tone has a waveform $\cos \omega t$ and an amplitude envelope $A(t)$ so that the tone's waveform is defined as $x(t) = A(t) \cos \omega t$. If the rate at which $A(t)$ changes

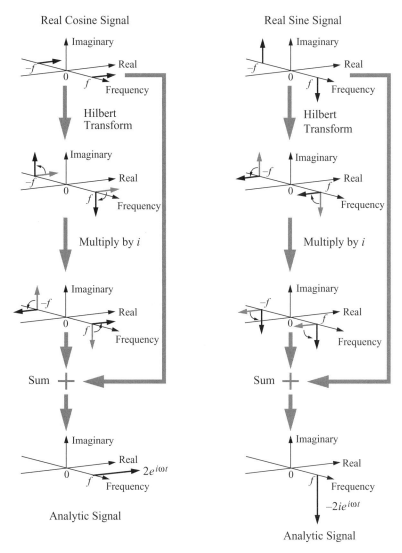

Figure 2.43
Creating analytic signals using the Hilbert transform.

is sufficiently slow compared to ω, then it's reasonably safe to say that the tone's Hilbert transform is approximately $y(t) \cong A(t) \sin \omega t$. Constructing the analytic signal,

$$z(t) = x(t) + iy(t)$$
$$= A(t)e^{i\omega t}.$$

and since $\left| e^{i\text{anything}} \right| = 1$,

$$A(t) = |z(t)| . \qquad\qquad\qquad\qquad \textit{Instantaneous Amplitude} \quad (2.64)$$

Equation 2.64 says that the amplitude envelope of a signal is the absolute value of its analytic signal through time if the rate at which $A(t)$ changes is sufficiently slow compared to ω. This is a relatively painless way of extracting the amplitude envelope of a signal though it works only for sinusoidal or quasi-sinusoidal signals. In fact, it's really too good to be true, because though it seems that we've managed to extract instantaneous amplitude for all times t in $A(t)$, in practice the best we can do is to get local amplitude, not true instantaneous amplitude. For non-sinusoidal signals, it is common to follow the Hilbert transform with a lowpass filter.

Frequency Detector   Using the definitions for $x(t)$, $y(t)$, and $z(t)$, we can express the instantaneous phase of a signal as

$$\psi(t) = \tan^{-1} \frac{y(t)}{x(t)}. \qquad\qquad\qquad \textit{Instantaneous Phase} \quad (2.65)$$

Equation 2.65 says that the phase angle at time t equals the arctangent of the ratio of the Hilbert-transformed input signal to the input signal. The instantaneous frequency is the derivative of instantaneous phase $\psi(t)$ with respect to time. We will learn about the derivative in chapter 6. Specifically, the instantaneous frequency is

$$f(t) = \frac{1}{2\pi} \frac{d}{dt} \psi(t). \qquad\qquad\qquad \textit{Instantaneous Frequency} \quad (2.66)$$

Once again, practical systems provide local frequency information, not true instantaneous frequency.

Other interesting effects such as frequency shifting are also possible using analytic signals (see chapter 9). The Hilbert transform is of fundamental importance to many disciplines as diverse as quantum mechanics and modern music-encoding technologies such as MP3 (see chapter 10).

Summary

The imaginary number i was invented to allow the square of a number to be negative. Complex numbers were created so that imaginary and real numbers could coexist in the same quantity.

This in turn required understanding how complex numbers can be manipulated arithmetically. Multiplication, for example, consists of multiplying the vector lengths of two complex numbers and adding their angles. Conjugation negates the sign of the imaginary part of a complex number. Complex numbers can be graphed easily by assigning the x-axis to the real part and the y-axis to the imaginary part. Graphing numbers on the complex plane shows that multiplying a number by i rotates it counterclockwise by $90°$ and dividing a number by i rotates it clockwise $90°$. Complex numbers provide a compact, powerful representation for sinusoids because we can treat complex numbers as vectors spinning on the complex plane.

Raising a complex number with unity magnitude to a power n is equivalent to multiplying the angle of the complex number by n (de Moivre's theorem). By combining the Taylor series for sine and cosine, and relating it to the series for e, we found the "most beautiful formula," $e^{i\pi} + 1 = 0$.

Adding a polar representation to complex numbers results in phasors. Adding time to a phasor makes it spin around the unit circle at a particular frequency. Positive-frequency phasors spin counterclockwise; negative-frequency phasors spin clockwise. Like the motion of a turntable, the phasor $e^{i\theta}$ embodies both the cosine and sine relations simultaneously. We observed a $90°$ relation between sine and cosine motion, called quadrature. Sinusoids are simply a generalization of phasor rotation, allowing us to project across the complex circle from an arbitrary position. Positive and negative frequencies result from reversing the direction of the phasor.

By investigating conjugate symmetrical phasors, we found that a real cosine is made up of the vector sum of two half-amplitude phasors of conjugate symmetry. Similarly, a real sine is made up of the vector difference of two half-amplitude phasors of conjugate symmetry.

The cosine function is called an even function, the sine function is called an odd function. With the exception of the zero function $f(x) = 0$, all functions are either even or odd, or a mixture of the two, and we found ways to break down any function into its even and odd functional components.

By injecting time into the phasor, we found that the frequency of a phasor in Hertz is the ratio of the angular velocity θ to the number of radians in a circle. Multiplying phasors together modulates their frequency, making them spin faster; demodulating them makes them spin slower.

By examining complex spectra, we found that the strengths of the frequency components of real signals are always balanced between negative and positive frequencies, have even symmetry around 0 Hz, and are in quadrature phase. If the real part of a signal is an even function, and its imaginary part is an odd function, its spectrum is said to be Hermitian. The spectrum of every real signal is Hermitian. The negative- and positive-frequency components of any real signal will, when summed, always cancel any imaginary magnitude, resulting in a signal that lies entirely along the real axis. The symmetry of a Hermitian spectrum allows us to discard all negative-frequency spectral information of a real signal because it is redundant.

A function is said to be analytic if it has no negative frequencies. Analytic signals provide a convenient spectral representation of real signals because they remove the redundant negative-frequency information in such a way that it can be restored if needed. The Hilbert transform of a signal is another signal whose frequency components are all phase-shifted by $90°$. To create an analytic signal, apply the Hilbert transform, then multiply the Hilbert transform output i, and add this signal to the input signal. The negative frequencies cancel, leaving just the positive frequencies.

3 Spectral Analysis and Synthesis

Mathematics and music, the most sharply contrasted fields of scientific activity, are yet so related as to reveal the secret connection binding together all the activities of our mind.
—Hermann von Helmholtz

3.1 Introduction to the Fourier Transform

Joseph Fourier[1] contributed a penetrating insight to our knowledge of waveforms in general and music in particular.

Any periodic vibration, no matter how complicated it seems, can be built up from sinusoids whose frequencies are integer multiples of a fundamental frequency, by choosing the proper amplitudes and phases.

Signals whose frequencies are integer multiples of some fundamental are the *harmonics* of that frequency. The process of building up a compound signal from simple sinusoids is known as *Fourier synthesis,* or *spectral synthesis*.

3.1.1 Fourier Synthesis

Fourier synthesis allows us to create a waveform from a specification of the strengths of its various harmonics. That is, in fact, all a spectrum really is: a specification (in the form of a function or a list) of the strengths of the harmonics of a waveform. If $f(t)$ represents a time domain waveform, where t is time, then its corresponding frequency domain spectrum is denoted $F(k)$, where k is frequency. Fourier synthesis starts with a spectrum $F(k)$ and interprets it as a recipe describing the strengths of harmonics that must be combined in order to produce the corresponding time domain signal $f(t)$.

Fourier's observation only applies to periodic vibrations, a subset of all possible vibrations, so Fourier synthesis does not apply universally. However, a great number of interesting vibrations are either periodic or quasi-periodic, including all pitched musical instruments. All sinusoids are periodic, and sinusoids are the building blocks out of which any arbitrary (periodic) vibration can be constructed by Fourier synthesis.

3.1.2 Fourier Analysis

Fourier also observed that the process works in reverse.

Any periodic vibration, no matter how complicated it seems, can be observed to be made up of a set of sinusoids whose frequencies are harmonics of a fundamental frequency, with particular amplitudes and phases.

This process is called *Fourier analysis,* or *spectrum analysis.* Fourier analysis provides a way to measure the strengths of the individual components of a harmonic signal. It starts with a time domain signal $f(t)$ and interprets it as a kind of recipe describing the spectral components and their strengths that must be combined in order to produce the corresponding frequency domain signal $F(k)$.

Like Fourier synthesis, Fourier analysis is also limited to periodic vibrations. It asserts that sinusoids are the irreducible elements for the analysis of all periodic waveforms.

3.1.3 Fourier Transform

The *Fourier transform* is the combination of Fourier analysis and Fourier synthesis. Fourier analysis and synthesis are called a *transform pair* because (ideally) the spectrum of a wave created by Fourier synthesis may be perfectly analyzed by Fourier analysis, and vice versa, with no loss of information.

A transform is really just a way to represent the same information in an equivalent form. Fourier analysis receives a time domain signal and converts it into an equivalent spectral representation; Fourier synthesis takes a spectrum and converts it into an equivalent time domain signal. The time doman and frequency domain representations are equivalent under Fourier transformation.

In recent times, the term *Fourier transform* has come to mean just Fourier analysis, while the term *inverse Fourier transform* means just Fourier synthesis. This can be a little confusing, so we must be careful to distinguish two meanings of the term *Fourier transform.* It can stand for both Fourier analysis and synthesis or only for Fourier analysis. In this chapter, I mostly mean Fourier analysis when I write "Fourier transform."

3.1.4 Additive Synthesis

Fourier synthesis states that we can create any periodic vibration by adding sinusoids together with frequencies that are integer multiples of a fundamental frequency (harmonics) at particular amplitudes and phases. For this reason, Fourier synthesis is also known as *additive synthesis.* Figure 3.1 shows an example of summing waveforms. The waveforms are added together point by point.

These rules govern additive synthesis:

- Only sinusoids may be combined.
- Frequencies of all sinusoids must be harmonically related.

Within these limitations, the choices of frequency, amplitude, and phase of the components are arbitrary.

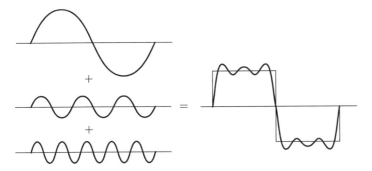

Figure 3.1
Adding odd-numbered overtones.

3.1.5 Example of Additive Synthesis

A clarinetlike tone can be constructed from the following rules:

- Only odd-numbered harmonics are present.

- Amplitudes of harmonics decrease as the harmonic number increases.

- The phase offset of each harmonic is 0. (This is not strictly required to obtain a clarinetlike tone, but it helps simplify the analysis.)

Equation (3.1) shows how to construct a waveform from these rules using sigma notation (see appendix section A.3).

$$s(t) = \sum_{n=1}^{\infty} \frac{1}{n} \sin(n\omega t + 0) \quad \text{such that } n \text{ is odd.} \tag{3.1}$$

The fundamental frequency of $s(t)$ is ω, and the frequency of harmonic n is $n\omega$. The amplitude of harmonic n is $1/n$, and its phase offset is 0. The waveform of each harmonic n separately is

$$\frac{1}{n} \sin(n\omega t + 0).$$

Here is the expansion of equation (3.1):

$$s(t) = \sin(\omega t + 0) + \frac{1}{3} \sin(3\omega t + 0) + \frac{1}{5} \sin(5\omega t + 0) + \cdots. \tag{3.2}$$

Equations (3.1) and (3.2) say, "Let there be a function $s(t)$, where t is time. The function $s(t)$ is the sum of an infinite number of terms, where each term is a harmonic waveform. The frequencies of the waveforms are all odd multiples of the fundamental frequency ω, the amplitudes are the inverse of the harmonic number n, and the phase offsets of the harmonics are all 0."

Figure 3.1 shows a graphical representation of the addition of the first three terms of equation (3.2).

If we kept adding higher-order harmonics according to equation (3.1), we would end up with a square wave in the limit at infinity (figure 3.1). Other waveforms such as triangular, impulse, and sawtooth can similarly be built up from combinations of sinusoids (see section 9.2.6).

3.1.6 Spectrum Analysis

As shown, we can build up compound waveforms by adding sinusoids together. Now let's take them apart.

The appeal of Fourier, or spectrum, analysis is that we can take *any* (periodic) waveform and break it down into its individual sinusoidal components. With this technique, we can

- Study musical timbre to understand why musical instruments sound the way they do
- Classify sounds by their spectral content and then identify them, for instance, to recognize speech and convert it to text
- Resynthesize sounds using Fourier synthesis based on Fourier analysis of sound (*analysis-based synthesis*)
- Synthesize hybrid sounds that are a mixture of previously analyzed sounds
- Create arbitrary mixtures of frequency components

For example, suppose in equation (3.1) we set the fundamental frequency to $\omega = 2\pi \cdot 100$, which corresponds to 100 Hz, and use it to create a waveform $s(t)$. If we then apply Fourier analysis to $s(t)$ and graph the resulting function, ideally we'd see a spectrum like that shown in figure 3.2.

3.1.7 How Spectrum Analysis Works

The key to spectrum analysis is to understand how the Fourier transform detects that there is energy at a particular frequency in a signal. Fortunately, there is a simple explanation. To explain what happens when signals are multiplied together, I'll need to review the rules of multiplication, then apply these rules to signals.

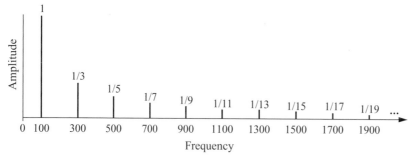

Figure 3.2
Fourier analysis of equation (3.1).

Multiplying Signals   There are four rules of multiplication:

1. $+x \cdot +y$ is a *one-quadrant multiply* (a positive number times a positive number is a positive number).

2. $+x \cdot \pm y$ and $\pm x \cdot +y$ are *two-quadrant multiplies* (a positive number times a signed number is a positive or negative number, depending upon the value of the signed number).

3. $\pm x \cdot \pm y$ is a *four-quadrant multiply* (a signed number times a signed number is a positive number if both are either positive or negative; otherwise the product is negative).

4. Anything times 0 is 0.

These rules are summarized in figure 3.3. Which quadrants the product can lie in depends upon the signs of the multipliers and multiplicands. For a one-quadrant multiply where $x \geq 0$ and $y \geq 0$, all possible products are bounded by quadrant 1. For a four-quadrant multiply, where x and y can be any positive or negative value, the product can be in any quadrant. There are four possibilities for the two-quadrant multiply. If, for example, $x \geq 0$ but y can be any positive or negative value, the product is bounded by quadrants 1 and 4.

To multiply two functions $s(t) = a(t) \cdot b(t)$, we find each point $a(t)$ and $b(t)$ and multiply them, using the quadrant rules. Some examples of multiplying two functions are shown in figure 3.4.

Squaring a Signal   If we multiply a signal by itself point by point, we square it. No matter whether the signal contains negative values, the resulting waveform will have all positive values, because (except for imaginary numbers) a number times itself is always positive. Figure 3.5 shows what happens when a sine wave is squared.

If there are two variables a and b, and $a = b$, then $a \cdot b = a^2$. The same goes for signals. If two signals are identical, their product is the same as the square of either one of them.

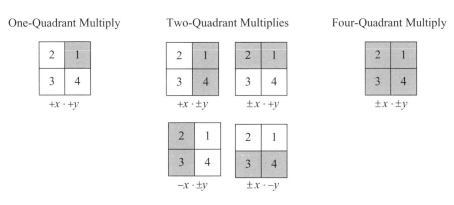

One-Quadrant Multiply   Two-Quadrant Multiplies   Four-Quadrant Multiply

$+x \cdot +y$ $+x \cdot \pm y$ $\pm x \cdot +y$ $\pm x \cdot \pm y$

$-x \cdot \pm y$ $\pm x \cdot -y$

Figure 3.3
Multiplication by quadrants.

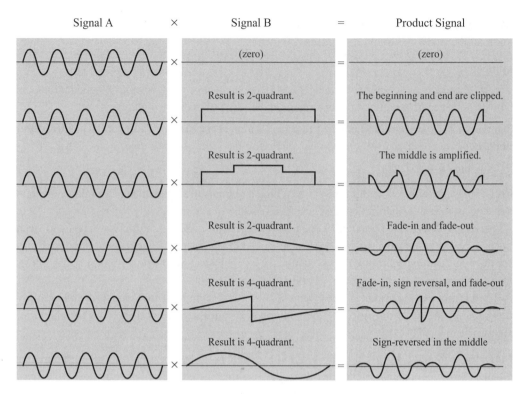

Figure 3.4
Examples of multiplying functions.

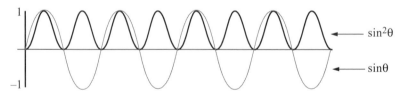

Figure 3.5
Squaring a signal.

Call the result of multiplying two arbitrary signals the product signal. Depending upon the exact values of the signals we are multiplying, sometimes the product signal will be positive, sometimes negative. However, for the special case where the signals are identical, *the product signal is always positive.* Here is the fundamental insight of the Fourier transform.

The more positive the product signal is, the closer the source signals are to being identical.

The more mixed positive and negative the product signal is, the less identical are the source signals.

We can adapt this to create a frequency detector, a device that indicates when the frequencies of two signals are equal. But first we have to figure out how to measure the positiveness of the product signal to know when signals are in fact identical.

Constructing a Frequency Detector Consider what would happen if we summed together all the points of a product signal. Figure 3.5 showed that the square of a signal, or equivalently, the product of two identical signals, is a signal with all positive values. (Zero is considered to be positive.) If we summed together all the values of this all-positive function, the result would be strongly positive because all the values being summed are positive.

If we multiplied signals that are not identical, their product would sometimes be positive and sometimes negative, depending upon the signs of the functions at each point. If we summed together all the values of such a mixed product signal, the result might be weakly positive or weakly negative.

Let's combine these ideas to construct a frequency detector. We begin by defining the following signals:

- Let $x(t)$ be a *test signal,* the signal to be analyzed. It is a sine wave with an unknown fixed frequency. Say it is a recording of a steady whistle tone.

- Let $y(t)$ be a *probe signal,* which we will use to determine the frequency of $x(t)$. It is a sine wave, provided by a variable-frequency oscillator.

- Let $c(t)$ be the product of the two signals.

The sum of the product signal $c(t)$ will be the most positive when the frequency of $y(t)$ is equal to the frequency of $x(t)$.

As shown in figure 3.6, we multiply the output of the oscillator and the recorded signal, then sum the result. The final output goes to a meter that measures the strength of the summation.

While the recorder plays the test signal, we watch the meter change as we turn the frequency dial of the oscillator. The frequency coming from the oscillator is identical to the frequency of the test signal when the meter achieves its most positive reading.

If $x(t)$ had been a compound signal containing several frequency components, the frequency detector would still work. We'd see positive bumps of the needle any time the oscillator frequency matched the frequency of any components of the compound signal.

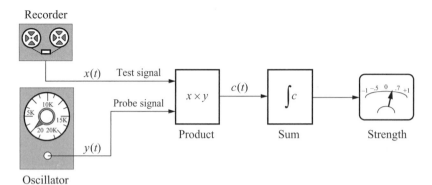

Figure 3.6
Frequency analyzer.

Formalizing the Frequency Detector   We can formalize the frequency detector as follows. First, represent the probe signal coming from the oscillator in figure 3.6 as a phasor, the *probe phasor*

$e^{-i2\pi ft}$.   *Probe Phasor* (3.3)

Because its exponent is negative, this phasor represents a single negative frequency component. The phasor spins clockwise for positive time t and positive frequency f (see section 2.6.3). Why use negative frequencies for the probe phasor? We could use positive frequencies, and some do. However, there will be a conceptual payoff later for using negative frequencies for the probe phasor, related to the ideas presented in section 2.9.

So, first multiply the input waveform and the probe phasor:

$x(t) \cdot e^{-i2\pi ft}$.

Next we need to understand the signal-summing device shown in figure 3.6. It is called an *integrator*. It does just what we want: at each moment, it adds the past value to the current input value and holds it for future use. Suppose we set the integrator's initial value to 0 and then feed it with a function equal to 1 at all points, $\{1, 1, 1, \ldots\}$. The output of the integrator will be $\{1, 2, 3, \ldots\}$. If we zero it and feed it a function such as $\{-1, -1, -1, \ldots\}$, the output of the integrator will be $\{-1, -2, -3, \ldots\}$. If we zero it and feed it a function that alternates, $\{1, -1, 1, -1, \ldots\}$, the output of the integrator will be $\{1, 0, 1, 0, \ldots\}$. The integrator must always be cleared to zero before use.

Given the way the integrator works, if we feed it a positive-only signal, its output will grow more positive with time because it's always adding positive values together. But if we feed it a mixed-sign signal (sometimes positive, sometimes negative) the integrator will tend to hover around 0. The mathematical symbol for integration is $\int c$ for some value c. So, adding integration to what we had before, the frequency detector can be expressed as

$\int x(t)e^{-i2\pi ft}$.   *Frequency Detector* (3.4)

Equation (3.4) says, "Integrate the product of the test signal and the probe phasor." Having come this far, it is a simple step to add the rest of the terms of the Fourier transform.

3.1.8 Continuous Fourier Transform

Here is the full definition of the continuous Fourier transform:

$$X(f) = \int_{-\infty}^{\infty} x(t)e^{-i2\pi ft}dt.$$
Fourier Transform (3.5)

It is continuous because it evaluates every moment of time and every possible frequency with no gaps or skips. When people refer to the Fourier transform, this formula is typically what they have in mind.

We've already seen most of the terms of equation (3.5):

- t is time, f is frequency. These are the two control variables in this equation.

- $x(t)$ is the test signal.

- $e^{-i2\pi ft}$ is the probe phasor. This signal is sometimes called the *kernel function* of the Fourier transform.

The new elements introduced in equation (3.5) are as follows:

- The result of the equation is $X(f)$, a function of frequency f called a *spectrum*. The value of $X(f)$ shows the amount of energy in the input signal at each frequency f.

- The term dt simply indicates that the integrator is operating on the time variable t rather than on the frequency variable f, that is to say, we are integrating through time.

- The equation says we are to integrate over the range $\pm\infty$. Because the integrator is operating with respect to time t, this requires us to integrate over *all time* in the interval $-\infty < t < \infty$ (see section 3.1.9).

Simply stated, equation (3.5) says that for some periodic function of time $x(t)$, there is an equivalent function of frequency $X(f)$. That's all a spectrum is: a function that reveals the strengths of the various frequency components it contains, ordered by frequency.

Here's a detailed look at the operation of equation (3.5). The spectrum $X(f)$ is built up as follows:

1. Assign a periodic signal to the test signal $x(t)$.

2. Fix f at frequency $-\infty$.

3. Let t vary over its entire range, calculating the value of the equation for every value of t. The final result is a single number indicating the amount of energy in $x(t)$ at frequency f. Store the result in function X at location f.

4. Choose a new value for f, and let t vary over its entire range again. Sum all the results of this new calculation, and store the result in the new location in $X(f)$.

5. Continue to fix f at all values in the range $-\infty < f < \infty$.

When we are done, $X(f)$ will have positive values where there was energy in the input signal $x(t)$; otherwise $X(f)$ will be at or near zero. However, we only get this ideal result when we integrate over all time. Because we can observe only local values of time, the resulting spectrum is less definitive, though still useful. Thus the Fourier transform acts as a frequency-dependent energy detector.

3.1.9 Limits on Fourier Analysis

The Fourier transform is restricted to periodic signals. Abstract signals such as sinusoids present no problem, but real-world signals such as music, speech, and other environmental sounds are not strictly periodic.

What Is Periodic?   Informally, we could say that a signal is periodic if it is the same every time it repeats. For instance, a pendulum is periodic if it visits the same points at the same times on each swing. In general,

If $f(t)$ is a periodic function with period τ, then for any reference point t, the value of $f(t + \tau)$ must equal $f(t)$, that is, the function must satisfy

$$f(t) = f(t + \tau), \qquad (-\infty < t < \infty) . \tag{3.6}$$

This is illustrated in figure 3.7. No matter what point t we choose, the value of $f(t)$ must equal $f(t + \tau)$ for the function to qualify as periodic. Although I've chosen to illustrate this with a sine wave, note that the definition does not restrict the shape of the curve except to say that all copies of it must be exactly alike.

The smallest possible nonzero value of τ is taken to be the *period*. If τ exists, the function is said to be periodic. If t represents time, it is said to be a *periodic function of time*.

Equation (3.6) must be true for all $-\infty < t < \infty$ for a function to qualify as periodic. But no physical signal can ever be periodic by this definition: a signal would have to exist for all time to qualify as periodic. If the Fourier transform requires infinite periodicity, then it is useless in practice. Fortunately, we can adapt the Fourier transform to deal with finite signals. What happens if we apply Fourier analysis to some time interval, say from t_1 to t_2? Theoretically, the Fourier transform does not exist if we

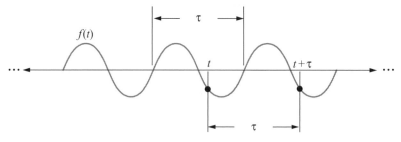

Figure 3.7
Periodic signal.

retrench in from infinity because it no longer allows conversion between the time and frequency domains without loss of information. Practically, however, it still works—sort of.

We live in a world with horizons, so perhaps it's not surprising that the level of detail provided by the Fourier transform should grow cruder as the time interval being analyzed shrinks. The practical consequence of this loss of information is that the Fourier transform of a finite signal is no longer *lossless*. Mathematically, a time-limited Fourier transform is not strictly a transform at all. But practically, a *lossy* Fourier analysis of a real-world signal is better than no analysis at all.

Other Conditions Some so-called pathological signals do not have Fourier transforms even if they are periodic. For example, although the Fourier transform exists for signals with an infinite number of harmonically related sinusoids, it does not exist if any of those sinusoids has an infinite amplitude. Additionally, signals must have a finite number of discontinuities and a finite number of maxima and minima. The rules that guarantee the existence of the Fourier transform of a signal are called the *Dirichlet conditions*. All real-world acoustical signals meet these conditions automatically, so we needn't be too concerned.

3.1.10   Inverse Fourier Transform

Whereas the Fourier transform starts with a function of time and produces a function of frequency, the inverse Fourier transform starts with a function of frequency and produces a function of time, that is, it synthesizes a signal from a specification given by its spectrum. Equation (3.7) gives the continuous form of the inverse Fourier transform.

$$x(t) = \int_{-\infty}^{\infty} X(f)e^{i2\pi ft} df. \qquad \textit{Inverse Fourier Transform} \ (3.7)$$

The symmetry between equations (3.5) and (3.7) is striking. The only things that have changed for the inverse Fourier transform are

- $x(t)$ is now the output function and $X(f)$ is the input function.
- The exponent of the probe phasor $e^{2\pi ift}$ is now positive; hence the phasor spins counterclockwise.
- The integration is with respect to f, not t. That is, dt is replaced with df, which says that the integrator is operating on the frequency variable f, not the time variable t.
- The integral operation is still constrained to operate over the range $\pm\infty$.

For some periodic function of frequency $X(f)$, this equation creates a corresponding periodic function of time $x(t)$. The actions of the inverse Fourier transform are as follows.

1. Assign a periodic function of frequency to $X(f)$.

2. Fix t at time $-\infty$.

3. Let f vary over its entire range, calculating the value of the equation for every value of f. The final result is a single number: the instantaneous value of the signal $x(t)$ at time t. Store the result in function x at point t.

4. Choose a new value for t, and let f vary over its entire range again. Sum all the results of this new calculation, and store the result in the new location in $x(t)$.

5. Continue to fix t at different values until we have covered all time.

When we are done, $x(t)$ will be a waveform with the spectrum specified by $X(f)$. Thus the inverse Fourier transform acts as a frequency-dependent signal generator.

3.2   Discrete Fourier Transform

The age of digital audio has led to a need for a version of the Fourier transform that operates on sampled signals. The *discrete Fourier transform* (DFT) operates on discrete sequences of time and frequency. The fast Fourier transform (FFT, see section 3.7) is a more computationally efficient version of the DFT. The FFT is widely used, and its name may be more familiar than the DFT, but it performs the same transform as the DFT. Since the DFT is easier to understand, we will start with it first.

3.2.1   Discrete Input and Output of the DFT

The input to the Fourier transform is a continuous signal $x_c(t)$, where t is continuous time. The input to the DFT is a *sampled* signal $x_s(n)$, where n is an integer that indexes individual samples. The sampled signal $x_s(n)$ is created by sampling $x_c(t)$ periodically according to the equation

$$x_s(n) = x_c(nT),$$

where T is the period between samples expressed as a real constant. The time interval $T = 1/R$, where R is the constant real sampling rate expressed as the number of samples per second. The product nT indicates a set of particular time instants indexed by n and separated by T on the continuous function $x_c(t)$ that is to be sampled.

The value of $x_c(nT)$ for each n is stored in the nth value of the sampled function $x_s(n)$. (See the discussion about sampled signals in chapter 1.) Figure 3.8 shows an example of a continuous function and what its discrete (sampled) equivalent might look like for some sample rate R.

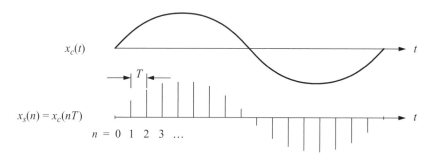

Figure 3.8
Continuous signal and corresponding discrete signal.

The same reasoning applies to the output of the DFT. Whereas the output of the Fourier transform is the continuous spectrum $X_c(f)$, where f is continuous frequency, the output of the DFT is the sampled spectrum $X_s(k)$, where k is an integer that indexes a particular harmonic in the spectrum.

The sampled spectrum is created by sampling $X_c(f)$ at periodic frequencies according to the equation

$$X_s(k) = X_c(kF),$$

where k indexes individual frequencies, and F is the frequency interval between harmonics. The product kF indicates a set of particular frequency values indexed by k and separated in frequency by F.

In the rest of this chapter, when I say $X(f)$, I mean the continuous form of the spectrum, and when I say $X(k)$, I mean the discrete form. Similarly, $x(t)$ refers to a continuous waveform, and $x(n)$ refers to a sampled waveform.

3.2.2 How Does the DFT Extract a Spectrum?

Here is the equation for the DFT:

$$X(k) = \frac{1}{N}\sum_{n=0}^{N-1} x(n)e^{-ik\omega n/N}, \qquad \textit{Discrete Fourier Transform (DFT)} \quad (3.8)$$

where $\omega = 2\pi$, N is the number of samples in the input signal $x(n)$, the variable n indexes the input signal, and k indexes the frequencies in the output spectrum $X(k)$. Both n and k are integers, so the input signal $x(n)$ and the output spectrum $X(k)$ are discrete functions of time and frequency, respectively. The probe phasor for the DFT is $e^{-ik\omega n/N}$. Because k, n, and N are integers, this probe phasor does not spin continuously but skips between discrete positions on the unit circle. The exact positions visited depend upon the values of k, n, and N. It will probe for energy in the input signal at discrete frequencies.

Example Let's run the test signal defined by equation (3.9) through the DFT and see what happens.

$$x(n) = A\sin\left(f\omega\frac{n}{N}\right). \qquad \textit{DFT Test Signal} \quad (3.9)$$

A is amplitude, $\omega = 2\pi$, f is frequency, and $0 \le n \le N-1$ for a signal of length N.

If we do not assign a value for A, and if we let $f = 1$ and $N = 8$, we generate eight samples for $x(n)$ numbered 0 through 7 (figure 3.9). For orientation the figure also shows the underlying continuous sine wave.

Simplifying the DFT To further simplify things, let's break down the DFT's complex probe phasor $e^{-ik\omega n/N}$ into its real and imaginary parts and handle them separately. Remembering Euler's formula (section 2.5), $e^{i\theta} = \cos\theta + i\sin\theta$, the form that corresponds to a negative frequency phasor is

$$e^{-i\theta} = \cos\theta - i\sin\theta.$$

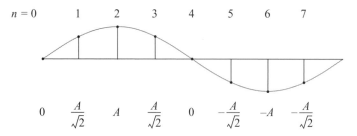

Figure 3.9
DFT test signal: eight samples of a sine wave.

Using this, we can break the DFT down into two simpler versions, the *discrete cosine transform* (DCT), containing only the real part of the probe phasor,

$$X(k) = \sum_{n=0}^{N-1} x(n) \cdot \cos\left(k\omega \frac{n}{N}\right). \qquad \textit{Discrete Cosine Transform (DCT)} \quad (3.10)$$

and the *discrete sine transform* (DST), containing the imaginary part,

$$X(k) = \sum_{n=0}^{N-1} x(n) \cdot -i\sin\left(k\omega \frac{n}{N}\right). \qquad \textit{Discrete Sine Transform (DST)} \quad (3.11)$$

The DST multiplies the input by i to make the result imaginary, as required by Euler's formula. However, for the moment, let's further simplify things by ignoring the negative imaginary component of the DST, and operate instead on a positive real-valued version of the discrete sine transform,

$$X(k) = \sum_{n=0}^{N-1} x(n) \sin\left(k\omega \frac{n}{N}\right). \qquad \textit{DST, Positive Real} \quad (3.12)$$

3.2.3   Analyzing a Frequency at Which Energy Is Present

According to equation (3.12), the DST multiplies each sample $x(n)$ by the appropriate value of the probe phasor, $\sin[k\omega(n/N)]$, sums the result, and stores it in $X(k)$ for each k. It then adds 1 to k and repeats the calculation, storing the result in $X(k+1)$, and so forth, N times.

Let's focus on that point in the analysis where the probe phasor is ready to detect the same frequency as the test signal. Recall that the test signal is $A\sin[f\omega(n/N)]$ and that the DST probe phasor from equation (3.12) is $\sin[k\omega(n/N)]$. Thus they have the same frequency when $k = f$. And since we set the test signal to $f = 1$, let's set $k = 1$ as well so we can watch it detect the input test frequency. That means, for $N = 8$, the DST multiplies the following two signals sample by sample:

Sample	0	1	2	3	4	5	6	7
Input	0	$\dfrac{A}{\sqrt{2}}$	A	$\dfrac{A}{\sqrt{2}}$	0	$-\dfrac{A}{\sqrt{2}}$	$-A$	$-\dfrac{A}{\sqrt{2}}$
Probe	0	$\dfrac{1}{\sqrt{2}}$	1	$\dfrac{1}{\sqrt{2}}$	0	$-\dfrac{1}{\sqrt{2}}$	-1	$-\dfrac{1}{\sqrt{2}}$
Product	0	$\dfrac{A}{2}$	A	$\dfrac{A}{2}$	0	$\dfrac{A}{2}$	A	$\dfrac{A}{2}$

We see that, for $k = f$,

$$\sum_{n=0}^{N-1} x(n)\sin\left(k\omega\frac{n}{N}\right) = 0 + \frac{A}{2} + A + \frac{A}{2} + 0 + \frac{A}{2} + A + \frac{A}{2}$$

$$= 4A.$$

When we sum all the terms with $N = 8$, we get the answer $4A$. Because all the product values are positive, the more of them we sum, the larger would be the positive result. For instance, if $N = 16$, we'd get the answer $8A$. Therefore, when the input signal and the probe phasor have identical frequency, we can write

$$\sum_{n=0}^{N-1} x(n)\sin\left(k\omega\frac{n}{N}\right) = N\frac{A}{2} \quad \text{if } x(n) = \sin\left(k\omega\frac{n}{N}\right).$$

So, if the input signal is a sine wave with frequency f, the amplitude of the DST when it is analyzing frequency f is $A/2$ scaled by N. The larger the analysis window (that is, the greater the value of N), the larger this result will be. This is unfortunate, because it would be better if the result did not vary with the length of the input signal so that we could compare the spectra of signals regardless of their length. We can eliminate N from the result by scaling the whole equation by $1/N$:

$$\frac{1}{N}\sum_{n=0}^{N-1} x(n)\sin\left(k\omega\frac{n}{N}\right) = \frac{A}{2} \quad \text{if } x(n) = \sin\left(k\omega\frac{n}{N}\right). \tag{3.13}$$

The $1/N$ term normalizes the amplitude of the spectrum so that when the test waveform matches the probe signal, the output will be $A/2$ instead of $N \cdot A/2$. From now on, I'll add this normalizing term to the DFT, DST, and DCT.

3.2.4 Analyzing a Frequency at Which Energy Is Absent

The previous section explained what happens when a probe phasor matches the frequency of the input signal. But what happens when a probe phasor is aimed at a place where the test signal has no energy? For instance, using the techniques developed previously, can we show that there is no energy in the test signal $x(n)$ at the frequency $k = 2f$? Clearly there shouldn't be. Using the DST (equation (3.11)) now we want to compute

$$\frac{1}{N}\sum_{n=0}^{N-1} x(n) \sin\left(2f\omega\frac{n}{N}\right).$$

If we set $N = 8$, we have the following sample values:

Sample	0	1	2	3	4	5	6	7
Input	0	$\dfrac{A}{\sqrt{2}}$	A	$\dfrac{A}{\sqrt{2}}$	0	$-\dfrac{A}{\sqrt{2}}$	$-A$	$-\dfrac{A}{\sqrt{2}}$
Probe	0	1	0	-1	0	1	0	-1
Product	0	$\dfrac{A}{\sqrt{2}}$	0	$-\dfrac{A}{\sqrt{2}}$	0	$-\dfrac{A}{\sqrt{2}}$	0	$\dfrac{A}{\sqrt{2}}$

Summing the product terms, we get

$$0 + \frac{A}{\sqrt{2}} + 0 - \frac{A}{\sqrt{2}} + 0 - \frac{A}{\sqrt{2}} + 0 + \frac{A}{\sqrt{2}} = 0$$

as expected. So the analyzer correctly reports that there is no energy at $k = 2f$.

3.2.5 Negative Frequencies in the DFT

Let's go back to the definition of the test signal in equation (3.9). When we probed for frequency $k = f$ in the test signal, why did we end up with a value of $A/2$? Why did we only detect half the amplitude of the input signal? Where is the other half of the input signal's amplitude?

I have a hunch that we should look for it among the negative frequencies; specifically, we should look at $k = -f$. We want to compute

$$\frac{1}{N}\sum_{n=0}^{N-1} x(n) \sin\left(-k\omega\frac{n}{N}\right).$$

If we set $N = 8$, we have the following sample values:

Sample	0	1	2	3	4	5	6	7
Input	0	$\dfrac{A}{\sqrt{2}}$	A	$\dfrac{A}{\sqrt{2}}$	0	$-\dfrac{A}{\sqrt{2}}$	$-A$	$-\dfrac{A}{\sqrt{2}}$
Probe	0	$-\dfrac{1}{\sqrt{2}}$	-1	$-\dfrac{1}{\sqrt{2}}$	0	$\dfrac{1}{\sqrt{2}}$	1	$\dfrac{1}{\sqrt{2}}$
Product	0	$-\dfrac{A}{2}$	$-A$	$-\dfrac{A}{2}$	0	$-\dfrac{A}{2}$	$-A$	$-\dfrac{A}{2}$

Summing the product terms, we get

$$0 - \frac{A}{2} - A - \frac{A}{2} + 0 - \frac{A}{2} - A - \frac{A}{2} = -4A$$

and normalizing by $1/N$, we get

$$\frac{1}{N}\sum_{n=0}^{N-1} x(n)\sin\left(-k\omega\frac{n}{N}\right) = -\frac{A}{2}.$$

So, according to the DST, half the signal is a negative frequency with a negative amplitude. What's going on?

Let's put the two halves together. To simplify, set the frequency of the test signal to $\theta = f\omega n/N$. Then it appears that according to the DST, the spectrum of our signal is

$$x(n) = \frac{A}{2}[(\sin\theta) - (\sin-\theta)]. \qquad (3.14)$$

The function $x(n)$ appears to be odd. Recall from section 2.6.12 that an odd function is of the form

$$f_o(x) = \frac{1}{2}[f(x) - f(-x)].$$

Could it be that we got this result because we've been using the DST? What if we redid the analysis using the DCT (equation (3.10))? Rather than going through the motions, let me just say that if we let $x(n) = A\cos(f\omega n/N)$ and set the probe phasor to $\cos(k\omega n/N)$, we would end up extracting:

$$x(n) = \frac{A}{2}[(\cos\theta) + (\cos-\theta)], \qquad (3.15)$$

which is an even function.

If we use the DST to look at an all-cosine waveform, it won't find any signal, and if we use the DCT to look at an all-sine waveform, it won't find any signal. The DST only detects odd

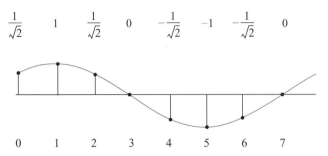

Figure 3.10
New test signal, offset in phase.

components of the spectrum, and the DCT only detects even components. Since the DFT combines the DST and DCT, it shows the entire spectrum.

3.2.6 DFT of Arbitrary Phase Signals

In the preceding section, I developed the use of the DST and DCT to detect the even (cosine) and odd (sine) components of the input signal separately. But real-world signals are seldom purely just one or the other, so it would be more desirable to have a frequency detector that could detect sinusoids of arbitrary phase. What would we have to do in order to analyze such signals?

Consider a new test signal $x(n)$, shown in figure 3.10, which is a mixture of sine and cosine. Its generating equation is

$$x(n) = a\cos\left(f\omega\frac{n}{N}\right) + b\sin\left(f\omega\frac{n}{N}\right), \qquad (3.16)$$

where, as before, $\omega = 2\pi$, f is frequency, and $0 \le n \le N - 1$ for signal length N. Let's set both the a and b coefficients of this test signal to $1/\sqrt{2}$, and set $f = 1$, and $N = 8$, as shown in figure 3.10.

To confirm that this equation generates the test signal sequence in figure 3.10, look at the eight values of the cosine and sine terms generated by equation (3.16):

Sample		0	1	2	3	4	5	6	7
$\frac{1}{\sqrt{2}}\cos(f\omega n/N)$		$\frac{1}{\sqrt{2}}$	$\frac{1}{2}$	0	$-\frac{1}{2}$	$-\frac{1}{\sqrt{2}}$	$-\frac{1}{2}$	0	$\frac{1}{2}$
	+								
$\frac{1}{\sqrt{2}}\sin(f\omega n/N)$		0	$\frac{1}{2}$	$\frac{1}{\sqrt{2}}$	$\frac{1}{2}$	0	$-\frac{1}{2}$	$-\frac{1}{\sqrt{2}}$	$-\frac{1}{2}$
$x(n)$	=	$\frac{1}{\sqrt{2}}$	1	$\frac{1}{\sqrt{2}}$	0	$-\frac{1}{\sqrt{2}}$	-1	$-\frac{1}{\sqrt{2}}$	0

Sure enough, there is our signal.

Figure 3.10 shows that $x(n)$ is a sinusoid with phase $\phi = 45°$. In order to properly detect this, the frequency of the probe phasors of the DST and DCT would have to be set to $k\omega(n/N) + 45°$. But this opens a Pandora's box: must we test each of the *infinitely many* possible phase angles in order to detect sinusoids of arbitrary phase? Fortunately not.

Finding a Sinusoid from a Sum of Sine and Cosine Remember from section 2.6.6 that a sinusoid of arbitrary phase ϕ can be constructed from the sum of a cosine wave of amplitude a plus a sine wave of amplitude b. It also works the other way around: for a cosine wave of amplitude a and a sine wave of amplitude b, we can determine the appropriate amplitude A and phase ϕ for a single equivalent sinusoid. In other words, for any frequency θ,

$$A \sin(\theta + \phi) = a \cos \theta + b \sin \theta \tag{3.17}$$

for suitable choices of A, ϕ, a, and b. (See appendix section A.4 for a proof.) We can use this knowledge to generalize the probe phasor in the DFT to detect signals of arbitrary amplitude and phase. What are the suitable choices of A, ϕ, a, and b?

Start with an unknown test sinusoid $x(n)$ of arbitrary amplitude and phase, such as equation (3.16). The method of discovering its amplitude A and phase ϕ goes back to the Pythagorean theorem:

1. Use a cosine probe phasor to get the DCT of the test signal:

$$a_k = \frac{1}{N} \sum_{n=0}^{N-1} x(n) \cos\left(k\omega\frac{n}{N}\right). \tag{3.18}$$

The values a_k are the even components of the spectrum.

2. Use a sine probe phasor to get the DST of the test signal:

$$b_k = \frac{1}{N} \sum_{n=0}^{N-1} x(n) \sin\left(k\omega\frac{n}{N}\right). \tag{3.19}$$

The values b_k are the odd components of the spectrum. Together, the a_k and b_k terms are called the *spectral coefficients* of the test signal.

3. By the Pythagorian theorem (figure 3.11), the amplitude of each component is

$$A(k) = \sqrt{a_k^2 + b_k^2}. \tag{3.20}$$

In this way we have detected the amplitudes of sinusoids with arbitrary phase.

4. To get the phases of each component ϕ_k, we must find the angle of the hypotenuse of the triangle for each component. This is defined in trigonometry as the arctangent of b/a, that is, $\text{atan}(b/a)$ (see appendix section A.2). It is also written $\tan^{-1}(b/a)$. So, the phase of each component is

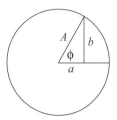

Figure 3.11
Pythagorean theorem.

$$\phi(k) = \tan^{-1}\frac{b_k}{a_k}. \tag{3.21}$$

In this way we have detected the phases of sinusoids with arbitrary phase.

Using this approach, we can create a frequency detector that can detect signals of arbitrary phase. The result gives both the amplitudes and phases of the detected spectral components.

Amplitude Spectrum, Phase Spectrum, and Power Spectrum   The preceding method is an effective if rather inelegant means of analyzing sinusoids of arbitrary phase. To restate, substituting equations (3.18) and (3.19) into equation (3.20) yields the amplitudes of the spectral components:

$$A(k) = \sqrt{\left[\frac{1}{N}\sum_{n=0}^{N-1}x(n)\cos\left(k\omega\frac{n}{N}\right)\right]^2 + \left[\frac{1}{N}\sum_{n=0}^{N-1}x(n)\sin\left(k\omega\frac{n}{N}\right)\right]^2}. \qquad \textit{Magnitude Spectrum} \quad (3.22)$$

Equation (3.22) is the *amplitude spectrum,* or *magnitude spectrum.* Each $A(k)$ represents the amplitude of the component at frequency k.

Substituting equations (3.18) and (3.19) into equation (3.21) yields the phases of the spectral components:

$$\phi(k) = \tan^{-1}\left[\frac{\dfrac{1}{N}\sum_{n=0}^{N-1}x(n)\sin\left(k\omega\frac{n}{N}\right)}{\dfrac{1}{N}\sum_{n=0}^{N-1}x(n)\cos\left(k\omega\frac{n}{N}\right)}\right]. \qquad \textit{Phase Spectrum} \quad (3.23)$$

$\phi(k)$ in equation (3.23) is called the *phase spectrum.* The numerator is the strength of sinusoidal energy at frequency k; the denominator is the strength of cosinusoidal energy. The arctangent of this ratio is an angle $\phi(k)$ that characterizes in one value the contribution of positive and negative cosinusoidal and sinusoidal energy occurring at frequency k.

The *power spectrum* is a variation on the magnitude spectrum. It is defined as

$$P(k) = a_k^2 + b_k^2, \qquad\qquad\qquad Power\ Spectrum \quad (3.24)$$

which is the same as equation (3.20) but without the square root operation, because power is proportional to the square of amplitude.

These equations allow us to find the amplitude, phase, and power of an arbitrary signal $x(n)$ using nothing more complicated than sines, cosines, and summation. But they look complicated. We can do better.

Complex Version of the Amplitude Spectrum   Fortunately, we can simplify equation (3.22) quite a bit by using Euler's formula $e^{i\theta} = \cos\theta + i\sin\theta$ to recombine the DCT and the DST into the DFT.

If we set $\theta = k\omega n/N$, the frequencies of the probe phasor, then Euler's formula allows us to combine the sine and cosine terms in equation (3.22) into a more compact and elegant representation:

$$X(k) = \frac{1}{N}\sum_{n=0}^{N-1} x(n)e^{-ik\omega n/N}, \qquad\qquad\qquad (3.25)$$

where $X(k)$ is now a complex function that contains both the amplitude spectrum $A(k)$ and the phase spectrum $\phi(k)$. This relatively tidy equation replaces equations (3.22) and (3.23). Thus it is written: *complex is simpler*.

3.2.7   Detecting Multiple Frequencies

Equation (3.25) is a powerful general-purpose frequency detector of sinusoids of any phase. But will it detect multiple components simultaneously? To get an answer, let's observe how the probe phasor $e^{-ik\omega n/N}$ behaves.

The magnitude of a phasor's exponent determines its rate of spin. Because the probe phasor's exponent in equation (3.25) is negative, the phasor spins clockwise. The spin rate of the probe phasor is directly proportional to i, k, ω, and n, and inversely proportional to N. But only k and n change during calculation. The variable k grows by 1 every time we evaluate the expression on the right side of equation (3.25). The variable n goes through all integers from 0 to $N-1$ every time k increments. Let's try some sample values for n and k to get a feel for how it works.

• When $k = 0$, the phasor is $e^{-ik\omega n/N} = e^0 = 1 + 0i$ regardless of the value of n. So when $k = 0$, we multiply every input sample by $(1 + 0i)$, sum the results, divide by N, and store the result in $X(0)$.

• When $k = 1$, the probe phasor is $e^{-i\omega n/N}$. During evaluation, n gradually increments by 1 from 0 to $N-1$. At first, when $n = 0$, the complex exponential equals $e^{-i\omega 0/N} = e^0 = 1 + 0i$. But as

n increases, the angle of the probe phasor will grow in a clockwise direction. Because n is divided by N, the phasor makes one clockwise rotation. At the end of evaluation, the complex exponential equals $e^{-i\omega(N-1)/N}$, which is one step short of a complete clockwise rotation.

- When $k = 2$, the phasor is $e^{-i2\omega n/N}$. It will spin clockwise twice as n goes from 0 to $N - 1$. When $k = 3$, the phasor spins three times, and so on.

Summarizing the roles of k and n,

- k selects the frequency on which to operate and determines where in the output function X to place the result.

- n steps through the samples of the input function x and also determines the phase value of the phasor for each calculation.

3.2.8   Frequency Range of the DFT

We've seen that n is bounded by $0 \le n \le N - 1$, but what are the bounds of k? Since frequency can be infinitely positive or infinitely negative, there are no bounds to the spectral function $X(k)$, and k can take any values in the range $\pm\infty$. But since the DFT operates on sampled functions, a curious thing happens when k reaches the Nyquist frequency. I illustrate with an example.

Take one period of a sine wave with $N = 8$, just like the test sample in figure 3.9, and play it repeatedly through a digital-to-analog converter at a rate of $R = 8000$ samples per second. When we reach the end ($n = 7$), we start over from the first sample ($n = 0$).

- Since the test signal is a simple sinusoid, the frequency we hear is $8000/8 = 1000\,\text{Hz}$.

- If we skip every other sample each time, we hear 2000 Hz.

- If we output the same sample each time, the output would be a constant function, and we hear silence.

- If we stepped backward through the samples, the frequency we'd hear is negative.

So, for $N = 8$ and $R = 8000\,\text{Hz}$, the frequencies we could get for skip value k (where k is an integer) are the positive frequencies

k	0	1	2	3
Frequency	0 Hz	1000 Hz	2000 Hz	3000 Hz

and the negative frequencies

k	-1	-2	-3
Frequency	-1000 Hz	-2000 Hz	-3000 Hz

Clearly, if we used skip values outside the range $-4 < k < 4$, all we would get is aliases of the frequencies inside this range (see section 1.3), given that the period contains only $N = 8$ samples.

Generalizing, we can say that if we have a discrete signal of length N, and k is its integer index, then the only valid frequencies we can generate are the integers in the interval $-N/2 < k < N/2$. For $N = 8$, we can express seven frequencies, three negative, three positive, and zero. If played back at sampling rate $R = 8000\,\text{Hz}$, the frequencies will be in increments of 1000 Hz. If we double N, the number of representable frequencies goes up to 15, and if R is still 8000 Hz, we halve the frequency distance between representable frequencies (to 500 Hz apart).

The same considerations apply to spectral analysis as well. The number of nonaliased frequencies that can be analyzed with the DFT is bounded by the same range, $-N/2 < k < N/2$, where k is the integer index.

The number of nonaliased frequencies that can be analyzed with the DFT is proportional to the number of input samples N.

The greater the number of input samples, the greater the number of analyzable frequencies and the finer the frequency distance between them.

The frequency distance between analyzable frequencies is inversely proportional to N.

3.3   Discrete Fourier Transform in Action

This section examines the mechanics of DFT operation, to see how the probe phasor's radian velocity relates to the analysis frequency. We'll look at a concrete analysis example and review the components of the DFT to define each precisely, so as to understand how they all relate.

3.3.1   Probe Phasor

In table 3.1, ϕ represents the angle of the probe phasor. In this example, $\phi = \angle e^{-i\omega kn/N}$, $N = 8$, and $R = 8000\,\text{Hz}$. The value of k determines the rate at which the phasor spins.

3.3.2   How the DFT Orders Frequencies

Notice from table 3.1 the rather curious order in which the radian velocity of the probe phasor detects frequencies. As the phase increment k goes from 0 to 7, the probe phasor detects frequencies in the following order:

	DC	Negative Frequencies (Increasing in Frequency)			Nyquist	Positive Frequencies (Decreasing in Frequency)		
k	0	1	2	3	4	5	6	7
Hz	0	−1000	−2000	−3000	±4000	3000	2000	1000

Table 3.1
Example of DFT Operation

k	Angular Velocity	Frequency	Comments
0		Radian $\phi = 0\pi$ Hertz $\frac{k}{N}R = 0$ Hz	0 Hz, called direct current (DC) in electronics. The probe phasor does not rotate.
1		Radian $\phi = -\frac{\pi}{4}$ Hertz $\frac{k}{N}R = -1000$ Hz	Smallest negative frequency. The probe phasor rotates once every N samples.
2		Radian $\phi = -\frac{\pi}{2}$ Hertz $\frac{k}{N}R = -2000$ Hz	The probe phasor rotates once every $N/2$ samples.
3		Radian $\phi = -\frac{3}{4}\pi$ Hertz $\frac{k}{N}R = -3000$ Hz	Largest representable negative frequency without aliasing.
4		Radian $\phi = \pm\pi$ Hertz $\frac{k}{N}R = \pm4000$ Hz	The Nyquist limit can be interpreted either as a positive or a negative frequency because either a clockwise or a counterclockwise spin would create the same sample sequence.
5		Radian $\phi = -\frac{5}{4}\pi = \frac{3}{4}\pi$ Hertz $\frac{k}{N}R = 3000$ Hz	Since, from here on, $k > N/2$, this and all greater frequencies alias to their equivalent positive frequencies. The "wheel" starts spinning the other way.

Table 3.1
(continued)

k	Angular Velocity	Frequency	Comments
6		Radian $\phi = -\frac{6}{4}\pi = \frac{\pi}{2}$ Hertz $\frac{k}{N}R = 2000$ Hz	Smaller positive frequency.
7		Radian $\phi = -\frac{7}{4}\pi = \frac{\pi}{4}$ Hertz $\frac{k}{N}R = 1000$ Hz	Smallest positive frequency.

This is because the phase increment hits the Nyquist frequency boundary at $k = 4$, and frequencies $k > 4$ alias to positive frequencies. While this frequency order follows rigorously from the logic of the Nyquist sampling theorem, it's not a particularly intuitive ordering for looking at spectra. When spectra are displayed in tables and graphs, it is common for the list of frequencies to be rotated by $N/2$ positions and then reversed so that they appear in the following order:

k	5	6	7	0	1	2	3	4
Hz	−3000	−2000	−1000	0	1000	2000	3000	±4000

This ordering puts 0 Hz in the middle. Rotating by $N/2$ positions and reversing the list again restores the frequencies to standard DFT frequency order. *It is important to be aware of which frequency order is being used to list or plot a spectrum.* I will always say explicitly which order is being used.

3.3.3 Analyzing a Sinusoid with the DFT

The best way to understand the DFT is to watch it in action.

Let's analyze this signal:

$$x(n) = \sin\frac{4\pi n}{N}, \qquad N = 8, n = 0, \ldots, N-1. \qquad \textit{DFT Example Signal} \quad (3.26)$$

The eight discrete samples numbered 0 through 7 are shown in figure 3.12. The continuous sine wave from which these samples are drawn is superimposed over it. The eight samples form two

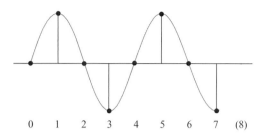

Figure 3.12
Test signal $x(n)$ shown as a bar graph.

complete sampled periods of a sine wave. Note that this is a periodic signal because sample 8 exactly repeats sample 0.

The definition of the probe phasor from equation (3.8) is

$$e^{-ik\omega n/N},$$

where n is the integer sample number, k is the integer frequency number, and N is the number of samples. We know from Euler's formula that we can break the probe phasor into a real cosine component and an imaginary sine component. This is how we can take the operation of the DFT apart. Table 3.2 shows the frequency analysis process of the DFT for the test signal $x(n)$, defined in (3.26). There are eight columns in table 3.2:

1. *Phase increment, k,* shows the values $k = 0, 1, 2, \ldots, 7$ of the probe phasor.

2. *Test signal* shows the test signal $x(n)$ for reference.

3. *Cosine probe phasor,* $\cos(k\omega n/N)$, increases in frequency because the phase increment k increases.

4. *Cosine product signal* shows the product of $x(n)$ and the real cosine probe phasor. When the cosine probe phasor equals a frequency component in $x(n)$, the sum of this signal will be nonzero. Then we normalize by multiplying the sum by $1/N$. Column 7 shows the (real) normalized sum of the samples of the cosine product signal.

5. *Sine probe phasor,* $-i\sin(k\omega n/N)$, increases in frequency according to the phase increment k.

6. *Sine product signal* shows the product of $x(n)$ and the sine probe phasor. Since the sine probe phasor contains i, but $x(n)$ is pure real, the result is pure imaginary. When the sine probe phasor equals a frequency component in $x(n)$, the sum of this signal will be nonzero. Then we normalize by multiplying the sum by $1/N$. Column 8 shows the (imaginary) normalized sum of the samples of the sine product signal.

7. *Cosine sum* shows the (real) normalized sum of the samples of the cosine product signal from column 4.

8. *Sine sum* shows the (imaginary) normalized sum of the samples of the sine product signal from column 6.

Table 3.2
Frequency Analysis Process of the DFT

(1)	(2)	(3)	(4)	(5)	(6)	(7)	(8)
k	Test Signal	Cosine Probe Phasor	Cosine Product Signal	Sine Probe Phasor	Sine Product Signal	Cosine Sum (Real)	Sine Sum (Imaginary)
0						0 +	$0i$
1						0 +	$0i$
2						0 +	$-0.5i$
3						0 +	$0i$
4						0 +	$0i$
5						0 +	$0i$
6						0 +	$0.5i$
7						0 +	$0i$

It's pretty easy to tell if the product signals sum to zero: if the samples are all zero, or if half the samples are above the line and half are symmetrically below, the sum will be zero. The sum of the product signals is nonzero at only two places: for the imaginary components at $k = 2$ and $k = 6$ in table 3.2. In the first case, the all-negative sum of the sine product signal at $k = 2$ results in a normalized value of $0 - 0.5i$ for that frequency. The all-positive sum of the sine product at $k = 6$ results in a normalized value of $0 + 0.5i$ for that frequency. The sums of all other product signals are zero. Notice that only imaginary components were found to have nonzero energy in the test waveform, which agrees with the fact that the test signal was a pure sine waveform and is odd. If the test signal had been a pure cosine waveform, the DFT would have detected only real components.

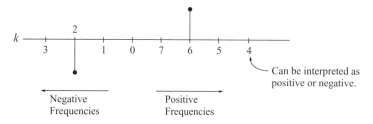

Figure 3.13
Imaginary spectrum from table 3.2 in rotated frequency order.

Representing the real and imaginary components of the DFT from table 3.2 in standard complex number notation and in standard DFT frequency order (not rotated order), we have as a final result the following complex spectrum:

	DC	Negative Frequencies			Nyquist	Positive Frequencies		
k	0	1	2	3	4	5	6	7
$X(k)$	$0 + 0i$	$0 + 0i$	$0 - \frac{1}{2}i$	$0 + 0i$	$0 + 0i$	$0 + 0i$	$0 + \frac{1}{2}i$	$0 + 0i$

Figure 3.13 is a plot of the imaginary spectrum from table 3.2 in rotated frequency order. Compare this to figure 2.29. (The signs of the components in 3.13 are flipped with respect to those in figure 2.29 as an artifact of the analysis method.)

3.3.4   Terms of the DFT

Here is a summary of what we know about the terms of the DFT:

- N determines the number of samples of the test waveform $x(n)$ that will be evaluated, and the number of divisions the spectrum will be broken into.
- k determines the location in $X(k)$ where the result of each measurement will be stored, and the frequency of the probe phasor as an integer from 0 to $N - 1$.
- n determines the instantaneous phase angle of the complex sinusoid during extraction of a particular frequency.
- $\omega = 2\pi$ causes the ratio n/N to indicate a fraction of a revolution of the unit circle.
- $-i$ makes the probe phasor spin clockwise.

3.3.5   The DFT Does Not Measure Absolute Frequency

Notice that R, the sampling rate, does not appear in the DFT equation. This means the DFT does not measure absolute frequency. Instead, it measures frequencies that are harmonics (integer multiples) of the *fundamental analysis frequency,* corresponding to the period of N samples, whatever the sampling rate. Let's call the fundamental analysis frequency f_N. For some real sampling rate R, the fundamental analysis frequency is

$$f_N = \frac{R}{N} \text{ Hz.} \qquad\qquad \textit{Fundamental Analysis Frequency} \quad (3.27)$$

We can use this to determine the analysis frequency f_k corresponding to the frequency index k as follows:

$$f_k = \begin{cases} k < \dfrac{N}{2}, & -k\dfrac{R}{N} \\[2ex] k \geq \dfrac{N}{2}, & (N-k)\dfrac{R}{N} \end{cases}$$

for $k = 0, 1, \ldots, N-1$. For example, if $R = 8000$, $N = 8$, then frequency order of the DFT is as we've seen before:

k:	0	1	2	3	4	5	6	7
f_k:	0	−1000	−2000	−3000	±4000	3000	2000	1000

This helps give a precise meaning to N and k: k indexes harmonics of the fundamental analysis frequency f_N.

3.3.6 Trade-off between Temporal and Spectral Resolution

N controls the amount of the input signal $x(n)$ that is analyzed and determines the number of harmonics into which the spectrum $X(k)$ is broken. A principle of information conservation is at work here.

We can obtain finer frequency resolution only by *decreasing* the fundamental analysis frequency f_N. Doing so requires that we *increase* N to analyze a larger chunk of the input signal. But what if $x(n)$ represents a signal with a time-varying spectrum, such as speech? If we increase N beyond a certain size, the analysis window may begin to cover multiple phonemes in the speech signal. Too large a value of N will cause the Fourier transform to blur the spectra of these phonemes together, making it impossible to distinguish them.

Conversely we can obtain finer time resolution only by *increasing* the fundamental analysis frequency f_N. Doing so requires that we *decrease* N to analyze a smaller chunk of the input signal. But as we decrease N, the frequency resolution of the Fourier transform suffers because we analyze the signal into fewer frequency bins that therefore cover increasingly large amounts of bandwidth. Below a certain value of N, the bandwidth of the analysis bins may grow so wide that they start to cover multiple frequency components in the input signal $x(n)$, blurring them together and making it impossible to distinguish them.

We might want to have fine frequency resolution in order to analyze a complex timbre such as a bell tone, which contains many inharmonic partials. We might want to have fine temporal resolution in order to analyze a transitory event such as a single phoneme of speech. However, if we want finer frequency resolution *and* finer temporal resolution, we are stuck, unless. . . .

3.3.7   Effect of Sampling Rate on the DFT

We can also change the sampling rate R at which the input signal $x(n)$ is recorded. If, for example, we double R and leave N the same, the fundamental analysis frequency is doubled. Doubling the fundamental analysis frequency means its period covers less temporal span, and the DFT is more sensitive to transient signals. The frequency resolution of the DFT remains the same in this case because N is unchanged. We have increased the temporal sensitivity of the DFT without decreasing its frequency resolution.

Conversely, doubling N and leaving R the same, the fundamental analysis frequency is halved. Halving the fundamental analysis frequency means its period covers greater temporal span, and the DFT is less sensitive to transient signals. However, the frequency resolution of the DFT doubles because N doubles. We have reduced the temporal sensitivity of the DFT and increased its frequency resolution.

Summarizing:

- Frequency resolution increases (and time resolution decreases) if N grows and R remains the same.
- Frequency resolution is unchanged (and time resolution increases) if R grows and N remains the same.

Table 3.3 shows the effect of a few selected sample rates R and window sizes N on the frequency resolution k.

Table 3.3
Frequency Resolution k for Various R and N

N	$R = 8{,}192$	$R = 16{,}384$	$R = 32$ K	$R = 64$ K
8	1,024	2,048	4,096	8,192
16	512	1,024	2,048	4,096
32	256	512	1,024	2,048
64	128	256	512	1,024
128	64	128	256	512
256	32	64	128	256
512	16	32	64	128
1,024	8	16	32	64
2,048	4	8	16	32
4,096	2	4	8	16
8,192	1	2	4	8
16,384	0.5	1	2	4
32K	0.25	0.5	1	2
64K	0.125	0.25	0.5	1

3.3.8   Using the DFT with Real and Complex Signals

Because the DFT's probe phasor is complex, the DFT performs a complex multiplication with $x(n)$. And $x(n)$ may itself be complex. However, $x(n)$ typically contains only real values when it is derived from the real world, for instance, via an analog-to-digital converter (see section 1.2). I've side-stepped this issue in the examples presented in this chapter by implicitly assigning $0i$ to the imaginary part of each sample of $x(n)$, which is a perfectly permissible way to make a real signal into a complex one.

However, it's worth keeping in mind that the DFT input $x(n)$ is expected to be a complex signal, and the DFT output is also a complex signal. Most software implementations provide two versions of the DFT. One assumes $x(n)$ is complex. The other assumes $x(n)$ is real and automatically pads each sample with $0i$ before passing it along to the version that assumes $x(n)$ is complex.

If the imaginary part of $x(n)$ is zero (that is, if it is a real signal), the following points apply:

• Magnitudes of the spectra of negative frequencies and positive frequencies of real signals are mirror images.

• Component amplitudes are 1/2 of their true values (except at 0 Hz and the Nyquist rate). To get the actual amplitude, the negative and positive frequency components must be added together.

• The phase of sine (odd) components are reversed between the positive and negative frequency sides of the spectrum (because $\sin -x = -\sin x$).

3.3.9   DFT in MUSIMAT

Following is a listing of a practical DFT program written in the MUSIMAT programming language (see volume 1, appendix section B.1).

This example introduces a new data type, `Complex`, which represents complex numbers in MUSIMAT. For example, we can define the complex value for the imaginary number $i = 0.0 + 1.0i$ in MUSIMAT as follows:[2]

```
Complex I(0.0, 1.0);// I has a zero real part and an imaginary part of 1.0
```

The `Complex` representation of e is

```
Complex E = Complex( 2.718281, 0.0 ); // the imaginary part of E is 0.0
```

Operations on data of type `Complex` perform complex arithmetic according to the rules laid out in chapter 2. For example, if we square `I`, we get the real number -1,

```
Complex iSquared = I * I;
```

and

```
Print( iSquared );
```

prints `(-1.0, 0.0)`.

This example also introduces a new function, Exp(), which raises the value of its first Complex argument to the power of its second Complex argument and returns the Complex result.

The code for the DFT in MUSIMAT is as follows:

```
Complex I = Complex( 0.0, 1.0 );          // define imaginary number i
Complex E = Complex( 2.718281, 0.0 );     // define e

ComplexList DFT( RealList x ) {
  // Function DFT takes a RealList argument and returns a ComplexList
  Integer N = Length( x );                // get the length of x
  ComplexList X;                          // a place to store result
  Complex mI2pidN = -I * 2.0 * Pi / N;    // combine constants

  For ( Integer k = 0; k < N; k++ ) {     // for every frequency...
    X[k] = 0.0;                           // prepare to accumulate
    For ( Integer n = 0; n < N; n++ ) {   // for every phase...
      Complex xn = Complex(x[n], 0.0);// cast Real input to Complex
      X[k] = X[k] + xn * Exp( E, mI2pidN * n * k );
    }
    X[k] = X[k] / N;                      // normalize
  }
  Return( X );
}
```

3.4   Inverse Discrete Fourier Transform

The *inverse discrete Fourier transform* (IDFT), given in equation (3.28), turns a spectrum created by the DFT back into a time domain signal.

$$x(n) = \sum_{k=0}^{N-1} X(k)e^{ik\omega n/N}. \qquad \textit{Inverse Discrete Fourier Transform (IDFT)} \quad (3.28)$$

Notice the symmetry between the DFT and IFDT:

- There is no $1/N$ scaler.

- $x(n)$ and $X(k)$ trade places.

- k and n are reversed: for each sample n, we iterate through all k.

- The probe phasor has a positive exponent, so it turns counterclockwise. Remembering Euler's formula, the probe phasor of the IDFT is

$$e^{i\theta} = \cos\theta + i\sin\theta.$$

The DFT can be thought of as a frequency detector, and the IDFT can be thought of as a frequency synthesizer. From this perspective, the IDFT uses the probe phasor to generate a set of k complex sinusoids, one for each harmonic in the input spectrum. Then it scales the amplitude and phases of the harmonics by multiplying each one by the corresponding value of the complex spectrum $X(k)$. The adjusted harmonics are then summed to create the output time domain waveform.

3.4.1 Matrix Form of the IDFT

We can alternatively express the IDFT in equation (3.28) as follows (see appendix section A.3):

$$x(n) = X(0)e^{i0\omega n/N} + X(1)e^{i1\omega n/N} + \cdots + X(N-1)e^{i(N-1)\omega n/N}. \tag{3.29}$$

This format makes clear all the calculations needed to make one sample of the output signal $x(n)$. To make computing examples of the IDFT easier, we can represent the terms of (3.29) using k and n, since these two variables uniquely identify each term. We define

$$\chi_n^k = X(k)e^{ik\omega n/N}.$$

With this definition the entire IDFT process for all $x(n)$ can be represented in matrix form:

$$\begin{bmatrix} \chi_0^{N-1} & \chi_1^{N-1} & \cdots & \chi_{N-1}^{N-1} \\ \vdots & \vdots & \ddots & \vdots \\ \chi_0^3 & \chi_1^3 & & \chi_{N-1}^3 \\ \chi_0^2 & \chi_1^2 & & \chi_{N-1}^2 \\ \chi_0^1 & \chi_1^1 & & \chi_{N-1}^1 \\ \chi_0^0 & \chi_1^0 & \cdots & \chi_{N-1}^0 \end{bmatrix}$$
$$\quad x(0) \quad x(1) \quad \cdots \quad x(N-1)$$

Each column computes one output sample $x(n)$ as given by equation (3.29). Each row calculates one harmonic of the fundamental analysis frequency f_N.

3.4.2 An Example of the IDFT

Let's run an example through the IDFT using the spectrum computed in section 3.3.3. This spectrum becomes the IDFT input, $X(k)$. Using standard DFT frequency order, the test spectrum is written

$$X(k) = \left\{ 0, 0, -\frac{i}{2}, 0, 0, 0, \frac{i}{2}, 0 \right\}. \qquad\qquad \textit{Test Spectrum} \ (3.30)$$

First, create a matrix of values for the probe phasor of the IDFT. Since each row of the matrix in section 3.4.1 computes one harmonic k of the IDFT, any value of $X(k)$ that is zero will fill its row k with zeros. Since for this example most values of $X(k)$ are zero, this should be a relatively simple calculation. The only nonzero rows are $k = 2$ and $k = 6$. So we need only compute probe phasor values for those two rows:

	n							
k	0	1	2	3	4	5	6	7
2	1	i	-1	$-i$	1	i	-1	$-i$
6	1	$-i$	-1	i	1	$-i$	-1	i

Next, scale the harmonics by their spectral strengths given in $X(k)$. We multiply rows 2 and 6 by $X(2) = -i/2$ and $X(6) = i/2$, respectively, and get this result:

	n							
k	0	1	2	3	4	5	6	7
2	$1\left(-\frac{i}{2}\right) = -\frac{i}{2}$	$i\left(-\frac{i}{2}\right) = \frac{1}{2}$	$-1\left(-\frac{i}{2}\right) = \frac{i}{2}$	$-i\left(-\frac{i}{2}\right) = -\frac{1}{2}$	$1\left(-\frac{i}{2}\right) = -\frac{i}{2}$	$i\left(-\frac{i}{2}\right) = \frac{1}{2}$	$-1\left(-\frac{i}{2}\right) = \frac{i}{2}$	$-i\left(-\frac{i}{2}\right) = -\frac{1}{2}$
6	$1\left(\frac{i}{2}\right) = \frac{i}{2}$	$-i\left(\frac{i}{2}\right) = \frac{1}{2}$	$-1\left(\frac{i}{2}\right) = -\frac{i}{2}$	$i\left(\frac{i}{2}\right) = -\frac{1}{2}$	$1\left(\frac{i}{2}\right) = \frac{i}{2}$	$-i\left(\frac{i}{2}\right) = \frac{1}{2}$	$-1\left(\frac{i}{2}\right) = -\frac{i}{2}$	$i\left(\frac{i}{2}\right) = -\frac{1}{2}$
Sum $x(n)$	0	1	0	-1	0	1	0	-1

Finally, by summing the columns of the preceding table (last row), we complete the analysis and retrieve the time domain signal, which is identical with equation (3.26), the example signal provided as input to the DFT. This demonstrates that the DFT and IDFT are a transform pair.

3.4.3 Complex Output from the IDFT

In the IDFT's defining equation (3.28), the output of the IDFT, $x(n)$, is complex because it is the result of complex operations. In the example above, I skirted around this problem because the imaginary parts of the output samples were all zero.

Digital-to-analog converters typically only operate on real-valued signals. If the imaginary part of $x(n)$ is zero, we just discard it and convert only the real part through the converter to hear a sound reconstructed by the IDFT. The imaginary part of $x(n)$ will be zero if the real part of $X(k)$ was even and the imaginary part was odd relative to frequency 0. This will automatically be the case if $X(k)$ was created by the DFT of a real signal. Thus, for most real-world applications such as recording audio, the imaginary component of the IDFT can be ignored.[3]

3.4.4 Separating Real and Imaginary Parts of the IDFT

Even if the output of the IDFT has a significant nonzero imaginary part, we can still meaningfully separate the real and imaginary output data. To see this, first remember that complex multiplication is defined as

$$(a + ib)(c + id) = (ac - bd) + i(bc + ad)$$

(see section 2.3.1). We can associate the a, b, c, and d terms of complex multiplication with the frequency synthesizer terms of the IDFT as follows. Recall that the function $\text{Re}\{x\}$ returns the real part of its complex argument x, and $\text{Im}\{x\}$ returns the (real-valued) imaginary part of its complex argument x.[4]

We can express the IDFT's complex input spectrum $X(k)$ as

$$X(k) = \underbrace{\text{Re}\{X(k)\}}_{a} + i\underbrace{(\text{Im}\{X(k)\})}_{b}$$

If we let $\theta = k\omega n/N$, we can express the complex probe phasor as

$$e^{i\theta} = \underbrace{\text{Re}\{\cos\theta\}}_{c} + \underbrace{i(\text{Im}\{\sin\theta\}}_{d}$$

Note that a, b, c, and d are all real values. Then, to compute just the real part of the IDFT,

$$\text{Re}\{x(n)\} = \sum_{k=0}^{N-1} ac - bd$$

$$= \sum_{k=0}^{N-1} \text{Re}\{X(k)\} \cos\theta - \text{Im}\{X(k)\} \sin\theta.$$

To compute just the imaginary part of the IDFT,

$$\text{Im}\{x(n)\} = \sum_{k=0}^{N-1} bc + ad$$

$$= \sum_{k=0}^{N-1} \text{Im}\{X(k)\} \cos\theta + \text{Re}\{X(k)\} \sin\theta.$$

(If the imaginary part of the input spectrum $X(k)$ is zero and the real part is even, then computing the imaginary part can be skipped.)

3.4.5 IDFT in MUSIMAT

The code for the IDFT in the MUSIMAT programming language is as follows:

```
Complex I = Complex( 0.0, 1.0 );       // define imaginary number i
Complex E = Complex( 2.718281, 0.0 ); // define e

ComplexList IDFT( ComplexList X ) {
  Integer N = Length( X );             // get length of X
  ComplexList x;                       // place to store result
  Complex I2pidN = I * 2.0 * Pi / N; // calculate constants once

  For ( Integer n = 0; n < N; n++ ) {
    x[n] = 0.0;                        // prepare to accumulate
    For ( Integer k = 0; k < N; k++ ) {
      x[n] = x[n] + X[k] * Exp( E, I2pidN * n * k );
    }
  }
  Return( x );
}
```

3.5 Analyzing Real-World Signals

So far, we've restricted the DFT and IDFT to periodic waveforms containing only harmonics (integer multiples) of the fundamental analysis frequency. This is fine as far as it goes, but most actual musical signals are not so well-behaved. What happens if we try to analyze a signal whose spectrum isn't aligned with the fundamental analysis frequency? Consider the following waveform:

$$x(n) = \sin\left(f 2\pi\frac{n}{N}\right), \qquad\qquad \textit{Nonintegral Test Signal} \;\; (3.31)$$

where $N = 16$ and frequency $f = 3/4$.

If the fundamental analysis period of the DFT is also $N = 16$, then $f = 3/4$ is clearly not an integer multiple of the fundamental analysis frequency. The DFT would receive input as shown in figure 3.14. What the DFT will do with this signal—what it does with every input signal—is interpret it as *one period of an infinitely repeating periodic function* at the fundamental analysis frequency. That is, the DFT operates on the example input function $x(n)$ as though it were like the one in figure 3.15. This is the *periodic extension* of the Fourier transform.

Notice the discontinuity in the waveform in figure 3.15. Periodic discontinuities in a waveform produce a spectrum with many high-frequency harmonics. The DFT "hears" a click in $x(n)$ when the waveform has such a discontinuity. This is revealed by looking at the magnitude spectrum that the DFT produces from this signal (figure 3.16). We see that there is some energy at all analyzed

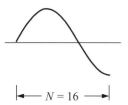

Figure 3.14
Nonharmonic test signal.

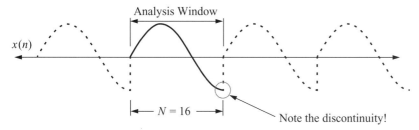

Figure 3.15
How the DFT interprets a nonperiodic input signal.

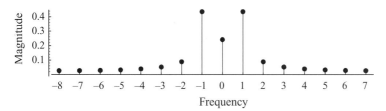

Figure 3.16
Magnitude spectrum of a nonintegral test signal.

frequencies. This is a problem, because we know from equation (3.31) that there is only one frequency component in the test signal at $f = 3/4$ Hz.

There are actually two problems here:

- *Picket Fence Effect.* We're trying to represent a frequency that is not an integer multiple of the fundamental analysis frequency f_N, so the results don't fit properly as harmonics of f_N. We are unable to view the underlying continuous spectrum because the DFT limits us to integer multiples of the fundamental analysis frequency f_N. This is analogous to trying to observe a row of evenly spaced trees through a picket fence.

- *Leakage.* Discontinuities at the edge of the analysis window spray noise throughout the rest of the spectrum. This phenomenon is called *leakage* because energy that should be in one spectral harmonic spreads away (leaks) into adjacent harmonics.

We must solve these problems because they arise whenever we try to analyze a signal with frequencies that are not locked to the rate of the analyzer.

3.5.1   Solving the Picket Fence Problem

Because of the picket fence phenomenon, we don't know whether two adjacent harmonics represent distinct partials or a single partial whose frequency is not a harmonic of the fundamental analysis frequency. There are a couple of things we can do to disambiguate these two cases.

If possible, resample the data, increasing the sampling rate R and/or the fundamental analysis frequency f_N until there are enough data points in the spectrum to disambiguate the two interpretations.

Alternatively, pad the signal to be analyzed with M additional zero-valued samples. This does not change its spectrum but increases its length. If we then increase the size of the analysis window to include the zero-valued padding samples in the DFT, we decrease the fundamental analysis frequency f_N, increasing the frequency resolution of the spectrum.[5] For example, if to a signal of length N samples we add $M = N$ additional zero-valued samples (doubling the signal's length) and take the DFT of $M + N$ samples, we increase the spectral frequency resolution by a factor of $(M + N)/N$. This is like adjusting the distance between the pickets in the fence until they line up with the evenly spaced trees.

3.5.2   Solving the Leakage Problem

The leakage problem is more serious. The best we can do is devise a work-around. We create a function exactly as long as the analysis window that gradually fades in and fades out at the edges. If we multiply the signal to be analyzed by this function, we decrease the effect of any discontinuities at the edges of the analysis window because the discontinuities are heavily attenuated at the analysis window edges. This helps reduce the impact of the discontinuities at the analysis window edges on the resulting spectrum.

But there's no free lunch because any alteration of the input signal will have some effect on the resulting spectrum. This is because the function that fades in and fades out is itself a signal, and it too has a spectrum. Later I describe several of these fade-in/fade-out functions and show their spectra.

To summarize, when analyzing waveforms that are not strictly harmonics of the fundamental analysis frequency:

• The spectrum of partials that are inharmonic to the fundamental analysis frequency can still be interpreted correctly because the DFT splits fractional frequencies proportionately into adjacent harmonics. We can interpolate between them to recover the fractional frequency components.

• The signal can be faded in and out at the edges of the analysis window to reduce the broadband spectral influence of discontinuities that would otherwise occur. The general term for this process is called *windowing*. The resulting spectrum will be a better approximation of the actual underlying signal, but we must account for the effects that windowing has on the resulting spectrum.

3.6   Windowing

To focus the DFT on part of the input signal, we can extract part of the signal or we can window it. So far, we've set the DFT summation limits to extract the desired portion of the signal (see equation (3.8)), covering the range of $n = 0$ to $n = N-1$. Alternatively, we can multiply the function by 0 everywhere except for the N samples we want to select, which we multiply by 1. This is windowing with a rectangular function, shown in figure 3.17 as the function $w(n, N)$, defined as

$$w(n, N) = \begin{cases} 1 & 0 \le n < N, \\ 0 & \text{otherwise}, \end{cases} \qquad \textit{Rectangular Window Function} \quad (3.32)$$

where N is the number of samples in the window and n indexes the window function.

3.6.1   Windowed DFT

We can rewrite the DFT to express windowing explicitly as follows:

$$X(k) = \sum_{n=-\infty}^{\infty} x(n)w(n, N)e^{-i2\pi kn/N}, \qquad \textit{Windowed DFT} \quad (3.33)$$

where $w(n, N)$ is the windowing function given in equation (3.32). Note that the sample index n now traverses all of time, but because of the windowing function, the result is the same as equation (3.8).

3.6.2   Tapering Functions

Windowing can introduce discontinuities at the edges of the analysis window if the underlying waveform is not a harmonic of the fundamental analysis frequency. The signal $x(n)$ selected by the window in figure 3.17 has such a discontinuity at its right edge. As shown in the figure, the underlying

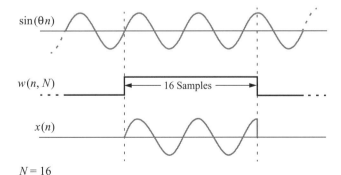

$N = 16$

Figure 3.17
Windowing with a rectangular function.

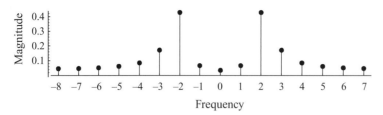

Figure 3.18
Spectrum of a rectangular function.

function $x(n)$ is 16 samples long, and by inspection we see that it contains 2.25 periods of a sine wave; hence the underlying function can be written as a periodic function:

$$\sin\frac{2\pi fn}{N},$$

where $N = 16$ and $f = 2.25$.

The magnitude spectrum of this signal (figure 3.18) shows that the DFT has added many spurious high-frequency harmonics, introduced by the discontinuity in the analyzed signal.

Analysis of nonharmonic signals via either sample extraction or windowing with a rectangular function can result in spurious energy estimates such as this unless we take steps to prevent it. We can diminish the impact of discontinuities at the edge of the analysis window by replacing the rectangular windowing function with a function that tapers to zero at its edges. Such functions are called *apodization functions,* or *tapering functions.* Tapering cannot be done with plain sample extraction; therefore windowing is necessary.

Triangular Window   The simplest tapering function is the *triangular window.* Its shape is shown in figure 3.19 as $w(n, N)$. It just consists of a complementary pair of slopes that make a tent function. Its equation is

$$w(n, N) = \begin{cases} 1 - \left|\dfrac{n - (N/2)}{N/2}\right|, & 0 \le n \le N, \\ 0 & \text{otherwise,} \end{cases} \qquad \textit{Triangular Window} \quad (3.34)$$

where N is the length of the window and n is the current sample. The operator $|\ |$ takes the absolute value of the expression it contains. The triangular window function w behaves differently depending upon whether the value of n is inside or outside the range of 0 to N. If it is outside, then w simply returns 0; otherwise it returns the appropriate point of the triangular function.

Figure 3.19 shows the result of windowing a sinusoid with a triangular window function. The resulting spectrum is shown in figure 3.20. The magnitude spectrum in figure 3.20 is much less noisy. Only frequencies near those that actually contain energy show significant strength. Thus, we have effectively enabled the DFT to be used with realistic signals because now we can remove much of the spurious noise that is introduced by discontinuities at the analysis window edges.

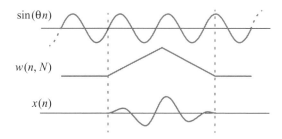

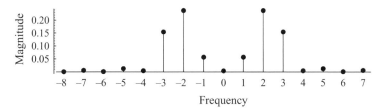

Figure 3.19
Windowing with a triangular function.

Figure 3.20
Spectrum of triangular windowing.

Whereas the highest peak in the rectangular-windowed DFT is about 0.43, the highest peak for the triangular-windowed DFT is only about 0.24. It is lower because applying the triangular function attenuates the signal wherever the triangular window function is less than 1.0 (everywhere but in the middle).

Other Window Functions   The triangular window is but one of many windowing functions. There seems to be a rather bewildering variety of them—the Bartlett window, Welch window, Parzen (triangular) window, Hann or hanning window, Hamming window, Blackman window, Lanczos window, Kaiser window, Gaussian window, and so on.

While each of these windows has a particular advantage in certain situations, any windowing function that reduces the discontinuities at the edges of the analysis window is a big improvement over the rectangular function. Here's a look at some of the standard window functions.

Hann (Hanning) Window   The equation for the *Hann window* (named after Julius von Hann and often referred to as the *hanning window*)[6] is

$$H(x, n, N) = \begin{cases} (1-a)\cos\left(2\pi\dfrac{n}{N} + \pi\right) + a, & 0 \le n < N, \\ 0 & \text{otherwise}, \end{cases} \tag{3.35}$$

where $a = 1/2$. It is shown in figure 3.21a.

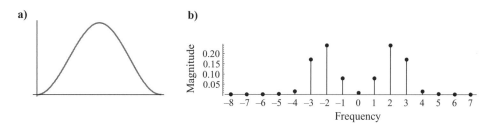

Figure 3.21
Hann window and resulting spectrum.

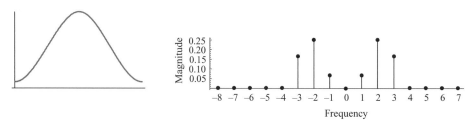

Figure 3.22
Hamming window and resulting spectrum.

Because it is just an inverted cosine wave scaled by 0.5 and offset by 0.5, this window is sometimes called the raised inverted cosine window. Its advantage is that, unlike the triangular window, it has no sharp edges (no sudden change in derivative) at all, so it is more effective at eliminating spurious artifacts from the analysis. Figure 3.21b shows the magnitude spectrum of the test signal, equation (3.31), windowed with the Hann window. Note that the peak amplitude is around 0.24, so there is some energy loss due to the overall attenuation of the signal, similar to what happened with the triangular window.

Hamming Window   The equation for the *Hamming window* (named after Richard W. Hamming) is the same as equation (3.35) except that $a = 0.54$ (see figure 3.22). The elevated value for a causes the Hamming window to let in more energy overall and, in particular, let in a little energy at the edges of the analysis window. The amplitude peak is a little higher than with the Hann window, but it also lets some of the broadband energy from the window edge back in.

Figure 3.23 shows the triangular, Hann, and Hamming windows superimposed. Note that both the Hann and Hamming windows emphasize the middle part of the input signal more than the triangular window does.

Bear in mind that windowing a signal always modifies its spectrum. If we perform an inverse transform on this spectrum, we will get back the *windowed version* of the signal. So, for example, performing the IDFT on the spectrum in figure 3.20 would reproduce the waveform $x(n)$, shown at the bottom of figure 3.19, not the original signal $\sin \theta n$ shown at the top of that figure.

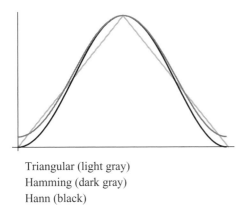

Triangular (light gray)
Hamming (dark gray)
Hann (black)

Figure 3.23
Triangular, Hann, and Hamming windows.

If we must window a signal $x(n)$ but need to get the original back after DFT/IDFT analysis/synthesis, we can divide the reproduced $x(n)$ by the windowing function to uncompensate $x(n)$. Any part of the signal windowed to zero is irretrievably lost (and very attenuated values might not be properly restored because of limited computer arithmetic precision). This means there is a problem with the triangular and Hann windows because they go to zero at their extremes, so we'd end up dividing by zero at those points, which is not meaningful. However, we could do this with the Hamming window because it does not reach zero at its extremities (which may have been one of the motivations for its development).

3.7   Fast Fourier Transform

With the addition of the technique of windowing, the DFT is able to handle real-world signals—almost. The next hurdle to its practical use is the sheer amount of computation required to analyze realistic-sized signals. If we want to perform a DFT of length N, then at each stage we must perform a complex multiplication of $w(n)$ against all N phasor frequencies. Since we must repeat this operation for all N input samples, the total number of stages is on the order of N^2. For signals beyond a certain size, we will quickly run out of computing power, patience, or both. The fast Fourier transform (FFT) reduces the number of computations from N^2 to $N \log_2 N$. For small values, the FFT does not have a big advantage, but as N grows, the FFT outperforms the DFT by enough to make a substantial difference, and it is widely used.

To take a practical example, say we wanted to perform the Fourier transform on just 1 second of a stereo audio signal, recorded at the conventional rate of $R = 44,100$ samples per second per channel. Since it's stereo, there are two channels, and we must calculate two DFTs, one for each of the two channels. Each channel has $N = 44,100$, for a combined total of $2N = 88,200$ samples. Further, say we perform the calculations on a computer that can perform 1,000,000 stages of the

Fourier transform per second (a fast computer by today's standards). For the FFT it would take about one and a half seconds to do the calculation, whereas for the DFT it would take a little over two hours.

	N	Order		Stages		Time
FFT	88,200	$N \log_2 N$	$=$	1.45×10^6	$=$	1.45 s
DFT	88,200	N^2	$=$	7.78×10^9	$=$	7779.24 s, or 2.16 hours

The FFT requires that the input signal length be a power of 2. But this requirement can be worked around easily: for length N choose the nearest power of 2 greater than N, and set the extra samples to zero. (All this does is slightly compress the resulting spectrum, exactly the same as padding the DFT with zeros.)

The FFT achieves its efficiency by reducing the number of stages that must be performed. The first step in reducing the computation, according to the Danielson-Lanczos (1942) lemma, is to rewrite the DFT into the sum of two smaller DFTs, each of length $N/2$. One DFT operates on just the $N/2$ even-numbered samples, and the other operates on just the $N/2$ odd-numbered samples. Here is a derivation. Let $W_N = e^{-i\omega/N}$ so that the DFT can be written as

$$X(k) = \sum_{n=0}^{N-1} x(n) W_N^{kn}, \qquad 0 \le k \le N-1,$$

(ignoring the normalization of the sum by $1/N$ for simplicity).

The order in which we sum the terms doesn't matter (that is, addition is commutative). We can add two summations, one indexing only the even terms, the other only the odd terms. Stepping through just the even terms is equivalent to indexing by $2n$, whereas for the odd terms it's equivalent to indexing by $2n + 1$. Each DFT must perform only $(N/2) - 1$ summations:

$$X(k) = \sum_{n=0}^{(N/2)-1} x(2n) W_N^{k(2n)} + \sum_{n=0}^{(N/2)-1} x(2n + 1) W_N^{k(2n+1)}, \tag{3.36}$$

where N is even.

Let's take a closer look at the probe phasor in the odd DFT:

$$W_N^{k(2n+1)} = e^{-i\omega k(2n+1)/N}.$$

Remembering that $x^{a+b} = x^a x^b$, we can expand this into

$$e^{-i\omega k(2n+1)/N} = e^{i\omega k(2n)/N} \cdot e^{-i\omega k/N}.$$

Notice that the term $e^{-i\omega k/N}$ does not depend upon n and that it appears in every term of the odd DFT. That means we can factor it out of the entire odd DFT summation. Since $e^{-i\omega k/N} = W_N^k$ we can rewrite equation (3.36) to read

$$X(k) = \sum_{n=0}^{(N/2)-1} x(2n) W_N^{k(2n)} + W_N^k \sum_{n=0}^{(N/2)-1} x(2n+1) W_N^{k(2n)}$$

$$= X_{\mathrm{E}}(k) + W_N^k \cdot X_{\mathrm{O}}(k),$$

where $X_{\mathrm{E}}(k)$ and $X_{\mathrm{O}}(k)$ are the even and odd DFTs. Notice that the probe phasors for the even and odd DFTs are now the same. The final summation of the odd DFT is additionally multiplied by W_N^k. This demonstrates that the DFT can be rewritten as a sum of two half-length DFTs.

This lemma can be recursively applied (so long as N is even), so we can progressively divide these two DFTs into $4, 8, \ldots$ DFTs of length $N/4, N/8, \ldots$ until we have subdivided all the way down to DFTs of length 1. What is a DFT of length 1? If we set $N = 1$ in equation (3.8), the DFT equation reduces to $X(0) = x(0)$. So the DFT of a signal of length 1 is just the value of the sample.

The recursive application of the Danielson-Lanczos lemma leads directly to the so-called radix 2 Cooley-Tukey (1965) fast Fourier transform. (A radix 4 FFT would partition the DFT into four subtransforms of length $N/4$.)

In outline, the FFT algorithm is as follows. Before the main FFT algorithm begins, the data are rearranged (by a technique called bit reversal) into a form that can be more efficiently accessed by the algorithm, and the values of the probe phasor are precomputed. The FFT algorithm itself consists of $\log_2 N$ stages in which successively longer subtransforms are computed from the previous stages. This process is repeated N times for a total of $N \log_2 N$ times.

Since the FFT implements exactly the same transform as the DFT, only more efficiently, I don't pursue the implementation of the FFT further, but this information should hopefully allow readers to make sense of other treatments of the subject (see Bracewell 1999; Smith 2003).

3.8 Properties of the Discrete Fourier Transform

Operations such as addition, multiplication, and shifting in the time domain have corresponding operations in the frequency domain, and vice versa. This section sets out some of the Fourier transform's most important properties.

3.8.1 Linearity of the Fourier Transform

The Fourier transform establishes a mathematical relation between periodic signals and their associated spectra. How the Fourier transform relates to the properties of superposition and proportionality determines whether it is a linear operation or not.

Superposition   As shown in figure 3.24, when two signals are added together, their spectra are added also. The figure shows the addition of the signals $f(t)$ and $g(t)$ and of their spectra $F(k)$ and

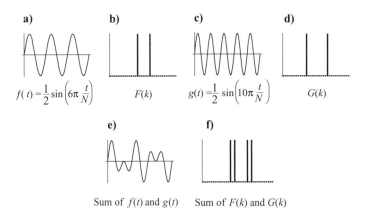

a) **b)** **c)** **d)**

$f(t) = \frac{1}{2}\sin\left(6\pi\,\frac{t}{N}\right)$ $F(k)$ $g(t) = \frac{1}{2}\sin\left(10\pi\,\frac{t}{N}\right)$ $G(k)$

e) **f)**

Sum of $f(t)$ and $g(t)$ Sum of $F(k)$ and $G(k)$

Figure 3.24
Addition of two spectra.

$G(k)$. It is clear that figure 3.24f is the same as figures 3.24b and 3.24d *superposed,* that is, added together point by point. We can express this as follows. Let $\mathcal{F}\{\ \}$ represent the Fourier transform. With $f(t)$ and $g(t)$ as defined, figure 3.24 demonstrates that

$$\mathcal{F}\{f(t) + g(t)\} = \mathcal{F}\{f(t)\} + \mathcal{F}\{g(t)\}$$

$$= F(k) + G(k),$$

where $F(k)$ is the Fourier transform of $f(t)$, and $G(k)$ is the Fourier transform of $g(t)$. \hfill (3.37)

The Fourier transform of the sum of two signals is the same as the sum of the transforms of each signal separately.

Proportionality   The same result can be achieved by either of the following methods:

• Scale a signal by a constant, then apply the Fourier transform.

• Transform the signal first, then scale the spectrum by the constant.

$$\mathcal{F}\{af(t)\} = a\,\mathcal{F}\{f(t)\}, \hfill (3.38)$$

where a is a constant. Intuitively, this makes sense. If we halve the strength of a signal, we expect it will scale all the spectral components by half as well. We wouldn't expect the transform to introduce or take away any energy by itself.

The strength of the transformed output is proportional to the strength of the input to the transform.

Linearity   The properties of proportionality and superposition demonstrate that the DFT is a linear transformation. Linear systems are relatively easy to characterize. Nonlinear systems are harder to characterize.

What do I mean by *linear*? Let there be a function f that represents a transform, such that for each input to the function x, the output y is given by

$$y = f(x) = ax + b. \tag{3.39}$$

The terms might be simple numbers or signals or anything else, depending upon the application of the function.

The function is said to be *linear* if the conditions of proportionality and superposition hold. Consider this proportionality:

$$f(cx) = c \cdot f(x). \qquad\qquad \textit{Proportionality} \quad (3.40)$$

For example, using equation (3.39), let $a = 1$, $b = 0$, $x = 3$, and $c = 0.5$; then

$$f(\tfrac{1}{2} \cdot 3) = \tfrac{1}{2} f(3).$$

Now consider this superposition:

$$f(x_1 + x_2) = f(x_1) + f(x_2), \qquad\qquad \textit{Superposition} \quad (3.41)$$

where x_1 and x_2 are any inputs. For example, let $a = 1$, $b = 0$, $x_1 = 2$, and $x_2 = 3$; then

$$f(2 + 3) = f(2) + f(3).$$

We can combine proportionality and superposition and write

$$f(c_1 x_1 + c_2 x_2) = c_1 f(x_1) + c_2 f(x_2). \tag{3.42}$$

Using induction, we can extend equation (3.42) to as many terms as we like:

$$f(c_1 x_1 + c_2 x_2 + \cdots + c_n x_n) = c_1 f(x_1) + c_2 f(x_2) + \cdots + c_n f(x_n).$$

Equation (3.40) says that if we multiply the input x by some constant c, then the output is multiplied by the same constant. Equation (3.41) is the more important of these two principles. It says that if f is linear, all we really need to know about f is how it responds individually to x_1 and x_2 and then we can automatically predict what it will do with the sum of x_1 and x_2 without even having to measure it. This is important. Consider the following example.

Value of Linearity   Suppose we were going to use the Fourier transform in a piece of medical apparatus with the expectation that someone's life might depend upon the result. We'd need to understand how the Fourier transform responds to all meaningful inputs so as to understand its performance in all meaningful circumstances. But must we examine every conceivable combination of inputs to accomplish this? No. Since the Fourier transform is a linear system, all we need to know is how the Fourier transform responds to each of its basic building blocks: the sinusoids.

If the outputs of the transform for each sinusoid are known, the output of the transform of any other signal can be immediately calculated, including without limit any possible combination of sinusoids.

We then have no need to make further measurements. The responses of the transform to any other input can be immediately calculated based on the measurements we have already made.

Contrast this with a nonlinear system. We might find that the response to a sum of two signals is much larger or smaller than would be expected based on the signals being taken separately. The response to the sum might be less even when strong responses are obtained separately.

A realistic example of such a nonlinear transform is the audio limiter, often found in the equipment racks of radio stations. Radio stations must be careful not to let the signals they send to the transmitter exceed a certain amplitude so that they will not interfere with other transmissions. Limiters allow signals that are below a threshold to pass linearly, but they progressively attenuate signals if their intensity exceeds the allowable threshold. Suppose two singers whose signals are $f(x_1)$ and $f(x_2)$ are on the air at a radio station. The microphone mixer combines the signals so that $f(x_1 + x_2) = f(x_1) + f(x_2)$ and passes $f(x_1 + x_2)$ to the limiter. The limiter has a threshold of τ. If $f(x_1 + x_2) > \tau$, the limiter reduces the gain before passing it to the transmitter. When the combined signal exceeds τ,

$$f(x_1 + x_2) \neq f(x_1) + f(x_2),$$

and the system is not linear.

For nonlinear systems, knowing the outputs of the system for all sinusoids is not enough. To fully understand a nonlinear transform, we must know not only how it responds to all sinusoids but also how it responds to all possible combinations of sinusoids—a much more daunting prospect. Thus, it is simpler to understand and predict linear systems than nonlinear ones.

Spectral Aspects of Linearity   Another important aspect of linearity for musical applications is that linear systems do not introduce any modifications to the spectrum of their inputs, whereas nonlinear systems typically do: they distort the timbre of signals that pass through them in some way. Of course, distortion may be desirable or undesirable, depending upon the context. For instance, electric guitarists often pay good money for effects boxes that artistically distort their sound.

What do guitar distortion boxes do? Consider the equation

$$y(n) = cx(n).$$

We can interpret this as a function that transforms a value on the x-axis to a proportional value on the y-axis. If $c = 1/1$, this operation is called the *identity function*. The $45°$ diagonal line in figure 3.25a shows a portion of the identity function, which stretches to infinity. If we think of the diagonal line inside the box as a mirror, then the value of $x(n)$ is reflected as $y(n)$.

The *inverse identity function* is shown in figure 3.25b. It exactly restores y-axis to x-axis values, giving the original function without loss of information. Figures 3.26a and 3.26b show the identity transform and its inverse for $c = 1/2$.

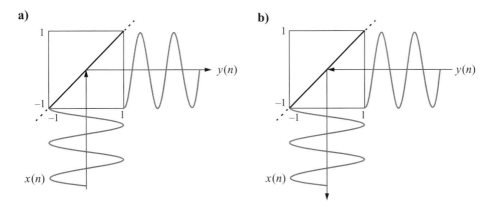

Figure 3.25
Identity transform and its inverse.

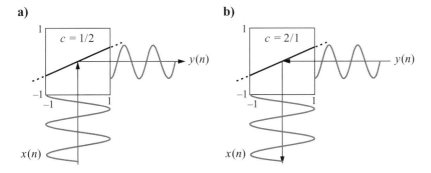

Figure 3.26
Scaled linear transform and its inverse.

This is nothing more than another demonstration of the law of proportionality given in equation (3.40): for a linear operation, it doesn't matter if the scaling takes place on the input side or the output side.

We can use the identity transform to demonstrate superposition as well:

$$g(x_1(n) + x_2(n)) = gx_1(n) + gx_2(n),$$

as shown in figure 3.27: for a linear operation, it doesn't matter whether we add and then transform (figure 3.27a), or transform and then add (figure 3.27b).

All the preceding transforms are linear because they are straight lines. Now consider the nonlinear transform:

$$y(x) = 2x^2 - 1, \tag{3.43}$$

a)

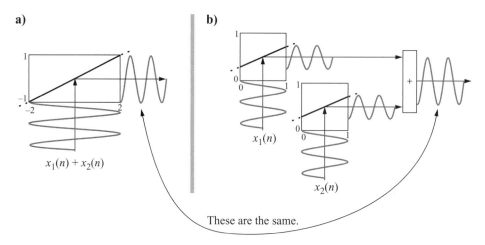

b)

$$x_1(n) + x_2(n)$$

$$x_1(n)$$

$$x_2(n)$$

These are the same.

Figure 3.27
Two superposed linear transforms combined.

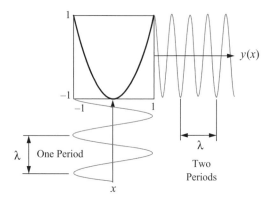

Figure 3.28
Nonlinear transform.

which is shown in figure 3.28. The transforming function is a parabola, and it has the interesting property that it doubles the frequency of its input. One period per unit of distance λ equals two periods per unit of distance on the output. Observe that as the input waveform goes through one quarter of a period, say from -1 to 0, it traverses the function from 1 to -1. During the next quarter period the input goes from 0 to 1 while the output goes from -1 to 1, thereby completing the first period of output while the input has only completed half a period. Thus, the output frequency is exactly twice the input frequency. To see this mathematically, we can substitute $\cos\theta$ for x in equation (3.43):

$$y(x) = 2x^2 - 1,$$

$$y(\cos\theta) = 2(\cos\theta)^2 - 1$$

$$= 2\left(\frac{\cos(\theta + \theta) + \cos(\theta - \theta)}{2}\right) - 1$$

$$= \cos 2\theta + \cos 0 - 1$$

$$= \cos 2\theta.$$

Curiously, the output is also a single sinusoid, but at twice the frequency of whatever signal it is driven by. This is an example of an unusual kind of distortion. For more about these interesting functions, see section 9.4.11.

Windowing is another example of a nonlinear transform. Windowing affects the spectrum of a signal—recall that this very property is the reason we use them (see section 3.6). We can treat a window function itself as a signal and take its DFT to reveal its spectrum. Figure 3.29 shows the magnitude spectra of a rectangular window and a triangular window. The rectangular window magnitude spectrum has a great deal more energy in its high frequencies than does the magnitude spectrum of the triangular window. This must certainly relate to why signals windowed with the rectangular function can have so much more high-frequency energy than triangular-windowed signals.

Recall that windowing involves multiplying a window function and an input signal, and it seems reasonable that its spectrum would somehow be superimposed on the input signal—but precisely how? I take this up in chapter 4. Meanwhile, note that almost any signal-processing operation that involves multiplying one signal by another or, in general, multiplying a signal by a curve of some sort is a nonlinear operation.

Time Invariance   A system that behaves the same at all times is *time-invariant*. For example, if we delay the input to a system by N samples and the output is also delayed by N samples, and there are no side effects due to the delay (that is, the output looks the same, just delayed), then the system is time-invariant.

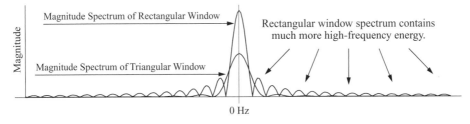

Figure 3.29
Magnitude spectra of rectangular and triangular window functions.

For instance, if we have an input function $x(n)$, and we apply a delaying function δ_D, where D is the amount of delay in samples, then the output function $y(n)$ will be delayed as well:

$$y(n-N) = \delta_{n-N} \cdot x(n). \tag{3.44}$$

Time-invariant systems are easier to analyze and understand than time-varying ones. If we know a system is time-invariant, we can be confident that if we know how it behaved yesterday, we know how it will behave today.

The Fourier transform is time-invariant by virtue of the fact that it analyzes its input signal for all time, as shown in equation (3.5). Think of the Discrete Fourier transform as a train traveling from right to left on an infinitely long track stretching from $-\infty$ to ∞, such that each car on the train covers one spectral sample. To delay the Fourier transform by N samples, all we do is shift the train N samples to the right.

Let's represent the Fourier transform at sample time n as $\mathcal{F}_n$, and let $X_n(k)$ be the output of the Fourier transform at sample time n. Represent the entire input signal as $x(\cdot)$, and say that it represents *all* input samples from $-\infty$ to ∞.[7] So now the Fourier transform reads $X_n(k) = \mathcal{F}_n\{x(\cdot)\}$. Think of the subscript n as selecting the Fourier transform beginning at time n. Now we can show the condition which defines time-invariance:

$$X_{n-N}(k) = \mathcal{F}_{n-N}\{x(\cdot)\} = \mathcal{F}_n\{x(\cdot-N)\}, \tag{3.45}$$

where $x(\cdot-N)$ means delaying the entire signal $x(\cdot)$ by N samples. This just says that it's equivalent whether we shift the Fourier transform over a stationary signal, or shift the signal over a stationary Fourier transform.

3.8.2 Limitations of the Fourier Transform

The DFT carves up the spectrum of a sound into bins of constant bandwidth. But the sensitivity of the ear to frequency is logarithmic, that is, the ear's sensitivity to pitch diminishes with increasing pitch. From the ear's perspective, the DFT overspecifies high frequencies and underspecifies low frequencies. To sufficiently match the ear's low-frequency spectral resolution therefore requires a value of N that provides unneeded high-frequency resolution.

We could make the encoding of musical signals more relevant to the ear if we could progressively increase the bandwidth of the higher frequency bins to more closely match the behavior of the ear. This would amount to having a *constant-Q* Fourier transform, where the bandwidth of the analysis bins increases in constant proportion as the center frequency of each band (see volume 1, sections 6.9.6 and 8.9.6). I say more about this in chapter 10.

3.9 A Practical Hilbert Transform

An *analytic signal* is a complex signal that has no negative frequencies. For example, $e^{i\omega}$ is analytic if $\omega \geq 0$ (see section 2.11).

Real signals (signals without an imaginary component) are never analytic because their energy is symmetrical around 0 Hz. But analytic signals can be constructed from real signals by using the Hilbert transform (see section 2.11.1). The Hilbert transform of a signal is another signal whose frequency components are all phase shifted by 90° ($-\pi/2$ radians for positive frequencies, $\pi/2$ for negative frequencies). If $x(t)$ is a real signal, then its analytic signal is

$$x_a(t) = x(t) + i\mathcal{H}\{x(t)\},$$

where $\mathcal{H}\{\ \}$ is the Hilbert transform.

In theory, we can create $\mathcal{H}\{x(t)\}$ simply by delaying all components in $x(t)$ by a quarter-cycle time shift. The problem is, the amount of time shift will be different for every component in $x(t)$. Thus implementing a practical Hilbert transform requires introducing the appropriate quarter-cycle phase delay for every frequency in the signal. This in turn requires that we analyze the signal into its constituent components so we can apply the appropriate phase offset to each one.

One way to create an analytic signal is to use the fast Fourier transform. Since an analytic signal is a real signal with zero negative frequencies, all we have to do is to take the FFT of the signal and zero out its negative frequencies. The inverse transform of this result is the (complex) analytic signal. There are limitations and artifacts associated with this procedure, but to a first approximation, we get what we expect. The steps are as follows:

1. Calculate the FFT of the real input signal $x(t)$ of length N, and store the result in $X(k)$. The spectrum $X(k)$ is also N elements long, and elements 0 and $N/2$ correspond to 0 Hz and the Nyquist frequency, respectively. The elements below $N/2$ are the negative frequencies, and those above are positive frequencies (see section 3.3.2).

2. Zero out the negative frequencies. The easiest way is to create an array of N multiplicands—call it $h(k)$—and form the elementwise product of $h(k)$ and $X(k)$. The formula for $h(k)$ is

$$h(k) = \begin{cases} 1, & k = 0, \\ 1, & k = N/2, \\ 0, & k = 2, 3, \ldots, (N/2)-1, \\ 2, & k = (N/2)+1, \ldots, N-1. \end{cases}$$

In MUSIMAT, this would be written

```
If ( k == 0 ) h[ k ] = 1;
Else If ( k == N/2 ) h[ k ] = 1;
Else If ( k >= 2 And k <= (N/2)-1) h[ k ] = 0;
Else If ( k >= (N/2)+1 And k <= N-1) h[ k ] = 2;
```

If all we have to do is zero out the negative frequencies, why must we scale some of the frequencies by 1 and others by 2? The most intuitive explanation I can think of has two parts.

First, we must double the magnitude of all positive frequencies to compensate for wiping out all the negative frequencies, so that we end up with the same amount of total energy. Hence, we multiply the magnitude of all positive frequencies by 2.

Second, the energy at 0 Hz and the energy at the Nyquist frequency really aren't positive *or* negative frequencies. Since there is no negative frequency corresponding to 0 Hz or the Nyquist frequency, we leave their strength the same, multiplying each by 1. Think of it this way: someone who sees a dot directly and also sees it in a mirror sees two dots. But if the dot is positioned *directly on the silver of the mirror,* only one dot is seen. It's as though energy at 0 Hz and the Nyquist frequency is on the silver of the mirror.

3. Form the elementwise product of $X(k)$ and $h(k)$:

$$X'(k) = h((k) \cdot X(k)).$$

The spectrum of $X'(k)$ has no negative frequencies.

4. Finally, take the inverse FFT:

$$x'(t) = \mathcal{F}^{-1}\{X'(k)\}.$$

The signal $x'(t)$ approximates the analytic form of the input signal $x(t)$.

Summary

Fourier analysis and synthesis are called a transform pair because the spectrum of a wave created by Fourier synthesis may be perfectly analyzed by Fourier analysis, and vice versa, with no loss of information.

Fourier analysis creates a spectrum whose frequencies are integer multiples of the fundamental analysis frequency. Fourier synthesis creates a periodic vibration from sinusoids whose frequencies are integer multiples of a fundamental frequency.

Fourier analysis is often called the Fourier transform, and Fourier synthesis is often called the inverse Fourier transform. Whereas the Fourier transform starts with a function of time and produces a function of frequency, the inverse Fourier transform starts with a function of frequency and produces a function of time.

The Fourier transform detects energy at a particular frequency by exploiting a property of multiplication. A signal multiplied by itself is squared. A squared waveform will have all positive values because (except for imaginary numbers) a number times itself is always positive. If we multiply two signals, one with a frequency we don't know times one with a frequency we do know, and the result is all positive, then the signals being multiplied must be identical. This provides a way to detect the presence or absence of particular frequency components: we can create a frequency detector by multiplying a known signal (called the probe phasor) by an unknown signal.

We can determine if the product signal is strictly positive by summing the multiplicands as we compute them. If the signals are identical, the multiplicands will all be positive, and the more of them we sum together, the larger their magnitude becomes. If the signals are unequal, the multiplicands will have mixed signs, and summing them together will not produce a large positive value over time. Thus, a large positive value from the frequency detector is an indication of frequency match.

The Fourier transform combines a frequency detector with an additional step that applies it to every frequency in a spectrum and records the result as a function of frequency.

The Fourier transform is restricted to periodic signals. By definition, periodic signals are of infinite length. By windowing the input signal, we can apply the Fourier transform to temporally limited signals. But the transform of a windowed signal is no longer lossless.

The Discrete Fourier Transform (DFT) operates on sampled signals. The DFT spectrum is one period of an infinitely repeating periodic function at the fundamental analysis frequency. The number of time domain samples analyzed determines the resolution of the frequency samples of the DFT spectrum. To increase spectral resolution, examine a larger swath of the input signal. But doing so decreases temporal resolution. Thus, we have a Heisenbergian contradiction: we can know either frequency content or temporal content in great detail, but not both.

The fast Fourier transform (FFT) accelerates the computation of the DFT. It reduces the number of computations from N^2 to $N \log_2 N$.

The Fourier transform is linear, and obeys the rules of superposition and proportionality.

Since an analytic signal is a real signal with zero negative frequencies, we can create one by taking the FFT of a signal and zeroing out its negative frequencies. The inverse transform of this result is the (complex) analytic signal.

4 Convolution

It is not surprising that the greatest mathematicians have again and again appealed to the arts in order to find some analogy to their own work. They have indeed found it in the most varied arts, in poetry, in painting, and in sculpture, although it would certainly seem that it is in music, the most abstract of all the arts, the art of number and of time, that we find the closest analogy.
—Havelock Ellis

4.1 Rolling Shutter Camera

Convolution lies at the heart of modern digital audio. The quickest and most intuitive introduction to convolution that I've found comes by way of an antique rolling shutter camera. When photography was first being developed early in the twentieth century, some cameras used a rolling shutter instead of an iris to control exposure of the film. A narrow slit was cut in a roll of opaque paper attached between two spring-loaded rollers that when released by the camera's trigger caused the slit to scroll quickly across the film plate between the lens and the film, exposing it to light and registering the image on the film (see figure 4.1).

Because the travel time of the shutter was not instantaneous, photographing fast-moving images led to image-skewing motion artifacts. In the famous photograph shown in figure 4.2, the slit moved from bottom to top while the camera operator swiveled to track the car going by. Because of motion artifacts and other problems, the rolling shutter was eventually supplanted by the iris shutter.

How is this relevant? The rolling shutter camera implemented a form of convolution. Suppose we had such a camera today and that we altered the loop of opaque paper so that there were two narrow slits, one some distance from the other. Now we retake the picture of the car driving horizontally past the camera. As the first slit passes across the film plate, it registers the car's image at one location on the film, and as the second slit passes over the film plate, it registers the car's image at a position proportional to how much farther the car has traveled in the meantime. The film registers the sum of the first image of the car and the second (time-shifted) image of the car—a double exposure with the exposures very close together in time. The slits are effectively sampling and time-shifting the image, depending upon when they pass between the film and the lens. This is basically what convolution is all about: time shifting and combining signals.

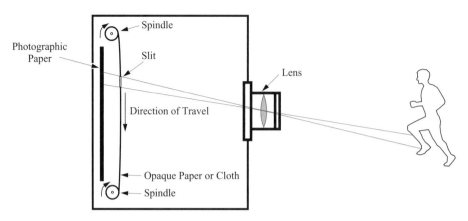

Figure 4.1
Camera with a slit to expose film.

Figure 4.2
"Voyage en auto; Papa à 80 km à l'heure, mars 1913." Photograph by Jacques Henri Lartigue. © Ministère de la Culture—France/AAJHL.

4.2 Defining Convolution

The formula for *convolution* is

$$h(n) = \sum_{m=0}^{n} f(m)g(n-m), \qquad \textit{Convolution} \quad (4.1)$$

where m and n are integers. The function g can be likened to the slit in the rolling shutter camera, and the function f can be likened to the image being recorded. The convolution operation scrolls function g past function f rather like the slit is scrolled past the film plate in the rolling shutter camera.

Let's try an example to see how equation (4.1) works. Say that the length of two functions f and g are both $N = 4$, so that their defined values lie in the range of 0 to $N - 1$. For this example, the actual defined values of the functions don't matter because we're just trying to get a feel for the abstract pattern of the convolution operation. Let's also say that values lying outside of the defined range of functions f and g are equal to zero. We can specify this mathematically for function f by writing

$$f(n) = \begin{cases} 0, & n \geq N, \\ f(n), & 0 \leq n < N, \\ 0, & n < 0. \end{cases} \qquad \textit{Convolution Range Rule} \quad (4.2)$$

Equation (4.2) says that values of f outside the defined range $0 \leq n < N$ are equal to zero, and values within this range are its defined values. We must also apply the convolution range rule to function g.

We start off by seeing what the convolution operation would calculate for $N = 0$, then continue by setting n to larger and larger values. Computing the first few elements of equation (4.1) for increasing n we have

$n = 0 \qquad h(0) = f(0)g(0)$

$n = 1 \qquad h(1) = f(0)g(1) + f(1)g(0)$

$n = 2 \qquad h(2) = f(0)g(2) + f(1)g(1) + f(2)g(0)$

$n = 3 \qquad h(3) = f(0)g(3) + f(1)g(2) + f(2)g(1) + f(3)g(0)$

Note that as n increases, convolution recruits increasing numbers of terms from f and g into the equation. Here's a visualization: suppose we mark out the two functions on the sides of two boards (figure 4.3), separated by a roller. The figure illustrates the computation of h for $n = 3$. As the boards roll past each other, we form the product of f and g, and then sum the products.

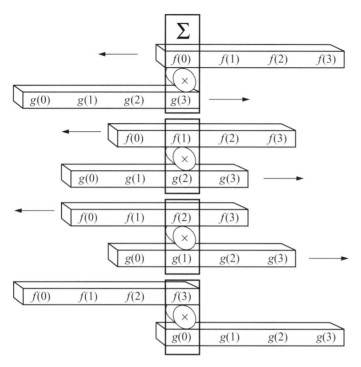

Figure 4.3
Convolution step $h(3)$.

But what happens when $n = 4$ or more? Since this is beyond the range of the functions, we invoke the convolution range rule and substitute zero for undefined values, which yields

$n = 4$ $\begin{aligned} h(4) &= f(0)g(4) + f(1)g(3) + f(2)g(2) + f(3)g(1) + f(4)g(0) \\ &= (f(0) \cdot 0) + f(1)g(3) + f(2)g(2) + f(3)g(1) + (0 \cdot g(0)) \\ &= f(1)g(3) + f(2)g(2) + f(3)g(1) \end{aligned}$

$n = 5$ $\begin{aligned} h(5) &= f(0)g(5) + f(1)g(4) + f(2)g(3) + f(3)g(2) + f(4)g(1) + f(5)g(0) \\ &= f(2)g(3) + f(3)g(2) \end{aligned}$

$n = 6$ $\begin{aligned} h(6) &= f(0)g(6) + f(1)g(5) + f(2)g(4) + f(3)g(3) + f(4)g(2) + f(5)g(1) + f(6)g(0) \\ &= f(3)g(3) \end{aligned}$

$n = 7$ $\begin{aligned} h(7) &= f(0)g(7) + f(1)g(6) + f(2)g(5) + f(3)g(4) + f(4)g(3) + f(5)g(2) + f(6)g(1) \\ &\quad + f(7)g(0) \\ &= 0 \end{aligned}$

For all values $n \geq 7$, $h(n) = 0$, so we might as well stop at $n = 6$.

In general if the length of f is N_f and the length of g is N_g, then the length of the convolved function h is $N_h = N_f + N_g - 1$. This rule works even when $N_f \neq N_g$ because we defined values that lie outside the range of functions f and g to be 0.

As a shorthand for equation (4.1), mathematicians notate convolution this way:

$$h(n) = f(\cdot) * g(\cdot), \hspace{3cm} \textit{Convolution Operator} \quad (4.3)$$

where $*$ is the convolution operator,[1] and the notation $f(\cdot)$ and $g(\cdot)$ can be translated to mean *all required values* of functions f and g. The centered dot notation, $f(\cdot)$ and $g(\cdot)$, is used in equation (4.3) to allow the convolution operation unlimited indexing range. Since $f(\cdot) * g(\cdot)$ is itself a function of n, sometimes convolution is written $(f * g)(n)$.

4.3   Numerical Examples of Convolution

The following numerical examples will be a lot clearer if the format for representing the steps of convolution is rotated to show terms that are being summed in columns, as in traditional arithmetic. Here is the convolution of functions f and g:

$f(0)g(0)$	$f(0)g(1)$	$f(0)g(2)$	$f(0)g(3)$	$\ldots$	$\ldots$	$\ldots$	$\ldots$
	$f(1)g(0)$	$f(1)g(1)$	$f(1)g(2)$	$f(1)g(3)$	$\ldots$	$\ldots$	$\ldots$
		$f(2)g(0)$	$f(2)g(1)$	$f(2)g(2)$	$f(2)g(3)$	$\ldots$	$\ldots$
			$f(3)g(0)$	$f(3)g(1)$	$f(3)g(2)$	$f(3)g(3)$	$\ldots$
			$\ldots$	$\ldots$	$\ldots$	$\ldots$	$\ldots$
$h(0)$	$h(1)$	$h(2)$	$h(3)$	$h(4)$	$h(5)$	$h(6)$	$\ldots$

In this format, increasing values of n run across the page instead of down the page. One value of the output function h is formed by summing each column. If the product of two functions would yield zero because of the application of the convolution range rule, the cell is marked with ellipses ($\ldots$) to indicate that its calculation is skipped.

Here is a concrete example of convolution.

$$h(n) = f(\cdot) * g(\cdot) = \{1, 2, 3, 4, 5\} * \{1, 2, 3\}. \hspace{3cm} (4.4)$$

	1	2	3				
		2	4	6			
			3	6	9		
				4	8	12	
					5	10	15
$h(n) =$	1	4	10	16	22	22	15

Here is the same example with the terms f and g transposed.

$$h(n) = g(\cdot) * f(\cdot) = \{1, 2, 3\} * \{1, 2, 3, 4, 5\}. \tag{4.5}$$

1	2	3	4	5		
	2	4	6	8	10	
		3	6	9	12	15
$h(n) =$ 1	4	10	16	22	22	15

Notice that equations (4.4) and (4.5) produce the same result, indicating that

Convolution is commutative.

4.3.1   Impulse Function

The next pair of examples reveals a useful insight into convolution. The *unit impulse function* $\delta(n)$ is defined as a single 1 located at $n = 0$, and all other indexed values are 0, written as follows:

$$\delta(n) = \begin{cases} 1, & n = 0, \\ 0, & n \neq 0, \end{cases} \qquad \textit{Unit Impulse Function} \tag{4.6}$$

where n is an integer.

In the first example, let $f(n) = \{1, 2, 3, 4, 5\}$, and let $g(n)$ be the unit impulse function $\delta(n)$. Then the convolution of $f(n)$ and $g(n)$ is as follows:

$$h(n) = f(\cdot) * g(\cdot) = \{1, 2, 3, 4, 5\} * \{1, 0, 0, 0, 0\}. \tag{4.7}$$

1	0	0	0	0				
	2	0	0	0	0			
		3	0	0	0	0		
			4	0	0	0	0	
				5	0	0	0	0
$h(n) = 1$	2	3	4	5	0	0	0	0

As we see, equation (4.7) simply copies f to h unchanged. In the next example, let $g(n) = \{0, 1, 0, 0, 0\}$. We have shifted the location of the 1 in the impulse function one element to the right.

$$h(n) = f(\cdot) * g(\cdot) = \{1, 2, 3, 4, 5\} * \{0, 1, 0, 0, 0\}. \tag{4.8}$$

0	1	0	0	0				
	0	2	0	0	0			
		0	3	0	0	0		
			0	4	0	0	0	
				0	5	0	0	0
$h(n) = 0$	1	2	3	4	5	0	0	0

Equation (4.8) copies f to h shifted right by one place. Last, we scale the impulse function g by $1/2$.

$$h(n) = f(\cdot) * g(\cdot) = \{1, 2, 3, 4, 5\} * \{0, 0.5, 0, 0, 0\}.$$ (4.9)

0	0.5	0	0	0	0			
	0	1	0	0	0			
		0	1.5	0	0	0		
			0	2	0	0	0	
				0	2.5	0	0	0

$h(n) = 0$	0.5	1	1.5	2	2.5	0	0	0

Convolving a function f with an impulse function just copies f.

Convolving f with a shifted impulse function makes a shifted copy of f.

Convolving f with a scaled impulse function scales f.

Convolving f with a scaled, shifted impulse function scales and shifts f.

Figures 4.4–4.9 illustrate convolutional shifting for some interesting wave types.

$f(x) * g(x) = h(x)$

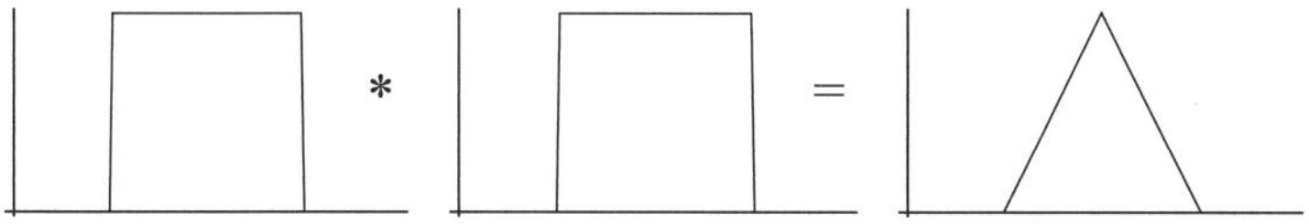

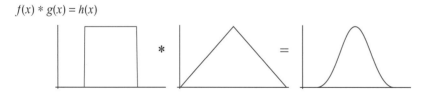

Figure 4.4
Convolution of two rectangular windows.

$f(x) * g(x) = h(x)$

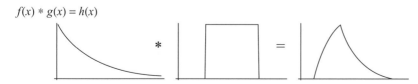

Figure 4.5
Convolution of window with triangular function.

$f(x) * g(x) = h(x)$

Figure 4.6
Convolution of exponential decay with rectangular window.

$f(x) * g(x) = h(x)$

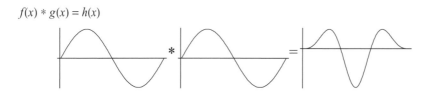

Figure 4.7
Convolution of two sine waves.

$f(x) * g(x) = h(x)$

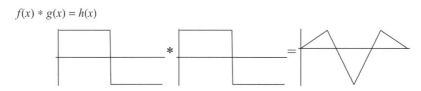

Figure 4.8
Convolution of two square waves.

$f(x) * g(x) = h(x)$

Figure 4.9
Convolution of impulse train with exponential decay.

4.3.2 Impulse Train

Consider the following example. An *impulse train function* is a periodic impulse function with period equal to the distance between impulses. Let f be an impulse train function with a period of $2, f = \{1, 0, 1, 0, 1\}$, and convolve it with $g = \{1, 2, 3, 4, 5\}$:

$$f(\cdot) * g(\cdot) = \{1, 0, 1, 0, 1\} * \{1, 2, 3, 4, 5\}. \tag{4.10}$$

1	2	3	4	5					
	0	0	0	0	0				
		1	2	3	4	5			
			0	0	0	0	0		
				1	2	3	4	5	
1	2	4	6	9	6	8	4	5	

The impulse train function adds a shifted copy of the signal for each impulse in the impulse train function (see figure 4.9). Recalling the rolling shutter camera example, when we added an extra slot to the opaque screen in the camera, we changed its function from a single impulse to an impulse train (with two pulses) and the camera produced the sum of the original and the delayed image, analogous to equation (4.10).

Let's modify the impulse train function so that each successive impulse is half its previous value: $f = \{1, 0, 0.5, 0, 0.25\}$. Then

$$f(\cdot) * g(\cdot) = \{1, 0, 0.5, 0, 0.25\} * \{1, 2, 3, 4, 5\}. \tag{4.11}$$

1	2	3	4	5				
	0	0	0	0	0			
		0.5	1	1.5	2	2.5		
			0	0	0	0	0	
				0.25	0.5	0.75	1	1.25
1	2	3.5	5	6.75	2.5	3.25	1	1.25

The scaled impulse train function sums a scaled and shifted copy of each impulse in the impulse train function.

Looking at this result, we can see why convolution is sometimes described as a smearing operation: by summing multiple delayed copies, convolution can render the original pattern less distinct. In fact, our ears are highly sensitive to this smearing effect: the walls of a room transmit delayed and scaled reflections to our ears from a sound source, so reverberation is a form of convolution. If we snap our fingers or create another impulsive sound in a room, the sum of all delayed and scaled reflections is called the room's *impulse response*. Speech is more intelligible in a bedroom than in a cathedral because the impulse response of a bedroom is shorter and less pronounced than the impulse response of a cathedral; hence there is less acoustical smearing in a bedroom.

Convolving with an impulse train adds a shifted original to the output for each impulse.

As many copies will be superimposed as there are nonzero impulses.

Copies will be shifted (delayed) by the position of the impulse in the impulse train function.

Shifted copies will be scaled by the amplitude of the impulses.

Shifted, scaled copies of the original are summed, smearing the result.

4.3.3   Convolution as Echo and Reverberation

Echoes are scaled and delayed copies of a source signal, and reverberation is a set of echoes generated recursively. So echoes and reverberation can be seen as forms of convolution (see volume 1, section 7.13).

If we convolve the impulse response of a good-sounding hall with a sound recorded elsewhere, it will be as if the sound had been recorded in that hall. Mathematically, the process is like equation (4.11): many scaled time-delayed copies of the sound are summed, based on the impulse response of the hall. So, convolving the impulse response of a room with another time domain signal applies the room's response to that signal. Thus, if one has the impulse response of a good concert hall, one can process recordings made in a nonreverberant room to sound as if they had been recorded in the concert hall.

This procedure works best if the sound being convolved is fairly dry (that is, with a short, mild impulse response), but it works surprisingly well even if the source signal is reverberant. Unfortunately, the convolution operation is computationally intensive, so using convolution to create artificial reverberation is prohibitively expensive on all but the fastest computers.[2] For more cost-effective approaches to creating artificial reverberation, see section 9.6.

4.4 Convolving Spectra

Explaining echoes and reverberation with convolution in the time domain begins to show the explanatory power of convolution, but there is much more to come. At the beginning of this chapter, I suggested that when two signals are multiplied, their spectra are convolved. Proof of this will lead in another interesting direction.

4.4.1 A Notational Convenience

We can simplify the following equations by defining the Fourier transform as the function $\mathcal{F}$, and the inverse Fourier transform as the function $\mathcal{F}^{-1}$. Then, if we apply the Fourier transform to a time domain function $s(t)$, we can write

$$\mathcal{F}\{s(t)\} = S(k),$$

indicating that the Fourier transform of signal $s(t)$ is the spectrum $S(k)$. Writing

$$s(t) = \mathcal{F}^{-1}\{S(k)\}$$

indicates that the inverse Fourier transform of spectrum $S(k)$ is signal $S(t)$.

We can use $\mathcal{F}$ and $\mathcal{F}^{-1}$ to stand for either the discrete or continuous Fourier transform, so long as we agree to apply the appropriate transform to the appropriate kind of data. If the spectrum and corresponding time domain function are continuous, $\mathcal{F}$ and $\mathcal{F}^{-1}$ refer to the continuous Fourier transform; otherwise they refer to the discrete Fourier transform.

4.4.2 Multiplying Signals Convolves Their Spectra

Say we have two discrete signals $f(n)$ and $g(n)$ that have discrete spectra $F(k)$ and $G(k)$, respectively. We wish to show that

$$\mathcal{F}\{f(n)g(n)\} = F(k) * G(k).$$

In order to prove that multiplying signals convolves their spectra, we can proceed as follows:

1. Compute the convolution of $F(k)$ and $G(k)$, that is, find $H_1(k) = F(k) * G(k)$.

2. Compute the spectrum of the product of signals $f(n)$ and $g(n)$, that is, find $H_2(k) = \mathcal{F}\{f(n)g(n)\}$.

3. Show that $H_1(k) = H_2(k)$.

Step 1 Say the spectra of $F(k)$ and $G(k)$ are as follows:

kHz	−4	−3	−2	−1	0	1	2	3	4
$F(k)$	0	0	0	0.5	0	0.5	0	0	0
$G(k)$	0	0.5	0	0	0	0	0	0.5	0

$F(k)$ and $G(k)$ each contain nine frequency samples: four for negative frequencies, four for positive frequencies, and one for 0 Hz. Since the table shows that the bandwidth of $F(k)$ and $G(k)$ is ±4 kHz, the bandwidth of each frequency sample must be 1.0 kHz. These spectra are shown in figure 4.10.

Now we convolve these two spectra to create $H_1(k)$:

$$H_1(k) = F(k) * G(k). \tag{4.12}$$

The computation is as follows. Refer to section 4.3 for help in following the convolution steps.

```
        0   0   0   0   0       0   0       0   0
            0   0   0   0       0   0       0   0   0
                0   0   0       0   0       0   0   0   0
                    0  0.25     0   0       0   0   0  0.25   0
                        0       0   0       0   0   0   0     0   0
                        0  0.25 0   0       0   0   0   0     0  0.25  0
                            0       0   0   0   0   0   0     0   0   0
                                0   0   0   0   0   0   0     0   0   0   0
                                    0   0   0   0   0   0     0   0   0   0   0
─────────────────────────────────────────────────────────────────────────────
H₁(k) =  0   0   0   0  0.25    0  0.25    0   0   0  0.25    0  0.25   0   0   0   0
```

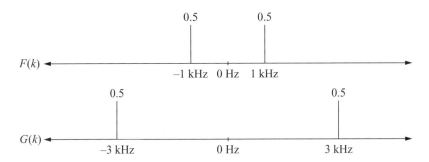

Figure 4.10
Spectra of $F(k)$ and $G(k)$.

The spectrum $H_1(k)$ contains 17 frequency samples, although there were but 9 samples each for $F(k)$ and $G(k)$. How do we interpret these extra frequency samples in $H_1(k)$? Each frequency sample in $H_1(k)$ covers just as much bandwidth as each sample of the input spectrum: 1.0 kHz each. So the spectral bandwidth of the convolution is $2N-1$ times the bandwidth of the source spectrum. But there is nothing in the spectrum outside of the range of ±4 kHz, the bandwidth of the input spectrum, and by the definition of convolution, there cannot be. So we ignore these extra samples. Here are the frequencies of $H_1(k)$ from the previous computation with the valid frequency samples shown shaded.

kHz	−8	−7	−6	−5	−4	−3	−2	−1	0	1	2	3	4	5	6	7	8
H1(k)	0	0	0	0	0.25	0	0.25	0	0	0	0.25	0	0.25	0	0	0	0

Figure 4.11 shows this result graphically.

We can interpret this result to say that we have ended up with copies of the spectrum of $F(k)$ placed with their centers where each component of $G(k)$ used to be. Equally validly we could say the same thing the other way around: copies of the spectrum of $G(k)$ are placed with their centers where each component of $F(k)$ used to be (figure 4.12). Both interpretations are correct.

We have conserved the total amount of energy in the two input spectra. Each of the four resulting spectral components has an amplitude of 0.25, so the total energy is the same.

Step 2 Having computed the convolution of $F(k)$ and $G(k)$ to find $H_1(k)$, we now compute $H_2(k)$, the spectrum of the product of signals $f(n)$ and $g(n)$ as follows:

1. Create the waveform for $f(n)$.

2. Create the waveform for $g(n)$.

3. Compute the Fourier transform of their product: $H_2(k) = \mathcal{F}\{f(n)g(n)\}$.

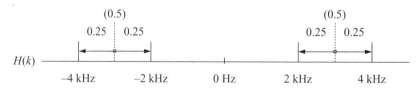

Figure 4.11
Convolved spectra of $F(k)$ and $G(k)$.

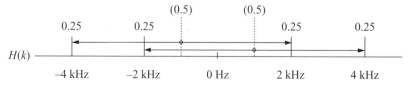

Figure 4.12
Convolved spectra of $G(k)$ and $F(k)$.

$f(n) = \{0, 0.36, 0.67, 0.9, 1., 0.96, 0.8, 0.53, 0.18, -0.18, -0.53, -0.8, -0.96, -1., -0.9, -0.67, -0.36\}$

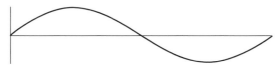

$g(n) = \{0, 0.9, 0.8, -0.18, -0.96, -0.67, 0.36, 1., 0.53, -0.53, -1., -0.36, 0.67, 0.96, 0.18, -0.8, -0.9\}$

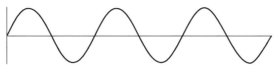

$f(n)\,g(n) = \{0, 0.32, 0.54, -0.16, -0.96, -0.65, 0.29, 0.52, 0.097, 0.097, 0.52, 0.29, -0.65, -0.96, -0.16, 0.54, 0.32\}$

$H_2(k) = \mathcal{F}\{f(n)\,g(n)\} = \{0, 0, 0, 0, 0.25, 0, 0.25, 0, \underline{0}, 0, 0.25, 0, 0.25, 0, 0, 0, 0\}$

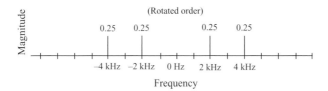

Figure 4.13
Calculation of DFT $\{f(n)\,g(n)\}$.

Step 3   Compare figure 4.11 with figure 4.13 to verify that $H_1(k) = H_2(k)$.

We have proved that indeed $H_1(k) = H_2(k)$, so multiplying signals convolves their spectra.

Multiplying in the time domain convolves in the frequency domain.

4.4.3   Convolving Spectra Multiplies Their Signals

If multiplying two signals convolves their spectra, then it ought to be the case that convolving two spectra multiplies their signals. We can prove this by the following steps.

1. Starting with $F(k)$ and $G(k)$, convolve them, as we did to create $H_1(k)$, and take the inverse Fourier transform of the result:

$$h_1(t) = f(t)g(t) = \mathcal{F}^{-1}\{F(k) * G(k)\}. \tag{4.13}$$

2. Get the product of the inverse Fourier transform of $F(k)$ and $G(k)$:

$$h_2(t) = f(t)g(t) = \mathcal{F}^{-1}\{F(k)\}\mathcal{F}^{-1}\{G(k)\}. \tag{4.14}$$

3. Observe that $h_1(t) = h_2(t)$.

This is left to the reader to check since it follows directly from the example above.

Convolving in the frequency domain multiplies signals in the time domain.

Some very interesting applications of this important rule are covered in later sections of this chapter and elsewhere in subsequent chapters.

4.5   Convolving Signals

The previous sections have established a relation between multiplication and convolution in the time domain and frequency domain that can be expressed this way:

Time domain   :   Frequency domain
Multiply $\rightarrow$ Convolve
Multiply $\leftarrow$ Convolve

In this section we consider the opposite approach:

Time domain   :   Frequency domain
Convolve $\rightarrow$ Multiply
Convolve $\leftarrow$ Multiply

So now we want to prove that

• When spectra are multiplied, their signals are convolved.

• When signals are convolved, their spectra are multiplied.

4.5.1   Multiplying Spectra Convolves Their Signals

Suppose we wish to reduce the high-frequency energy of a recorded signal. One way to proceed would be as follows:

1. Take the Fourier transform of the signal.

2. Window (multiply) the resulting spectrum with a function that shapes the spectrum as desired, for example, by using a decreasing exponential function (figure 4.14) to attenuate the high frequencies. In this case, we interpret the exponential function as a spectrum ranging from high energy at low frequencies to low energy at high frequencies.

3. Take the inverse Fourier transform of the product spectrum. On listening to the result, we would hear a signal with attenuated high-frequency energy compared to the original. From this, we see that *multiplication of spectra is equivalent to filtering*.

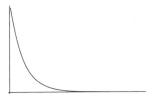

Figure 4.14
Decreasing exponential function.

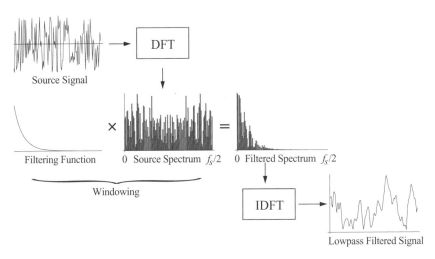

Figure 4.15
Windowing a spectrum as a form of filtering.

Figure 4.15 sketches the general process of filtering by multiplying spectra. We take the Fourier transform of the source signal, then multiply its spectrum with the desired filtering function. The inverse Fourier transform of the result produces a signal with frequency content shaped by the filtering function.

4.5.2   Windowing as a Form of Filtering

Figure 4.15 shows filtering performed by multiplication in the frequency domain. Using the notation $\mathcal{F}^{-1}\{\ \}$ for the Fourier transform, we can express this operation as

$$\mathcal{F}^{-1}\{F(k)G(k)\} = f(t) * g(t). \tag{4.15}$$

The product of spectra, for instance, those in figure 4.15, produces the convolution of the corresponding time domain signals. This subject is elaborated in chapter 5.

4.5.3   Generating Fractal Noise by Multiplying Spectra

We can generate noise with an arbitrary spectral tendency by modifying the power spectrum of uniform noise. In fact, completely arbitrary spectral functions of any kind, fractal and otherwise, can be obtained this way (see volume 1, section 9.17.2). The method is to compute the Fourier transform of a signal, scale its power spectrum as desired, and retransform with the inverse Fourier transform.

Below is a code listing in MUSIMAT that performs filtering using the DFT. It is based on the DFT described in section 3.3.9. The process can be made more efficient through the use of the fast Fourier transform (F. R. Moore 1990, 447–448). First we define some needed functions.

```
// Return the real part of a Complex list
RealList RealPart( ComplexList x ) {
  Integer N = Length( x );          // find its length
  RealList t;                       // place to store the result
  For ( Integer i = 0; i < N; i = i + 1 )
    t[ i ] = Re( x[ i ] );          // Take just the real part
  Return( t );
}
```

The function Re( ) takes a list of complex numbers and returns a list of just their real parts. This function iteratively calls the built-in MUSIMAT function Re( ) to take the real part of each complex number on the list.

```
// Normalize a Real list to lie between L and U
RealList Normalize( RealList x, Real L, Real U ) {
  Integer N = Length( x );              // find the length of the list
  Real max = Max( x );                  // find the maximum
  Real min = Min( x );                  // find the minimum
  For ( Integer k = 0; k < N; k = k + 1 )
    x[ k ] = (U - L) * ( x[ k ] - min) / (max - min) + L;
  Return( x );
}
```

The function Normalize( ) scales a list of real values to lie within a specified range. The ShapedNoise( ) function in the following code listing first creates a broadband noise signal of a length specified by argument N and then takes the Fourier transform of the noise signal. Then it applies a scaling function by $1/f^B$, filtering the signal by multiplication in the frequency domain. To allow the signal to be real after conversion to the time domain, the spectrum is forced to be conjugate-symmetrical around 0 Hz. Then the function takes the IDFT of the shaped noise spectrum, and finally it normalizes and returns the resulting time domain noise sequence.

```
// Shape N samples with spectral tendency B, values lie between L and R
RealList ShapedNoise( Real B, Integer N, Real L, Real U) {
```

```
RealList x;
For ( Integer k = 0; k < N; k = k + 1 )// start with broadband noise
  x[k] = Random( -1.0, 1.0 );
ComplexList X = DFT( x );              // take the Fourier transform of x
For ( k = 1; k <= N/2; k = k + 1 ) { // scale the noise by 1/f^B
  Real r = Re( X[k] );                // take the real part of a sample
  Real i = Im( X[k] );                // take its imaginary part
  Real power = r * r + i * i;         // square amplitude to get power
  power = power / Pow(Real(k), B);    // apply the spectral tendency
  Real amplitude = Sqrt( power );     // convert back to amplitude
  Real phase = Atan2( i, r );         // calculate the phase
  RealSet( X[k], amplitude * cos(phase) ); // reconstruct
  ImagSet( X[k], amplitude * sin(phase) ); // the spectrum
}
// Force the spectrum to be conjugate-symmetrical around 0 Hz
For ( k = 1; k <= N/2; k = k + 1 ) {
  RealSet( X[N - k], Re(X[k]) );
  RealSet( X[N - k], -Im(X[k]) );
}
ComplexList cx = IDFT( X );           // take inverse Fourier transform
RealList cr = Re( cx );               // discard the imaginary part
RealList result = Normalize( cr, L, U ); // normalize
Return( result );
}
```

We can invoke ShapedNoise( ), for example, as follows:

```
Print("B=0.0",  ShapedNoise( 0.0, 512, -1.0, 1.0 ) );
Print("B=1.0",  ShapedNoise( 1.0, 512, -1.0, 1.0 ) );
Print("B=2.0",  ShapedNoise( 2.0, 512, -1.0, 1.0 ) );
Print("B=3.0",  ShapedNoise( 3.0, 512, -1.0, 1.0 ) );
```

Figure 4.16 shows the results of calling ShapedNoise( ). To help with comparison, the same noise sequence was used for all plots (unlike in the preceding, code, which selects a different random sequence each time the function is invoked). Figure 4.16a shows time domain noise signals that were generated with $b = \{0, 1, 2, 3\}$. As b increases, high frequencies are more and more sharply attenuated. Figure 4.16b shows the log-log power magnitude spectral plots of the corresponding signals. The upper diagonal line in the spectral plots corresponds to $1/f$, and the lower diagonal line to $1/f^2$. When $b = 0$, the spectrum is relatively flat, and when $b = 1$, the spectrum has a spectral tendency of approximately $1/f$. For $b = 2$, the spectral tendency is closer to $1/f^2$.

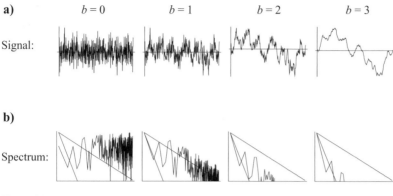

Figure 4.16
Shaped noise examples.

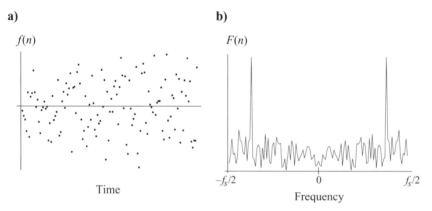

Figure 4.17
Noisy signal and its Fourier transform.

4.5.4   Data Smoothing

We can use noise shaping to extract periodic signals from noise. Say we have recorded a signal $f(n)$, whose samples are shown in figure 4.17a. Just looking at it, it's hard to know that there's a periodic signal embedded within the noise. However, the Fourier transform $F(k)$ shown in figure 4.17b clearly indicates the presence of a sinusoid. How can we extract the time domain signal from the noise?

Inspecting the Fourier-transformed signal in figure 4.17b, we observe that the noise is broadband, whereas the signal itself is at a specific frequency. This suggests that if we filtered out all but the regions around this frequency, we'd reduce the noise accompanying the signal and get a clearer picture of the signal.

If we multiplied the Fourier transform of the noisy signal by a spectrum that is nonzero only in the vicinity of the frequency band of interest, we'd remove much of the noise. This is *bandpass filtering*.

We can create the filtering spectrum in any way, but the following is a useful technique. The first step is to create a function, a *kernel*, to extract the interesting portion of the spectrum of $f(n)$. The kernel must be exactly as long as the Fourier transform $F(k)$. Suppose that $F(k)$ has length L. While the kernel can be any function, let's use of the following exponential function for the kernel:

$$G(k) = e^{-k/Q}, \qquad -\frac{L}{2} < k < \frac{L}{2} - 1. \tag{4.16}$$

The parameter Q, called the *quality factor*, controls how quickly the skirts of the exponential function fall toward zero as k departs from the vicinity of the origin (see volume 1, equation (8.26)). For example, if we set $Q = L/4$, we get the function in figure 4.18a.

But before we can use it, we must transform it so that it has two peaks, one each over the positive frequency and negative frequency bands of $F(k)$ in which we are interested. The transformation is effected simply by rotating a copy of the kernel to the left and rotating a copy to the right, then adding them together (figure 4.18b).

Now we form the product of $F(k) G(k)$ (figure 4.19a). We see that though there is still a little noise in the skirts of the signal, elsewhere the noise is virtually gone.

a)   b)

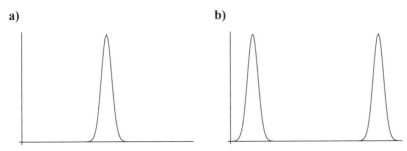

Figure 4.18
Bandpass filter kernels.

a)   b)

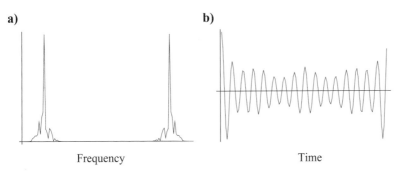

Frequency   Time

Figure 4.19
Product of lowpass spectrum and the signal under test.

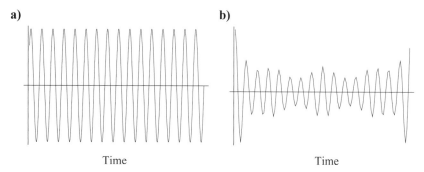

Figure 4.20
Comparison of original and reconstructed signals.

Last, we take the inverse Fourier transform of the foregoing, and the time domain signal is revealed, stripped of a good deal of its noise (figure 4.19b).

Figure 4.20 shows the original signal (figure 4.20a) next to the reconstructed signal (figure 4.20b). Although the reconstructed signal still contains artifacts of its contamination with noise, if we didn't have the original signal by itself, this would be better than the hash we started with. More sophisticated noise reduction techniques are discussed in chapter 10.

4.5.5 Convolving Signals Multiplies Their Spectra

Suppose you are in an anechoic chamber (so sound reflections in the room can be ignored), and a sound arrives at your ears. Depending upon the direction of the sound source, some sound energy—the *direct signal*—goes straight to your tympanum without bouncing off any other surface. The rest of the sound energy—the *reflected signals*—bounce off the folds of your pinnae (the protuberances on either side of your head that enfold your ear canal) before reaching your tympanum. There are as many reflected signals as there are indirect paths from the sound source to your tympanum. Every reflected path is longer than the direct path, so the reflected signals arrive at your tympanum ever so slightly after the direct signal. The intensity of the reflected signals will be attenuated in comparison to the direct signal because of their longer path and the consequences of the inverse square law. They are attenuated even more because the pinnae absorb some amount of sound energy.

The direct signal and the reflected signals interact to change the spectrum of the sound you hear in subtle ways. Some frequencies of the direct signal will arrive at the tympanum out of phase with respect to the same frequencies arriving along a longer path and cancel each other at the tympanum; other frequencies will arrive in phase and reenforce each other. Sound arriving from other directions will cancel and intensify different frequencies because the pinnae will reflect them into the ear differently.

Thus, the delayed and attenuated reflections introduced by the pinnae modify the spectrum of the arriving signal in characteristic ways, depending upon the direction of the sound and the location and shape of the pinnae. We can say that the pinnae *filter* the spectrum of the arriving signal and that the spectral shape of the pinnae depends upon the direction of the arriving sound.

Our auditory system uses this cue (among others) to determine a sound's direction (see volume 1, chapter 6). The pinnae effectively convolve the direct signal with multiple time-shifted and amplitude-scaled copies created by reflections on the pinnae. Since convolving in the time domain is multiplying in the frequency domain, the spectrum of the source signal is multiplied by the spectrum of the pinnae; our ears use this spectral modification as a cue to sense the direction of an arriving sound.

For instance, suppose a motorcycle is passing you: your auditory system performs at least the following minor miracles:

· It notices that the motorcycle sound spectrum is being modified in a way that is characteristic of spectral changes due to the filtering properties of your pinnae.

· It observes that your head is stationary and determines that this means the sound source must be moving.

· It converts the spectral changes into distance and direction information, and passes this information on to your awareness, which synthesizes the percept of the motorcycle passing you.

Figure 4.21 shows an example of the spectrum of the pinnae adapted from a spectral plot of a test subject's ear. It was created by placing a small microphone at the outer end of the ear canal, placing a loudspeaker at some angle to the head (in this case, facing directly into the left ear), and recording the intensity at each frequency over the range of hearing. It's quite a complicated curve. There is a lot of variation in this type of curve from individual to individual. There is a running controversy over just how the ear uses this (and other) information to determine sound location. Nonetheless, pinna filtering seems to provide most of the auditory cues we use to determine sound direction. For example, consider the following experiment. We measure an individual's pinna filters for several different directions and use each one in turn to filter an arbitrary recorded sound, such as from an automobile engine. When we play each filtered automobile sound to the individual through headphones, the subject will quite reliably indicate the correct direction corresponding to each filter.

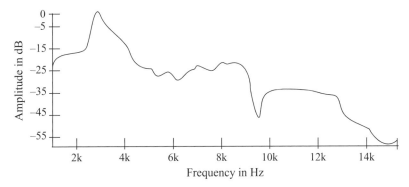

Figure 4.21
Pinna filter for a particular source angle and individual.

4.6 Convolution and the Fourier Transform

Take another look at the equation for convolution, equation (4.1):

$$h(n) = \sum_{m=0}^{\infty} f(m)g(n-m).$$

The Fourier transform can be related to convolution as follows.

If $g(n)$ is the phasor $e^{i\omega n}$, where $\omega = 2\pi$, then (4.1) becomes

$$h(n) = \sum_{m=0}^{\infty} f(m)e^{i\omega(n-m)}.$$

By the rule $x^{a-b} = x^a x^{-b}$ we can rewrite this as

$$h(n) = \sum_{m=0}^{\infty} f(m)e^{i\omega n}e^{-i\omega m}.$$

Notice that the term $e^{i\omega n}$ does not depend on the index variable m. Since its value doesn't vary through each summation, it is a constant in each term of the summation.

Just as $ab + ac$ can be factored into $a(b+c)$, $e^{i\omega n}$ can be factored out of the summation, yielding

$$h(n) = e^{i\omega n} \sum_{m=0}^{\infty} f(m)e^{-i\omega m}.$$

But the resulting summation is just the Fourier transform of $f(m)$, which can be simplified as

$$h(n) = e^{i\omega n} \, \mathcal{F}\{f(m)\}. \tag{4.17}$$

where $\mathcal{F}\{f(m)\}$ is the one-sided Fourier transform of $f(m)$, defined as

$$\mathcal{F}\{f(\omega)\} = \sum_{m=0}^{\infty} f(m)e^{-i\omega m}. \tag{4.18}$$

Equation (4.17) says that convolution of a signal $f(m)$ with a phasor $e^{i\omega n}$ is the same as the Fourier transform of $f(m)$ modulated by the phasor $e^{i\omega n}$.

4.7 Domain Symmetry between Signals and Spectra

By now we have a pretty good arsenal of tools to examine and modify spectra, including addition, multiplication, convolution, the Fourier transform, and the inverse Fourier transform, so we are in a good position to summarize the many symmetries between the time domain and the frequency domain.

It is important to keep in mind the layout of spectral frequencies in order to make sense of the figures in this section. I've used three common arrangements in this book:

- Positive frequencies only, from 0 Hz to $f_s/2$, the Nyquist frequency, shown left to right
- Positive and negative frequencies, from $-f_s/2$ to $f_s/2$, left to right with 0 Hz in the middle
- Standard DFT order, from left to right, 0 Hz followed by increasing negative frequencies up to $\pm f_s/2$, then decreasing positive frequencies until nearly 0 Hz. This is sometimes called wrap-around order.

The spectra in this section are all shown using standard DFT order. It may be a little harder to scan visually, but it will give a clearer picture of the underlying mathematics.

Complex spectra are represented here using overlapping bar charts: empty bars show real components, solid bars show imaginary components. That way we can see the imaginary and real components of a spectrum side by side. I sometimes also supply the magnitude spectrum.

4.7.1 Cosine Symmetry

Figure 4.22 shows a cosine signal and its Fourier transform. The complex spectrum shows that there are two real components at equal positions on either side of 0 Hz. The individual values of the complex spectrum in figure 4.22 are as follows:

Frequency	0	2 1	2 2	2 3	2 4	2 5	2 6	2 7	2 8	8	7	6	5	4	3	2	1
Real	0	0.5	0	0	0	0	0	0	0	0	0	0	0	0	0	0	0.5
Imaginary	0	0	0	0	0	0	0	0	0	0	0	0	0	0	0	0	0

What would happen if we interpreted the complex spectrum from figure 4.22 as a time domain signal and took its DFT? Since its imaginary part is zero, we can ignore that and focus on the real part, interpreting it as an impulse train consisting of two impulses.

Figure 4.23 shows that we recover exactly the same real cosine we started with. This shows the symmetry of the Fourier transform:

$$\mathcal{F}\{\mathcal{F}\{\cos\theta\}\} = \cos\theta, \qquad (4.19)$$

in other words, the Fourier transform of the Fourier transform of a signal is the identical signal. (For simplicity, we ignore any scaling asymmetries that may arise, depending on which Fourier transform is utilized.)

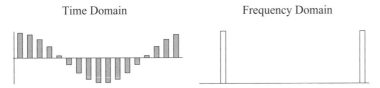

Figure 4.22
Fourier transform of a cosine wave.

Time Domain   Frequency Domain

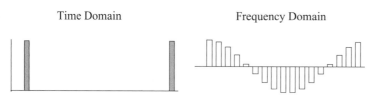

Figure 4.23
Fourier transform of a cosine wave spectrum.

Time Domain   Frequency Domain

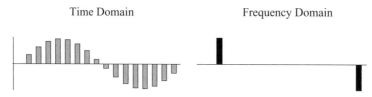

Figure 4.24
Fourier transform of a sine wave.

Time Domain   Frequency Domain

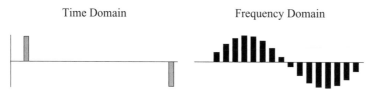

Figure 4.25
Fourier transform of a sine wave spectrum.

This also works for sine waves. Figure 4.24 shows a real sine wave and its Fourier transform. The complex spectrum is entirely imaginary. The values of the complex spectrum from figure 4.24 are as follows:

Frequency	0	−1	−2	−3	−4	−5	−6	−7	−8	8	7	6	5	4	3	2	1
Real	0	0	0	0	0	0	0	0	0	0	0	0	0	0	0	0	0
Imaginary	0	0.5	0	0	0	0	0	0	0	0	0	0	0	0	0	0	−0.5

If we take the Fourier transform of the resulting complex spectrum, we recover the sine wave (figure 4.25). This shows the symmetry of the Fourier transform for the sine function:

$$\mathcal{F}\{\mathcal{F}\{\sin\theta\}\} = \sin\theta. \tag{4.20}$$

This also works for the IDFT, that is,

$$\mathcal{F}^{-1}\{\mathcal{F}^{-1}\{\sin\theta\}\} = \sin\theta.$$

I leave the demonstration to the reader.

Wavelength Symmetry   It is a basic premise of physics that period and frequency are inverse, so the frequency f corresponding to a time interval T can be expressed as

$$f = \frac{1}{T \sec} = \frac{1}{T} \sec^{-1}. \tag{4.21}$$

Equation 4.21 suggests that if we were to shorten the time domain features of a signal, we would expand the corresponding frequency domain features, and vice versa. Figure 4.26 demonstrates this effect for doubling the frequency of a cosine wave. Shorter distances in the time domain correspond to longer distances, in the frequency domain.

Impulse/Sinusoidal Symmetry   It is remarkable that sinusoids (the quintessence of smoothness and continuity) in the time domain are represented by functions that are spikes (the quintessence of jaggedness and discontinuity) in the frequency domain (see for example, figure 4.26). Of course, it works both ways: the inverse Fourier transform of the spiky spectral function produces a continuous sine function in the time domain.

Impulse and Constant Function Symmetry   Figure 4.23 showed that an impulse train consisting of two impulses positioned as shown is converted into a cosine wave by the DFT. What is a single impulse transformed into? Let's create an impulse function that has a single imaginary impulse at position 0, with zeros at all other positions. The Fourier transform of this single impulse is shown in figure 4.27.

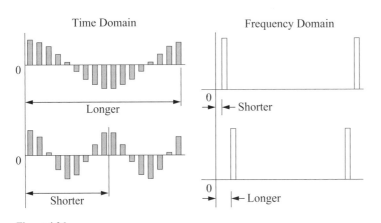

Shorter distances in the time domain correspond to longer distances in the frequency domain.

Figure 4.26
Effect of doubling the frequency of a cosine on its spectrum.

Figure 4.27
Transforming an impulse.

The spectrum of a single impulse turns out to be a uniform constant-valued real function. That's why impulsive sounds such as clicks have spectra that contain equal energy at all frequencies. Clicks in a signal splatter energy broadly across the spectrum because they share the spectral characteristics of an impulse function.

What is the spectrum of a uniform constant-valued signal? In particular, if the constant-valued imaginary spectrum in figure 4.27 is interpreted as a signal, what would its spectrum be? Figure 4.28 shows that we get the single impulse function back.

The frequency of a constant function of time is 0 Hz.

This gives some substance to the otherwise rather abstract notion of 0 Hz, "the frequency of no frequency."

A flow that reverses direction is an *alternating current* (AC). A flow that does not reverse direction is a *direct current* (DC). The flow of a river at its source is a direct current because the water never flows uphill. However, where it flows into the sea, the river's current is alternating because the tides push the water in and out of the river's mouth. A flow that varies without reversal is still a direct current: the flow at a river's source is just as much a direct current during a drought as it is after a rainstorm.

Impulse Train Signals   So far we've observed that

- One impulse transforms into a spectrum with equal energy at all frequencies.
- Two impulses symmetrically placed transform into a cosine wave.
- Two impulses, one negative, symmetrically placed transform into a sine wave.

In the time domain, an impulse train signal consists of a multitude of equally spaced pulses (figure 4.29). Its spectrum is also an impulse train. I've chosen values for figure 4.29 so that in

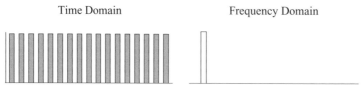

Figure 4.28
Fourier transform of a constant-valued function.

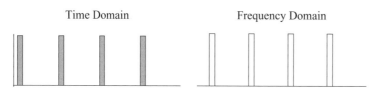

Figure 4.29
Impulse train.

this case the two signals are identical. The time domain signal and its Fourier transform are as follows:

Time domain	1	0	0	0	1	0	0	0	1	0	0	0	1	0	0	0
Frequency	0	-1	-2	-3	-4	-5	-6	-7	8	7	6	5	4	3	2	1
Real	0.25	0	0	0	0.25	0	0	0	0.25	0	0	0	0.25	0	0	0
Imaginary	0	0	0	0	0	0	0	0	0	0	0	0	0	0	0	0

If the impulse train spectrum from figure 4.29 is transformed again, the resulting spectrum is identical to the original time domain signal. Notice that the spectrum of the impulse train signal in figure 4.29 does not change in amplitude with varying frequency. Also, the frequency domain impulses become more densely packed as the time domain pulses become more sparsely packed, and vice versa.

4.8   Convolution and Sampling Theory

So far, I've only provided an intuitive sense of the effects of sampling on the spectrum of a signal being digitized. We are now in a position to formalize these intuitions.

I've described *sampling* as recording instantaneous, periodic measurements of a continuous time domain function. But there's another way to think about digitization that allows us to leverage our knowledge of convolution.

In this view of sampling, we first construct a *sampling function $f(t)$* that is a continuous impulse train function in the time domain. This means it is not a list of discrete sample values but a continuous-valued function consisting of positive impulses at some periodic interval, ranging in time from minus to plus infinity. The impulses are all of unit area and infinitesimal duration.[3]

Let's say $g(t)$ is a continuous signal that we wish to record digitally. If we multiply the sampling function $f(t)$ and the signal $g(t)$, the result will be an impulse train function whose heights correspond to the instantaneous amplitudes of the input signal at each point where the sampling function was nonzero. Call this $h(t)$, the sampled signal. This process is shown in figure 4.30. Note that $h(t)$ is also continuous, even though it looks like a sampled function, because it is the product of continuous functions.

Let the spectrum of $h(t)$ be $H(k)$. What will $H(k)$ look like? Since we are multiplying the sampling function $f(t)$ and the input signal $g(t)$ to create $h(t)$, we know $H(k)$ must be the convolution of the spectra of the sampling function $F(k)$ and input signal $G(k)$.

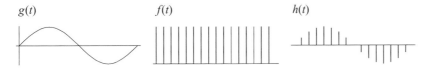

Figure 4.30
Sampling via an impulse train function.

But what is the spectrum of the sampling function? Since $f(t)$ is an impulse train, its spectrum $F(k)$ will be like that shown in figure 4.29: a sequence of impulses. Importantly, the amplitude of these spectral components does not decrease with increasing frequency. Each spectral component of an impulse train is of constant height. Furthermore, since the sampling function is of infinite length, its Fourier transform will have an infinite number of spectral components, all the same height. If the impulses of the sampling function are spaced T seconds apart, the spectral components will be spaced at frequencies that are harmonics (integer multiples) of the sampling frequency $f_s = 1/T$.

As discussed in section 4.4.2, we end up with copies of the whole spectrum of $G(k)$ placed with their centers on each component of $F(k)$.

The spectrum of the sampled signal H(k) contains an infinite number of copies of the input spectrum G(k) spaced at intervals of the sampling spectrum F(k).

Said another way, copies of the sampled spectrum are centered on each harmonic of the sampling rate. But since there are an infinite number of these harmonics, the spectrum of a digitized signal is *infinitely periodic*. Figure 4.31 shows the periodic spectrum of a digitized signal. Though these spectral copies have an identical shape, they are not the same because each copy is centered on a different component of the spectrum $F(k)$.

If $G(k)$ has no spectral components outside the interval $\pm f_s/2$, the Nyquist frequency range, all is well. But if the spectrum of $G(k)$ is wider than $\pm f_s/2$, the components that lie outside of the Nyquist frequency range spill over into the adjacent spectral periods. Figure 4.32 shows an example of this, highlighting the areas of overlap.

The components that overlap each other are *aliased components*. These components typically bear no harmonic relationship to the underlying spectrum of $G(k)$, so they tend to introduce a very objectionable form of distortion into the digitized signal. To avoid aliasing, we must apply an anti-aliasing filter to $g(t)$ to eliminate any components in its spectrum that may lie outside of the Nyquist limit prior to digitization (see section 1.3).

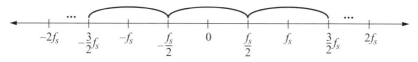

Figure 4.31
Periodic spectrum of a digitized signal.

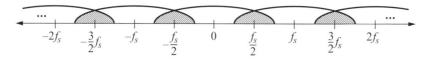

Figure 4.32
Spectrum of a digitized signal with overlapping components.

Ideally, we'd eliminate the undesirable aliases by constructing a "brick wall" filter that passes all frequency components within the Nyquist frequency range equally and utterly rejects all others. But in reality there are compromises, and a less-than-ideal filter must be used. Depending upon what compromises are made, the filter may not entirely reject all aliased components (bad), or it may cut out a bit of the desired spectrum (better). It may also distort the spectrum by amplifying some regions of the selected frequency band more than others. Proper design of optimal anti-aliasing systems go beyond the scope of this book.[4]

All digitized signals contain an infinite periodic spectrum with a period equal to the sampling frequency.

What happens when we try to play back the sampled signal $h(t)$? We must apply another anti-aliasing filter, called a reconstruction filter, to eliminate the multiple spectral copies, leaving only spectral components within the Nyquist frequency range. This subject is discussed in section 1.3 and chapter 5.

4.9   Convolution and Windowing

In the discussion of windowing (section 3.6), I used an intuitive approach to explain its impact on the resulting spectrum. Now, with convolution, this can be made more formal.

4.9.1   Rectangular Functions and the Sinc Function

Figure 4.33a shows a continuous rectangular function. The *duty cycle* of a rectangular function is the ratio of the time it is nonzero to the duration of its period. By inspection, the duty cycle of this rectangular function is 1/2. The real part of the spectrum of the rectangular function is shown in figure 4.33b. This is the *sinc function,* defined as

$$\operatorname{sinc} x \;=\; \frac{\sin x}{x}\,. \hspace{4cm} \textit{Sinc Function (Cardinal Sine)} \quad (4.22)$$

a) Continuous Rectangular Function

0

b) Frequency Domain, Real Part Only, Close-up

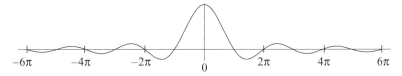

-6π -4π -2π 0 2π 4π 6π

Figure 4.33
Rectangle function, sinc function.

Since $\sin 0 = 0$, $\text{sinc}(0) = 0/0 = 1$ by the rule that the ratio of two equal quantities is 1. Note, however, that computers and calculators typically can't divide by 0, so for these applications the sinc function can be defined as

$$\text{sinc}\, x = \begin{cases} (\sin x)/x, & x \neq 0, \\ 1, & x = 0. \end{cases}$$

The zero-crossing points on the sinc function (where $\sin x = 0$) correspond to values of $x = n\pi$, where n is an integer and $n \neq 0$. Notice that when $n = 0$, $\text{sinc}(0) = (\sin 0)/0 = 1$.

4.9.2   Windowing with a Rectangular Function

If we select a portion of a signal by windowing it with a rectangular function, we multiply the signal point for point with a rectangular function set to 1.0 for the portion we wish to extract and 0.0 elsewhere (see section 3.6). Since this constitutes multiplying two time functions together, we know this means the spectra of the two functions are convolved.

To see this at work, start with the following definitions:

- A real rectangular windowing function $f(t)$ has the spectrum $F(k)$.
- A real signal to be windowed $g(t) = \sin(2\pi ft)$, has the spectrum $G(k)$.
- The windowed signal—the product of $f(t)$ and $g(t)$—is $h(t)$, and its spectrum is $H(k)$.

We know from figure 4.33 that the real part of the spectrum $F(k)$ is a sinc function.

Since we are multiplying $f(t)$ and $g(t)$ in the time domain, we are convolving $F(k)$ and $G(k)$ in the spectral domain. We end up with copies of the whole spectrum of $F(k)$ placed with its center on each component of $G(k)$. In this example, $G(k)$ is the spectrum of a sine wave, shown in figure 4.34a. The convolution of $G(k)$ with $F(k)$ is shown in figure 4.34b. The central lobe of the copies of the sinc function end up centered on the components of $G(k)$.

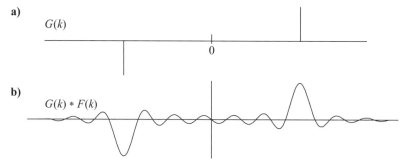

Figure 4.34
Sine spectrum and its convolution with a sinc function.

4.9.3 Spectral Interpretation of the Fourier Transform

We can use figure 4.34 to gain a valuable insight into the workings of the Fourier transform. With the discrete Fourier transform, as in the preceding example, we are windowing a phasor with a rectangular function. Thus figure 4.34b is essentially the same as the sine portion of the probe phasor's spectrum because both were created by multiplying a sine wave and a rectangular function.

This suggests that from a spectral perspective, the Fourier transform generates a probe phasor with a spectrum like figure 4.34b for each harmonic under test. For each such spectrum in turn, the Fourier transform then multiplies the spectrum of the whole input signal by the spectrum of the probe phasor. Nonzero values in the result indicate the presence of energy at those frequencies.

This in turn provides insight into why the Fourier transform works perfectly for harmonic spectra but imperfectly for inharmonic spectra. Notice that the spectrum in figure 4.34b is mostly nonzero. But it is *exactly zero at all integer multiples of the frequency of the main lobe*. The amplitude at the center of the main lobe is 1.0, and the amplitude at all other harmonics is 0.0. Thus, when multiplied against the input spectrum, the spectrum of the probe phasor does not change energy at the center of its main lobe, and it zeros out energy that is exactly harmonic. *But inharmonic energy is imperfectly suppressed.* This is the source of leakage when we try to analyze real-world signals that are not aligned with the fundamental analysis frequency of the Fourier transform. Concisely, the side lobes of the sinc function will only zero energy in the input spectrum if that energy is strictly harmonic. Otherwise the other components of the input signal contribute energy to the spectrum at each harmonic, skewing the resulting measurement.

Figure 4.34b has been structured so that the period of the windowing function $f(t)$ is an integer multiple of the period of $g(t)$, which causes the zero crossings of the window function's spectrum $F(k)$ to coincide with any other harmonics $G(k)$ might have. This fact doesn't matter much in this example because a pure sine wave has no harmonics, but if the input spectrum contained only harmonics of this frequency, this would zero out all harmonics except the one under the main lobe of $F(k)$, allowing us to isolate it and measure its energy.

4.9.4 Frequency-Based Interpretation of the Fourier Transform

The previous section provides a stepping-stone to a frequency-based interpretation of the Fourier transform. Remember that the DFT examines the entire input signal one harmonic at a time, and reports the amount of energy present at each harmonic. Unless we supply a different window, the probe phasor is windowed by a rectangular function with a duty cycle of N samples, the length of the Fourier transform. The spectrum of the probe phasor is therefore a complex version of figure 4.34b: a sinc function convolved with the complex spectrum of the phasor.

Because the Fourier transform multiplies the input signal by a probe phasor at the frequency of the harmonic currently being measured, it effectively constructs a temporary spectrum that is the convolution of the input spectrum with the spectrum of the probe phasor at that frequency.

• If a component in the input spectrum has the same frequency as the current harmonic being measured, its amplitude is not changed by the analysis (figure 4.35a) and it contributes fully to the measurement.

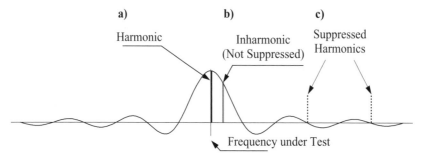

Figure 4.35
Components under test.

• Any nonharmonic components in the input spectrum are attenuated, because they fall on the skirts of the sinc function, but they are not eliminated entirely (figure 4.35b).

• All other harmonics fall on zero crossings of the sinc function (figure 4.35c); they will not contribute to the measurement.

The Fourier transform is only good at testing harmonic components, and any component in the analysis spectrum that does not fall on a zero crossing will add its energy into the measurement of this harmonic, thereby distorting that harmonic's true amplitude.

4.9.5 Filter Bank Interpretation of the Fourier Transform

The preceding discussion suggests that it is possible to think of the Fourier transform as constructed out of a tunable filter. In this view, the Fourier transform filters the spectrum of the analysis signal at each harmonic of the fundamental analysis frequency to measure its energy. The main lobe of the sinc function resonates with energy in the analysis signal at the frequency under test.

However, if there is energy in the analysis spectrum at any frequency under a nonzero section of the sinc function, that component contributes to the current measurement in proportion to its magnitude. The amplitude of the side lobes of the sinc function falls off very slowly, meaning that there's lots of opportunity for measurement contamination from regions covered by these side lobes. For instance, consider the amplitude of the first upper and lower side lobes. Since the sinc function is sinusoidal (see equation (4.22)), the zero crossings occur when x has moved through a semicircle. Thinking along these lines, the first side lobe must have a peak where $x = 3\pi/2$. Converting that to dB of amplitude, we have

$$20 \log_{10} \operatorname{sinc} \frac{3\pi}{2} \cong -13.5 \text{ dB}$$

with reference to the height of the central lobe (0 dB). The significance of this is that any components that are located in the vicinity of these lobes will only be suppressed by about −13.5 dB. Looking at the rest of the lobes, we see that they decline in amplitude relatively slowly, so any energy that lies under their "umbrella" will also contribute—erroneously—to the spectral measurement.

Figure 4.36
Sinc squared function and its triangular waveform.

4.9.6   Sinc Squared Function

If we could make the amplitude of the side lobes fall off more quickly, there would be less opportunity for distant components in the analysis spectrum to contaminate the measurement. We could accomplish this if we squared the sinc function. The central lobe of the sinc squared function is slightly narrower (though it is still just as tall), and the amplitude of the side lobes falls toward zero much more quickly. The time domain function corresponding to the frequency domain sinc squared function is the triangle function (figure 4.36). This explains from a frequency domain perspective why a triangular window is better at rejecting spurious frequency content than a rectangular window.

4.10   Correlation Functions

If two functions are similarly shaped, we say they are *correlated*. If they are dissimilar, we say they are *uncorrelated*. If $f(t)$ and $g(t)$ are sampled functions, their discrete correlation $h(n)$ is denoted

$$h(n) = f \star g = \sum_{m=(-\infty)}^{\infty} f(m)g(n+m). \qquad\qquad \textit{Correlation} \quad (4.23)$$

Correlation and convolution are close mathematical relatives because this formula is almost identical to convolution (see equation 4.1) except that the function g is not reversed in time; that is, we have $g(n+m)$ for correlation and $g(n-m)$ for convolution. To memorialize this distinction, we use star ($\star$) for correlation and asterisk ($*$) for convolution.

Correlation is also sometimes denoted as corr(f, g). If two functions can be made to show a high degree of correlation by shifting one of them in time t, we say they are correlated with lag t.

We can create a *correlation function* that compares the amount of correlation as one function is shifted to the left or right of the other by varying the time lag t. Occasionally lag time is indicated with the notation corr(f, g)(t). If high correlation is achieved with positive t, then the first function $f(n)$ lags (appears to the right of) the second function $g(n)$. Otherwise, if high correlation is achieved with negative t, then the first function leads (appears to the left of) the second function. This can be summarized by writing

$$\text{corr}(f, g)(t) = \text{corr}(g, f)(-t).$$

4.10.1   Correlation and Spectra

Spectrally, correlation is also similar to convolution. The Fourier transform $\mathcal{F}\{\ \}$ of the discrete correlation of two real functions $f(n)$ and $g(n)$ can be written

$$\mathcal{F}\{\operatorname{corr}(f, g)\} = F(\omega)\overline{G(\omega)}, \tag{4.24}$$

where the bar over $G(\omega)$ indicates the complex conjugate.[5] This says that the Fourier transform of the correlation of two real signals equals the product of the spectrum of the one by the complex conjugate spectrum of the other.

4.10.2   Fast Correlation

The fact that the two sides of equation (4.24) are related by the Fourier transform suggests that correlation can be computed efficiently with the fast Fourier transform (see section 3.7). First, take the FFT of both functions, then take the complex conjugate of one of them. The inverse FFT of the product of these signals is their correlation. Though the result is complex, the imaginary parts will be zero if both input functions are real. The elements of the inverse FFT correspond to the correlation at various lags, with the value of FFT(0) corresponding to zero lag, and so on.

4.10.3   Autocorrelation

The correlation of a function with itself, $\operatorname{corr}(f, f)$, is its *autocorrelation*

$$\phi_n = \operatorname{corr}(f, f)(n) = f \star f = \frac{1}{N+1-n}\sum_{m=0}^{N-n} f(m)f(n+m), \qquad n = 0, 1, \ldots, N-1.$$

$$\textit{Autocorrelation} \tag{4.25}$$

The term $1/(N+1-n)$ normalizes the sum for the number of elements in the function that are being autocorrelated for a particular n.

By equation (4.24), the autocorrelation of a function f is equal to the power spectrum of that function:

$$\mathcal{F}\{\operatorname{corr}(f, f)\} = F(\omega)\overline{F(\omega)} = |F(\omega)|^2, \qquad \textit{Wiener-Khinchin Theorem} \tag{4.26}$$

where the power spectrum is the square of the magnitude spectrum (see equation (3.24)).

4.10.4   Uses of Correlation

Correlation can be used, for example, to determine the time delay between when a sound is transmitted and when it is received. If the propagation delay is known, correlation can be used to determine the distance to a reflecting surface, or even the geometry of a room. If $g(t)$ is a signal reflected from a wall and $f(t)$ is the original signal, then their correlation function $h(n)$ will show a spike for the value of n corresponding to the time delay in samples between the two signals.

Autocorrelation can help detect deterministic signals that are masked by a random background signal. The autocorrelation of a deterministic signal, such as a sine wave, persists over all time displacements, whereas the autocorrelation of random processes tends towards zero for large displacements.

More generally, correlation can be used to determine the similarity between any two functions. For instance, it can be used to help identify sounds, such as phonemes of speech: a phoneme from a speaker is correlated against a codebook of sample phonemes; the highest degree of correlation is taken to indicate the phoneme being uttered.

Summary

In section 3.8.1, we saw demonstrated the following principles of symmetry between signals and spectra.[6]

- When signals are added together, so are their spectra.
- When spectra are added together, so are their signals.
- When a signal is multiplied by a constant, its spectrum is multiplied by the same constant.
- When a spectrum is multiplied by a constant, its signal is multiplied by the same constant.

We can now add two new rules:

1. When two signals are multiplied, their spectra are convolved.
2. When two spectra are multiplied, their signals are convolved.

And two new corollaries:

3. When two spectra are convolved, their signals are multiplied.
4. When two signals are convolved, their spectra are multiplied.

Corollary (3) is just rule (1) starting in the frequency domain. Corollary (4) is just rule (2) starting in the time domain. We can combine all these observations as follows:

Rule	Time Domain	Direction	Frequency Domain
1	multiply	$\rightarrow$	convolve
2	convolve	$\leftarrow$	multiply
3	convolve	$\rightarrow$	multiply
4	multiply	$\leftarrow$	convolve

Multiplying or convolving in one domain convolves or multiplies in the other domain, respectively.

Convolution certainly travels in good company, appearing in the same ranks as addition and multiplication in importance. It is fundamental to the symmetry between signals and spectra. It explains the spectral consequences of digital sampling theory. It unlocks such subjects as filtering and reverberation, among many other things.

Suggested Reading

McClellan, Jim, Ron Schafer, and Mark Yoder. 1997. *DSP First: A Multimedia Approach*. Upper Saddle River, N.J.: Prentice-Hall.

Steiglitz, Ken. 1996. *A Digital Signal Processing Primer: With Applications to Digital Audio and Computers Music*. Boston: Addison-Wesley.

5 Filtering

The science of pure mathematics, in its modern development, may claim to be the most original creation of the human spirit. Another claimant for this position is music.
—Alfred North Whitehead

Just as an antique camera technology provided an intuitive introduction for convolution (chapter 4), analog tape recorders provide a very effective model for understanding filtering.

5.1   Tape Recorder as a Model of Filtering

Analog tape recorders operate by dragging a magnetic tape past erase, record, and playback heads at a constant rate (figure 5.1). The tape is pinched between the motor-driven capstan and the pinch roller to pull it along. The reel on the left supplies tape at the rate determined by the angular velocity of the capstan; the motor-driven reel on the right takes up the tape after it passes the capstan.

When recording, the tape first passes under the erase head, which removes any previous recording. A moment later, the erased tape passes under the record head, which converts the incoming electrical audio signal into magnetic fluctuations on the tape. Then the newly recorded tape passes under the playback head, which picks up the magnetic fluctuations, converts them back to electrical signals, and sends them to an amplifier and loudspeaker for reproduction. This arrangement allows the recordist to monitor the recording in progress. (If the erase and record heads are switched off, a previously recorded tape can be played.)

Because of the delay introduced by the passage of the tape from record to playback head during recording, a performer playing into this setup will hear a delayed version of the performance in the loudspeaker. The delay introduced by the tape recorder is what makes it a kind of filter. The study of filtering can be likened to the ways that a tape recorder can be configured.

5.1.1   A Simplified Echoplex Machine

A variant of the analog tape recorder called the Echoplex was manufactured in the mid-twentieth century. Like the standard tape recorder, it had an erase head and a record head, but it had multiple playback heads spaced at adjustable intervals. Figure 5.2 shows its essential elements.

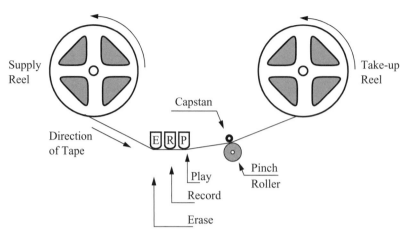

Figure 5.1
Transport of an analog tape recorder.

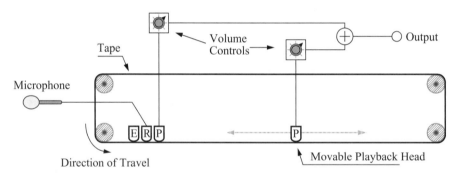

Figure 5.2
Simplified Echoplex machine.

This simplified Echoplex machine has just two playback heads, marked P in the figure. The first one is fixed in place, but the distance of the second one is adjustable. The machine provides separate volume knobs for the two playback heads, then sums their outputs. (A real Echoplex has many more playback heads.)

Suppose I connect a microphone to the record head (through an amplifier, not shown) and attach headphones to the output (again, through an amplifier, not shown). Speaking into the microphone, I'd hear my voice after a moment's delay while the sound travels from the record head to the first playback head, then I'd hear the sound again when it reaches the second playback head some time later, depending upon how far away the second playback head is and how fast the tape is running. If I snapped my fingers into the microphone, I'd hear two snaps separated by some time distance—in other words, I'd hear two echoes.

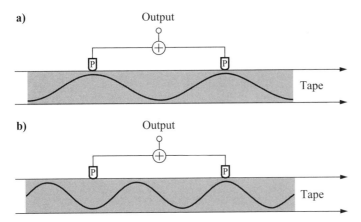

Figure 5.3
Playback heads sum different frequencies differently.

If I sped up the tape or put the playback heads closer together (or did both), the echoes would come closer together until at some point, as I continued to increase the tape speed (or decrease the distance, or both), the two echoes would seem to fuse together, but I'd hear a kind of hollowness in the sound that wasn't there before. As I continued to speed up the tape, the pitch of the hollowness would appear to change.

Without getting too analytical (yet), what is happening is that the microphone signal is being filtered by the tape recorder. The hollow sound is created by the way the two playback heads combine the waveforms on the tape. They sample the sound recorded on the tape at different points.

Consider the sinusoidal waveforms recorded on the magnetic tape shown in figure 5.3. The playback heads are summed together, as they are in figure 5.2. First, let's fix a constant distance between the playback heads. Then let's connect a sine wave oscillator to the record head and record different frequencies onto the tape while we measure the strength of the output signal. We'll observe that some frequencies are loud but others virtually disappear, even though the output strength of the oscillator does not change. Why?

The answer is that some frequencies picked up by the two playback heads are in phase and therefore sum together, while other frequencies are out of phase and therefore cancel.

In figure 5.3a, the distance between the playback heads is exactly one period of the sinusoid recorded on the tape. When the output of the playback heads is summed for this frequency, the two signals reinforce each other.

In figure 5.3b, the distance between the playback heads is 1.5 periods of the sinusoid recorded on the tape. When the output of the playback heads is summed for this frequency, the two signals cancel each other.

Thus, the distance between the playback heads strengthens some frequencies and weakens or even cancels others. That's what filtering is: frequency-selective amplification.

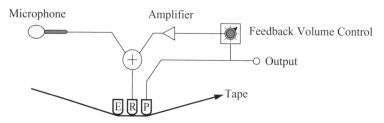

Figure 5.4
Tape recorder feedback.

5.1.2   Tape Recorder with Feedback

For this example, let's go back to just three heads, erase, record and play, but this time let's feed some of the output signal from the playback head back into the input of the record head mixing it together with the input signal (figure 5.4).

The feedback volume control determines how much of the signal received by the playback head is fed back to the record head again. If this volume control is all the way off, no signal is fed back, and we hear a brief delay between the signal presented at the microphone and the signal presented at the output. But as the feedback volume control is turned up, at some point we hear the familiar echo . . . echo . . . echo . . . echo of a recirculating delay line in operation that reminds us of the sound dying away as it reflects between the walls of a room.

If I vary the tape speed (or the distance between record and playback head, or both) the echo rate varies. If I turn up the amount of playback signal fed back to the record head, at first the echoes become more persistent, then they seem to fuse into a continuous wash of sound. Beyond a certain critical feedback volume setting, sounds become louder and louder all by themselves even if I don't touch the feedback volume control, and the sound quickly becomes highly distorted.

This is a kind of filter too, called an *infinite impulse response filter*. If I record an impulsive sound such as a hand clap or finger snap, the sound could keep recirculating in a loop forever, depending upon the setting of the feedback volume control. Its critical feature is that it recirculates its output back to its input, yielding potentially infinite repeats.

The type of filter shown in figure 5.2 is called a *finite impulse response filter*. It is finite because no matter what we do, we'll only get back as many impulses as we have playback heads. Hence its response to impulses is finite.

Tape recorder analogies are very useful in a variety of settings. We've already seen their usefulness to describe echoes and two kinds of filters. They can also be used to describe Doppler shift. For example, suppose we modify an Echoplex so that we can dynamically vary the displacement of the playback head. As we slide the playback head closer and farther away from the record head during recording, we hear changes in pitch and tempo of the recorded sound that are exactly the same as those discussed for the Doppler effect in volume 1, section 7.12 (for the case of a stationary sound source and varying listener position).

Tape recorders also serve as models for a discussion of the technology of time dilation and pitch shifting in chapter 9.

5.2 Introduction to Filtering

In chemistry, a filter is something that removes foreign matter from a fluid. In optics, a colored gel selects a band of frequencies from a spectrum of light corresponding to the color of the gel. As I use the term in this chapter, a filter can remove, reduce, or even strengthen certain bands of frequencies in a signal.

5.2.1 Frequency Response

The *frequency response* of a filter shows which frequencies a filter passes and which it rejects.

Filters are classified by the kind of frequencies they pass. Simple categories of filters include

- *Lowpass*   Low frequencies are allowed to pass through.
- *Highpass*   High frequencies are allowed to pass through.
- *Bandpass*   A band of frequencies between a lower and an upper limit are allowed to pass through.
- *Bandreject*   A band of frequencies between a lower and upper limit are blocked.

For frequencies where the curves in figure 5.5 are above the horizontal line, the filter makes the signal stronger in that frequency range. Where the curve is below the line, the filter weakens the signal by a corresponding amount in that frequency range.

For completeness, we must include one more kind of filter:

- *Allpass*   All frequencies are allowed to pass equally.

What good is the allpass filter? This filter changes only the phase of the signal. Whereas all filters have some effect on phase, the allpass filter *only* has an effect on phase.

More complex filters can be built up out of combinations of these simpler filters. Any complex filter can be broken down into some combination of these simple filters.

In addition to modifying the amplitude of signals based on their frequency, filters typically also modify the phase of the signals, delaying the phases of some frequencies more than others. Even though an allpass filter may not change the amplitudes of any frequencies, it may have an effect on their relative phases. We call the effect of a filter on the phase of a signal its *phase response*. Though the influence on phase is important to the theory of filters, the amplitude response is generally a more salient measure of a filter's operation because our ears are generally more sensitive to amplitude than phase.

The frequency response of a filter is defined as the ratio of the filter's output amplitude to its input amplitude. This ratio is called the *gain* of the filter. Say the frequency response of a filter is $H(k)$, where k is frequency; then its gain is

$$H(k) = \frac{O(k)}{I(k)}, \qquad\qquad\qquad \textit{Gain} \;\; (5.1)$$

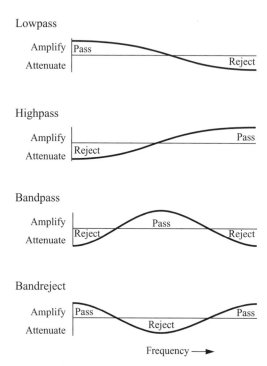

Figure 5.5
Frequency responses of various filters.

where $O(k)$ and $I(k)$ are the amplitudes at frequency k of the output and input signals, respectively. A gain ratio greater than 1 is *amplification,* and a ratio less than 1 is *attenuation.*

The *passband* is the frequency region which the filter lets pass, and the *stopband* is the region the filter blocks. In real-world filters, the amount of amplification or attenuation changes gradually with frequency, and so the basic filters in figure 5.5 show the transition from passband to stopband as gentle curves. However, ideally, we might like to have a filter that passes all frequencies within its passband with equal amplification and perfectly rejects all frequencies in its stopband. Figure 5.6 shows the frequency response of an ideal lowpass filter.

The end of the passband is indicated by f_c, the cutoff frequency. In figure 5.6, the gain is unity in the passband and minus infinity in the stopband. Filters that approximate the frequency response shown in figure 5.6 are sometimes called "brick wall" filters because of the sudden change in gain between passband and stopband.

5.2.2   Filter as Spectral Multiplication

In section 4.5.4, I characterized the operation of a filter as spectral multiplication:

$$Y(k) = H(k)X(k), \tag{5.2}$$

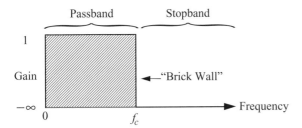

Figure 5.6
Ideal lowpass filter frequency response.

where $X(k)$ is the spectrum of an input to the filter, $Y(k)$ is the spectrum of the output of the filter, and $H(k)$ is the filter response. Since filtering can be seen as spectral multiplication, that means there are two ways to implement any filter: as multiplication in the frequency domain, and as convolution in the time domain. Theoretically, both perspectives are equivalent. However, in real-world filters, the implementation choice really depends on some combination of the design aims and such practicalities as the kinds and amount of processing we can afford. In the following discussion I consider filtering from both the time domain and frequency domain perspectives.

5.2.3 Discrete-Time and Sampled Signals

Filtering requires operating on multiple points of a signal simultaneously. In order to study filters mathematically, we must have a flexible way to represent multiple discrete moments of a signal. If we have a sampled signal $x(n)$, and say that sample number n corresponds to *now,* then $n-1$ indexes the immediately previous sample, and $n+1$ indexes the immediately subsequent sample. If time is graphed on the x-axis with *now* at n, the past $(n-1, n-2, n-3, \ldots)$ is generally shown stretching to the left, and the future $(n+1, n+2, n+3, \ldots)$ to the right.

5.3 A Simple Filter

The simple *lowpass filter* is the best place to begin. It can be expressed as the following difference equation:

$$y(n) = x(n) + x(n-1), \qquad n = 0, 1, 2, 3, \ldots. \tag{5.3}$$

A *difference equation* references multiple points at different locations on the same function. In this equation we observe the values of the same input function x at different moments of time. Equation (5.3) says that the output $y(n)$ is equal to the current input $x(n)$ plus the previous input $x(n-1)$ of the discrete-time signal x.

What does *previous* mean? What is the elapsed time T between now, $x(n)$, and then, $x(n-1)$? In order to know how long the delay is, we must insert T into equation (5.3).

Suppose we have a continuous-time function $x(t)$. If we discretize it by sampling it at instants $t = nT$ (where n is an integer), then we can express the discrete-time filter over continuous time as

$$y(nT) = x(nT) + x[(n-1)T], \qquad n = 0, 1, 2, 3. \ldots$$

Establishing the time period T allows us to *realize* the filter, that is, to tie it to a particular stretch of real time, because $t = nT$ can select any moment of time by suitable choices of n and T. But it's better to leave T out of the picture when discussing filter theory. The equations are simpler, and we can always add time back in if we need to realize a filter in a particular context. So I generally use the notation $x(n)$ rather than $x(nT)$ unless it is required.

Consider a working example. In the system diagram of the lowpass filter (figure 5.7), the successive values of the discrete-time input signal $x(n)$ are supplied as input, and the filter generates a successive discrete-time output signal $y(n)$.

The output signal $y(n)$ is the sum of the input $x(n)$ and the contents of a box with the mysterious label z^{-1} (explained later). The function of this box is to act as a delay for one sample time T. The delay box holds the current input sample $x(n)$ for the duration of one sample time causing it to become sample $x(n-1)$ and then supplies it to the adder, labeled $\oplus$ in the figure. The adder adds together samples $x(n)$ and $x(n-1)$.

Let's make up a discrete-time test signal and pass it through the filter to see how this works. Suppose we have a sampled-time signal

$$x(n) = \{1, 2, 3, 4, 5, 6, 7, 8\}.$$

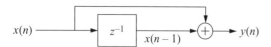

Figure 5.7
System diagram of a simple lowpass filter.

Table 5.1
Filtering a Test Signal

Time n	Current Input $x(n)$	+	Delayed Input $x(n-1)$	=	Output $y(n)$
0	1		–		–
1	2		1		3
2	3		2		5
3	4		3		7
4	5		4		9
5	6		5		11
6	7		6		13
7	8		7		15

Table 5.1 shows the filter in action. Notice that every current input becomes the next delayed input. The value $y(0)$ is not defined because $y(0) = x(0) + x(-1)$, but there is no $x(-1)$. So the first valid filter output is $y(1)$. This means there are only seven valid output values for eight input values. In general, for L inputs, there are $L-1$ valid outputs for this simple filter.

5.4 Finding the Frequency Response

I've said that the filter in figure 5.7 is a lowpass filter, but can I prove it? What is its frequency response? We could try to find out in various ways, for instance, experimentally—by applying various frequencies to the filter's input and plotting the filter's response to each frequency; analytically—by inserting particular sinusoids into the difference equation; and theoretically—by deriving a formula that characterizes the frequency response of the filter directly from equation (5.3).

5.4.1 Determining the Frequency Response Experimentally

To sample the operation of a filter to determine its effect experimentally, we could feed sine waves of various frequencies into the filter and use the magnitude Fourier transform, developed in chapter 3, to show the frequency response of the filter. We could set up a test system as shown in figure 5.8.

The sine wave generator produces a signal at a constant amplitude and adjustable frequency. If the system is running at sampling rate f_s, then by the Nyquist sampling theorem (see section 1.3.1), the frequencies the sine wave generator must be able to produce are in the range $\pm f_s/2$. The generator's signal drives the filter, and the filter's output is passed to a spectrum analyzer, which computes the magnitude Fourier transform of the signal.

Figure 5.9 shows the magnitude frequency response of the filter at eight test frequencies. Because the input signal is real, the magnitude spectrum is symmetrical: half of the energy of each test frequency appears as negative frequencies, so we see 16 half-amplitude components ranging in frequency from $-f_s/2$ to $f_s/2$. The frequency response is highest near 0 Hz and diminishes smoothly toward zero at the Nyquist frequency. This would seem to confirm that this is indeed a lowpass filter. Note that the overall shape of the frequency response looks like it could be sinusoidal. The components between the test frequencies are the result of artifacts of the analysis method (windowing of the transform with a nonharmonic test signal).

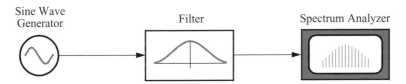

Figure 5.8
Test system for filter frequency response.

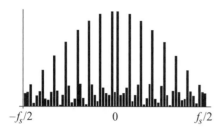

$-f_s/2$ 0 $f_s/2$

Figure 5.9
Magnitude frequency response of a simple lowpass filter.

The advantage of this experimental approach is that we don't need to observe the inner state of the filter in order to understand how it works; this is a *black-box analysis*. We have experimentally established that the filter passes only low frequencies, so it is reasonable to assume that it is a lowpass filter. However, we only have proof for the frequencies we studied. Technically, we shouldn't assume anything about the filter's response to other frequencies. What if we just happened to pick frequencies that tended to confirm our hypothesis? Since there's a possibility of doubt, we would have to test all frequencies to be sure, but this would mean testing an infinite number of frequencies.

Using the experimental approach, we have no definitive proof that we can generalize from the sample frequencies we've tested to cover all frequencies. In contrast, a *white-box analysis* of the filter would provide an understanding of the filter's behavior under all possible circumstances.

5.4.2 Determining the Frequency Response Analytically

The derivations in the rest of this chapter generally follow Julius Smith's (1985) introduction to digital filter theory. Using the experimental approach, we plugged sine waves at various frequencies into an implementation of the filter and observed the resulting frequency response. We can perform a comparable experiment analytically by plugging sinusoids into the difference equation. The advantage here is that we'll know for sure the exact values the filter would produce, and the results won't be subject to experimental error. Using cosines will provide a tactical advantage later in the analysis, so let's use the general cosine signal, $A\cos(2\pi fnT + \phi)$, as the test signal.

Here is equation (5.3), the filter equation, for reference:

$$y(n) = x(n) + x(n-1).$$

If we let $x(n) = A\cos(2\pi fnT + \phi)$, we can rewrite it as

$$y(n) = A\cos(2\pi fnT + \phi) + A\cos[2\pi f(n-1)T + \phi], \qquad (5.4)$$

where A is amplitude, f is frequency, T is the sample period, and $n = 0, 1, 2, 3, \ldots$.

So (5.4) is the test equation for the lowpass filter. There are two ways to look at the values of $y(n)$ in (5.4): one at a time—we can fix n at a particular value and evaluate $y(n)$ for that value; or all at once—we can say that the expression $n = 0, 1, 2, 3, \ldots$ means that n takes on *all possible integer values at once*. Centered dot notation is used to signify this, so $x(\cdot) = \cos(\cdot)$ means x is defined over all possible cosine values.

Now let's see what happens. Starting with the frequency $f = 0$,

$$y(n) = A\cos\phi + A\cos\phi$$

$$= 2A\cos\phi.$$

If we set $A = 1$ and $\phi = 0$, we are left with just

$$y(n) = 2,$$

which makes sense, because in this case both the input signal and the delayed input are both just the unity function:

$$x(n) = \cos 0$$

$$= \{1, 1, 1, \ldots, 1\}.$$

Using the "one-at-a-time" rule, we see that if $f = 0$, $A = 1$, and $\phi = 0$, then

$$y(n) = 1 + 1$$

$$= 2$$

(5.5)

for all n. That takes care of the bottom end of the frequency response. What about the top end?

Let's think about the relation between f and T. Remember that the sampling period T is the reciprocal of the sampling frequency, that is, $T = 1/f_s$. If the frequency f that we're plugging into the filter were operating at the sampling rate, then we'd have $f = f_s = 1/T$. Therefore, if we set f at the Nyquist frequency, then $f = f_s/2 = 1/2T$. So the value of f corresponding to the Nyquist frequency is $1/2T$.

What is the input $x(n)$ when $f = 1/2T$? It is

$$x(n) = A\cos\left(2\pi\frac{1}{2T}nT + \phi\right)$$

$$= A\cos(\pi n + \phi).$$

A handy trigonometric identity (see appendix section A.2) shows that

$$\cos(a + b) = \cos a \cos b - \sin a \sin b.$$

So we can write

$$x(n) = A\cos(\pi n)\cos\phi - A\sin(\pi n)\sin\phi,$$

(5.6)

but (remembering $n = 0, 1, 2, \ldots$) we can substitute

$$A\cos(\pi n) = A\{1, -1, 1, -1, \ldots\}$$

$$= A(-1)^n.$$

Furthermore we can get rid of the second half of equation (5.6) by observing that for $n = 0, 1, 2, \ldots, A \sin(\pi n) \sin \phi = 0$ because $\sin \pi n$ is always zero. So we can simplify $x(n)$:

$$x(n) = A(-1)^n \cos \phi, \quad n = 0, 1, 2, \ldots$$

$$= \{A \cos \phi, -A \cos \phi, A \cos \phi, \ldots\}. \tag{5.7}$$

Finally, using equation (5.7) for $x(n)$ we can write

$$y(n) = x(n) + x(n-1)$$

$$= A(-1)^n \cos \phi + A(-1)^{n-1} \cos \phi$$

$$= A(-1)^n \cos \phi - A(-1)^n \cos \phi \tag{5.8}$$

$$= 0.$$

Now, during the derivation of equation (5.8), how did I figure out that if $x(n) = A(-1)^n$ it's okay to write $x(n-1) = A(-1)^{n-1} \cos \phi$? Where is the proof that this step is valid? Clearly,

$$(-1)^n = \{-1^0, -1^1, -1^2, -1^3, \ldots\}$$

$$= \{1, -1, 1, -1, \ldots\} \tag{5.9}$$

for $n = 0, 1, 2, \ldots$. But what does $(-1)^{n-1}$ equal? Start with just the first term and set $n = 0$. Then we have

$$(-1)^{0-1} = (-1)^{-1}. \tag{5.10}$$

Remember from algebra that a value raised to a negative exponent is the same as the reciprocal raised to a positive exponent, that is

$$a^{-n} = \frac{1}{a^n}.$$

Armed with this insight, we can rewrite equation (5.10) to read

$$(-1)^{-1} = \frac{1}{-1^1} = -1. \tag{5.11}$$

Reasoning along these lines, we can show that for $n = 0, 1, 2, \ldots,$

$$(-1)^{n-1} = \{-1^{-1}, -1^0, -1^1, -1^2, \ldots\}$$

$$= \{-1, 1, -1, 1, \ldots\}. \tag{5.12}$$

Notice that the sequences of values in equations (5.9) and (5.12) have opposite sequences of signs. This provides an alternative explanation for why the terms of equation (5.8) cancel to equal zero. Rather than taking this rather obscure step, I relied on another equality:

$-(-1)^n = (-1)^{n-1}.$

Here's an example: let $n = 3$, then $-(-1)^3 = 1 = (-1)^2 = (-1)^{3-1}.$

In summary, we've plugged an analytical signal into the lowpass filter with frequencies of $f = 0$ Hz and $f = f_s$ Hz, and observed that the output of the filter is 2 (see equation (5.5)) and 0 (see equation (5.8)), respectively. We have proof that for those two frequencies at least the filter operates as a lowpass filter. We could re-do the computation for any frequencies by plugging in other values of f, but testing just one frequency at a time still will not prove that for every frequency the response is lowpass. The next sections will show a more powerful way of determining filter response.

5.4.3 Moving Average Filter

Another view of the simple lowpass filter in equation (5.3) is to think of it as calculating a running two-point average. The average of two values is $(A + B)/2$. Equation (5.3) does not have the divide-by-2, but if we implemented it, then the output of the lowpass filter would be 1 at 0 Hz instead of 2. The filter is defined as follows:

$$y(n) = \frac{x(n) + x(n-1)}{2}. \qquad \qquad \textit{Moving Average Filter} \quad (5.13)$$

This is a *moving average filter* because the average at sample n is based on a moving window of adjacent values that are used to calculate the average of the input sequence.

Equation (5.13) uses a window of only two samples. We can generalize the moving average filter as follows. Given a sequence of samples $x(n)$ in the range from 0 to N, a moving average of window size w is a new sequence $y_w(n)$ in the range from 0 to $N - w - 1$ defined by the average of w terms:

$$y_w(n) = \frac{1}{w} \sum_{j=n}^{j+n-1} x(j). \qquad \qquad (5.14)$$

Averaging tends to round off abrupt changes, that is, to inhibit high frequencies.

• Adjacent samples of a low-frequency signal more often have the same sign, so they tend to grow in strength when added.

• Adjacent samples of a high-frequency signal more often have opposite signs, so they tend to decrease in strength when added.

5.4.4 Determining the Frequency Response Theoretically

The flaw in the previous two attempts to prove the frequency response of the lowpass filter is that we were testing it one frequency at a time. But since there is an infinity of frequencies to choose from, proving that it is a lowpass filter for all of them is impossible. But if we could plug them all in at once, we would be in a position to prove the filter's lowpass character once and for all. This would be tantamount to plugging the general sinusoid into the difference equation. Here is equation (5.4) for reference:

$$y(n) = x(n) + x(n-1)$$
$$= A\cos(2\pi fnT + \phi) + A\cos[2\pi f(n-1)T + \phi].$$

Let $\omega = 2\pi f$, $\phi = 0$, and $A = 1$. Then we have, equivalently,

$$y(n) = \cos \omega nT + \cos[\omega(n-1)T]. \tag{5.15}$$

All that remains is to reduce this to a single function for $y(n)$. The final function we get will have frequency-dependent amplitude and phase shift properties that will indicate exactly how the filter works at all frequencies.

Tuck the following fact away for future reference: For any two sinusoids having the same frequency but possibly different amplitudes, we can find one sinusoid at that frequency but with another phase. That is the meaning of the equation

$$a\sin \omega n + b\cos \omega n = \sin(\omega n + \phi), \tag{5.16}$$

for appropriate values of a, b, and ϕ.

Clearly, the output of the filter $y(n)$ is a single sinusoid, but what formula expresses it? I use complex frequencies to simplify the derivation. Say that the input signal $x(n)$ is now complex, and

$$x(n) = Ae^{i(2\pi fnT + \phi)}. \tag{5.17}$$

Let's make all the previous simplifications again with this new definition: let $\omega = 2\pi f$, $\phi = 0$, and $A = 1$. Then we have, equivalently,

$$x(n) = e^{i\omega nT} = \cos \omega nT + i\sin \omega nT. \tag{5.18}$$

Now we plug this complex definition of $x(n)$ into the difference equation:

$$y(n) = x(n) + x(n-1)$$
$$= e^{i\omega nT} + e^{i\omega(n-1)T}.$$

Because $x^{a-b} = x^a x^{-b}$, we can write

$$y(n) = e^{i\omega nT} + e^{i\omega nT}e^{-i\omega T}.$$

Because $a + ab = (1+b)a$, we can write

$$y(n) = (1 + e^{-i\omega T})e^{i\omega nT}.$$

Since by definition $x(n) = e^{i\omega nT}$, we can write

$$y(n) = (1 + e^{-i\omega T})x(n). \tag{5.19}$$

In equation (5.19), we've managed to equate y to a scaled version of x, which is what we've been looking for. Let's isolate the scaling terms from equation (5.19) and put them into their own function:

$$H(e^{i\omega T}) = (1 + e^{-i\omega T}).$$ *Lowpass Filter Transfer Function* (5.20)

We can rewrite equation (5.19) with this new definition:

$$y(n) = H(e^{i\omega T})x(n).$$ (5.21)

Equation (5.21) says that the output signal y is equal to the input signal x times a scaling function H. The scaling function supplies a scaling coefficient that *depends upon the frequency of the input signal*. The definition of $H(e^{i\omega T})$ says that as ω grows (that is, as the frequency increases), the value of $H(e^{i\omega T})$ follows the semicircle shown in figure 5.10.

Just as the definition of H consists of two terms, figure 5.10 consists of two vectors, a positive real unit vector of length 1 corresponding to the first term in the definition of H, and a vector corresponding to the complex exponential $e^{-i\omega T}$. Because the exponent is negative, this complex vector spins clockwise around the circle as ω grows. When the frequency of the input signal $\omega = 0$, then $e^{-i\omega T} = e^0 = 1$, which is also a positive unit vector on the real axis. Since according to the definition of H, the two vectors are added, the frequency response of the filter to 0 Hz is just $1 + 1 = 2$. Correspondingly, when the frequency of the input signal equals the Nyquist frequency, $\omega = f_s/2 = \pi$ radians, then the complex vector will have rotated clockwise 180° so that it is now a negative real unit vector, and the combination of the two vectors cancels. Thus, the frequency response of the filter at the Nyquist frequency is $1 - 1 = 0$.

So, for $\omega = 2\pi f$ and $T = 1$, if we let $f = 0$, we can write

$$H(e^{i\omega T}) = H(e^0) = (1 + e^{-0}) = (1 + 1) = 2,$$

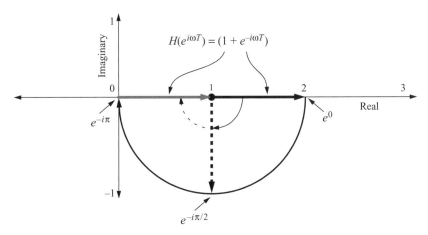

Figure 5.10
Lowpass filter transfer function.

but if we let $f = 0.5$ (which corresponds to the Nyquist frequency $f_s/2$), then we can write

$$H(e^{i\omega T}) = H(e^{i\pi}) = (1 + e^{-i\pi}) = (1 - 1) = 0.$$

5.4.5   Summary of Frequency Response Determination

It's true that so far we've still only analyzed the same two frequencies as before, but there's an important difference between this analysis and the previous two: now we have a function that describes *for any input* what the frequency response will be. We can plug any frequency into H and see how the filter will react. We see that H describes a complex vector path from 2 to 0 as frequency goes from 0 Hz to the Nyquist frequency. The function $H(e^{i\omega T})$ given in equation (5.20) is called the *transfer function* of the filter because it shows how much of x is transferred to y depending on frequency. Note that the function does not depend upon the passage of time; the parameter n does not appear in it anywhere. Hence, the transfer function is *time-invariant*. The function $H(e^{i\omega T})$ only depends upon ω, the frequency of the input (assuming that the sampling period T remains constant—usually a safe assumption).

While the transfer function $H(e^{i\omega T})$ is a complete description of the filter's frequency response, it is not very user-friendly. At only two places on the transfer function—0 Hz and the Nyquist frequency—are the solutions real numbers: the rest of the frequency values between 0 and π require four dimensions to chart. While mathematically that's no problem, it's not particularly intuitive. Why does the transfer function require four dimensions? Let's take a brief digression into mapping functions.

5.4.6   Mapping Functions

As shown in figure 5.11a, a real-valued function of a real variable like $y = f(x) = x^2$ requires a plane, consisting of the dimensions x and y. As shown in figure 5.11b, a complex-valued function of a real variable like $f(x) = \cos x + i \sin x$ requires three dimensions: the complex plane (real and imaginary axes) and x. We can envision three dimensions as two dimensions stacked up like a deck of cards or as a threaded series of planes.

a) $y = f(x) = x^2$ **b)** $z = f(x) = \cos x + i \sin x$

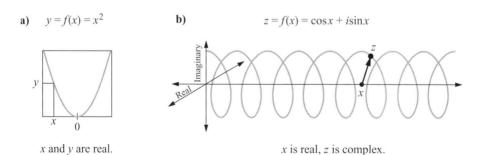

x and y are real.   x is real, z is complex.

Figure 5.11
Function dimensions.

a)   b)

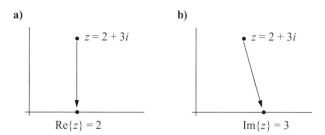

Figure 5.12
Real function of a complex variable.

In general, functions can be said to *map* the coordinates from one domain or one space into another. There can be any number of dimensions involved in either domain or space; the only requirement is that there be an unambiguous mapping. In figure 5.11a, we're mapping from one dimension to one dimension (for a total of 2 dimensions), and in figure 5.11b, we're mapping from one dimension to two (for a total of 3 dimensions).

Also, we can map from higher dimensions to lower ones. For example, the functions Re{ } and Im{ } presented in section 2.3.1 take a complex number in two dimensions and return a one-dimensional real number (figure 5.12). (Recall that Im{ } returns the imaginary part of its argument as a real number.)

The number of dimensions required by a function is the sum of the number required for the function value(s) and the function variable(s). So a complex function of a complex variable requires four dimensions. One way to visualize it is to imagine that each point in the complex plane maps to another point in the complex plane. There are methods to circumvent the need for four-dimensional graphics by showing how the function transforms sets of lines that lie in the complex plane. Each line is mapped into a corresponding curve in the complex plane, and these can be represented in two dimensions. This is the approach taken in figure 5.10, but this is a simple filter, and this approach will not be powerful enough for more complicated functions. It would be nice to reduce the complexity of the transfer function to facilitate understanding. Is there a way to decompose the transfer function into meaningful lower-dimension functions?

5.4.7   Frequency Response and Phase Response

As we've seen in preceding sections, there are only two things a filter can do to a signal: it can perform frequency-dependent amplitude change and frequency-dependent phase change, and nothing more. If we break the transfer function down into these attributes, we'll have two two-dimensional functions instead of one four-dimensional one.

Let's start by defining the attributes we hope to separate out.

The frequency response is the frequency-dependent gain attribute of the transfer function.

The phase response is the frequency-dependent delay attribute of the transfer function.

Remembering that the magnitude of a complex number is the real-valued length of a vector drawn to it from the origin of the complex plane, we define the frequency response for frequency ω as the magnitude of the transfer function:

$$G(\omega) \equiv \left| H(e^{i\omega T}) \right|.$$ *Magnitude of the Transfer Function* (5.22)

$G(\omega)$ is a real-valued function of the real frequency variable ω.

Remembering that taking the angle of a complex number reveals its phase, let's define the phase response for frequency ω as the angle of the transfer function:

$$\Theta(\omega) \equiv \angle H(e^{i\omega T}).$$ *Angle of the Transfer Function* (5.23)

$\Theta(\omega)$ is a real-valued function of the real variable frequency ω.

5.4.8   Transfer Function

Finally, we must show that the transfer function consists of the combination of the frequency response and the phase response. The proper way to write the combination is perhaps a little surprising:

$$H(e^{i\omega T}) = \underbrace{G(\omega)}_{\text{Amplitude}} \underbrace{e^{i\Theta(\omega)}}_{\text{Phase}}.$$ (5.24)

Why is the phase function written as an imaginary exponent of e? Remember that what we're trying to do is to put Humpty-Dumpty back together again: we need to recombine the frequency response and the phase response to equal the original transfer function, which is complex. But $G(\omega)$ and $\Theta(\omega)$ are not complex. Making $\Theta(\omega)$ an exponent of e is just the right thing because $e^{i\Theta(\omega)}$ is a phasor with an angle corresponding to the delay specified by the real-valued function $\Theta(\omega)$. The length of this phasor is scaled by the real-valued function $G(\omega)$. Thus we recreate H by appropriate recombining of the frequency response and the phase response.

5.4.9   Using the Transfer Function

Equation (5.24) is the mother of all transfer functions in the sense that we should be able to take the transfer function of any actual filter and idealize it into the form shown in that equation. An equation such as this, which summarizes other equations in a fundamental way or reduces specific equations to general ones, is called a *canonical* equation. We can use (5.24) as a general tool to separate the amplitude and phase responses of any filter, thereby splitting the transfer function from four dimensions into two two-dimensional functions.

For example, let's take the transfer function of the lowpass filter and idealize it into the canonical transfer function, thereby breaking it down into its separate frequency response and phase response. This will provide an intuitive model of the operation of the lowpass filter that covers all possible values of the frequency ω. Here is the lowpass filter transfer function, equation (5.20), for reference:

$$H(e^{i\omega T}) = (1 + e^{-i\omega T}).$$

Because $e^0 = 1$, we can substitute as follows:

$$H(e^{i\omega T}) = (1 + e^{-i\omega T}) \tag{5.25}$$

$$= (e^0 + e^{-i\omega T}).$$

5.4.10 Balancing the Exponents

The next step involves manipulating exponents. Since by elementary algebra we know that $x^a \cdot x^b = x^{a+b}$, it should come as no surprise that $x^{-1} \cdot x^{+1} = x^0$. With this in mind, consider the following equation:

$$x^0 + x^{2a} = (x^{-a} + x^a)x^a,$$

which depends upon the additive property of exponents. Now consider this equation, based on the same principle:

$$x^0 + x^a = (x^{-a/2} + x^{a/2})x^{a/2}.$$

Julius Smith (1985) calls this *balancing the exponents*. We could balance the exponents of equation (5.25) by writing

$$H(e^{i\omega T}) = (e^0 + e^{-i\omega T}) \tag{5.26}$$

$$= (e^{i\omega T/2} + e^{-i\omega T/2})e^{-i\omega T/2}.$$

With this step we have regularized the exponents so that they are all $\pm i\omega T/2$. Notice that the expression $e^{i\omega T/2} + e^{-i\omega T/2}$ in equation (5.26) is now remarkably similar to the trigonometric cosine identity (see equation (2.52)):

$$\cos\theta = \frac{e^{i\theta} + e^{-i\theta}}{2}.$$

We can make these terms match the cosine identity as follows. If we let $\theta = \omega T/2$, then the cosine identity becomes

$$\cos\frac{\omega T}{2} = \frac{e^{i\omega T/2} + e^{-i\omega T/2}}{2}.$$

If we multiply both sides by 2, it becomes

$$2\cos\frac{\omega T}{2} = e^{i\omega T/2} + e^{-i\omega T/2}.$$

And since by equation (5.26), $H(e^{i\omega T}) = (e^{i\omega T/2} + e^{-i\omega T/2})e^{-i\omega T/2}$, we can write

$$H(e^{i\omega T}) = 2\cos\frac{\omega T}{2}e^{-i\omega T/2}. \tag{5.27}$$

At this point the transfer function for the lowpass filter has started to look like the canonical transfer function given in equation (5.24). We can now have a go at extracting $G(\omega)$ and $\Theta(\omega)$.

5.4.11   Extracting the Frequency Response

Remembering that we defined $G(\omega) \equiv |H(e^{i\omega T})|$, we can write

$$G(\omega) = \left|2\cos\frac{\omega T}{2}e^{-i\omega T/2}\right|.$$

We can simplify this by remembering that $\left|e^{i \cdot \text{anything}}\right| = 1$. Also, $|2| = 2$, so

$$G(\omega) = 2\left|\cos\frac{\omega T}{2}\right|. \tag{5.28}$$

We can simplify some more by eliminating the absolute value operation from equation (5.28). If we restricted ourselves just to the region of the cosine function that lies in the interval $[-\pi, +\pi]$ where the value of the cosine function is all positive, we can drop the absolute value because in that range

$$\left|\cos\frac{\omega T}{2}\right| = \cos\frac{\omega T}{2}.$$

Restricting ourselves to this region is okay because that covers the frequency range of interest, $\pm f_s/2$, the Nyquist range (see section 1.3.3). So finally we can write

$$G(\omega) = 2\cos\frac{\omega T}{2}, \qquad -\pi \le \omega \le \pi. \tag{5.29}$$

Equation (5.29) is the frequency response of a lowpass filter. Its plot is shown in figure 5.13.

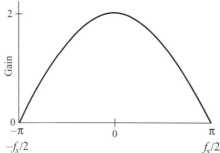

Figure 5.13
Frequency response of a lowpass filter.

5.4.12 Extracting the Phase Response

Now let's tackle the phase response. Remember from equation (5.24) that the canonical transfer function H is the product of the frequency response and the phase response. Here it is again for reference:

$$H(e^{i\omega T}) = G(\omega)e^{i\Theta(\omega)}.$$

We have determined that the transfer function of the lowpass filter is

$$H(e^{i\omega T}) = 2\cos\frac{\omega T}{2}e^{-i\omega T/2}. \tag{5.30}$$

We see from equation (5.29) that the frequency response is governed by the term $2\cos(\omega T/2)$. So the phase response must be determined by the only other term in that equation, $e^{-i\omega T/2}$. By comparing equation (5.24) and equation (5.30), we see that the term $-\omega T/2$ corresponds to the phase response, $\Theta(\omega)$. So we can now define

$$\Theta(\omega) = -\frac{\omega T}{2}, \qquad -\pi \leq \omega \leq \pi. \tag{5.31}$$

Here are a few strategic values of this function:

$$\Theta(0) = 0, \qquad \Theta\left(\frac{\pi}{2}\right) = -\frac{(\pi/2)T}{2} = -\frac{\pi}{4}T, \quad \Theta(\pi) = -\frac{\pi}{2}T.$$

If we set $T = 1$, the plot looks like figure 5.14.

The lowpass filter evidently has no phase delay at 0 Hz and a maximum delay of $-(\pi/2)T$ at the Nyquist frequency.

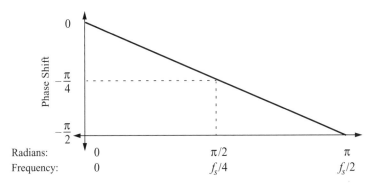

Figure 5.14
Phase response of a lowpass filter.

Note that equation (5.31) expresses delay as a function of the frequency ω and the sample period T. This means that the retardation of the output of the filter is measured in units of the sampling period. For instance, if the radian velocity ω of the input is doubled from 1 to 2, the phase response goes from $\Theta(1) = -T/2$ to $\Theta(2) = -T$. That is, doubling ω causes the retardation to grow by $T/2$, which is one half of a sampling period. The filter progressively delays the output signal as a linear function of input frequency.

However, this frequency-dependent delay is counteracted because doubling the frequency of the input halves its period, balancing the growth in retardation introduced by the filter. For example, as the radian velocity ω of the input goes from 1 to 2, the wavelength shrinks by half. As a consequence, the phase response of a simple lowpass filter is actually constant and equals one-half sample, and the delay through the filter is the same at all frequencies.

If the phase response of the filter is a linear (straight-line) function of frequency (excluding phase wraps at ±180 degrees), the filter is said to be *linear phase*. Filters that are not linear phase can introduce *phase distortion*.

5.4.13 Summary of Lowpass Filter Analysis

Here's a summary of what we know about lowpass filters. First, remembering that we defined $\omega = 2\pi f$, we can rewrite the frequency response $G(\omega) = 2\cos(\omega T/2)$ as

$$G(f) = 2\cos \pi fT. \hspace{3cm} \textit{Real Frequency Response} \quad (5.32)$$

We can similarly rewrite the phase response $\Theta(\omega) = -\omega T/2$ as

$$\Theta(f) = -\pi fT. \hspace{3cm} \textit{Real Phase Response} \quad (5.33)$$

What we have demonstrated is that if we plug a general-purpose sinusoid of the form

$$x(n) = A\cos(2\pi fnT + \phi) \hspace{5cm} (5.34)$$

into a simple lowpass filter,

$$y(n) = H(e^{i2\pi fT})x(n), \hspace{5cm} (5.35)$$

where H is the transfer function defined in equation (5.30), we produce the output

$$y(n) = 2A\cos(\pi fT)\cos(2\pi fnT + \phi - \pi fT). \hspace{2.5cm} (5.36)$$

The first term, $2A\cos \pi fT$, applies a frequency-controlled amplitude scaling to the second term, $\cos(2\pi fnT + \phi - \pi fT)$, which is a sinusoid at frequency f with frequency-controlled phase response governed by $-\pi fT$.

The total frequency response is the product of the input amplitude A and the filter's frequency response $G(f) = 2\cos \pi fT$. Similarly, the total phase response is the sum of the input phase delay ϕ and the filter's phase response $\Theta(f) = -\pi fT$.

5.4.14 General Form of a Linear Time-Invariant Filter

Summarizing, a filter may be expressed in general as follows:

$y(n)$ = Transfer function · Circular motion

$$= G(\omega)e^{i\Theta(\omega)} \cdot A e^{i\omega nT + \phi} \qquad (5.37)$$

$$= G(\omega)A \cdot e^{i(\omega nT + \Theta(\omega) + \phi)}.$$

Thus, a filter can be characterized as circular motion with radius $G(\omega)A$ and phase $\Theta(\omega) + \phi$. The particular kind of filter implemented depends only on the definitions of $G(\omega)$ and $\Theta(\omega)$. This analysis only applies to linear time-invariant filters, however.

5.5 Linearity and Time Invariance of Filters

In section 3.8.1, superposition, proportionality, and linearity were discussed for the Fourier transform. This section is a condensed version of this same topic for filters.

5.5.1 Linearity

A filter is linear if the following two rules apply.

Linearity *A filter is linear if (1) its input is proportional to its output, and (2) its output is the same whether the input signals are combined and then filtered, or filtered and then combined.*

Consider $f(x) = x$, the identity function. If $x = 3$ and $g = 0.5$, then proportionality means

$$f(gx) = gf(x) = f(0.5 \cdot 3) = 0.5f(3) = 1.5.$$

If $x = 2$ and $y = 3$, superposition means

$$f(x + y) = f(x) + f(y) = f(2 + 3) = f(2) + f(3). \qquad (5.38)$$

If f is linear, it remains linear even if the values supplied to the function are also functions:

$$f(x(p) + y(q)) = f(x(p)) + f(y(q)).$$

Note that even if f is a linear function, if we let g be a function of x as well, we lose linearity. That is, $y(x) = f(gx) = gf(x)$ is linear, but $y(x) = f(g(x)x) = g(x)f(x)$ is nonlinear, even if g and f are linear. For instance, let $g(x) = f(x) = x$, the identity function; then $y(x) = g(x)f(x) = f^2(x)$, which is an exponential, not a linear, function. Suppose that $g(x)$ is a function such as an automatic gain control, expander, limiter, or compressor of some kind that scales the signal depending upon its content. In that case, it matters whether $g(x)$ sees the signal before or after it is processed by $f(x)$, and therefore

$$f(g(x)x) \neq g(x)f(x).$$

5.5.2   Time Invariance

A filter is time-invariant if the following is true.

Time Invariance   *A filter is time-invariant if no parameter of the filter transfer function changes with time.*

For instance, the lowpass filter $y(n) = H(e^{i2\pi f T})x(n)$ is time-invariant because the transfer function $H(e^{i2\pi f T})$ does not include the time parameter n. The phrase linear time-invariant is often abbreviated LTI.

The human larynx is an example of a time-varying filter, which varies its filtering properties in order to produce the different vowel sounds. A time-varying filter can be linear, but not if any time-varying parameters depend upon the signal (as is the case, for example, of an automatic gain control).

5.6   FIR Filters

The simple lowpass filter analyzed in previous sections is an example of a linear time-invariant (LTI) filter. It is a member of the class of filters characterized by the difference equation

$$y(n) = \text{FIR}(a, x, M) = \sum_{r=0}^{M} a_r x(n-r). \tag{5.39}$$

The *order* of a filter is determined by the value of M; parameter a is a set of M coefficients, and x is the input signal. For example, the simple lowpass filter is first-order, so $M = 1$, and equation (5.39) becomes $y(n) = a_0 x(n) + a_1 x(n-1)$. Compare this to equation (5.3). All filters described by equation (5.39) sum scaled delayed copies of the *input* signal to produce the output signal. Filters described by equation (5.39) are called finite impulse response (FIR) filters.

5.7   IIR Filters

There is another class of filters characterized by the difference equation:

$$y(n) = \text{IIR}(b, y, N) = \sum_{r=1}^{N} b_r y(n-r), \tag{5.40}$$

where N is the order, b is a set of N coefficients, and y is the output signal.

These filters sum scaled delayed copies of the filter's *output* signal y. Filters described by equation (5.40) are called infinite impulse response (IIR) filters. The only problem with filters defined by equation (5.40) is that there's no way to give them any input; but in order to be useful, they obviously must receive inputs from somewhere. As a consequence, an IIR filter must be coupled to an FIR filter of some kind to supply some input. So, how are these filters combined?

5.8   Canonical Filter

The general difference equation of the causal, finite-order, linear, time-invariant filter combines the FIR and IIR structures as follows.

$$y(n) \ = \ \text{FIR}(a, x, M) - \text{IIR}(b, y, N)$$

Causal, Finite-Order, Linear, Time-Invariant Filter   (5.41)

$$= \ \sum_{r=0}^{M} a_r x(n - r) - \sum_{s=1}^{N} b_s y(n - s).$$

The canonical filter defined by equation (5.41) is *causal* because it references only current and past input (but not future input), and past output (but not current or future output). The FIR filter section accesses only current and past input because it can only retrieve input in the range $x(n - 0)$ to $x(n - M)$. The IIR filter section accesses only past output because it can only access output in the range $y(n - 1)$ to $y(n - N)$. The reason the IIR filter section starts from $s = 1$ instead of from 0 is because it must not reference its current output, $y(n - 0)$, whose value is undefined until the current operation is complete, whereupon it becomes a past output.

There is no reason in principle why we couldn't implement an *acausal* version of this filter by setting $M < 0$ and/or $N < 1$, in which case the filter would access future input and/or future output, which is tantamount to reversing the direction of time. Of course, acausal filters cannot exist in real time; however they are perfectly legal in nonreal time, such as, for instance, when operating on a prerecorded signal. There, acausal filtering can be used to undo the phase effects of a filter. Suppose we filtered a signal in the forward direction, then filtered it again in the reverse direction: we'd undo the phase delay introduced by the forward filtering, and we'd also square the effect of the filter on the amplitude response.

Equation (5.41) is called *finite-order* because $M < \infty$ and $N < \infty$. The *order* of a filter is defined as the maximum of M and N. It is linear and time-invariant (LTI) because it embodies the properties discussed in section 5.5.

If $N > 1$ (that is, if any past output is used by the filter as input), the filter is said to be *recursive*, which means that its output is recycled back into the filter.

5.9   Time Domain Behavior of Filters

We have focused so far on frequency-domain characterizations of filters. But we can characterize them in the time domain as well. In section 5.3, we saw how to plug sinusoids of various frequencies into a filter to determine its frequency response. We can use the same basic procedure to discover the time response of a filter. For instance, if we assign values to the coefficients a_r, we can observe the output y of the FIR filter

$$y(n) \ = \ a_0 x(n) + a_1 x(n - 1) + a_2 x(n - 2) \tag{5.42}$$

by applying some input signal x.

But what input signal shall we use? While we could proceed as before and try sinusoids at various frequencies, the test would be more definitive if we had just one signal that contained every frequency in equal proportion. Happily, the *unit impulse function,* also called the *delta function,* is such a signal. It is defined as

$$\delta(n) = \begin{cases} 1, & n = 0, \\ 0, & n \neq 0. \end{cases} \qquad\qquad \textit{Unit Impulse (Delta) Function} \quad (5.43)$$

Since the unit impulse function contains every frequency at once, the filter's response to it characterizes how the filter will respond to any and all frequencies (see figure 4.27). The response of a filter to the impulse function is called its *impulse response.* So the impulse response, denoted $h(r)$, is the filter's signature in the time domain, just as the transfer function $H(e^{i\omega T})$ is its signature in the frequency domain. The impulse function can be approximated by sudden, sharp sounds, such as those made by an electrical spark, a hammer blow, a gun firing, or a balloon bursting.

5.9.1 Impulse Response of an FIR Filter

Let's examine the impulse response of the FIR filter given in equation (5.42). Let $a_0 = 1$, $a_1 = 0.6$, and $a_2 = 0.3$, and let the input signal be the impulse function $\delta(n)$, so that we have the function $y(n) = a_0\delta(n) + a_1\delta(n-1) + a_2\delta(n-2)$. An FIR filter simply sums the time-shifted gain-scaled copies of its input signal. The first three outputs of the filter's response to the impulse signal are shown in table 5.2.

As n increases with time, the impulse function $\delta(n)$ is shifted through the filter, isolating the value of each coefficient one at a time as it goes. The impulse response simply shows a picture through time of the successive values of the coefficients.

We see in table 5.2 that the impulse response of this filter dies out after $M = 3$ samples, that is, the impulse response is finite. All filters that only process their inputs are *finite impulse response* (FIR) filters. The length of the impulse response equals the order M of the filter. We can write the impulse response for this filter as

$$h(r) = \{a_0, a_1, a_2\}. \qquad\qquad (5.44)$$

Values of $h(r)$ outside the range $0 \leq r < M$ are defined to be zero.

Table 5.2
Impulse Response of a Third-Order FIR Filter

n				a_0	a_1	a_2			
0	$\ldots$	0	0	1	0	0	$\ldots$		
1	$\ldots$	0	0	1	0	0	$\ldots$		
2	$\ldots$	0	0	1	0	0	$\ldots$		
Output	$\ldots$	0	0	1	0.6	0.3	0	0	$\ldots$

5.9.2 Impulse Response of an IIR Filter

The impulse response can be used to characterize recursive filters as well. For recursive filters, $N > 1$, so they reuse their past output. However, the impulse response length of recursive filters is not tied to the order of the filter. The impulse response may die out, remain constant, or even grow over time. Consider the IIR filter

$$y(n) = x(n) + by(n-1). \tag{5.45}$$

Let x be the impulse function, and $b = 0.9$. The impulse response of this filter is shown in figure 5.15. After the initial impulse, each output equals the previous output times 0.9, resulting in the sequence $\{1, 0.9, 0.81, 0.729, 0.6561, 0.59049, \ldots\}$. The output is an exponential function that never goes to zero, so the impulse response is infinite. All filters that have at least one output term fed back into their input are IIR filters. If the b coefficient is set to a small value, the impulse response dies out quickly. But the impulse response is still infinite in length. Even though values quickly become insignificant, technically they reach zero only in the limit at infinity.

If $b = 1$, the impulse response is an impulse train that does not die away and is of infinite length. If $b > 1$, the impulse response grows in amplitude without bounds, indefinitely. If $b \geq 1$ the filter is *unstable* because its output grows without bounds. Note that only recursive filters can be unstable because only they recirculate their output. Every finite-order nonrecursive filter is stable (so long as its coefficients are not infinite).

As with the FIR filter, the impulse response of the IIR filter—also denoted $h(r)$—characterizes the filter in time just as the transfer function $H(e^{i\omega T})$ characterizes it in frequency. Although the IIR filter impulse response is always infinite in length, if $b < 1$, the impulse response eventually becomes insignificant and from that point on can either be ignored or estimated.

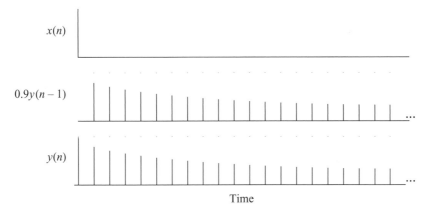

Figure 5.15
Impulse response of an IIR filter.

5.10 Filtering as Convolution

Another way to view a sampled waveform $x(n)$ is to think of it as a *linear combination of delayed scaled impulse functions.* For example, consider the sampled sine wave

$$x(n) = \{0, 0.707, 1, 0.707, 0, -0.707, -1, -0.707, \ldots\},$$

which can equivalently be expressed as

$$x(n) = [0 \cdot \delta(n)] + [0.707 \cdot \delta(n-1)] + [1 \cdot \delta(n-2)] + [0.707 \cdot \delta(n-3)] + \cdots,$$

as shown pictorially in figure 5.16.

Observing the pattern, we can express $x(n)$ as a linear combination of scaled and shifted impulses as follows:

$$y(n) = \sum_{k}^{n} x(k)\delta(n-k)$$

$$= x(n) * \delta(n).$$

Convolution is commutative, that is, it doesn't matter whether we scroll x past δ, or vice versa: we still get the same summation. Therefore, we can write

$$y(n) = \sum_{k}^{n} \delta(k)x(n-k)$$

$$= \delta(n) * x(n).$$

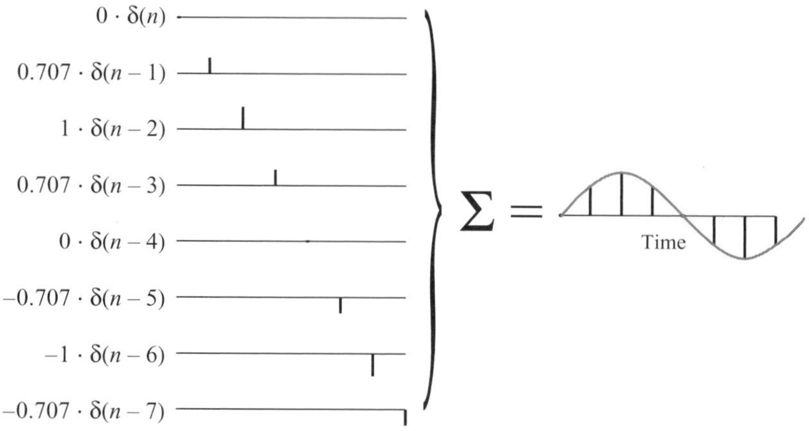

Figure 5.16
Sampled signal as a combination of impulses.

Each sample of any signal may be viewed as an impulse arriving at some amplitude and time. Each impulse arriving at the filter's input will cause the filter to produce its entire impulse response. If another impulse arrives before the first has died away, they will be superimposed. For example, we can equally well interpret the second-order FIR filter in equation (5.42) as adding three delayed copies of $x(n)$, the first copy being scaled by a_0, the second by a_1, and the third by a_2:

...	$a_0x(n-2)$	$a_0x(n-1)$	$a_0x(n-0)$	...		
	...	$a_1x(n-2)$	$a_1x(n-1)$	$a_1x(n-0)$	...	
+		...	$a_2x(n-2)$	$a_2x(n-1)$	$a_2x(n-0)$	...
...	$y(n-2)$	$y(n-1)$	$y(n)$	$y(n+1)$	$y(n+2)$	...

The shaded column shows the computation of $y(n)$, and the columns to the left and right show the computation of previous and future outputs, respectively. But observe that this is exactly the same operation as convolving $x(n)$ with the filter impulse response. Remembering that we defined the impulse response for this filter as $h(r) = \{a_0, a_1, a_2\}$ in equation (5.44), we could recreate the previous sequence with the expression

$$y(n) = \sum_{r=0}^{n} h(r)x(n-r) \tag{5.46}$$

$$= h(n) * x(n),$$

recalling that nonexistent values of $h(r)$ are defined as zero.

Equation (5.46) shows that

Filtering, seen previously to be multiplication in the frequency domain, can also be seen as convolution with a filter's impulse response h in the time domain.

Notice how similar equation (5.46) is to the FIR portion of the canonical filter in equation (5.41) and also to equation (5.39). All we have done is to substitute $h(r)$ for a_r. This allows us to apply our knowledge of convolution, and so we see that a filter operates by summing weighted echoes of the input signal.

The output of any LTI filter can be computed by convolving the input x with the impulse response h.

The impulse response $h(r)$ of an FIR filter is finite, whereas the impulse response of an IIR filter is theoretically infinite. However, if an IIR filter is stable (that is, $b < 1$), then for all practical purposes its impulse response eventually becomes insignificant. In practical applications, an IIR impulse response may be truncated after the amplitude goes below a chosen threshold.

5.11   Z Transform

The convolution theorem states that convolving functions p and q in time is the same as multiplying their spectra $P(e^{i\theta})$ and $Q(e^{i\theta})$ in frequency:

$$p * q \Leftrightarrow P(e^{i\theta})Q(e^{i\theta}).$$

The double-arrow notation $\Leftrightarrow$ indicates that the terms on either side constitute a transform pair. In this case, they are a Fourier transform pair, with the time domain shown on one side and the frequency domain shown on the other.

Let's define $y = p * q$. Then by the convolution theorem, $Y(e^{i\theta}) = P(e^{i\theta})Q(e^{i\theta})$ defines the spectrum of y. Now let's apply the same reasoning to filtering.

We know from equation (5.46) that the output of a filter y is equal to the impulse response h convolved with input signal x. In other words,

$$y = h * x.$$

By the convolution theorem, the spectrum of y is

$$Y(e^{i\theta}) = H(e^{i\theta})X(e^{i\theta}).$$

Rearranging this equation to isolate H, we discover:

$$H(e^{i\theta}) = \frac{Y(e^{i\theta})}{X(e^{i\theta})}. \tag{5.47}$$

The frequency response of a filter is the ratio of the frequency response of the output to the frequency response of the input.

Thus the action of a filter in the frequency domain can be inferred by measuring the ratio of the output spectrum to the input spectrum, frequency by frequency. But how can we make practical sense out of this? How can we characterize the action of the filter in producing this effect?

First, let's review some facts.

- A filter's frequency response H is a complex-valued function of complex frequency (see section 5.4.5).

- A filter's magnitude spectrum is the magnitude of its complex frequency response. So, if we take the magnitudes of the complex functions equated in equation (5.47), the results are the magnitude spectra of the frequency responses. That is,

$$\left|H(e^{i\theta})\right| = \frac{\left|Y(e^{i\theta})\right|}{\left|X(e^{i\theta})\right|},$$

where each of the terms $|H|$, $|Y|$, and $|X|$ are magnitude spectra.

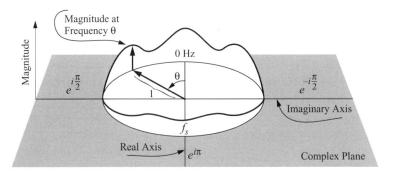

Figure 5.17
Complex frequency response.

- The argument to the frequency responses, $e^{i\theta}$, is a vector that makes a counterclockwise traversal of the unit circle in the complex plane as θ goes from 0 to 2π.

- Spectra of sampled signals are periodic in frequency, with a period of f_s Hz.

Taking this collection of facts, we can envision the magnitude of the complex frequency response $|H(e^{i\theta})|$ as a function erected upon the circumference of the unit circle. A hypothetical magnitude spectrum is shown in figure 5.17 to provide a visualization. The impulse response is shown like an oddly carved cylinder raised above the complex plane. The cylinder is in the shape of the magnitude spectrum of the signal being represented. The real axis is shown receding into the distance so as to reveal that the positive and negative halves of the hypothetical spectrum are conjugate symmetrical, which means it was derived from a real (not complex) signal. The vector shown at angle θ points to the base of another vector that is orthogonal to the complex plane and whose length is the amplitude of the spectrum at that frequency.

From here on I adopt a term from signal processing and call the complex plane the *z plane*.

We see that figure 5.17 defines the spectrum as occurring only where $e^{i\theta}$ touches the *z* plane, that is, on or above the unit circle on the complex plane. It isn't necessary to show a spectrum this way; we could perfectly well unroll it and show it as a two-dimensional function. But showing it in its native cylindrical form draws attention to an important fact about the frequency response, $|H(e^{i\theta})|$. The frequency response of a digital filter is a *periodic function* with a period of f_s Hz. If we were to show H unwrapped, we'd have to show an infinite number of periods to adequately express its circularity in a linear manner (see figure 4.31).

Figure 5.17 defines the spectrum as occurring only where $e^{i\theta}$ touches the *z* plane. What about all the other points on the *z* plane? Do they have any significance with respect to the spectrum being shown? We know we have three dimensions: the real axis, the imaginary axis, and amplitude. But three-dimensional functions can be represented as surfaces. Is there significance to the idea of extending the frequency response beyond only those points touched by the unit circle $e^{i\theta}$—extending it to define a surface?

How can we identify the meaning of the rest of the points in the complex plane that are not on the unit circle? The discussion has been restricted to the points on the unit circle because the argument to the frequency response function has been constrained to be the complex variable $e^{i\theta}$, which is a unit vector that only varies by angle θ. So $e^{i\theta}$ only allows us to examine the frequency response at points on the unit circle. What we are looking for is a representation of the frequency response that is not restricted to values on the unit circle $e^{i\theta}$ but is able to take on any value z in the complex plane.

To go forward, we must first review how to convert a time domain signal into a spectrum via the Fourier transform, but this time we must generalize the method to evaluate the frequency at every point on the z plane, not just the unit circle. Let's start with a complex definition of the discrete Fourier transform:

$$X(e^{i\omega T}) = \sum_{n=0}^{N-1} x(n)e^{-i\omega nT}, \tag{5.48}$$

where N is the number of samples in the complex input waveform $x(n)$ that are summed, n is an integer, $\omega = 2\pi$, and $e^{-i\omega nT}$ indexes the spectrum in the complex plane. Then we generalize equation (5.48) so it is not restricted to evaluating points on the unit circle but can evaluate any point on the complex plane. We do so by substituting the general complex variable z in equation (5.48) for all terms $e^{i\omega T}$:

$$X(z) = \sum_{n=0}^{N-1} x(n)z^{-n}, \qquad\qquad\qquad Z \; Transform \tag{5.49}$$

where n and N are integers. Because z is not restricted to the unit circle, we are operating not just on points on the unit circle of the complex plane but on all points z in the complex plane. What is the behavior of this equation? In particular, what is the effect of raising a point z in the complex plane to successive powers? What happens when z^{-n} is multiplied by the input $x(n)$? What happens when the product is summed?

5.11.1   What Happens When a Point in the Complex Plane Is Raised to Successive Powers?

Complex number z can be represented as $z = re^{i\theta}$ with magnitude r and an angle θ (see equation 2.46). Raising z to successive powers means finding $z^n = (re^{i\theta})^n$. This makes z's magnitude r grow geometrically if it started out being greater than 1, shrink geometrically if it started out being less than 1, and remain the same if it equaled exactly 1. Raising z to successive powers causes it to spin counterclockwise if $\theta > 0$ and clockwise if $\theta < 0$. The amount of spin depends on the initial angle θ. Small values of θ spin slowly; larger (positive or negative) values of θ spin faster. See figure 5.18a.

Consider the sequence

$$z^{-n}, \qquad n = 0, 1, 2, 3, \ldots,$$

$$= 1, \frac{1}{z^1}, \frac{1}{z^2}, \frac{1}{z^3}, \ldots. \tag{5.50}$$

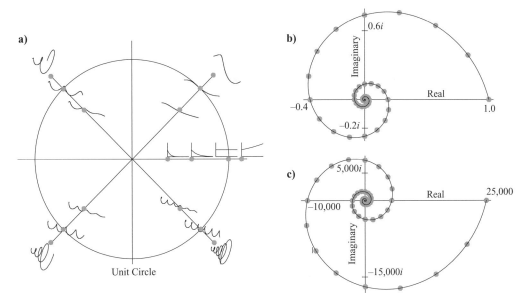

Figure 5.18
Complex exponentials.

If we set $z = re^{-i\omega}$, set $\omega = 2\pi/16$, and let r be a value greater than 1 (for example, $r = 1.1$), the sequence $z^{-n} = (re^{-i\omega})^{-n}, n = 0, 1, 2, 3, \ldots$, will shrink geometrically toward zero while the angle spirals counterclockwise (figure 5.18b). Dots on the spiral in the figure show the successive integer values of n.

If we set $z = 0.9e^{-i\omega}$, the sequence $z^{-n} = (re^{-i\omega})^{-n}, n = 0, 1, 2, 3, \ldots$, grows as it spins counterclockwise (figure 5.18c).

Raising a complex number to positive powers of n will just invert these two sequences. That is, setting $z^{n} = (re^{-i\omega})^{n}$ will spiral clockwise, and will grow if $r > 1$ and will shrink if $r < 1$. We can summarize this by observing that when we raise a complex value with nonzero imaginary part to successive powers of n, the following behaviors result:

$\left|z^{-n}\right| < 1$	Expanding counterclockwise spiral		$\left|z^{n}\right| < 1$	Contracting clockwise spiral
$\left|z^{-n}\right| = 1$	Unit circle		$\left|z^{n}\right| = 1$	Unit circle
$\left|z^{-n}\right| > 1$	Contracting counterclockwise spiral		$\left|z^{n}\right| > 1$	Expanding clockwise spiral

Of course, a complex value with a zero imaginary part is just a real number, so all complex values that lie on the real number line obey the normal rules for raising real numbers to successive powers.

5.11.2 What Happens When z^{-n} Is Multiplied by $x(n)$?

In equation (5.49) the spiral trajectory of z^{-n} is multiplied sample by sample against $x(n)$. The input $x(n)$ could itself be a complex signal, or if it is derived from a natural signal, its imaginary part will be zero.

5.11.3 What Happens When the Product of z^{-n} and $x(n)$ Is Summed?

Recall how the Fourier transform frequency extractor works: for each frequency under test, we multiply the analysis signal by the input signal and sum the result. Where the analysis signal and input signal are identical or nearly identical, the sum will be a large positive number. All that has changed between equation (5.48) and equation (5.49) is that the analysis signal can now be a complex spiral. If we sum the sequence shown in figure 5.18b, each new addition will add a smaller and smaller amount, so the rate at which the sum grows will slow progressively, and in the limit (at infinity) the sum will converge to a particular value. On the other hand, the sum of the sequence in figure 5.18c diverges, because we always add larger and larger values.

Let's now extend the modified DFT to an infinite number of summations. Equation (5.51) is the same as equation (5.49), but with the limit of summation set to infinity.

$$X(z) = \sum_{n=0}^{\infty} x(n)z^{-n}. \qquad\qquad \textit{Z Transform, Infinite} \quad (5.51)$$

It is called *one-sided* because the index variable n begins at 0 instead of $-\infty$.

What is the significance of extending the number of summations to infinity? As with the DFT, as N grows, we sum a larger and larger portion of the input waveform, providing finer and finer frequency resolution. Setting $N = \infty$ simply means we evaluate the entire input signal, which is presumed to be infinite in length, in order to determine to an infinitesimal degree what each point of its Z transform is.

For every value of z, equation (5.51) sums the infinite geometric progression $\{x(n)z^{-n}\}$, $n = 0, 1, 2, \ldots$, using the following steps:

For each value of z,

1. Compute the geometric progression $\{x(n)z^{-n}\}$, $n = 0, 1, 2, \ldots$:

$$\{x(0)z^{-0}, x(1)z^{-1}, x(2)z^{-2}, \ldots\}.$$

2. Sum the progression:

$$x(0)z^{-0} + x(1)z^{-1} + x(2)z^{-2}, \ldots.$$

3. Store the result in $X(z)$.

But what good is this? How could one usefully calculate an infinite progression of any kind, let alone do it over and over for every possible value of z? Fortunately, this is not required, and we can use this equation without having to form sums of infinite series because there is a way to directly express the value of the nth sum of a general geometric progression.

5.11.4 Sums of Geometric Progressions

The nth sum S_n of the general geometric progression $\{ar^n\}$ can be written as

$$S_n = \frac{a}{1-r} - \left(\frac{a}{1-r}\right)r^n . \tag{5.52}$$

This can be verified by plugging in any real numbers a, r, and integer n. For instance, let $a = 1$, $r = 1/2$. Then, the sequence $\{ar^n\}$ for $n = 0, 1, 2, \ldots$ is

$$\left\{1, \frac{1}{2}, \frac{1}{4}, \frac{1}{8}, \frac{1}{16}, \ldots \right\}. \tag{5.53}$$

We can use equation (5.52) to find the sum of the first n terms of this progression. For $n = 1, 2, 3, \ldots$, the sum of all previous terms of the example forms the series

$$\left\{1, \frac{3}{2}, \frac{7}{4}, \frac{15}{8}, \frac{31}{16} \right\}.$$

For example, the third term, 7/4, is the sum of the first three terms of equation (5.53): $1 + 1/2 + 1/4$.

If we make sure that $|r| < 1$, we can make the right-hand term of equation (5.52) as close to zero as we wish by making n large enough. It follows that if $|r| < 1$, then $a/(1-r)$ is closely approximated by S_n when n is a large enough number. In fact, the larger the number n, the closer the approximation. To indicate that we can make the difference between S_n and $a/(1-r)$ as small as we please by making n large enough, we write

$$\lim_{n \to \infty} S_n = \frac{a}{1-r},$$

which reads, "The limit of S_n as n increases without bound is equal to $a/(1-r)$."

Because S_n is the nth sum of the sequence, if we set n to infinity, then S_n must be the sum of the infinite series. That is,

$$ar^0 + ar^1 + ar^2 + \cdots = \frac{a}{1-r} - \left(\frac{a}{1-r}\right)r^\infty, \qquad |r| < 1,$$

$$= \frac{a}{1-r}.$$

Or, if we put this in *closed form*,

$$\sum_{n=0}^{\infty} ar^n = \frac{a}{1-r}, \qquad |r| < 1, \tag{5.54}$$

which reads, "The sum of the infinite geometric progression ar^n converges at $a/(1-r)$ for $|r| < 1$."

Remember that this works only if $|r| < 1$ because then the second term of the right-hand side of equation (5.52) goes to zero as n grows. If instead of taking r^n we take r^{-n}, then we must make sure $|r| > 1$ so that the series will still converge. Thus we'd write

$$\sum_{n=0}^{\infty} ar^{-n} = \frac{a}{1 - r^{-1}}, \qquad |r| > 1. \tag{5.55}$$

5.11.5 Z Transform of the Unit Step Function

Equations (5.54) and (5.55) provide very convenient tools to understand filter frequency response in general and IIR filter frequency response in particular. Let's examine the Z transform of the unit step function:

$$u(n) = \begin{cases} 0, & n < 0, \\ 1, & n \geq 0. \end{cases}$$

We can use the unit step function to understand the Z transform of IIR filters because the unit step function is equivalent to the output of a simple first-order IIR filter with unity gain driven with a single impulse.

If we let $x(n)$ be the unit step function, then we would use equation (5.54) to write its one-sided Z transform as follows:

$$X(z) = \sum_{n=0}^{\infty} 1 \cdot z^{-n} = \frac{1}{1 - z^{-1}}, \qquad |z| > 1. \tag{5.56}$$

Examining equation (5.56), we see that if we set $z = 1$, then $X(z)$ would be positive infinity (because then we're dividing by zero). Thus, the value of $X(1) = \infty$. The function $X(z)$ is said to have a *pole* at $z = 1$.

By multiplying the numerator and denominator of equation (5.56) by z, we may also write it as

$$X(z) = \frac{z}{z - 1}, \qquad |z| > 1. \tag{5.57}$$

We see that if we set $z = 0$, then $X(z)$ would be zero (because the numerator is zero). Thus, the value of $X(0) = 0$. The function $X(z)$ is said to have a *zero* at $z = 0$.

What about all the other points in the complex plane? $X(z)$ is a complex function of a complex variable, requiring four dimensions to represent, so it is hard to visualize. However, if we take the magnitude of $X(z)$, it becomes a three-dimensional function that we can graph and study. If we computed $|X(z)|$ for all z, we'd see it is like a 3-D relief map of the Z transform of the unit step function. The surface of this 3-D function resembles a tent with the zeros acting as tent pegs and the poles acting as . . . tent poles. In between, the values would be a function of $|X(z)|$. For example, the magnitude Z transform of the unit step function, equation (5.57), is shown in figure 5.19. The plot shows the unit circle with a pole at $1 + 0i$ and a zero at $0 + 0i$. As shown, the function takes on relatively small nonzero values elsewhere. Keep in mind that poles are infinitely high and infinitely narrow at the top; the representation of the pole in figure 5.19 is truncated to fit on the page.

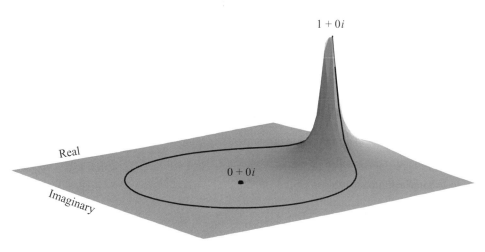

Figure 5.19
Magnitude Z transform of a unit step function.

The magnitude of the Z transform provides a 3-D representation of the complex frequency response $X(z)$. The Z transform is sometimes expressed as $X(z) \Leftrightarrow x(n)$ because the Z transform of a time domain signal $x(n)$ is its complex spectrum, and vice versa.

5.11.6 Convolution and the Z Transform

We can restate the convolution theorem with the more general notion of frequency embodied in the Z transform. Remember that filters convolve in the time domain or multiply in the frequency domain:

$$x(n) * y(n) \Leftrightarrow X(z)Y(z).$$

We can characterize the output of the filter as the input convolved with its impulse response:

$$y(n) = h(n) * x(n),\tag{5.58}$$

where h is the impulse response. Since equation (5.58) convolves h and x, we can equivalently multiply their complex spectra:

$$Y(z) = H(z)X(z).$$

$Y(z)$ is the spectrum corresponding to the impulse response $y(n)$. Rearranging, we get

$$H(z) = \frac{Y(z)}{X(z)}.\tag{5.59}$$

Equation (5.59) says that the Z transform of the filter's frequency response is equal to the Z transform of the output signal divided by the Z transform of the input signal. $H(z)$ is called a transfer function because it shows how the input spectrum is changed while being transferred to the output.

The transfer function H(z) of a linear time-invariant filter (LTI) is defined as the Z transform of the impulse response h(n).

5.12   Z Transform of the General Difference Equation

According to equation (5.59), the transfer function of a filter is simply the ratio of its complex output spectrum to its complex input spectrum. Remember that the general difference equation (5.41) is just another way to describe filtering as convolution. What would we see if we took the Z transform of the general difference equation? Before we can do that, we must understand two more properties of the Z transform.

5.12.1   Shift Theorem of the Z Transform

The shift theorem states that

$$x(n-k) \Leftrightarrow z^{-k}X(z).$$ *Shift Theorem of the Z Transform* (5.60)

In other words, a real signal $x(n)$ delayed by k samples has the Z transform $z^{-k}X(z)$, which is the Z transform of the original unshifted signal $x(n)$ multiplied by z^{-k}. Here's how it works. Start with the equation for the one-sided Z transform,

$$X(z) = \sum_{n=0}^{\infty} x(n)z^{-n}.$$ (5.61)

Define a new function, $w(n) = x(n-k)$. This new function equals $x(n)$ delayed by k samples. Expressing this mathematically,

$$W(z) = \sum_{n=0}^{\infty} w(n)z^{-n} = \sum_{n=0}^{\infty} x(n-k)z^{-n},$$

which says that $W(z)$ equals the Z transform of $w(n)$, and $W(z)$ also equals the Z transform of $x(n)$ delayed by k samples.

Now, if we require that $x(n-k) = 0$ for all $n < k$ (that is, if $x(n)$ is itself one-sided), then we can safely ignore any values of $x(n)$ for $n < 0$. If so, we can relate $W(z)$ directly to $x(n-k)$ by restricting the index of summation to begin on the kth sample:

$$W(z) = \sum_{n=k}^{\infty} x(n-k)z^{-n}.$$ (5.62)

No matter what value we give k, we're guaranteed to always start at $x(0)$, but z^{-n} will have different values depending upon k.

Now define a new relation $m = n - k$. This implies $n = m + k$. With this relation, we can substitute m as the index of summation instead of n so long as we start the summation at 0 instead of k. Substituting these definitions into equation (5.62) produces

$$W(z) = \sum_{m=0}^{\infty} x(m) z^{-(m+k)}. \tag{5.63}$$

One way to verify that equation (5.63) is equal to equation (5.62) is to set k to some value in each equation and observe that they form identical sequences.

Now, by the law of exponents, $x^{a+b} = x^a x^b$, we can rewrite equation (5.63):

$$W(z) = \sum_{m=0}^{\infty} x(m) z^{-m} z^{-k}.$$

The term z^{-k} is a constant since it does not depend upon the index of summation. Since $ab + ac = a(b + c)$, we can extract the term z^{-k} and put it outside the summation:

$$W(z) = z^{-k} \underbrace{\sum_{m=0}^{\infty} x(m) z^{-m}}_{X(z)}.$$

What's left inside the summation is equivalent to the definition we started with for $X(z)$, given in equation (5.61). Because of this, we can relate $W(z)$ to $X(z)$ directly as follows:

$$W(z) = z^{-k} X(z), \tag{5.64}$$

where $x(n) = 0$ for all $n < 0$.

This shows how to find the Z transform of a signal delayed by k samples, and further, that delaying a signal by k samples in the time domain corresponds to multiplying its spectrum by z^{-k} in the frequency domain. From equation (5.64) we see that a signal delayed by one sample can be represented as its Z transform multiplied by z^{-1}. We can interpret the quantity z^{-1} as the *unit sample delay operator* because $x(n) z^{-1} = x(n - 1)$.

5.12.2 Convolution Theorem

We can immediately use the shift theorem to prove that convolution in the time domain corresponds to multiplication in the frequency domain. The Z transforms of signals $x(n)$ and $y(n)$ are $X(z) = Z\{x(n)\}$ and $Y(z) = Z\{y(n)\}$, respectively, where $Z\{\ \}$ is the Z transform defined by equation (5.51). We want to show that the Z transform of the convolution of x and y is equal to the product of X and Y, or

$$Z\{x(n) * y(n)\} = X(z) Y(z).$$

By the definition of the Z transform, equation (5.51), we are seeking to demonstrate

$$Z\{x(n) * y(n)\} = \sum_{n=0}^{\infty} [x(n) * y(n)]z^{-n}.$$

Recall equation (4.1), which defines convolution

$$x(n) * y(n) = \sum_{m=0}^{\infty} x(m)y(n-m).$$

Substituting the right-hand side of the convolution equation into the previous equation, we obtain

$$Z\{x(n) * y(n)\} = \sum_{n=0}^{\infty} \sum_{m=0}^{\infty} x(m)y(n-m)z^{-n}.$$

It doesn't matter in which order we perform the summations, so we can switch their order.

$$Z\{x(n) * y(n)\} = \sum_{m=0}^{\infty} \sum_{n=0}^{\infty} x(m)y(n-m)z^{-n}.$$

Also, since $x(m)$ does not depend on n, it can be extracted to the outer summation. Notice that the inner summation (shaded) is now in the form of the Z transform of y delayed by m samples.

$$Z\{x(n) * y(n)\} = \sum_{m=0}^{\infty} x(m) \sum_{n=0}^{\infty} y(n-m)z^{-n}.$$

By the shift theorem of the Z transform, we can rewrite the shaded term as $z^{-m}Y(z)$. Then we have

$$Z\{x(n) * y(n)\} = \sum_{m=0}^{\infty} x(m)z^{-m}Y(z),$$

and since $Y(z)$ is a constant within the summation, we can factor it out:

$$Z\{x(n) * y(n)\} = \left(\sum_{m=0}^{\infty} x(m)z^{-m} \right) Y(z).$$

Notice that the shaded term is the Z transform of x. Finally we write

$$Z\{x(n) * y(n)\} = X(z)Y(z).$$

Reexpressing this using the $\Leftrightarrow$ operator, we have

$$x(n) * y(n) \Leftrightarrow X(z)Y(z) \qquad \text{Convolution Theorem} \quad (5.65)$$

In other words, convolution in the time domain corresponds to multiplication in the frequency domain, and vice versa.

5.12.3 Linearity

Since the Z transform is derived from the Fourier transform, and the Fourier transform is linear (see section 3.8.1), the Z transform is also linear. All linear transforms $\mathcal{L}\{\ldots\}$ share the following two properties:

- Gain scaling can be applied before or after applying the transform:

$$\mathcal{L}\{gx(\cdot)\} = g\mathcal{L}\{x(\cdot)\}. \tag{5.66}$$

For instance, gain-scaling a signal scales its spectrum.

- Addition can be performed before or after applying the transform:

$$\mathcal{L}\{x(\cdot) + y(\cdot)\} = \mathcal{L}\{x(\cdot)\} + \mathcal{L}\{y(\cdot)\}. \tag{5.67}$$

For instance, the spectrum of the sum of two signals is the same as the sum of their spectra.

5.12.4 Z Transform of the General LTI Filter

Now we can apply the Z transform to the general LTI filter equation.

If we let $\mathcal{Z}\{\ \}$ denote the Z transform, and

$$y(n) = a_0 x(n) + a_1 x(n-1) + \cdots + a_M x(n-M) - b_1 y(n-1) - \cdots - b_N y(n-N),$$

then

$$\mathcal{Z}\{y(n)\} = \mathcal{Z}\{a_0 x(n) + a_1 x(n-1) + \cdots + a_M x(n-M) - b_1 y(n-1) - \cdots - b_N y(n-N)\}.$$

By equation (5.67),

$$\mathcal{Z}\{y(n)\} = \mathcal{Z}\{a_0 x(n)\} + \mathcal{Z}\{a_1 x(n-1)\} + \cdots + \mathcal{Z}\{a_M x(n-M)\} - \mathcal{Z}\{b_1 y(n-1)\} - \cdots$$
$$-\mathcal{Z}\{b_N y(n-N)\},$$

and by equation (5.66),

$$\mathcal{Z}\{y(n)\} = a_0 \mathcal{Z}\{x(n)\} + a_1 \mathcal{Z}\{x(n-1)\} + \cdots + a_M \mathcal{Z}\{x(n-M)\} - b_1 \mathcal{Z}\{y(n-1)\} - \cdots$$
$$- b_N \mathcal{Z}\{y(n-N)\}.$$

By the shift theorem,

$$Z\{y(n)\} = a_0 Z\{x(n)\} + a_1 z^{-1} Z\{x(n)\} + \cdots + a_M z^{-M} Z\{x(n)\} - b_1 z^{-1} Z\{y(n)\} - \cdots$$

$$-b_N z^{-N} Z\{y(n)\}.$$

Replacing $Z\{y(n)\} = Y(z)$ and $Z\{x(n)\} = X(z)$, we have

$$Y(z) = a_0 X(z) + a_1 z^{-1} X(z) + \cdots + a_M z^{-M} X(z) - b_1 z^{-1} Y(z) - \cdots - b_N z^{-N} Y(z).$$

Putting all $Y(z)$ to one side, we have

$$Y(z) + b_1 z^{-1} Y(z) + \cdots + b_N z^{-N} Y(z) = a_0 X(z) + a_1 z^{-1} X(z) + \cdots + a_M z^{-M} X(z).$$

Factoring by common terms,

$$Y(z)[1 + b_1 z^{-1} + \cdots + b_N z^{-N}] = X(z)[a_0 + a_1 z^{-1} + \cdots + a_M z^{-M}].$$

Rearranging this equation to isolate $X(z)$ and $Y(z)$, and by equation (5.59), we have

$$H(z) = \frac{Y(z)}{X(z)} = \frac{a_0 + a_1 z^{-1} + \cdots + a_M z^{-M}}{1 + b_1 z^{-1} + \cdots + b_N z^{-N}} . \tag{5.68}$$

Equation (5.68) gives us a way to represent $H(z)$ as a ratio of input and output polynomials expressing the transfer function as the ratio of gains at different delays. We are now in possession of a general tool to map the location of the poles and zeros of a transfer function.

5.12.5   Finding the Roots

Any value of z that causes $Y(z)$ to be zero will cause $H(z)$ to be zero as well. Any value of z that causes $X(z)$ to be zero will cause $H(z)$ to become infinite and hence indicates the location of a pole. The closer an input frequency comes to a pole, the more it is amplified by the filter; the closer to a zero, the more attenuated. The values of the independent variable that make a polynomial zero are called the *roots* of that polynomial. For a polynomial of degree N, there are no more than N roots, which may or may not be distinct and may or may not be complex.[1] We can take the two polynomials for $Y(z)$ and $X(z)$, and since we know their degrees are N and M, respectively, we can write an equation that reveals the roots of these polynomials.

We start with equation (5.68). It would be convenient if the numerator and denominator of equation (5.68) were more alike, but the numerator polynomial starts with a_0, and the denominator polynomial starts with b_1. This can be rectified by factoring a_0 from the numerator. Here is just the numerator from equation (5.68): $a_0 + a_1 z^{-1} + \cdots + a_M z^{-M}$. We can divide this by any amount so long as we also multiply it by the same amount. So divide and multiply it by a_0:

$$a_0 \left(\frac{a_0 + a_1 z^{-1} + \cdots + a_M z^{-M}}{a_0} \right) = a_0 \left(\frac{a_0}{a_0} + \frac{a_1}{a_0} z^{-1} + \cdots + \frac{a_M}{a_0} z^{-M} \right)$$

$$= a_0\left(1 + \frac{a_1}{a_0}z^{-1} + \cdots + \frac{a_M}{a_0}z^{-M}\right).$$

Now, for cosmetic reasons, let $\alpha_i = a_i/a_0$, and let $g = a_0$, and the numerator looks like this.

$$g(1 + \alpha_1 z^{-1} + \cdots + \alpha_M z^{-M}).$$

Recombining the numerator with the denominator in equation (5.68), we have

$$H(z) = g\frac{1 + \alpha_1 z^{-1} + \cdots + \alpha_M z^{-M}}{1 + b_1 z^{-1} + \cdots + b_N z^{-N}} = \frac{N(z)}{D(z)}. \tag{5.69}$$

We know that the numerator polynomial $N(z)$ has M roots, and the denominator polynomial $D(z)$ has N roots, and we are in striking distance of showing them explicitly, but we need one more tool first.

5.12.6 Factoring Polynomials

In equation (5.69), I've identified the numerator $N(z)$ and denominator $D(z)$ polynomials separately so we can handle them individually. We see that they are expressed as polynomials in expanded form. What would they look like in factored form? Let's take a simpler example of a general polynomial for a thought experiment. Remember from algebra that $(x + ay)(x + ay)$ can be expanded to

$$x^2 + axy$$
$$\frac{+\ axy\ +\ a^2 y^2}{x^2 + 2axy + a^2 y^2}$$

If we let $x = 1$, then the expanded polynomial would be $1 + 2ay + a^2y^2$, and its factored form would be $(1 + ay)(1 + ay)$. Clearly, we could extend the factored form to any degree we wanted, $(1 + ay)(1 + ay)\ldots$, and we'd end up with a corresponding polynomial expansion. To express the factored form of large polynomials, we can use the notation

$$(1 + ay)(1 + ay)\ldots = \prod(1 + ay).$$

We can also set a limit for how many terms to multiply by decorating the product symbol with expressions above and below it to indicate how many products to form. We can express the polynomial $N(z)$ as

$$1 + \alpha_1 z^{-1} + \cdots + \alpha_M z^{-M} = \prod_{n=1}^{M}(1 - Q_n z^{-1}),$$

where Q_n are the zeros of the polynomial.

Now we are in a position to write equation (5.69) in its factored form:

$$H(z) = \frac{N(z)}{D(z)} = g\,\frac{\displaystyle\prod_{n=1}^{M}1 - Q_n z^{-1}}{\displaystyle\prod_{n=1}^{N}1 - P_n z^{-1}}, \tag{5.70}$$

where Q_n are the zeros of $H(z)$, and hence P_n are the poles of $H(z)$.

Now consider what happens when we take equation (5.70), which represents the factored form of $H(z)$, and set $z = e^{i\omega T}$ to get its frequency response. Limiting z to lie only on the complex unit circle means we only evaluate H on the unit circle. With this value of z, we have

$$H(e^{-i\omega T}) = \frac{N(e^{-i\omega T})}{D(e^{-i\omega T})} = g\,\frac{\displaystyle\prod_{n=1}^{M}1 - Q_n e^{-i\omega T}}{\displaystyle\prod_{n=1}^{N}1 - P_n e^{-i\omega T}}, \tag{5.71}$$

which expresses how the transfer function of H responds to a phasor with frequency ω in terms of its poles and zeros.

5.12.7 Determining the Gain of a Filter

In equation (5.71), $H(e^{-i\omega T})$ is still a complex function of a complex variable, but if we take its magnitude, it reduces to a three-dimensional function. Going back to figure 5.19, the frequency response of the filter corresponds only to those points touched by the unit circle as it is pushed up by the poles and drawn down by the zeros. So we can express the gain of the filter as

$$\left|H(e^{-i\omega T})\right| = g\,\frac{\left|\displaystyle\prod_{n=1}^{M}1 - Q_n e^{-i\omega T}\right|}{\left|\displaystyle\prod_{n=1}^{N}1 - P_n e^{-i\omega T}\right|}.$$

Next we want to modify this so that the effects of P and Q are more obvious. In algebra, the value of a fraction is unchanged if the numerator and denominator are scaled by the same amount, so we can scale it by

$$\frac{\left|e^{iM\omega T}\right|}{\left|e^{iN\omega T}\right|} = \frac{1}{1}.$$

Does that only work if M and N are the same? No, because we are taking the magnitudes of the numerator and denominator separately, and $\left| e^{i \cdot \text{anything}} \right| = 1$. So now we can say,

$$
\left| H(e^{-i\omega T}) \right| \;=\; g \frac{\left| e^{iM\omega T} \right|}{\left| e^{iN\omega T} \right|} \frac{\left| \displaystyle\prod_{n=1}^{M} 1 - Q_n e^{-i\omega T} \right|}{\left| \displaystyle\prod_{n=1}^{N} 1 - P_n e^{-i\omega T} \right|} . \tag{5.72}
$$

Notice that $e^{iM\omega T}$ can be expressed as

$$
\prod_{n=1}^{M} e^{i\omega T},
$$

and the same for $e^{iN\omega T}$. This means there is exactly one term $e^{i\omega T}$ for every term in the numerator and the denominator of equation (5.72), and when we substitute them in, we have

$$
\left| H(e^{-i\omega T}) \right| \;=\; g \frac{\left| \displaystyle\prod_{n=1}^{M} e^{i\omega T}(1 - Q_n e^{-i\omega T}) \right|}{\left| \displaystyle\prod_{n=1}^{N} e^{i\omega T}(1 - P_n e^{-i\omega T}) \right|} \;=\; g \frac{\left| \displaystyle\prod_{n=1}^{M} e^{i\omega T} - Q_n \right|}{\left| \displaystyle\prod_{n=1}^{N} e^{i\omega T} - P_n \right|} .
$$

because $e^{i\omega T} \cdot 1 = e^{i\omega T}$, and $e^{i\omega T} \cdot Q_n e^{-i\omega T} = Q_n$.

Clearing away the rubble, we're left with

$$
\left| H(e^{-i\omega T}) \right| \;=\; g \frac{\left| \displaystyle\prod_{n=1}^{M} e^{i\omega T} - Q_n \right|}{\left| \displaystyle\prod_{n=1}^{N} e^{i\omega T} - P_n \right|} . \tag{5.73}
$$

Equation (5.73) clearly shows the zeros Q and the poles P. But it's a little awkward to speak of $\left| H(e^{-i\omega T}) \right|$ as "the magnitude of the transfer function evaluated on the unit circle" when what we're really saying is much simpler: equation (5.73) shows the gain of the filter as a function of frequency ω. Since we know that the gain of the filter is a function of real frequency, we can define $G(\omega) = \left| H(e^{-i\omega T}) \right|$. Now we can expand equation (5.73) to

$$
\begin{aligned}
G(\omega) &= \left| H(e^{-i\omega T}) \right| \\
&= \frac{\left| (e^{i\omega T} - Q_1) \right| \left| (e^{i\omega T} - Q_2) \right| \ldots \left| (e^{i\omega T} - Q_M) \right|}{\left| (e^{i\omega T} - P_1) \right| \left| (e^{i\omega T} - P_2) \right| \ldots \left| (e^{i\omega T} - P_N) \right|} .
\end{aligned} \tag{5.74}
$$

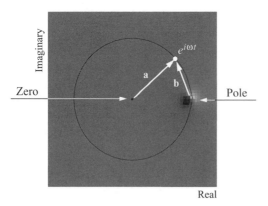

Figure 5.20
Gain as a ratio of zeros and poles.

This can be interpreted as follows: $e^{i\omega T}$ is a point on the unit circle corresponding to some frequency ω. Each Q_n is a point corresponding to the nth zero, and each P_n is the nth pole. The expression $\left| e^{i\omega T} - Q_n \right|$ is the magnitude (length) of a vector drawn from the nth zero to the point $e^{i\omega T}$ representing the frequency under examination, ω.

__The gain G(ω) of a filter at frequency ω is the product of the magnitudes of the vectors drawn from the zeros to the point $e^{i\omega T}$ divided by the product of the magnitudes of the vectors drawn from the poles to the point $e^{i\omega T}$.__

Figure 5.20 is a view of figure 5.19 taken from directly overhead. The gain of the filter corresponds only to the height of those points above the unit circle. Since in this case there are only one zero and one pole, the gain of the filter at frequency ω is a function of the ratio of the magnitude of the vector drawn from the zero to the point $e^{i\omega T}$ (call it a) divided by the magnitude of the vector drawn from the pole to the point $e^{i\omega T}$ (call it b).

Figure 5.21 shows the height of the function above the unit circle, as though the unit circle were unwrapped like an orange peel. As $e^{i\omega T}$ approaches 0, the ratio a/b goes to infinity, and as $e^{i\omega T}$ approaches π, the ratio a/b goes to $1/2$.

5.12.8   Determining the Phase Response of the Transfer Function

There is a story for phase response comparable to the preceding description of frequency response. The phase response is a function of sums and differences of angles:

$$\angle H(e^{i\omega T}) = \sum_{n=1}^{M} \angle(e^{i\omega T} - Q_n) - \sum_{n=1}^{N} \angle(e^{i\omega T} - P_n) + (N-M)\omega T. \tag{5.75}$$

But why should we have to say "the angle of the transfer function evaluated on the unit circle" when what we're really saying is much simpler: equation (5.75) shows the phase response of the filter

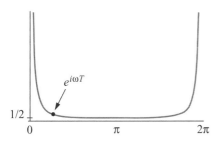

Figure 5.21
Height of function on a unit circle.

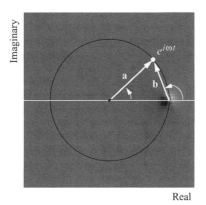

Real

Figure 5.22
Phase as a sum and difference of zeros and poles.

as a function of frequency ω. Since we know that the phase response of the filter is a function of real frequency ω, we can define $\Theta(\omega) = \angle H(e^{i\omega T})$. Now, if we let $\theta_n = \angle(e^{i\omega T} - Q_n)$, and $\phi_n = \angle(e^{i\omega T} - P_n)$, then we can write

$$\Theta(\omega) = \sum_{n=1}^{M} \theta_n - \sum_{n=1}^{N} \phi_n + (N - M)\omega T. \tag{5.76}$$

For some frequency ω, the phase response $\Theta(\omega)$ corresponds to the sum of the angles of the vectors from the zeros to the point $e^{i\omega T}$ minus the sum of the angles of the vectors from the poles to the point $e^{i\omega T}$.

The term $(N - M)\omega T$ is an offset based on the number of poles and zeros and the frequency ω. But observe that this is a linear phase term and thus represents pure delay. If the numbers of zeros M and poles N are equal (they usually are), this term cancels out, so it is usually insignificant in practice and can be ignored.

Figure 5.22 shows how to interpret equation (5.76). At the indicated frequency, the phase response is just $\angle a - \angle b$.

5.12.9 A Better Filter

We have come to an understanding of the Z transform by examining the unit step function (see section 5.11.5). Although it may be a good example to ponder, in practice it requires the summation of an infinite number of terms, and its gain becomes infinite as the frequency goes toward 0 Hz. The problem, of course, is the location of the pole right on top of $z = 1$. The pole is there because the Z transform of the unit step function is $1/(1 - z^{-1})$, and when the value of $z = 1$, the value of its frequency response is infinite. If we could simply move the location of the pole away from right on top of the unit circle, we'd have a more usable filter. Suppose we introduce a variable g to scale the position of the pole:

$$X(z) = \frac{1}{1 - gz^{-1}}, \qquad |z| > |g|. \tag{5.77}$$

Now $X(z)$ has a pole at $z = g$. For instance, if we set $g = 1/2$, then the Z transform has a pole at $z = 1/2$. As soon as the tip of the pole moves off $z = 1$, even just an iota, the frequency response of the filter at 0 Hz is no longer infinite, merely very large, because the frequency response of the filter is defined as the magnitude of the Z transform *on the unit circle*. The further we move the pole away from the unit circle, the less severe will be the gain of the filter. The Z transform described by equation (5.77) still has a zero at $z = 0$, as before.

We have in equation (5.77) a description of the Z transform of a filter we'd like to implement. We used equation (5.55) to go from the filter sequence to the Z transform. We've seen that it is possible to derive a Z transform from a filter sequence, but here we want to derive a filter sequence from a Z transform. So we need to reverse-engineer equation (5.77). Fortunately, equation (5.55) shows us how to do it. Here is (5.55) for convenience:

$$\sum_{n=0}^{\infty} ar^{-n} = \frac{a}{1 - r^{-1}}, \qquad |r| > 1.$$

Setting $a = 1$ and substituting $r = gz$, we get

$$\sum_{n=0}^{\infty} gz^{-n} = \frac{1}{1 - gz^{-1}}, \qquad |z| > |g|$$

$$= \sum_{n=0}^{\infty} (gz^{-1})^{n} \tag{5.78}$$

$$= \sum_{n=0}^{\infty} g^{n}z^{-n}$$

$$= g^{0}z^{-0} + g^{1}z^{-1} + g^{2}z^{-2} + \cdots + g^{n}z^{-n} + \cdots.$$

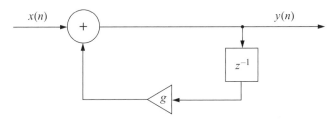

Figure 5.23
Simple recursive filter.

By the convolution theorem, we know that equation (5.78) is equivalent to the sequence

$$\{g^0 z^{-0}, g^1 z^{-1}, g^2 z^{-2}, \ldots, g^n z^{-n}, \ldots\}$$

(see section 5.10). This shows that the filter is an infinite exponential sequence g^n delayed by a corresponding z^{-n} for each successive power n.

Suppose we set $g = 0.9$. Then the sequence would be $\{1, 0.9 + 0.81 + 0.729 + \cdots\}$. At each stage in the sequence, we could compute the next value g^{n+1} by multiplying the current value g^n by g, then delay it by z^{-n}. Figure 5.23 shows a simple apparatus that could perform this process recursively. If, for example, we set $g = 0.9$ and drive the input of this filter with an impulse, $\delta(n) = \{1, 0, 0, \ldots\}$, the output $y(n)$ would be

$$y(n) = 1, 0.9 + 0.81 + 0.729 + \cdots.$$

The box labeled z^{-1} stores the value of the previous sample, and g scales the previous sample before adding it to the current sample. The output of the filter is the sum of the scaled previous sample and the current sample. Expressing this mathematically, we have the difference equation

$$y(n) = x(n) + gy(n-1). \tag{5.79}$$

It can be shown that the corresponding system function for this filter is

$$H(z) = \frac{1}{1 - gz^{-1}},$$

as we expect. The filter is causal, and there is a pole at $z = g$, which means that the region of convergence must be of the form $|z| > |g|$. The impulse response of the filter is $h(n) = g^n u(n)$. This simple recursive filter smooths sample sequences, as does the simple averaging filter considered earlier in this chapter, but note that $h(n)$ for positive n is nonzero for an infinite duration, and this is therefore an infinite impulse response (IIR) filter.

We could also think of this filter structure as an *accumulator,* which is a computational device that adds its current value and its input to obtain its new value. (For example, the M+ key on some common pocket calcuators acts as an accumulator.) We can think of this filter structure as accumulating the energy of the input signal over time. We would be slightly closer to the mark if we described it as a "leaky" accumulator because, though it accumulates successive sums through time, the coefficient g bleeds away some of the energy stored in the filter at a constant rate, effectively dissipating energy out of the accumulator over time.

Note that for stability, $|g| < 1$. Remember the tape recorder example where the output is fed back to the input. In fact, figure 5.23 is equivalent to a tape recorder with its output fed back to its input through a volume control. For the tape recorder, the equivalent to having $|g| = 1$ is when the volume control is set so that the output only grows in volume when new sound is input to the tape loop; otherwise the volume remains constant. This is called *regeneration.* If $|g| = 1$, then whatever energy goes into the filter remains forever—it is not bled away—and energy arriving from the input signal is accumulated. Positive signal values drive the level in the accumulator more positive, and negative values drive it more negative. So long as the average value fed to the filter is zero, the average level in the filter accumulator remains the same, but if it is unbalanced, either more often positive than negative or vice versa, then the level in the filter will accumulate in the direction of the imbalance. If it grows indefinitely, it may exceed the capacity of the filter to store energy, and the filter will become unstable. When the input stops, the filter will still contain all the energy, and it will never die away.

If $|g| > 1$, then even a single nonzero impulse will cause the filter's energy to grow without bounds as long as the filter runs. Using the tape recorder example, if the volume control is $|g| > 1$, then any energy in the system—a click or even just the hum and tape hiss of the equipment—will cause the energy in the tape to grow until the magnetic particles on the tape cannot store any stronger flux, and the sound starts to distort.

5.13   Filter Families

Previous sections have covered filter theory in general and have presented simple FIR and IIR structures as thought experiments. Another common classification of filters is to group them by the structure of their Z transforms.

5.13.1   One-Zero Filter

Let's take the simple averaging filter and add gain coefficients a_0 and a_1:

$$y(n) = a_0 x(n) + a_1 x(n-1), \qquad n = 0, 1, 2, 3, \ldots. \tag{5.80}$$

Its system diagram is shown in figure 5.24.

Its Z transform is

$$Y(z) = a_0 X(z) z^0 + a_1 X(z) z^{-1}.$$

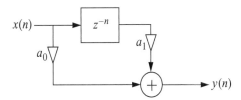

Figure 5.24
System diagram of a one-zero filter.

Its transfer function is

$$H(z) = \frac{Y(z)}{X(z)} = \frac{a_0 X(z)z^0 + a_1 X(z)z^{-1}}{X(z)}$$

$$= \frac{X(z)(a_0 z^0 + a_1 z^{-1})}{X(z)} \tag{5.81}$$

$$= a_0 + a_1 z^{-1}.$$

We obtain the frequency response by setting $z = e^{i\omega}$ so that $H(z) = a_0 + a_1 e^{-i\omega}$, that is, by evaluating the Z transform on the unit circle. This says that the frequency response of the filter is "the value of a_0 plus a_1 times the complex sinusoid $e^{-i\omega}$." For instance, if we set $\omega = 0$, $a_0 = 1$, and $a_1 = 1$, then the response H is $1 + 1 \cdot e^0 = 2$, that is, the filter amplifies energy at 0 Hz by a gain factor of 2. For $a_0 = 1$, $a_1 = 1$, and $\omega = \pi$ (corresponding to half the sampling rate) the frequency response is $1 + 1 \cdot e^{-i\pi} = 1 + 1 (-1) = 0$. So this is a lowpass filter. Values of ω in between move gradually from 2 to 0 as ω goes from 0 to π.

Notice that we can control the lowpass effect by changing a_1. If we set $a_1 = 0$ and $a_0 = 1$, then the two test values of ω are

$$\omega = 0 \qquad 1 + (0e^0) = 1$$

$$\omega = \pi \qquad 1 + [0(-1)] = 1$$

We note that the gain is unity regardless of the frequency. This is easy to see just looking at the difference equation, equation (5.80). If we set $a_1 = 0$, we effectively eliminate the delayed sample $x(n-1)$, and the filter is reduced to $y(n) = a_0 x(n)$. In this case we have a *zeroth-order filter*, otherwise known as a volume control, where the volume is controlled by a_0.

What would happen if we let $a_1 < 0$? Let's try setting $a_1 = -1$, and $a_0 = 1$. Now as we vary ω from 0 to π we have

$$\omega = 0 \qquad 1 + [(-1)e^0] = 0 \qquad \text{0 Hz is canceled}$$

$$\omega = \pi \qquad 1 + [(-1)(-1)] = 2 \qquad \pi \text{ Hz is doubled}$$

We see that we have created a highpass filter. So this filter can act as a lowpass or a highpass filter, depending upon the sign of a_1.

But we've only looked at two frequencies, and what we should do now is to derive the transfer function of the filter to see how it would behave for any value of a_0 and a_1 for any frequency. Returning to equation (5.81), we saw that $H(z) = a_0 + a_1 z^{-1}$, and if we restrict z to points on the unit circle, we know we can say $H(e^{i\omega}) = a_0 + a_1 e^{-i\omega}$. Then, by Euler's formula, we can also say

$$H(e^{i\omega}) = (a_0 + a_1 \cos\omega) + i(a_1 \sin\omega). \tag{5.82}$$

Now we see that H is the complex response to a phasor input at all frequencies ω. H is a four-dimensional function. It would be more convenient to view its magnitude because this yields a real function of the gain of the filter. Let's call the gain of the filter $G(\omega)$. To take the magnitude of a complex function, we apply the Pythagorean theorem:

$$
\begin{aligned}
G(\omega) &= \sqrt{(\text{Real part})^2 + (\text{Imaginary part})^2} \\[2mm]
&= \sqrt{(a_0 + a_1 \cos\omega)^2 + (a_1 \sin\omega)^2} \\[2mm]
&= \sqrt{(a_0 + a_1 \cos\omega)(a_0 + a_1 \cos\omega) + (a_1 \sin\omega)^2} \quad \text{Expansion} \\[2mm]
&= \sqrt{a_0^2 + 2a_0 a_1 \cos\omega + a_1^2 \cos^2\omega + a_1^2 \sin^2\omega} \quad \text{Cross-product} \\[2mm]
&= \sqrt{a_0^2 + 2a_0 a_1 \cos\omega + a_1^2(\cos^2\omega + \sin^2\omega)}. \quad \text{Factor}
\end{aligned}
$$

But $\cos^2\omega + \sin^2\omega = 1$ for any ω (see appendix, section A.4.8), so

$$G(\omega) = \sqrt{a_0^2 + a_1^2 + 2a_0 a_1 \cos\omega}. \tag{5.83}$$

So equation (5.83) describes the frequency response of the one-zero filter for all frequencies ω and all values of the coefficients a_0 and a_1. Now, if we let $a_0 = a_1 = 1$, we get

$$G(\omega) = \sqrt{2 + 2\cos\omega}.$$

Plotting this, we see that for $1 \leq \omega \leq$. we have the cosine curve shown in figure 5.25. Thus, for these settings of a_0 and a_1, the frequency response is a lowpass filter.

Now if we set $a_0 = 1$, and $a_1 = -1$, we get

$$G(\omega) = \sqrt{2 - 2\cos\omega}.$$

Plotting this, we get the cosine curve shown in figure 5.26. Thus, for these settings of a_0 and a_1, the frequency response is a highpass filter.

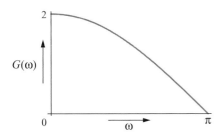

Figure 5.25
Lowpass filter response.

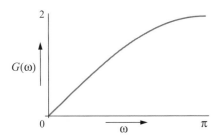

Figure 5.26
Highpass filter response.

We can plot equation (5.83), the general frequency response of a one-zero filter, by applying appropriate settings of ω, a_0, and a_1. For instance, figure 5.27a shows equation (5.83) for the values $a_0 = 1$, $-1 \leq a_1 \leq 1$, and $0 \leq \omega \leq \pi$. We can see that when $0 < a_1 \leq 1$ the filter is lowpass, but when $-1 \leq a_1 < 0$ the filter is highpass. When $a_1 = 0$, the filter is unity gain at all frequencies. This can be seen in figure 5.27b, which shows a perspective as though sighting along the arrow in figure 5.27a.

We can also use equation (5.82) to determine the phase response.

$$\Theta(\omega) = \tan^{-1}\left(-\frac{\text{Imaginary part}}{\text{Real part}}\right)$$

$$= \tan^{-1}\frac{-a_1 \sin\omega}{a_0 + a_1 \cos\omega}.$$

(5.84)

For $a_1 = -1$, $a_0 = 1$, we have

$$\Theta(\omega) = \tan^{-1}\frac{\sin\omega}{1 - \cos\omega},$$

a)

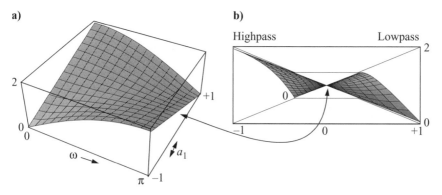

b)

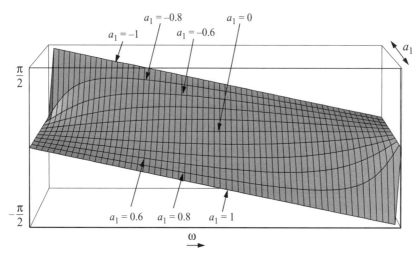

Figure 5.27
Frequency response of a general one-zero filter.

Figure 5.28
Phase response of a one-zero filter.

and for $a_1 = 1$, $a_0 = 1$, we have

$$\Theta(\omega) = \tan^{-1}\frac{-\sin\omega}{1 + \cos\omega}.$$

A plot of equation (5.84) for $-1 \le a_1 \le 1$ and $0 \le \omega \le \pi$ is given in figure 5.28.

Where is the zero in this one-zero filter? We find it by looking where the transfer function $H(z) = 0$. For this filter, the transfer function is

$$H(z) = a_0 + a_1 z^{-1}.$$

What values will cause $H(z)$ to be zero? If we simplify by setting $a_0 = 1$, then the transfer function becomes $H(z) = 1 + a_1 z^{-1}$. Now, if we could get the term $a_1 z^{-1} = -1$, we'd have $H(z) = 1 - 1 = 0$. This can be accomplished by solving for z, which yields $z = -a_1$. Sure enough, this causes $H(z)$ to be zero because

$$H(z) = 1 + a_1(-a_1)^{-1}$$

$$= 1 + a_1\left(-\frac{1}{a_1}\right)$$

$$= 1 - \frac{a_1}{a_1}$$

$$= 1 - 1$$

$$= 0.$$

To summarize: the root of the polynomial $H(z) = a_0 + a_1 z^{-1}$ is $z = -a_1$ because it is at this value that the filter has a *zero of transmission*. That is, the filter *blocks the input from being transmitted to the output* when $z = -a_1$. This means the frequency response of the filter depends upon the value we assign to a_1.

If we set $a_1 = 1$, the zero of transmission is at $z = -a_1 = -1 + 0i$, which corresponds to the point on the unit circle of $e^{i\pi} = -1 + 0i$. Remembering that $e^{i\pi}$ corresponds to the frequency $f_s/2$, this means that the filter has a zero of transmission at the Nyquist frequency, that is, at the highest representable frequency; therefore it is a lowpass filter.

If we set $a_1 = 0$, the transfer function is reduced to just $H(z) = a_0 = 1$, and no filtering takes place—the input signal is just copied from the input to the output. (Remember, we are just considering $a_0 = 1$ in this discussion. If a_0 were a different value, the gain of the signal would be scaled accordingly, but the scaling would not be frequency-dependent, that is, it would not be filtered.)

If we set $a_1 = -1$, the zero of transmission is at $z = -a_1 = 1 + 0i$, which corresponds to the point on the unit circle of $e^0 = 1 + 0i$. Remembering that e^0 corresponds to the frequency 0 Hz, this means that the filter has a zero of transmission at 0 Hz, that is, at the lowest representable frequency; therefore it is a highpass filter.

We have considered just a couple of important points on the transfer function. But what about all other values of $H(z)$? Figures 5.29 and 5.30 show 3-D graphics of the magnitudes of transfer functions of one-zero filters to answer this question. But there is one other important point on the transfer function we must first consider.

For a one-zero filter, the transfer function $H(z) = a_0 + a_1 z^{-1}$ also has a pole at $z = 0$. As $z \to 0$, the term $a_1 z^{-1}$ will become large, no matter what a_1 is (so long as it is not zero). Thus, when $z = 0$, $H(z) = \infty$, and it will also be large in the vicinity of zero. Because the pole is centered on $z = 0$, it influences all frequencies alike, so we can ignore it, which is why we can

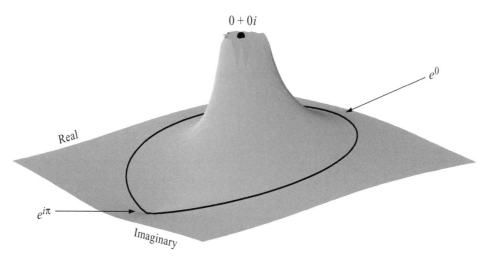

Figure 5.29
Magnitude of the transfer function of a one-zero lowpass filter.

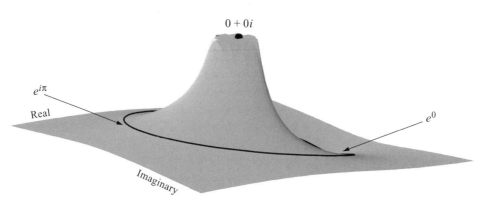

Figure 5.30
Magnitude of the transfer function of a one-zero highpass filter.

get away with calling this a one-zero filter (instead of having to call it a one-zero, one-pole filter). But we have to know that it is there in order to make sense out of the surface plots of the transfer functions shown in figures 5.29 and 5.30.

The transfer function is a complex function of a complex variable, requiring four dimensions, so the magnitude of the transfer function is plotted instead. Figure 5.29 shows the magnitude for a one-zero lowpass filter, and figure 5.30, for the one-zero highpass filter.

In each figure the pole is shown as a truncated tabletop mountain. The location of $z = 0$ is shown in the middle of the tabletop by a black dot. (In reality the pole is infinitely high; it is truncated to show it on the page.) From this plot we can observe the magnitude of the transfer function—the gain—for all complex values of z in the range $z = \pm(1.5 + 1.5i)$. The unit circle is superimposed upon the surface of the plot for reference, and the positions of 0 Hz and the Nyquist frequency are indicated. The height of the unit circle above the z plane indicates the amount of gain the filter will apply at that frequency.

5.13.2 One-Pole Filter

Whereas the one-zero filter forms the weighted average of two input samples to produce its output, the one-pole filter forms its output by computing the weighted sum of the current input sample and the most recent output sample. Its difference equation is

$$y(n) = ax(n) + by(n-1).$$ (5.85)

Current output   Current input   Previous output

Since it refers to its own past output, its next state depends upon its previous state, and it is *recursive*. Let's look at what this means for this filter's response to an impulse. First we construct the delta function according to equation (5.43) for all $n \geq 0$. Recall that the zeroth value of the delta function is 1 and the rest are all 0. If we set $a = 1$, and $x = \delta$, and compute the first few terms of the filter, we'll have the sequence shown in table 5.3.

The impulse response of the filter is

$$h(n) = b^n, \qquad n \geq 0.$$ (5.86)

Since we let n take any positive number (including ∞), that means the impulse response is infinite in length,[2] asymptotically approaching zero without ever reaching it (so long as $b \neq 0$). So this is an infinite impulse response (IIR) filter.

Table 5.3
One-Pole Impulse Response

n	$x(n) + by(n-1)$	$=$	$y(n)$
0	$1 + b \cdot 0$		1
1	$0 + b \cdot 1$		b
2	$0 + b \cdot b$		b^2
3	$0 + b \cdot b^2$		b^3
4	$0 + b \cdot b^3$		b^4
.	.		.
.	.		.
.	.		.

In order to derive the frequency response of the one-pole filter we would like to perform a Z transform of its impulse response. But the impulse response is infinite, and therefore the Z transform is also infinite:

$$H(z) = h(n)z^{-n}$$
$$= b^0 z^{-0} + b^1 z^{-1} + b^2 z^{-2} + \cdots + b^\infty z^{-\infty}. \tag{5.87}$$

We didn't face this problem with the one-zero filter because, as an FIR filter, it had an impulse response that died out after only two samples. A way out of the dilemma is to recognize that equation (5.87) is a geometric series. If we keep b and z less than ± 1, the values of later terms in this summation will become vanishingly small as n grows large. As demonstrated in section 5.11.4, such a progression will converge, and so we know that we can express equation (5.87) in closed form as follows. Remember that, in general,

$$\sum_{n=0}^{\infty} ar^n = \frac{a}{1-r}, \qquad |r| < 1,$$

and if we want to compute with r^{-n}, we have to switch it around:

$$\sum_{n=0}^{\infty} ar^{-n} = \frac{a}{1-r^{-1}}, \qquad |r| > 1. \tag{5.88}$$

Applying equation (5.88) to the infinite Z transform, we get

$$H(z) = h(n)z^{-n} = \frac{1}{1 - bz^{-1}}, \qquad |bz| > 1. \tag{5.89}$$

Another, swifter but less elucidating, way to derive the same thing is to write

$$y(n) = ax(n) + by(n-1),$$

$$Y(z) = aX(z) + bz^{-1}Y(z),$$

$$H(z) = \frac{X(z)}{Y(z)} = \frac{a}{1 - bz^{-1}}.$$

In any case, we now have the Z transform of the one-pole filter. If we set $z = e^{i\omega}$, we get the complex frequency response:

$$H(e^{i\omega}) = \frac{a}{1 - be^{-i\omega}}. \tag{5.90}$$

In order to get the gain response $G(\omega)$ of the filter, we take the magnitude of the complex frequency response. We do this by finding the square root of the sum of the squares of the real and imaginary parts of the complex frequency response.

We need to isolate the real and imaginary parts of equation (5.90). Say we have a complex variable z such that $z = p + iq$. Defining $\text{Re}\{z\} = p$, and $\text{Im}\{z\} = q$, allows us to extract the real and imaginary parts. Then, by the Pythagorean theorem, we could express gain as the magnitude of the complex frequency response:

$$G(\omega) = \left|H(e^{i\omega})\right| = \sqrt{[\text{Re}\{H(e^{i\omega})\}]^2 + [\text{Im}\{H(e^{i\omega})\}]^2}. \tag{5.91}$$

Now we can separate the real and imaginary parts of the complex frequency response equation (5.90) and plug them into equation (5.91). If we start with the complex frequency response, then by Euler's formula we can write

$$H(e^{i\omega}) = \frac{a}{1 - be^{-i\omega}} \tag{5.92}$$

$$= a\underbrace{\frac{1}{1 - b\cos\omega}}_{\text{Real}} + \underbrace{bi\sin\omega}_{\text{Imaginary}}.$$

Having separated the real and imaginary parts, we can now use the Pythagorean theorem to obtain the frequency response of the filter:

$$G(\omega) = a\frac{1}{\sqrt{(1 - b\cos\omega)^2 + (b\sin\omega)^2}}$$

$$= \frac{a}{\sqrt{1 - 2b\cos\omega + b^2\cos^2\omega + b^2\sin^2\omega}} \tag{5.93}$$

$$= \frac{a}{\sqrt{1 - 2b\cos\omega + b^2(\cos^2\omega + \sin^2\omega)}}$$

$$= \frac{a}{\sqrt{1 - 2b\cos\omega + b^2}}.$$

Similarly, working from equation (5.92), we can express the phase response as:

$$\Theta(\omega) = \tan^{-1}\frac{b\sin\omega}{1 - b\cos\omega}. \tag{5.94}$$

Where is the pole in the one-pole filter? Remember the transfer function is

$$H(z) = h(n)z^{-n} = \frac{a}{1 - bz^{-1}}, \qquad |bz| > 1,$$

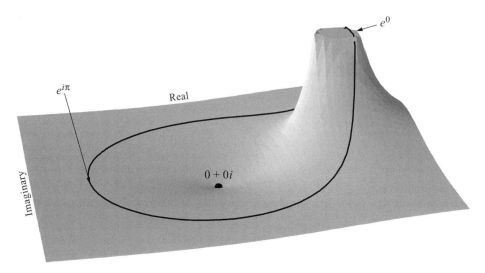

Figure 5.31
Magnitude of the transfer function of a one-pole lowpass filter.

and we're looking for what makes the denominator zero, which means we want to solve $1 - bz^{-1} = 0$ for b. Interestingly, the answer is $b = z$. To see it in action, we can set z to b and write

$$H(b) = \frac{a}{1 - b(1/b)} = \frac{a}{0} = \infty.$$

So there will be a pole wherever b is. Since b must be a real number, the pole will always be on the real axis. It makes sense that the sign of the b coefficient determines if the filter is highpass or lowpass: when $b > 0$, the pole will amplify frequencies near 0 Hz because it will be on the low-frequency side of the unit circle, toward e^0. When $b < 0$, the pole will amplify high frequencies because it will be on the side of the unit circle toward the Nyquist frequency, $e^{i\pi}$. The closer the peak of the pole comes to the unit circle, the greater the degree of amplification. Figure 5.31 shows the magnitude of the transfer function of the one-pole filter as a lowpass with $b = 0.9$, and figure 5.32 shows it as a highpass with $b = -0.8$.

Comparing the one-pole filter to the one-zero filter, we see that the one-pole filter boosts frequencies in the passband rather than attenuating the frequencies in the stopband, as does the one-zero filter.

Notice that there is a hidden zero in the one-pole filter at $z = 0$. It is hidden in the same sense that a pole at $z = 0$ is hidden in the middle of the one-zero filter: because the zero is in the center of the unit circle, it influences all frequencies equally, hence it can be ignored. We can prove the zero is there by solving $H(z)$ for $z = 0$:

$$H(0) = \frac{a}{1 - b(1/0)} = \frac{a}{1 - \infty}, \qquad b \neq 0.$$

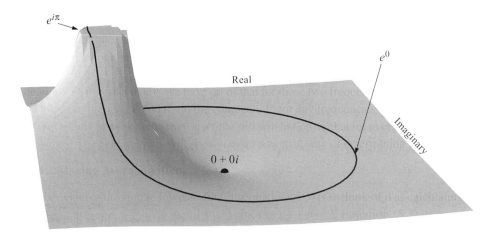

Figure 5.32
Magnitude of the transfer function of a one-pole highpass filter.

Because of the fact that the one-pole filter boosts gain, its use must be very carefully managed. It is customary to set a to $1 - |b|$ so that the peak gain is normalized to be approximately 1. In any event, a one-pole filter will only be stable if $|b| < 1$. Otherwise, the impulse response will be a geometrically increasing function that will eventually overflow the capacity of any real system. Consider the tape recorder in echo mode as an example of this system at work.

5.13.3   Two-Pole Filter

In the case of the one-pole filter, we saw that the pole slides along on the real axis. The closer it comes to the unit circle, the more pronounced its effect, but we are stuck with its one-dimensional mobility: we can boost low or high frequencies, but we can't single out any particular frequency for emphasis.

The two-pole filter introduces this desired capability. We can move two poles in *conjugate symmetry* to any point in the z plane to produce a resonance at that frequency. Saying that the poles move in conjugate symmetry means that they are mirror images above and below the real axis. The conjugate of a complex variable z is written $\bar{z}$, and if $z = a + bi$, then its conjugate is $\bar{z} = a - bi$. As with the one-pole filter, when the poles are far from the unit circle, their effect is muted: they affect a broader set of frequencies but do not boost any of them very much. As they move closer to the unit circle, the circle effectively cuts the poles high on their slopes, and the gain is correspondingly greater for a narrower band of frequencies. Thus, the magnitude of the slope where the poles intersect the unit circle determines the gain, while their angle θ determines the frequency of resonance.

The difference equation of the two-pole filter is

$$y(n) = ax(n) - b_1 y(n-1) - b_2 y(n-2). \tag{5.95}$$

Recalling the transfer function of the general filter equation,

$$H(z) = \frac{Y(z)}{X(z)} = \frac{a_0 + a_1 z^{-1} + \cdots + a_M z^{-M}}{1 + b_1 z^{-1} + \cdots + b_N z^{-N}},$$

we can see that the transfer function of the two-pole filter is

$$H(z) = \frac{Y(z)}{X(z)} = \frac{a_0}{1 + b_1 z^{-1} + b_2 z^{-2}}.$$

It is a two-pole filter because the denominator polynomial is of order 2. It will have two roots because it is an exponential function.

As long as b_1 and b_2 are real numbers, we know that if the two roots are complex, they will always be conjugates of each other. It's also entirely possible that for some values of b_1 and b_2 the roots will just be reals, but if they are complex, they will always be conjugates.

Therefore, we can factor the denominator polynomial like this:

$$X(z) = 1 + b_1 z^{-1} + b_2 z^{-2} = (1 - P_1 z^{-1})(1 - P_2 z^{-1})$$

for some values of P_1 and P_2 so long as they are conjugates of each other. P_1 and P_2 are in general complex. They must be conjugates of each other if their product is to produce b_1 and b_2 because b_1 and b_2 are reals, and the product of two conjugates is a real.

While we can see that the points P_1 and P_2 indicate the locations of poles, they do not provide much insight into how one would go about setting them to particular values in order to realize a filter with a particular resonance. It would be nice if we could find a way to set the location of the poles more intuitively. The one-pole and one-zero filters were relatively simple in this regard: all we had was a radius indicating how far the pole or zero was from the center of the z plane along the real axis. The two-pole filter requires another degree of freedom. We must be able to indicate

• The strength of the resonance in terms of the proximity of the poles to the unit circle

• The frequency of the resonance in terms of its angle on the unit circle

But these are the properties of polar coordinates. Perhaps if we recast the transfer function in terms of polar coordinates, we'd have an easier time understanding how to control this filter.

Let's represent points P_1 and P_2 in polar coordinates, such that $P_1 = Re^{i\theta}$, and $P_2 = Re^{-i\theta}$. Here we define R to be the common radius of both poles, and θ and $-\theta$ are their respective angles. Substituting these definitions of P_1 and P_2 into the transfer function yields:

$$H(z) = \frac{a_0}{(1 - P_1 z^{-1})(1 - P_2 z^{-1})} = \frac{a_0}{(1 - Re^{i\theta} z^{-1})(1 - Re^{-i\theta} z^{-1})}$$

$$= \frac{a_0}{1 - Re^{-i\theta}z^{-1} - Re^{i\theta}z^{-1} + R^2z^{-2}} = \frac{a_0}{\underbrace{1 - 2R(\cos\theta)\,z^{-1}}_{b_1} + \underbrace{R^2z^{-2}}_{b_2}}.$$

We can see from this that $b_1 = -2R\cos\theta$, and $b_2 = R^2$.

Note that to change the resonant frequency, we only need to change b_1 because that's the only term that has θ in it. But if we have to change the radius, we must change both b_1 and b_2.

Armed with this knowledge, we can rewrite the difference equation, substituting for b_1 and b_2:

$$y(n) = ax(n) - b_1 y(n-1) - b_2 y(n-2)$$

$$= ax(n) + 2R(\cos\theta)y(n-1) - R^2y(n-2).$$

The transfer function in terms of a radius R and an angle θ is:

$$H(z) = \frac{a_0}{1 - 2R(\cos\theta)\,z^{-1} + R^2z^{-2}}, \qquad \textit{Two-Pole Filter Transfer Function} \quad (5.96)$$

where R controls the bandwidth of the resonance, and θ determines its resonant frequency. Figure 5.33 shows the transfer function of the two-pole filter with $a_0 = 1$, $R = 0.8$, and $\theta = \pi/4$. As we can see, there is also a zero at $0 + 0i$, although its role is insignificant because it affects all frequencies alike.

Now, if we take the transfer function equation (5.96) and substitute $z = e^{i\theta}$, we can obtain the gain at the resonant frequency θ for different values of R:

$$H(e^{i\theta}) = \frac{a_0}{[(1-R)^2(1 - 2R(\cos 2\theta) + R^2)]^{1/2}}. \qquad (5.97)$$

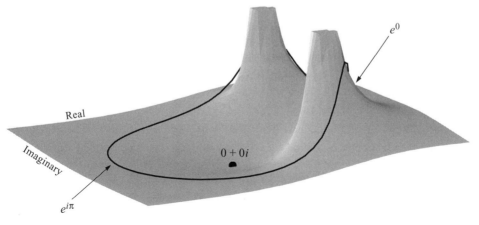

Figure 5.33
Transfer function of a two-pole filter.

If the location of the poles is complex, the two poles will be at conjugate locations, as shown in figure 5.33, and this is a bandpass filter. But if the poles are located on the real axis, the filter devolves into a simple lowpass filter. For instance, if we set $\theta = 0$, the two poles merge into one, and the frequency response for $\theta = 0$ is

$$
\begin{aligned}
H(e^{i\theta}) &= \frac{a_0}{[(1-R)^2(1-2R(\cos 2\theta)R^2)]^{1/2}} \\[2mm]
&= \frac{a_0}{[(1-R)^2(1-2R+R^2)]^{1/2}} \\[2mm]
&= \frac{a_0}{[(1-R)^2(1-R)^2]^{1/2}} \\[2mm]
&= \frac{a_0}{(1-R)^2}.
\end{aligned}
$$

So the upper bound of the gain of the resonant peak is $a_0/(1-R)^2$ when the poles lie on top of each other at $\theta = 0$. In this case, as R approaches 1, the gain approaches infinity at 0 Hz. It so happens that we get the complementary frequency response if we set $\theta = \pi$. In that case, as R approaches 1, the gain approaches infinity at the Nyquist frequency. So we see that when the poles are on the real axis, this filter behaves similarly to the one-pole filter.

If $R = 0$, the transfer function degenerates to $H(z) = a_0$, which is a zeroth-order filter, otherwise known as a frequency-independent gain control.

Now let's look at what happens as we vary the distance of the poles from the unit circle. We can intuitively see that the closer a pole is to the unit circle, the higher the unit circle will be lifted by that pole as it runs like a highway over it. When the pole is far from the unit circle, it is as though the circle runs over the "foothills" of the pole. The gain will be relatively broad and weak. As the peak of the pole approaches the unit circle, it becomes relatively taller and narrower.

The bandwidth of a filter is a measure of how broadly it amplifies various frequencies. The *bandwidth of a resonance* is the distance in frequency between two points that have half as much power as the peak (see figure 5.34). Scaling power by 1/2 is the same as attenuating amplitude by approximately -3 dB. If the gain of a filter at the peak is g, then half-power points are the frequencies surrounding the peak where the gain is $g/\sqrt{2}$.

Note that for the bandwidth of the filter to change, the proximity of the poles to the unit circle must change, which requires changing both b_1 and b_2.

The frequency response for the two-pole filter is derived as follows. Starting with the transfer function,

$$
H(z) = \frac{a}{1 + b_1 z^{-1} + b_2 z^{-2}},
$$

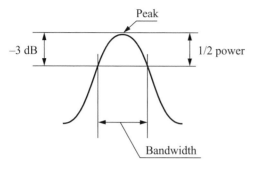

Figure 5.34
Bandwidth of a resonance.

we evaluate this for $e^{i\omega}$:

$$H(e^{i\omega}) = \frac{a}{1 + b_1 e^{-i\omega} + b_2 e^{-i2\omega}}$$

$$= \frac{a}{1 + b_1(\cos\omega) - ib_1(\sin\omega) + b_2(\cos 2\omega) - ib_2(\sin 2\omega)}$$

then apply the Pythagorean theorem to obtain the magnitude of the transfer function, which is $G(\omega)$, the frequency response of the filter:

$$G(\omega) = \frac{a}{\sqrt{[1 + b_1(\cos\omega) + b_2(\cos 2\omega)]^2 + [-b_1(\sin\omega) - b_2(\sin 2\omega)]^2}},$$

where $b_1 = -2R\cos\omega$, and $b_2 = R^2$.

The phase response of the filter is

$$\Theta(\omega) = -\tan^{-1}\frac{-b_1(\sin\omega) - b_2(\sin 2\omega)}{1 + b_1(\cos\omega) + b_2(\cos 2\omega)}. \tag{5.98}$$

This process can be extended to N-pole filters by observing the pattern of the transfer function of the two-pole filter:

$$H(z) = \frac{a}{1 + b_1 z^{-1} + b_2 z^{-2} + \cdots + b_n z^{-n}} = \frac{a}{1 + \sum_{n=1}^{N} b_n z^{-n}}. \quad \textit{N-Pole Filter Transfer Function} \tag{5.99}$$

5.13.4 Allpass Filter

It might seem strange to think that a filter having unity gain at every frequency could be useful for anything, but in fact *allpass filters* have important uses in modeling concert hall acoustics and

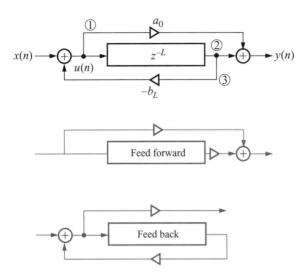

Figure 5.35
Allpass filter.

music synthesis, among other things. Their usefulness comes from their phase response, which can be arbitrarily designed to introduce frequency-based signal delays without changing amplitude.

Figure 5.35 shows that an allpass filter is the combination of a feedforward filter network and a feedback filter network sharing the same delay line. There are three contributions to the output of this filter, as marked in the figure:

- The feedforward path $y(n) \propto a_0 x(n)$
- The delay path, $y(n) \propto x(n-L)$
- Because of feedback, $y(n)$ receives a contribution from itself from z^{-L} samples ago, so $y(n) \propto b_L y(n-L)$.

Since these paths are all summed by the filter, we can write its difference equation by inspection as

$$y(n) = a_0 x(n) + x(n-L) + b_L y(n-L). \tag{5.100}$$

The transfer function of this filter is

$$H(z) = \frac{Y(z)}{X(z)} = \frac{a_0 + z^{-L}}{1 + b_L z^{-L}}. \tag{5.101}$$

By setting $z = e^{i\omega T}$, the magnitude frequency response for sample period T can be expressed as

$$\left| H(e^{i\omega T}) \right| = \left| \frac{a_0 + e^{-i\omega LT}}{1 + b_L e^{-i\omega LT}} \right|. \tag{5.102}$$

The allpass condition arises when the two coefficients are complex conjugates, $a_0 = \overline{b_L}$, or if the coefficients are real, when $a_0 = b_L$. If we let $\alpha = a_0 = \overline{b_L}$, we can write

$$\left| H(e^{i\omega T}) \right| = \left| \frac{\overline{\alpha} + e^{-i\omega LT}}{1 + \alpha e^{-i\omega LT}} \right|.$$

Since $1 + \alpha e^{-i\omega LT} = e^{i\omega LT} + \alpha$,

$$\left| H(e^{i\omega T}) \right| = \left| \frac{\overline{\alpha} + e^{-i\omega LT}}{e^{i\omega LT} + \alpha} \right|.$$

Finally, $\overline{\alpha} + e^{-i\omega LT} = \overline{\alpha + e^{i\omega LT}}$, and $|\overline{x}/x| = 1$, so

$$\left| H(e^{i\omega T}) \right| = \left| \frac{\overline{\alpha + e^{i\omega LT}}}{\alpha + e^{i\omega LT}} \right| = 1.$$

In general, if the frequency response of a filter is unity for all frequencies ω, it is said to be *loss-less* because it transfers all input signal energy to the output. By this criterion, the allpass filter with unity gain is lossless.

Summary

The frequency response of a filter shows which frequencies it passes and which it rejects. Characteristic patterns include highpass, lowpass, bandpass, band-reject, and allpass. Filters have an effect on the frequency and phase of components in a signal.

Filtering can be viewed in the frequency domain as multiplication of the spectrum of the input signal by the transfer function of a filter. It can be viewed in the time domain as convolution of a signal with the impulse response of a filter.

The output of a simple lowpass filter is equal to the current input plus the previous input. The remoteness in time of the previous input is a function of the sampling period of the input signal and filtering system. An experimental approach to finding the frequency response is to drive the filter with a variable-frequency oscillator and examine its output on a spectrum analyzer. With this black-box approach we observe that the filter passes low frequencies, but we don't know why.

By plugging the general sinusoid into the filter equation, we can discover a function that describes for any input what the frequency response will be.

Filters affect the amplitude and phase of the input signal in a frequency-selective manner. The transfer function of a filter is the combination of its frequency response and phase response. By factoring the transfer function, we can separate out the frequency response of a filter from its phase response, to observe them separately.

A filter is linear if its output is proportional to its input, and the output is the same whether the input signals are combined and then filtered, or filtered and then combined. A filter is time-invariant if no parameter of the filter transfer function changes with time. A filter is causal if it does not depend upon future inputs or outputs.

The time domain behavior of a filter is called its impulse response. Filters can only be of two kinds: finite impulse response (FIR) filters sum scaled delayed copies of their input to produce the output signal. The impulse response of such filters is therefore finite. Infinite impulse response (IIR) filters sum scaled delayed copies of their input and scaled delayed copies of their output to produce the output signal. The impulse response of such filters is therefore infinite. The canonical filter is the combination of an FIR filter and an IIR filter.

We can break down a filter's impulse response into a linear combination of delayed scaled impulse functions. Doing so, we observe that the filter convolves the input signal with the filter's impulse response.

The frequency response of a filter is the ratio of the frequency response of the output to the frequency response of the input.

The Z transform is an extension of the Fourier transform that evaluates the frequency response of a filter at every point on the complex plane (the z plane). The magnitude of the Z transform provides a three-dimensional representation of the complex frequency response of the input signal. This function resembles a tent with the zeros acting as tent pegs and the poles acting as tent poles. It characterizes how a filter responds to all frequencies. The transfer function of a linear time-invariant filter is defined as the Z transform of its impulse response.

The shift theorem of the Z transform states that a real signal delayed by k samples has the Z transform of the original unshifted signal multiplied by z^{-k}. From the shift theorem, we see that convolution in the time domain corresponds to multiplication in the frequency domain.

The transfer function of a filter can be seen as a ratio of input and output polynomials expressing the transfer function as the ratio of gains at different delays. Any value of the transfer function that causes its numerator to be zero will cause its output to be zero, indicating the location of a zero, while any value that causes the denominator to be zero will cause its output to become infinite, indicating the location of a pole. The closer an input frequency comes to a pole, the more it is amplified by the filter; the closer to a zero, the more attenuated.

The gain of a filter at some frequency is equal to the product of the magnitudes of the vectors drawn from the zeros to its corresponding point on the unit circle, divided by the product of the magnitudes of the vectors drawn from the poles to that same point. The phase response corresponds to the sum of the angles of the vectors from the zeros to the corresponding point on the unit circle, minus the sum of the angles of the vectors from the poles to the same point.

A common classification of filters is to group them by the structure of their Z transforms: one-zero filter (gradual lowpass or highpass filter), one-pole filter (potentially steeper lowpass or highpass filter), two-pole filter (resonator, can be bandpass or band-reject). The allpass filter passes all frequencies uniformly but with a frequency-dependent phase shift.

6 Resonance

There is geometry in the humming of the strings. There is music in the spacing of the spheres.
—Pythagoras

The study of resonance unlocks the deepest understanding of musical instrument sounds. It explains how sound propagates in rooms, why microphones and loudspeakers sound the way they do, and how our ears work. The subject can be boiled down to just three characteristics of vibration: displacement, velocity, and acceleration. Furthermore, these three qualities can be seen as just different aspects of the same underlying phenomenon. The unifying perspective is called the *derivative*. It provides a simpler and more coherent way to understand resonance than was given in volume 1, section 8.9, and it unifies a broad range of musical phenomena.

6.1 The Derivative

The velocity v and acceleration a of an object are determined by observing how the object's displacement varies with time (see volume 1, chapter 4). This suggests that displacement and time are the basic properties of motion, whereas velocity and acceleration are derivative properties. The basic English meaning of *derivative* is a fact that can be gleaned from other more basic facts. This simple definition actually agrees quite well with the mathematical interpretation of the term. Derivatives have appeared in various disguises in previous chapters; this chapter finally introduces them by name.

If a function $y(t)$ describes an object's displacement through time, then we can *derive* its velocity at some instant t by observing the slope of $y(t)$ at that instant (see volume 1, section 4.8). If $y(x)$ is a function that can be differentiated,[1]

*The **derivative** of a function y(x) is the slope of a line tangent to the function at x.*

This definition implies that we are selecting a *particular value* for x and finding the derivative (the slope of its tangent line) at that point. While that is a completely reasonable interpretation, we can also view the variable x as a place that could store *the range of all possible values*. For instance, we can think of a mailbox as a place to store a particular envelope or as an empty space that can store any envelope that fits. Using this broader view, if we take the derivative of $y(x)$ across all possible values of x, we will have created a function describing how the slope changes for all x.

For instance, say that $y(t)$ describes an object's displacement through time. If we take the derivative of $y(t)$ at every instant, we create a function describing how the object's velocity changes through time. If we let $v(t)$ be the object's velocity through time that we have so derived, then we can derive the object's acceleration by creating another function $a(t)$, which describes how the slope of $v(t)$ changes through time.

This process of taking derivatives could continue indefinitely. For instance, we can derive the object's rate of change of acceleration by observing how the slope of $a(t)$ changes through time. The phrase *rate of change* holds the key to understanding the derivative:

- Velocity measures the *time rate of change* of an object's *position*.
- Acceleration measures the *time rate of change* of an object's *velocity*.

The *rate of change* is defined as the slope that is tangent to the function at the point of interest. And if we look at how the slope changes at all points of interest along a curve, we create another curve that is its derivative.

Of course, we could apply the same operation to the derivative function just created. We could go on creating derivatives of derivatives forever. Because there can be any number of derivatives of a function, we must be able to uniquely identify them. Let's call *velocity* the *first derivative of displacement* with respect to time. Similarly, *acceleration* is the *first derivative of velocity* with respect to time. Now, since velocity is the first derivative of displacement, and acceleration is the first derivative of velocity, it stands to reason that acceleration is the *second derivative* of displacement with respect to time. This suggests that there are an infinite number of derivatives for any function including third derivatives, and so forth (figure 6.1). For example, the third derivative of displacement is called *jerk*.[2] Although there are no standardized terms for derivatives beyond the third, some have been suggested, such as *snap, crackle,* and *pop* for the fourth, fifth, and sixth derivatives, respectively, though one assumes the tongue must be firmly planted in the cheek to pronounce them properly.

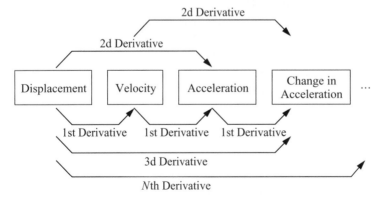

Figure 6.1
Chain of derivatives.

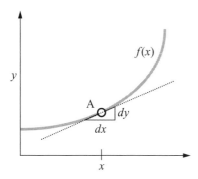

Figure 6.2
True tangent to a curve.

The derivative is a concept from *differential calculus*. It is used to study rates of change and to find the instantaneous rate of change of one quantity relative to another. *Integral calculus* is a related discipline used to study the accumulation of quantities, such as areas under a curve. We used integration in chapter 3 to study the Fourier transform. Integral and differential calculus are mirror images of each other.

6.1.1 Finding the Derivative

We can determine the derivative of a function by measuring the slope of a line tangent to the function at the point of interest. To make this more concrete, say we have a curve such as shown in figure 6.2, and we wish to find its derivative at x. Say that A $= f(x)$. So the task is to find the tangent of the curve at A and then measure its slope.

Whatever the slope is, we know it can be expressed as the ratio of some distance along the y-axis to a corresponding distance along the x-axis—the so-called "rise over run" of a slope. If we call the distance along the x-axis dx and the distance along the y-axis dy, then we can say that the slope is the ratio dy/dx.

But how in practice can we actually find the tangent of an arbitrary function? And then how do we determine its slope? Let's start with an *approximation* to the tangent and see where it leads.

Suppose that the line marked τ in figure 6.3 is the true tangent at point A, and the task is to discover this fact. First, we pick a point B somewhere farther up the curve, and connect a line between the two points such that A $= f(x)$ and B $= f(x + \Delta x)$, where Δx is an arbitrary-sized interval along the x-axis. Set Δx to a big enough value so we can get a good look at it. If we set $\Delta y = f(x + \Delta x) - f(x)$, then the ratio $\Delta y/\Delta x$ is not the same as dy/dx, the true tangent of the function at A, but it's close. Now, if we make Δx progressively smaller, then point B moves closer to point A along the curve, and the line between B and A becomes a better and better approximation of the true tangent at A.

In the limit as Δx gets *infinitely close* to zero (but doesn't reach zero), B also moves infinitely close to A (but doesn't reach A). As Δx shrinks, the difference between the slope of the line AB and the true tangent of A vanishes, and in the limit, they are equal.

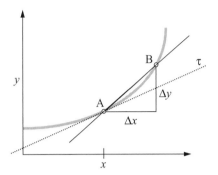

Figure 6.3
Approximating the tangent to a curve.

We have equated the true tangent dy/dx with the ratio we discovered by taking the approximation $\Delta y/\Delta x$ to the limit. Therefore we can write

$$\frac{dy}{dx} = \lim_{\Delta x \to 0} \frac{f(x + \Delta x) - f(x)}{\Delta x}.$$ *Derivative* (6.1)

6.1.2 Derivative Notation

We have accomplished a couple of things with equation (6.1). First, we discovered a way to find the derivative of an arbitrary differentiable function. Second, we can now use the ratio dy/dx as a convenient notation for the derivative.

The ratio dy/dx is not some strange mathematical jargon designed to perplex; dy is really just a distance—an infinitesimal distance—on the y-axis; and the same for dx on the x-axis. Though their lengths are infinitesimal, their *ratio* is not, because the size of their ratio is a function of their relative proportions, not their absolute size. dy/dx is actually just a fraction that expresses a ratio of two distances. And we discover its value by using the limit process on the right-hand side of equation (6.1).

Mathematicians commonly use two other notations for derivatives. Say we have a function $g(x)$. We can express the derivative as the slope of the function with respect to *some particular value* of x by writing

$$\frac{dg}{dx} = \dot{g} = g'.$$

Or, looking at *every possible value* of x at once, we can express the derivative as

$$\frac{d}{dx}g(x) = \dot{g}(x) = g'(x),$$

where both $\dot{g}(x)$ and $g'(x)$ are equivalently the derivative *functions* of $g(x)$.

Similarly, the second derivative of $g(x)$ at some particular value can be written

$$\frac{d^2 g}{dx^2} = \ddot{g} = g''.$$

The second derivative *function* can be expressed as

$$\frac{d^2}{dx^2} g(x) = \ddot{g}(x) = g''(x).$$

Sometimes it will be more convenient to express derivatives in terms of a so-called dummy variable that represents the value of the function. For instance, if $y = g(x)$, we can write the derivative with respect to x variously as

$$\frac{dy}{dx}, \qquad \frac{dg}{dx}, \quad \text{or} \quad \frac{d}{dx} g(x),$$

depending on what works best in context.

6.1.3 Velocity and Acceleration as Derivatives of Displacement

The notion of derivatives simplifies and clarifies the relations between displacement, velocity, and acceleration in an important way that I exploit in the discussion of resonance. If a function $s(t)$ represents the displacement s of a car through time t, the car's *velocity* with respect to t can be written

$$\frac{ds}{dt} = \dot{s} = s'. \qquad\qquad\qquad\qquad\qquad\qquad \textit{Velocity} \; (6.2)$$

(The three representations are equivalent.) Velocity is just the *time rate of change of displacement*.

Using the same function, we can express the *acceleration* of $s(t)$ with respect to t as

$$\frac{d^2 s}{dt^2} = \ddot{s} = s''. \qquad\qquad\qquad\qquad\qquad\qquad \textit{Acceleration} \; (6.3)$$

Acceleration is just the time rate of change of velocity, or the *time rate of change of the time rate of change* of displacement.

Suppose you are standing at the top of an incline, and I roll a ball up the incline toward you in an arc, so that it rolls smoothly to a stop at your feet, then accelerates smoothly away back down the incline. Say it covers a distance up and back of 1 meter in the span of 2 seconds. Suppose we observe from measurements that its displacement s from your feet at time t is the square of the elapsed time, that is, $s(t) = t^2$ (figure 6.4a).

If we track how the slope of this function changes over time, we obtain the function of its first derivative, velocity (figure 6.4b). Clearly, the velocity of the ball is zero when it is at your feet $t = 0$ and is decreasingly negative (decelerating) while it is rolling toward you and increasingly positive (accelerating) while it is rolling away. Thus, overall, the velocity grows steadily more

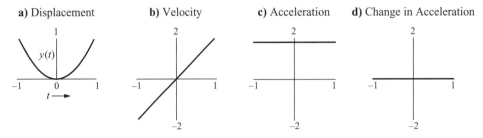

Figure 6.4
Deriving velocity and acceleration from displacement.

positive through time. We can also observe from measurements that the velocity has a constant slope. In this case, we observe the velocity is $2t$.

If we track how the slope of velocity changes over time (it's a straight line, so the slope doesn't change), we obtain acceleration (figure 6.4c). Since the slope of velocity doesn't change, the acceleration is constant through time. In this case, the acceleration is 2. And if we track how the slope of acceleration changes through time (it also doesn't change), we obtain the rate of change of acceleration (the jerk, figure 6.4d), which is 0.

Thus, if $s(t) = t^2$, then $\dot{s} = 2t$, $\ddot{s} = 2$, and $\dddot{s} = 0$ with respect to t. Notice the pattern of 2, 2, 0 throughout this sequence. We can generalize from this example as follows: if a is any constant, then

$$\frac{dx^a}{dx} = ax^{a-1}. \qquad\qquad Power\ Rule\ \ (6.4)$$

The derivative of x^a with respect to x is ax^{a-1}, where a is a positive integer.

Applying this to the example of the ball, if $a = 2$, then $s(t) = t^a$, and the velocity of the ball can be given as

$$\frac{d}{dt}s(t) = \frac{d}{dt}t^a = \frac{d}{dt}t^2 = at^{a-1} = 2t^{2-1} = 2t.$$

6.1.4 Derivative of the Sine Function

The derivative of the sine and cosine functions will be handy in the discussion of the mathematics of resonance (section 6.4), where the displacement, velocity, and acceleration of air, or the vibrating parts of a musical instrument, must be compared. What, for instance, is the derivative of $\sin\theta$?

Recalling that the derivative is the slope of a line tangent to the function at every point (section 6.1), let's look at some selected points on the sine wave shown in figure 6.5. A steep upward slope is strongly positive; a steep downward slope is strongly negative; a flat line has zero slope. If we plot the values of the slopes, we observe that they are points on a cosine wave. Thus, the derivative of $\sin\theta$ appears to be $\cos\theta$. How can we prove this?

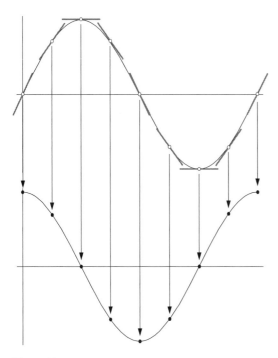

Figure 6.5
Finding the derivative of a sine wave.

We can express the equation $y = \sin\theta$ in terms of the definition of the derivative given in equation (6.1) by simple substitution:

$$\frac{dy}{d\theta} = \frac{d}{d\theta}\sin\theta = \lim_{\Delta\theta\to 0}\frac{\sin(\theta+\Delta\theta)-\sin\theta}{\Delta\theta}. \tag{6.5}$$

As before, we are taking the difference of two points on a function separated by some distance, and dividing the result by that distance so as to approximate a tangent to the function, then letting the distance between the points become infinitely small. The only change from before is that now the function is a sine wave and the distance between the points is an angular distance, $\Delta\theta$, instead of a linear distance.

But what does the function described by equation (6.5) actually look like? We can seek the answer by breaking equation (6.5) up into its terms, setting $\Delta\theta$ to a relatively large value, and plotting all the terms for various values of θ. Figure 6.6 shows a plot of equation (6.5) with $\Delta\theta = 15°$ and θ plotted from 0 to 2π. Curve (a) shows the value of the term $\sin(\theta+\Delta\theta)/\Delta\theta$, and curve (b) shows the term $-\sin(\theta)/\Delta\theta$, which is almost the opposite of (a). Curve (c) shows the sum of curves (a) and (b), which looks like a cosine function and has a range of approximately ±1. In fact,

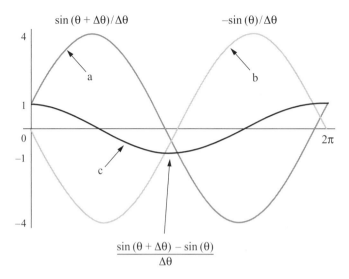

Figure 6.6
Sine derivative in separated terms.

if we plot equation (6.5) setting $\Delta\theta$ to ever smaller values, curve (c) approaches the cosine function ever more closely, and in the limit it equals the cosine function.

We see that the derivative of a sine function is the cosine function, and we can write

$$\frac{d}{d\theta}\sin\theta = \cos\theta. \qquad\qquad\qquad\qquad\qquad \textit{Sine Derivative} \ \ (6.6)$$

6.1.5   Derivative of the Cosine Function

As with the sine function just described, the equation $y = \cos\theta$ can be expressed in terms of the definition of the derivative given in equation (6.1) by simple substitution:

$$\frac{dy}{d\theta} = \frac{d}{d\theta}\cos\theta = \lim_{\Delta\theta\to 0}\frac{\cos(\theta+\Delta\theta)-\cos\theta}{\Delta\theta}. \qquad\qquad\qquad (6.7)$$

Using the same analysis as for the sine function, figure 6.7 shows a plot of equation (6.7) with $\Delta\theta = 15°$ and θ plotted from 0 to 2π. Curve (a) shows the value of $\cos(\theta+\Delta\theta)/\Delta\theta$, and curve (b) shows $-\cos(\theta)/\Delta\theta$, which is almost the opposite of (a). Curve (c) shows the sum of curves (a) and (b), which looks like a negated sine function and has a range of approximately ± 1. If we plot equation (6.7) setting $\Delta\theta$ to ever smaller values, curve (c) approaches ever more closely to a negated sine function, and in the limit it equals a negated sine function.

We see that the derivative of a cosine function is a negated sine function, and we can write

$$\frac{d}{d\theta}\cos\theta = -\sin\theta. \qquad\qquad\qquad\qquad\qquad \textit{Cosine Derivative} \ \ (6.8)$$

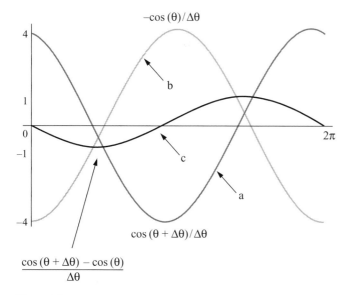

Figure 6.7
Cosine derivative in separated terms.

6.1.6 Derivative of *e*

The derivative of *e*, the base of the natural logarithms, when raised to a power, is identically *e* to the same power.

$$\frac{d}{dx}e^x = e^x$$

A plot of *e* is shown in figure 6.8. (See section 2.4.2 for more about *e*.)
The derivative of e^{ax} is given by

$$\frac{d}{dx}e^{ax} = ae^{ax},$$

where *a* is a constant.

For example, find the derivative of $f(x) = 7e^{-3x}$. By the formula just given, we have

$$f'(x) = -21e^{-3x}.$$

By the definition of the derivative (equation 6.5), we have

$$\frac{d}{dx}e^x = \lim_{\Delta x \to 0}\frac{e^{x+\Delta x} - e^x}{\Delta x} = \lim_{\Delta x \to 0}\frac{e^x - 1}{\Delta x}.$$

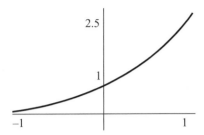

Figure 6.8
Plot of e^x.

One way to think about e is that it is the number that satisfies

$$\lim_{\Delta x \to 0} \frac{e^x - 1}{\Delta x} = 1.$$

Going the other way,

$$\frac{d}{dx} \ln ax = \frac{1}{x}.$$

6.1.7   Derivative Rules

Here are some other useful rules about derivatives that will come in handy.

Constant Multiplier Rule   The derivative of a constant times a function is the constant times the derivative of the function:

$$[af(x)]' = a[f(x)]'. \hspace{3cm} \textit{Constant Multiplier Rule} \hspace{0.3cm} (6.9)$$

This means that we can either extract a constant factor out of a derivative or put one in, depending upon what we are trying to do. This often can be used to simplify derivative expressions.

Sum Rule   The derivative of the sum of two functions is the same as the sum of the derivatives of the functions:

$$\frac{d}{dx}(f(x) + g(x)) = \frac{d}{dx}f(x) + \frac{d}{dx}g(x). \hspace{2cm} \textit{Sum Rule} \hspace{0.3cm} (6.10)$$

Product Rule   The derivative of the product of two functions is the same as the sum of each function times the derivative of the other:

$$\frac{d}{dx}(f(x)g(x)) = g(x)\frac{d}{dx}f(x) + f(x)\frac{d}{dx}g(x). \hspace{1.5cm} \textit{Product Rule} \hspace{0.3cm} (6.11)$$

Here's an immediate use for the product rule. If we set $f(x) = \sin x$, and $g(x) = \cos x$, then by the product rule we have

$$\frac{d}{dx}(\sin x \cos x) = \cos x \frac{d}{dx}\sin x + \sin x \frac{d}{dx}\cos x$$

$$= (\cos x)(\cos x) + (\sin x)(-\sin x)$$

$$= \cos^2 x - \sin^2 x.$$

The middle step in this derivation was made by remembering the derivatives of sine and cosine.

Chain Rule Suppose we have two functions $f(x)$ and $g(x)$. We can compose them like this: $f(g(x))$, which simply means we take the output of $g(x)$ as the input of $f(x)$. We can take the derivative of such a composite function. Say we let $y = f(g(x))$ and $u = g(x)$; then we can express the chain rule in terms of u, x, and y as follows:

$$\frac{dy}{dx} = \frac{dy}{du} \cdot \frac{du}{dx}. \qquad\qquad\qquad\qquad\qquad \textit{Chain Rule} \text{ (6.12)}$$

The derivative of f(g(x)) is the derivative of f with respect to g(x) times the derivative of g(x).

Equivalently we can write the chain rule as

$$\frac{d}{dx}f(g(x)) = f'(g(x))g'(x) \quad \text{or as} \quad \frac{d}{dx}f(u) = f'(u)\frac{du}{dx}.$$

For example, to differentiate $\sin(5q + 3)$ we use the chain rule. Let $y = \sin x$ and $u = 5q + 3$. Then

$$\frac{d}{dx}f(g(x)) = \frac{d}{du}\sin u \cdot \frac{d}{dq}(5q + 3) \quad \text{Chain rule}$$

$$= \cos u \left(\frac{d}{dq}5q + \frac{d}{dq}3 \right) \quad \text{Sum rule}$$

$$= \cos u \frac{d}{dq}5q \qquad\qquad \text{Derivative of a constant is 0.}$$

$$= 5\cos u \frac{dq}{dq} \qquad\qquad \text{Power rule, and substitute } u.$$

$$= 5\cos(5q + 3). \qquad\quad \text{Result}$$

The easiest way to understand the step "power rule, and substitute u" is to realize that dq is just an interval along an axis, and whatever that interval is, if we divide it by itself, the result is 1, and the term drops out. But we can also make it drop out by applying the power rule to dq/dq:

$$\frac{d}{dq}(q^1) = 1 \cdot q^0 = 1 \cdot 1 = 1.$$

Here is another relevant application of the chain rule:

$$\frac{d}{dx}\ln\sin x = \frac{1}{\sin x}\cdot\frac{d}{dx}\sin x = \frac{\cos x}{\sin x}\,. \tag{6.13}$$

6.1.8 Second Derivatives of Sine and Cosine

Recall that the first derivative of the sine is the cosine, and the first derivative of the cosine is the negation of the sine. Since the second derivative is just the derivative of the derivative, the second derivative of the sine is the negated sine:

$$\frac{d^2}{d\theta^2}\sin\theta = -\sin\theta. \qquad\qquad\textit{Sine Second Derivative} \tag{6.14}$$

If the first derivative of the cosine is the negated sine, and the first derivative of the negated sine is the negated cosine, then the second derivative of the cosine is the negated cosine.

$$\frac{d^2}{d\theta^2}\cos\theta = -\cos\theta. \qquad\qquad\textit{Cosine Second Derivative} \tag{6.15}$$

6.1.9 Derivative of Sinusoid with Varying Frequency

What is the derivative of a sinusoid when its frequency is varied? For instance, if we have $\sin\theta$, what is the derivative of twice its frequency, $\sin 2\theta$? In general, we want to solve

$$\frac{d}{d\theta}\sin a\theta \tag{6.16}$$

where a is a constant. We know that the effect of the constant a on the sinusoid in equation (6.16) is to vary its frequency (see volume 1, section 5.2.8). Let $u = a\theta$. Then

$$\frac{d}{d\theta}\sin a\theta = \frac{d}{du}\sin u\frac{du}{d\theta} \qquad \text{Chain rule}$$

$$= \cos a\theta\frac{d}{d\theta}a\theta \quad \text{Derivative of sine is cosine}$$

$$= a\cos a\theta\frac{d\theta}{d\theta} \qquad \text{Constant multiplier rule}$$

$$= a\cos a\theta. \qquad \text{Power rule}$$

Summarizing the result,

$$\frac{d}{d\theta}\sin a\theta = a\cos a\theta. \tag{6.17}$$

Scaling the frequency of a sinusoid proportionately scales both the frequency and the amplitude of its derivative.

To take a practical example, consider a small packet of air particles that is being displaced by a sound wave represented by

$$x = \sin a\theta. \tag{6.18}$$

By equation (6.17), the *velocity* of the air packet is

$$a \cos a\theta. \tag{6.19}$$

and the *acceleration* of the air packet is the second derivative of $\sin a\theta$. We can find the second derivative by taking the derivative of equation (6.17) again. Let $u = a\theta$ and find the first derivative. By (6.17),

$$\frac{d}{d\theta}\sin a\theta = a \cos a\theta.$$

Then find the second derivative.

$$\frac{d}{d\theta}a \cos a\theta = a\frac{d}{d\theta}\cos a\theta \qquad \text{Constant multiplier rule}$$

$$= a\frac{d}{d\theta}\cos a\theta \frac{d}{d\theta}a\theta \quad \text{Chain rule}$$

$$= -a\sin a\theta\frac{d}{d\theta}a\theta \qquad \text{Cosine derivative, equation (6.8)}$$

$$= -a^2\sin a\theta\frac{d\theta}{d\theta} \qquad \text{Result}$$

Finally, we can eliminate $d\theta/d\theta$, either because it is unity or by applying the power rule, so we end up with

$$\frac{d^2}{d\theta^2}\sin a\theta = \frac{d}{d\theta}a \cos a\theta = -a^2 \sin a\theta. \tag{6.20}$$

Thus the acceleration of the air packet is $-a^2 \sin a\theta$.

Scaling the frequency of a sinusoid proportionately scales the frequency, and negates and squares the amplitude of its second derivative (acceleration).

In equations (6.18), (6.19), and (6.20) we have an elegant representation for the displacement, velocity, and acceleration of the harmonic motion of vibrating objects, such as a packet of air particles, the basilar membrane, and many kinds of musical instruments.

6.2   Differential Equations

Differential equations are expressions relating a primary quantity such as displacement to derived quantities such as velocity and acceleration. They allow us to relate physical properties of vibrating systems such as musical instruments to mathematical representations of their vibration, providing insight into the very essence of what makes musical instruments sound the way they do.

6.2.1   First-Order Differential Equations

First-order differential equations cover physical systems where displacement is proportional to velocity. This characterizes, for example, the way energy from a plucked string dissipates through time.

Figure 6.9a shows a waveform that is representative of a plucked string instrument note. Figure 6.9b shows how the total energy of this waveform dissipates through time. This curve is called the envelope of the signal (see volume 1, sections 2.8.4 and 8.10). The slopes of the lines at points a, b, and c in figure 6.9b are tangent to the envelope and therefore indicate the rate at which energy is being dissipated by the instrument at moments t_a, t_b, and t_c, respectively. Dissipated energy is radiated from the string as sound and heat. Say that at time t_0 the string has 2 arbitrary units of acoustical energy.

The beginning of the note has the greatest total energy and also the steepest decay rate. The end of the note has less total energy and a less steep decay rate. In fact, it appears that throughout the note's duration, total energy is directly related to decay rate. Distilling these observations, we see that

- The magnitudes of the waveform's *displacements* from the equilibrium state of points a, b, and c are in the order $a > b > c$.

- The magnitudes of the energy envelope's *slopes* at points a, b, and c are also in the order $a > b > c$.

We see that the amount of energy that the string can dissipate at any moment is related to the amount of energy it contains.

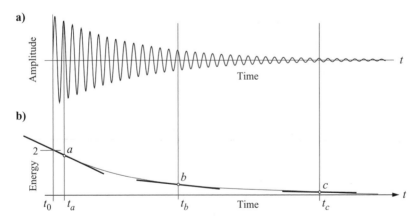

Figure 6.9
Damped waveform of a plucked musical instrument.

At any moment, the rate of total energy loss is proportional to the remaining energy.

Since it is proportional, we should be able to find some constant of proportionality that will allow us to equate the displacement and slope, which in turn will allow us to study the evolution of such envelopes.

Suppose a string is plucked with a certain force at time t_0. What will be the intensity of its sound at some particular later time? In order to address this question, we proceed in three stages:

1. Describe the physical system.

2. Relate the mathematical description to the problem at hand.

3. Determine a particular solution.

The next three subsections follow this agenda. Finally, we check to see that the solution relates to the question asked.

Describing the Physical System   Let the total energy in the vibrating string at time t be denoted by the function $y(t)$. By definition, the energy rate of change is dy/dt. Assuming that the string can be modeled as a linear system, the rate of energy change is proportional to the remaining energy:

$$\frac{dy}{dt} = ky, \tag{6.21}$$

where k is a constant of proportionality that can be determined for particular musical instruments. The constant k specifies the steepness of the slope. For example, a plucked violin note with a rapid decay rate would have a relatively large value of k, whereas a low-frequency piano note would have a small value of k, reflecting its slow decay rate.[3] Since the amount of energy initially in the string is positive and decreases with time, it stands to reason that both dy/dt and k must be negative because the slope of the energy curve is negative at all times (that is, the string can only dissipate energy).

Relating the Description to the Problem at Hand   We know that equation (6.21) describes the *amount of energy* in the vibrating string shown in figure 6.9b as a function of $y(t)$. It says that for every time t, the *rate of energy change* (dy/dt) is related by the constant k to the *remaining amount of energy* (y). However, we don't know what the function $y(t)$ is. In fact, about all we really know is that (if there is any solution to equation (6.21) at all) the derivative of $y(t)$ must be proportional to y, but this is just restating what we already know. How can we identify what the function $y(t)$ is?

While some systematic methods of solving equation (6.21) exist, they go beyond the scope of this book. However, a method available to us is the educated guess. This may seem counterintuitive, but mathematicians do sometimes guess at solutions. Educated guessing is common enough in mathematics to have been given a name: *ansatz,* a German word used to describe a solution to a problem that is guessed. If we look at the shape of the curve in figure 6.9b, it looks like an exponential curve. For comparison, a family of exponential curves is shown in figure 6.10, defined as ce^{-at}, where c is a constant that determines the initial magnitude of the function, and a is a constant that determines the rate at which the function moves toward zero as time t increases. In figure 6.10,

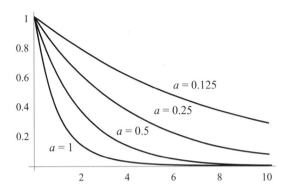

Figure 6.10
Exponential curves.

$c = 1$, and a takes on the values shown. By inspection, it appears that k in equation (6.21) performs the same function as a does here.

Let's try out the exponential function as a solution to equation (6.21) by substituting k for a and defining

$$y(t) = ce^{-kt}. \tag{6.22}$$

If we substitute this definition of the function y back into equation (6.21), we get (by the power rule)

$$\frac{dy}{dt} = ky = \frac{d}{dt}ce^{-kt} = kce^{-kt}. \tag{6.23}$$

Equation (6.22) is called a *general solution* of equation (6.21) because, for instance, we can use it to describe the exponential decay of any plucked or struck musical instrument. The next step is to find a *particular solution* that matches the curve to the appropriate variables for some particular instrument.

Determining a Particular Solution   Recall that the constant c introduced in equation (6.22) determines the magnitude of the exponential function and that the string has 2 arbitrary units of acoustical energy at time $t = 0$. Clearly, the amount of energy remaining at time t is related to the initial energy when $t = 0$, and so $y(0) = 2$ can be described as the *initial condition* of this system. Therefore, we must specify the value of c so that $y = 2$ when $t = 0$. If we evaluate equation (6.22) with this initial condition

$$y(0) = ce^0$$

$$= c \cdot 1,$$

we see that if we set c equal to 2, it will properly reflect the initial condition. If we use this value of c, then equation (6.22) becomes

$$y(t) = 2e^{-kt}. \tag{6.24}$$

This equation is certainly true for $y(0)$ because in that case the right side of equation (6.24) equals 2. But is it true for other values of $t > 0$?

Checking the Solution   Using equation (6.24), we can rewrite equation (6.23) as

$$\frac{dy}{dt} = 2ke^{-kt} = ky,$$

and we also know that $y(0) = 2e^0 = 2$, so equation (6.22) satisfies equation (6.21) as well as satisfying the initial condition.

Order of the Equation   Equation (6.21) is called an *ordinary differential equation of the first order*. The *order* of a differential equation is determined by the highest derivative used in the equation. Since the first derivative is the highest derivative used in equation (6.21), it is a first-order differential equation. In general, physical systems involving a rate of change will lead to a differential equation of some kind.

Making the Equation Homogeneous   In general, something is homogeneous if it is of a uniform nature. Macroscopically, air is homogeneous in this sense.

A polynomial is homogeneous if all its terms are of the same degree. For example $x^2 + y^2$ is a homogeneous polynomial.

If a homogeneous polynomial is set equal to zero, it is said to be a *homogeneous equation*. For example, $x^2 + y^2 = 0$ is a homogeneous equation, whereas $x^2 + y^2 = 3$ is not.

Let's rewrite equation (6.21) as

$$\frac{dy}{dt} - ky = 0.$$

Putting it in this form makes it a homogeneous equation (Nelson 2003). Generally, the solution to homogeneous equations is simpler than the solution of nonhomogeneous equations.

Linearity of the Equation   Equation (6.21) is *linear* because each term has only one order of derivative in it. We do not see, for instance, a term containing $\dot{y} \cdot y$. Also, we do not see $\dot{y} \cdot \dot{y}$ (no derivatives are raised to a power).

6.2.2 Second-Order Equations

We've seen that first-order differential equations such as equation (6.21) can be solved by the exponential function $y(x) = ce^{-kx}$. First-order differential equations cover physical systems such as a plucked string, where displacement (corresponding to the total energy) is proportional to velocity (corresponding to the rate at which the energy dissipates). What about physical systems that include acceleration, the second derivative of displacement? Second-order differential equations cover physical systems that include acceleration. This describes, for example, instruments with a driven vibrating system, producing a sustained tone, such as the bowed string, woodwinds, and brasses.

Consider solving a second-order homogeneous differential equation of the form

$$\ddot{y} + a\dot{y} + by = 0, \tag{6.25}$$

where a, b, and y are real constants. In the previous section, we took an educated guess to discover that the exponential function solved equation (6.21). Could a suitable variation of the exponential function also solve equation (6.25)? To keep things simple, let's take a generalized exponential function,

$$y(x) = ce^{\kappa x}. \tag{6.26}$$

By the power rule, the first and second derivatives are $\dot{y} = \kappa c e^{\kappa x}$ and $\ddot{y} = \kappa^2 c e^{\kappa x}$, respectively. If we substitute these into equation (6.25), we get

$$\kappa^2 c e^{\kappa x} + a\kappa c e^{\kappa x} + bc e^{\kappa x} = 0.$$

Factoring yields

$$e^{\kappa x}(\kappa^2 + a\kappa + b) = 0.$$

This suggests that equation (6.26) is a solution of equation (6.25) if equation (6.26) is a solution to the quadratic equation

$$\kappa^2 + a\kappa + b = 0. \qquad\qquad \textit{Auxiliary Equation} \quad (6.27)$$

This equation is such a common stepping stone that it has not one but two names: the *auxiliary equation* and the *characteristic equation*.

Recalling the quadratic formula from algebra, we know that there are two roots of equation (6.27), which we can define as

$$\kappa_1 = \frac{-a + \sqrt{a^2 - 4b}}{2} \quad \text{and} \quad \kappa_2 = \frac{-a - \sqrt{a^2 - 4b}}{2}. \tag{6.28}$$

This suggests that there are two solutions to equation (6.25):

$$y_1(x) = e^{\kappa_1 x} \quad \text{and} \quad y_2(x) = e^{\kappa_2 x}. \tag{6.29}$$

I leave to the reader the process of checking this by substituting equation (6.29) into equation (6.25) and applying the power rule to obtain the derivative terms. The result should yield $0 = 0$.

6.3 Mathematics of Resonance

In volume 1, section 8.7.1, we saw that every vibration can be shown to be a linear weighted combination of its normal modes. The approach developed in the present section will draw a similar conclusion but from a much more fundamental perspective based on an instrument's transient (time-varying) behavior. I construct physical models of the vibrating system and examine its dynamic behavior. The physical basis of the normal modes of vibration will become clear from this analysis and will generalize to all vibrating systems.

The first step is to examine the transient behavior of a system vibrating in one-dimensional simple harmonic motion, first without taking friction into account, then with friction included (section 6.3.2).

All vibration arises from the interaction of elastic forces and inertial forces (see volume 1, section 8.4). Elasticity provides a restoring force, and inertia causes the restoring force to overshoot its equilibrium point (in a relatively friction-free system), thereby perpetuating the vibration. The inertial force and elastic force balance at all points. These facts suggest an equation describing how simple harmonic motion is governed by the sum of the elastic force and inertial force:

$$ma(t) + kx(t) = 0, \tag{6.30}$$

where $x(t)$ is displacement of a mass m tethered by a spring with spring constant k and accelerating according to $a(t)$.

6.3.1 Simple Harmonic Oscillator

How does equation (6.30) represent an oscillatory vibration? It's not immediately obvious just from looking at it. Essentially, what we'd like to know is how the system evolves as we vary time t. We'd especially like to find an algebraic solution for equation (6.30) because then we could determine the state of the system at arbitrary moments of time.

Intuition suggests that equation (6.30) should describe a sinusoidal motion (see volume 1, section 8.4). Actually, it should describe *every possible sinusoidal motion*. That's because even such a simple spring/weight system is theoretically capable of creating an infinite number of different sinusoidal motions with different initial phases, amplitudes, and frequencies. They should all be embodied in equation (6.30).

We can eliminate the acceleration function $a(t)$ by remembering from the discussion of derivatives (section 6.1) that acceleration is the second derivative of displacement, $a(t) = \ddot{x}(t)$. Thus, we can rewrite equation (6.30) as

$$m\ddot{x}(t) + kx(t) = 0. \qquad \text{\textit{Harmonic Equation}} \tag{6.31}$$

Equation (6.31) is sometimes called the *harmonic equation*. Although we've simplified it to just displacement $x(t)$, elasticity k, and inertia m, it still isn't clear how to solve it for $x(t)$. Equation (6.31) is a second-order differential equation, and such equations are typically more difficult to solve than algebraic ones because there is no universal body of techniques for solving such equations. Realistically, an educated guess is our only option. Since we observe that physical systems constructed like equation (6.31) vibrate with sinusoidal motion, let's guess that the solution to equation (6.31) is the general sinusoid

$$x(t) = A \sin(\omega t + \phi), \qquad \text{\textit{Trial Solution}} \tag{6.32}$$

where $\omega = 2\pi f$, f is the frequency of the sinusoid, A is amplitude, ϕ is phase offset, and t is time.

Applying the trial solution to the term $kx(t)$ yields $kA \sin(\omega t + \phi)$. Then, applying it to the other term $m\ddot{x}(t)$ yields $-m\omega^2 A \sin(\omega t + \phi)$. Equation (6.31) is just the sum of these two terms:

$$-m\omega^2 A \sin(\omega t + \phi) + kA \sin(\omega t + \phi) = 0. \tag{6.33}$$

There's a pattern in this equation that would be clearer if it weren't so cluttered. Let's make it simpler by temporarily setting $A = 1$ and letting $\alpha = \sin(\omega t + \phi)$. Switching the position of the two terms, we have

$$k\alpha - m\omega^2\alpha = 0. \tag{6.34}$$

Then, if we set $\omega^2 = k/m$:

$$k\alpha - m\omega^2\alpha = 0,$$

$$k\alpha - m\frac{k}{m}\alpha = 0,$$

$$k\alpha - k\alpha = 0.$$

This demonstrates that $x(t) = A\sin(\omega t + \phi)$ is indeed a solution to equation (6.31) so long as we let $\omega^2 = k/m$.

Parameters A and ϕ determine the *initial conditions* of amplitude and phase of the oscillator. Because there are no dissipative forces in the simple harmonic oscillator, any nonzero initial conditions will result in unending sinusoidal vibration.

Of course, if $\omega^2 = k/m$, we can also write $\omega = \sqrt{k/m}$ (by picking the positive solution).

It is apparent from this that ω is the natural vibrating frequency of the simple harmonic oscillator because it represents the angular velocity of the harmonic oscillator's sinusoidal motion based on the values of m and k. In honor of this important fact, let's give this use of the term ω its own name and call it

$$\omega_r = \sqrt{k/m}, \qquad\qquad\qquad\qquad\qquad \textit{Resonant Frequency} \tag{6.35}$$

the resonant frequency of the simple harmonic oscillator.

If we solve $\omega = 2\pi f$ for f, we have

$$f = \frac{\omega}{2\pi}.$$

Substituting the definition for ω_r in (6.35) into this, we arrive at

$$f = \frac{\omega_r}{2\pi} = \frac{\sqrt{k/m}}{2\pi}, \qquad\qquad \textit{Natural Frequency of the Harmonic Oscillator} \tag{6.36}$$

which is the formula for the natural frequency, characteristic frequency, or *eigenfrequency* of the harmonic oscillator.

Other Solutions   So, we've guessed correctly that $A\sin(\omega t + \phi)$ is a solution of equation (6.31), and the angular frequency ω and phase offset ϕ can be any constants.

There are other solutions as well. If we set $\phi = \pi/2$, then $\sin(\omega t + \phi) = \cos(\omega t + 0)$, which shows that $B\cos(\omega t + \phi)$ is also a solution (where B is a constant) because a cosine is just a sine wave started with a different phase offset.

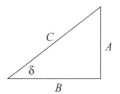

Figure 6.11
Right triangle.

Solutions of linear equations such as equation (6.31) can be added together to produce yet more valid solutions; thus the combination of the preceding two solutions is also a solution:

$$x = A \sin \omega t + B \cos \omega t. \tag{6.37}$$

The values of A and B depend upon the initial conditions of displacement and velocity of the harmonic oscillator at time $t = 0$. For instance, if the displacement is initially zero but the velocity is not, then it must be that $B = 0$ and $A \neq 0$, and the motion is sinusoidal. If the displacement is initially nonzero and velocity is zero, then it must be that $A = 0$ and $B \neq 0$, and the motion is cosinusoidal. If displacement and velocity are both nonzero, then neither A nor B is zero.

From elementary trigonometry, we know that the A and B terms in equation (6.37) can also be interpreted as the sides of a right triangle (figure 6.11). If we call the hypotenuse of such a triangle C, then we can define its length as $C = \sqrt{A^2 + B^2}$. And if we call the slope of the hypotenuse δ, then we can define its slope as $\tan \delta = A/B$.

Thus we can rewrite equation (6.37) as

$$x = C \cos(\omega t - \delta). \tag{6.38}$$

Although equation (6.32) shows that amplitude is a factor in defining harmonic motion in equation (6.31), the frequency of the oscillation does not depend upon amplitude because (6.36), which defines the frequency, does not have term A in it. This means the mass can vibrate at any amplitude, and its resonant frequency will be unchanged. The amplitude only determines the extent of the displacement.

Peak Displacement   There is a little more to say about how the constants A and ϕ interact in the solution for $x(t)$. These values are determined by the position and velocity of the harmonic oscillator at any fixed time t. To see this, let's look at the position and velocity of the mass at the beginning of time. Applying $t = 0$ to the definition of $x(t)$ in equation (6.32) yields the *initial displacement:*

$$x(0) = A \sin \phi. \qquad\qquad \textit{Initial Displacement} \tag{6.39}$$

The velocity $v(t)$ is just the first derivative of equation (6.32):

$$v(t) = A \omega \cos(\omega t + \phi) \text{ with respect to } t.$$

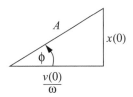

Figure 6.12
Calculating peak displacement.

Setting $t = 0$, we have the *initial velocity:*

$$v(0) = A\omega\cos\phi. \hspace{4cm} \textit{Initial Velocity} \quad (6.40)$$

Solving (6.39) for $\sin\phi$ and solving (6.40) for $\cos\phi$, we get $\sin\phi = x(0)/A$, and $\cos\phi = v(0)/(A\omega)$. According to the rules of trigonometry, we can interpret A as the hypotenuse of a right triangle whose sides are $x(0)$ and $v(0)/\omega$ (see figure 6.12), which can be written as

$$A = \sqrt{(x(0))^2 + \left(\frac{v(0)}{\omega}\right)^2}. \hspace{2cm} \textit{Peak Displacement} \quad (6.41)$$

A is the peak displacement that the mass will achieve, expressed in terms of the three parameters which determine it: initial position $x(0)$, initial velocity $v(0)$, and angular frequency ω of the mass.

6.3.2   Damped Harmonic Oscillator

The discussion of the harmonic oscillator so far has left out friction and other nonconservative forces, but in the natural world such forces are always present to some degree. The traditional way to study the effects of friction on vibrating systems is to add a dashpot to the spring mass system (figure 6.13). A *dashpot* is just a reservoir of fluid with some viscosity. A paddle attached to the mass is dipped into the fluid. The viscosity of the fluid resists the vertical vibrating motion of the mass in proportion to the rate at which the paddle is drawn through the fluid—the faster its velocity, the greater the drag.

Linear vs. Nonlinear Dissipation   The counterforce exerted by the dashpot on the vibrating system is linear because it is always proportional to the velocity of the system. However, dissipative forces in musical instruments are often nonlinear. For example, the force of friction between a bow and a violin string is not a linear function of pressure applied by the violinist because the bow has a tendency alternately to stick and slip on the string as it is dragged along. This motion is what creates the characteristic sawtooth pattern of violin string vibration (also called Helmholtz motion; see section 8.3.1). So studying linear dissipation is not sufficient to understand important classes of musical instrument vibration. However, modeling dissipation as a linear force has a great advantage: relatively simple solutions exist for the differential equations using linear dissipation, whereas solutions for nonlinear dissipative systems tend to be less tractable.

Figure 6.13
Spring, weight, and dashpot.

Damping   Since the motion of the damped harmonic oscillator in figure 6.13 is along the y-axis, let's represent displacement of the mass with the variable y. Its velocity is $\dot{y}$ and its acceleration is $\ddot{y}$.

If we call the counterforce exerted by the dashpot F_c, and say that it is proportional to the velocity of the mass, then

$$F_c = -(c)\dot{y}, \hspace{4cm} \textit{Viscous Resistance} \quad (6.42)$$

where c is a constant of proportionality called the *damping factor*. The minus sign reminds us that this is a counterforce to the direction of motion. (Though the expression surrounding it is negative, the damping factor c is itself always a positive quantity.) The greater the damping factor, the more rapidly energy is dissipated from the vibrating system. Since we're considering an undriven system (that is, a system that receives energy only at the onset of vibration), greater dissipation causes the vibration to die away more quickly. We can reduce the damping factor of the dashpot by making the liquid less viscous or by decreasing the size of the paddle, or both. Doing either decreases c and lengthens the time required for the vibration to die away.

When dissipation is small compared to the other forces, the effect is called *underdamping*. As we gradually increase the damping factor c of the dashpot, the oscillation dies away more rapidly, as we'd expect. But at a certain point, increasing the damping factor causes the system to stop oscillating—that is, it no longer crosses the equilibrium point and reverses direction; it just heads gradually toward equilibrium along an exponential trajectory. This effect is called *critical damping*. As we continue to increase the damping factor c, the system's progress toward equilibrium slows, and it takes longer and longer to achieve equilibrium. This effect is called *overdamping*. It is important for the study of oscillation, developed in section 6.3.3, to shed some light on this behavior.

Combining Forces   We have identified three forces at work on a damped harmonic oscillator system:

- Inertial force, $F_m = m\ddot{y}$
- Restoring force, $F_k = -ky$
- Dissipative force, $F_c = -c\dot{y}$

The restoring force (elasticity or tension) and the dissipative force (friction and radiation) both oppose the inertial force, so we write $F_m = F_k + F_c$, and after substituting we have

$$m\ddot{y} = -c\dot{y} - ky.$$

Rearranging, we have:

$$m\ddot{y} + c\dot{y} + ky = 0, \qquad\qquad\qquad \textit{Damped Harmonic Motion} \ (6.43)$$

which is a second-order homogeneous differential equation describing damped harmonic motion.

Again, we must guess at a solution. If we study natural systems with the features of this equation, such as the plucked string shown in figure 6.9, we observe that they dissipate energy at an exponential rate. Since this matches the general behavior of an exponential function such as

$$x(t) = e^{\omega t}, \tag{6.44}$$

why not try this as a solution?

To substitute the trial solution (6.44) into (6.43), we must find its first and second derivatives with respect to time t, which are $\dot{x}(t) = \omega e^{\omega t}$ and $\ddot{x}(t) = \omega^2 e^{\omega t}$, respectively. Substituting the appropriate derivative of the trial solution into equation (6.43) yields

$$m\omega^2 e^{\omega t} + c\omega e^{\omega t} + ke^{\omega t} = 0. \tag{6.45}$$

Dividing both sides by $e^{\omega t}$ causes this term to drop out, and equation (6.45) reduces to

$$m\omega^2 + c\omega + k = 0.$$

We can use the quadratic formula from algebra to solve this equation for ω. Doing so yields two valid solutions that we can name ω_+ and ω_-, respectively:

$$\omega_+ = \frac{-c + \sqrt{c^2 - 4km}}{2m} \quad \text{and} \quad \omega_- = \frac{-c - \sqrt{c^2 - 4km}}{2m}. \tag{6.46}$$

Since both solutions are valid, but we only need one, we can take our pick. We'll have a small tactical advantage later if we choose the negative solution and let $\omega = \omega_-$. The expansion,

$$\omega = -\frac{c}{2m} - \frac{\sqrt{c^2 - 4km}}{2m}, \tag{6.47}$$

reveals the formula for angular velocity of the damped harmonic oscillator. Now that we have a definition of ω, we can plug it back into the trial solution equation (6.44), which becomes:

$$x(t) = e^{\omega t} = e^{\left(-\frac{c}{2m} - \frac{\sqrt{c^2 - 4km}}{2m}\right)t}.$$ (6.48)

We can simplify that gnarly[4] exponent of e by remembering from algebra that $x^{(\alpha + \beta)} = x^\alpha x^\beta$, allowing us to rewrite equation (6.48) as

$$x(t) = e^{\omega t} = e^{-\frac{c}{2m}t} e^{-\frac{\sqrt{c^2 - 4km}}{2m}t}.$$ (6.49)

If we let

$$\alpha = \frac{c}{2m} \quad \text{and} \quad \beta = \frac{\sqrt{c^2 - 4\,km}}{2\,m},$$ (6.50)

then the derivative becomes

$$x(t) = e^{-\alpha t} e^{-\beta t}.$$ (6.51)

We see from equation (6.51) that $x(t)$ can be interpreted as the product of two exponents of e. Assuming we use positive physical constants for c and m, the first term $e^{-\alpha t}$ is always a simple exponential decay function. But the second term $e^{-\beta t}$ is a little trickier because it contains a square root in its definition. It may be a simple exponential decay function, or it may be a phasor, depending upon the settings of its parameters.

Does It Vibrate? Depending upon the values we choose for m, c, and k, there may not be any oscillation at all. Let's look again at the definition of β in equation (6.50). The term under the radical sign $c^2 - 4km$ is called the *discriminant*. It gets a special name because (in this case) it discriminates between the conditions that cause vibration and those that do not. If the discriminant is positive, then β is real because then we're taking the square root of a positive number. If the discriminant is negative, then β is complex because we're taking the square root of a negative number.

- If β is complex, then $e^{-\beta t}$ in equation (6.51) is a phasor, which vibrates.
- If β is real, then $e^{-\beta t}$ is a simple exponential decay function, which doesn't vibrate.

The discriminant is positive if $c^2 - 4km > 0$. Assuming we use positive physical constants, the system does not vibrate when dissipation dominates over the other forces. Thus, the relative strength of the damping factor over the other forces determines the behavior of the oscillator. There are three possibilities for an undriven, damped harmonic oscillator:

- If $c^2 - 4km > 0$, the system is said to be *overdamped*. It does not vibrate.
- If $c^2 - 4km = 0$, the system is said to be *critically damped*. It does not vibrate, and it goes to zero with a pure exponential decay.
- If $c^2 - 4km < 0$, the system is said to be *underdamped,* and it vibrates.

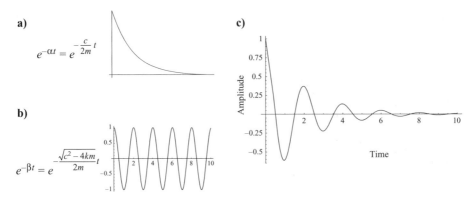

Figure 6.14
Damped harmonic motion.

Overdamping   Assuming positive physical values for m, c, and k, and time $t \geq 0$, the exponents of e in equation (6.51) are both real and negative. In this case, (6.51) is the product of two declining exponential curves, one of which goes toward zero a little faster than the other. The more the damping factor dominates the system, the faster the oscillator's energy dissipates.

Critical Damping   If $c^2 - 4km = 0$, then $\beta = 0$. In this case, equation (6.51) reduces to

$$x(t) = e^{-\alpha t} e^{-\beta t} = e^{-\alpha t} e^0 = e^{-\alpha t},$$

so the result is a pure exponential decay term.

Underdamping   This is the interesting case. For positive values of time ($t \geq 0$), equation (6.51) is the product of a declining exponential curve ($e^{-\alpha t}$) and the phasor $e^{-\beta t}$. It stands to reason that the result is an exponentially decaying phasor. For example, with the values $m = 1$ kg, $c = 1$ Ns/m, and $k = 10$ N/m, the plot of $e^{-\alpha t}$ is shown in figure 6.14a. Since the discriminant is complex for these values, we project $e^{-\beta t}$ onto the real axis, observing the harmonic motion shown in figure 6.14b. The product of these two curves yields the plot shown in figure 6.14c, which is an exponentially decaying sinusoid.

Note that increasing c causes $e^{-\alpha t}$ to decay more quickly, and increasing m causes it to decay more slowly. These facts are in line with the intuition that increasing friction should make a note go silent more quickly, and adding mass should make it last longer, all else being equal.

Now consider the term β. It holds some secrets that are revealed when it is simplified. Taking its definition from equation (6.50), first we square it, then expand it:

$$\beta^2 = \frac{c^2 - 4km}{4m^2} = \frac{c^2}{4m^2} - \frac{4km}{4m^2}$$

$$= \frac{c^2}{4m^2} - \frac{k}{m}.$$

Then we restore it by taking its square root again:

$$\beta = \sqrt{\frac{c^2}{4m^2} - \frac{k}{m}}.$$

This formula shows that β is the natural vibrating frequency of the damped harmonic oscillator because (when underdamped) it regulates the angular velocity of the oscillator's phasor. In honor of this fact, let's rename β and call it the *resonant frequency of the damped harmonic oscillator*.

$$\omega_{rd} = \beta = \sqrt{\frac{c^2}{4m^2} - \frac{k}{m}}, \qquad \textit{Resonant Frequency of the Damped Harmonic Oscillator} \quad (6.52)$$

It is interesting to compare the resonant frequency ω_{rd} of the damped harmonic oscillator with the resonant frequency ω_r of the simple harmonic oscillator given in equation (6.35). The resonant frequency ω_r is based only on the values of m and k, whereas the resonant frequency ω_{rd} depends also on c. That means that as the amount of dissipation c *increases*, the resonant frequency of the damped harmonic oscillator *decreases,* and the resonant frequency goes to zero when the system becomes critically damped.

This is shown progressively in figure 6.15. A harmonic oscillator's constant values of $m = 1$ kg and $k = 4$ N/m are plotted against decreasing values of the damping factor c ranging from 6 Ns/m

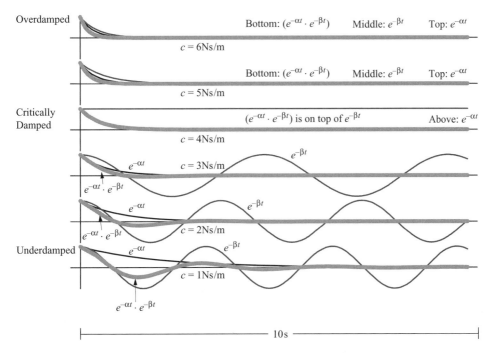

Figure 6.15
Progression from overdamped to underdamped oscillator.

to 1 Ns/m, plotted over 10 seconds. The darker line is $e^{-\alpha t}$, the lighter line is $e^{-\beta t}$, and the thickest line is their product. The first plot is heavily overdamped, the second, less so. The third, $\beta = 0$ is critically damped. The bottom three are progressively more underdamped.

Note that the frequency of vibration goes up as the damping factor goes down.

6.3.3   Driven Harmonic Oscillator

When we apply a driving force to a simple harmonic oscillator, we have two frequencies to keep track of: the natural vibrating frequency of the oscillator ω_0, and the frequency of the driving force ω.

The response amplitude of the system tends to grow when the frequency of the driving force ω is close to the natural frequency ω_0 of the oscillator. The tendency of a system to vibrate more strongly at a particular frequency is called *resonance*.

Recall from equation (6.43) that the equation of motion for a damped harmonic oscillator is

$$m\ddot{y}(t) + c\dot{y}(t) + ky(t) = 0. \tag{6.53}$$

To study this as a *driven* harmonic oscillator, we add a driving force $s(t)$, which represents any function of time. As before, we simply want to combine this new force with the ones studied previously such that now the oscillator consists of an inertial force m, a dissipative force c, an elastic force k, and a driving force $s(t)$. The driving force injects energy into the oscillator, and the other forces respond in opposition to it, so we write

$$m\ddot{y}(t) + c\dot{y}(t) + ky(t) = s(t). \tag{6.54}$$

The *driving force*—also sometimes called the *forcing function*—is the input to the system, and the corresponding solution describing the resulting vibration is called the output or the *response* of the system to the driving force.

Although $s(t)$ can be any function, for the moment we are particularly interested in sinusoids because then we can focus on *forced harmonic motion,* which is mathematically more tractable and is also responsible for the behavior of many musical instruments as well as the motions of the basilar membrane.

In the introduction to complex numbers, I quoted the mathematician Jacques Hadamard, who said, "The shortest path between two truths in the real domain passes through the complex domain." The derivation to follow is a case in point. We can simplify the impending calculations a great deal by the use of a phasor as a forcing function instead of a sinusoid. So, let

$$s(t) = Fe^{i(\omega t + \phi)}, \qquad\qquad\qquad\qquad \textit{Forcing Function} \tag{6.55}$$

which is a phasor with amplitude F, time t, angular velocity ω, and phase offset ϕ. Thus we have

$$m\ddot{y}(t) + c\dot{y}(t) + ky(t) = Fe^{i(\omega t + \phi)}. \tag{6.56}$$

Dividing both sides of (6.56) by m puts it into standard second-order differential equation format:

$$\ddot{y}(t) + \frac{c}{m}\dot{y}(t) + \frac{k}{m}y(t) = \frac{F}{m}e^{i(\omega t + \phi)}. \tag{6.57}$$

Solutions for Transient and Steady States   The states of a driven harmonic oscillator that we must find solutions for include

• Its *transient response*—how it behaves when the driving force changes, such as when it is switched on and off. When the driving force is stopped, vibration dies out exponentially, and when it is started, its motion depends upon initial conditions.

• Its *steady-state response*—how it behaves when the driving force is applied for a long time. We can observe the steady-state response by waiting until any contribution of the transient response has died away. Then the amplitude settles down to a constant that depends solely upon the response of the resonator to the driving function and no longer depends upon the initial conditions.

When the driving force is switched on, the system starts vibrating immediately, and the amplitude of the response grows exponentially through time. After this transient effect dies out, we are left with the steady-state vibration of the system. To study resonance, we are particularly interested in the amplitude of the steady state for various frequencies of the driving function because the amplitude the oscillator achieves depends upon the proximity of the resonant frequency ω_0 to the frequency of the driving function ω. When it is switched off again, the vibration continues but dies away exponentially.

Equation (6.57) must somehow embody all transient and steady-state behavior, and its solutions must include all possible combinations of resonant properties, initial conditions, and driving functions. We can characterize the solutions of the driven harmonic oscillator separately, as follows.

Transient Response   If we set $F = 0$, then equation (6.57) becomes identical to equation (6.43) and its duplicate (6.53), which is the homogeneous equation of motion for the damped harmonic oscillator (section 6.3.2).

Driven harmonic motion is identical to damped harmonic motion in the absence of a driving function.

Recall that the solution to equation (6.43) is equation (6.44): $y(t) = e^{i\omega t}$.

Since (6.43) is a homogeneous equation, let's call (6.44) the *homogeneous solution* $y_h(t)$ for driven harmonic motion and write

$$y(t) = y_h(t) = e^{i\omega t}. \hspace{3cm} \textit{Homogeneous Solution} \ (6.58)$$

This solution characterizes the transient response of the resonant system.

Steady-State Response   We must find a *particular solution* to the full differential equation (6.57). Let's call the solution $y_p(t)$ and say that it will characterize the steady-state response of the system.

We should expect that any steady-state solution we find will resemble undamped harmonic motion. That's because the forcing function continuously injects energy into the system, compensating for the energy lost to dissipation. Properly, an undamped vibration can only arise in the absence of dissipating forces, but if we inject energy into a vibrating system at a rate that exactly matches its rate of dissipation, it will be indistinguishable from an undamped vibration. The amplitude of the resulting oscillation will reveal a balance between energy injected by the forcing function and that lost to dissipation over time.

Combining Transient and Steady-State Solutions   During and immediately after transitions in the driving function, a harmonic oscillator contains both transient and steady-state motions. But eventually—if there are no further changes in the driving function—the contribution of $y_h(t)$ becomes insignificant and we are left only with $y_p(t)$.

Thinking along these lines, we see that the full solution to equation (6.57) must be the sum of the two solutions:

$$y(t) = y_h(t) + y_p(t).$$

Since we already have a solution for the transient component of a resonator's motion $y(t) = e^{i\omega t}$, all we need to find is the particular solution $y_p(t)$.

Finding the Particular Solution   As always, a degree of luck is required to guess at a solution to equation (6.57). We used the phasor $e^{i\omega t}$ to solve the equation of motion for the damped harmonic oscillator, and its motion resembles driven harmonic motion. Let's use it again with the slight twist of adding an amplitude term:

$$y(t) = Ae^{i(\omega t + \phi)}, \tag{6.59}$$

where we intend A to represent the amplitude of the solution, but right now we have no idea what the equation for amplitude should be. Let's leave A undetermined with the expectation that we will uncover its meaning as part of the particular solution. Proceeding this way is called the *method of undetermined coefficients*.

In order to apply the trial solution, equation (6.59), to equation (6.57), we must provide the two derivatives of $y(t)$ used in (6.57). They are $\dot{y}(t) = i\omega Ae^{i(\omega t + \phi)}$ and $\ddot{y}(t) = -\omega^2 Ae^{i(\omega t + \phi)}$. Substituting these into (6.57) yields

$$-\omega^2 mAe^{i(\omega t + \phi)} + i\omega cAe^{i(\omega t + \phi)} + kAe^{i(\omega t + \phi)} = Fe^{i(\omega t + \phi)}.$$

The common term $e^{i(\omega t + \phi)}$ drops out of both sides. Factoring A out from the common terms reduces it further to just $A(-\omega^2 m + i\omega c + k) = F$. Solving this for A yields

$$A = \frac{F}{-\omega^2 m + i\omega c + k}. \tag{6.60}$$

We have found an equation for the previously undetermined coefficient A.

Now we can apply this formula for A in the definition of $y(t)$ given in equation (6.59), which yields the particular solution:

$$\begin{aligned} y(t) = y_p(t) &= \frac{F}{-\omega^2 m + i\omega c + k} \cdot e^{i(\omega t + \phi)} \\[2mm] &= \frac{Fe^{i(\omega t + \phi)}}{-\omega^2 m + i\omega c + k}. \end{aligned} \tag{6.61}$$

Note that this equation for $y(t)$ incorporates all the parameters involved in the equation of motion given in equation (6.57). The numerator is a phasor with amplitude F, angular velocity ω, and phase offset ϕ. The denominator of equation (6.57) reveals the interaction of all the parameters affecting the amplitude of this motion. Understanding the denominator reveals the inner workings of resonance.

If, as is typically the case, the values for m, c, and k are constants, then observe that the amplitude of $y(t)$ is modulated by the angular velocity parameter ω and the amplitude F of the driving function. For a musical instrument such as a flute, F corresponds to the amplitude of the oscillating air current, and ω embodies frequency f via the relation $\omega = 2\pi f$.

As $y(t)$ spins around complex zero in the complex plane, its real part is a sinusoid corresponding to the displacement of the driven harmonic oscillator. For any value of the driving frequency ω, $y(t)$ encodes the amplitude and phase of the oscillator's steady-state response in the manner dictated by the interaction of its terms. Equation (6.61) represents the *steady-state solution* $y_p(t)$ with a phasor whose complex amplitude is modulated by the frequency of the driving function.

6.3.4 How Amplitude Varies with Frequency

What is the phenomenon responsible for a resonator's having greater amplitude at some frequencies than others? In order to understand this, we must study how the amplitude of $y(t)$ responds to the frequency ω of the driving function.

The formula for the amplitude response, equation (6.61), is a phasor. So, to talk about its amplitude, we must talk about the magnitude of the phasor, which is just the length of its vector (see section 2.6). Let's reason through the terms in the numerator and denominator of equation (6.61) separately to obtain its magnitude.

Numerator   First, the magnitude of a positive real is a positive real, so $|F| = F$. Next, recall that the magnitude of an unscaled phasor is always unity, therefore $\left|e^{i(\omega t+\phi)}\right| = 1$ (see deMoivre's theorem, section 2.4).

Denominator   Now, since the denominator of equation (6.61) is complex, let's rearrange its terms into standard complex number format: $(k - \omega^2 m) + i\omega c$. Recalling that the magnitude of a complex number is the square root of the sum of its squares (and remembering to treat *both* components of the complex number as real values for the square root operation), the magnitude of the denominator is $\sqrt{(k - \omega^2 m)^2 + \omega^2 c^2}$.

Combining Numerator and Denominator   Putting the pieces together, the magnitude of equation (6.61) can be expressed as

$$|y(t)| = \frac{F}{\sqrt{(k - \omega^2 m)^2 + \omega^2 c^2}}. \tag{6.62}$$

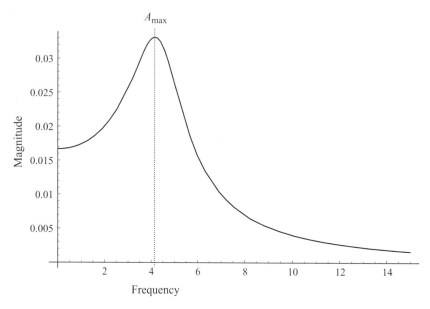

Figure 6.16
Response spectrum of a driven harmonic oscillator.

It's time for an example. Setting $m = 3$ kg, $c = 7$ Ns/m, $k = 60$ N/m, and $F = 1$ N, and varying ω over the range 0 to 15 Hz, equation (6.62) produces the classic response spectrum of a resonator (figure 6.16). Though I refer to it as a *response spectrum,* it is technically the magnitude spectrum of the response of a driven harmonic oscillator to a sinusoidal driving function.

Finding the Resonant Frequency   Figure 6.16 shows a resonant peak amplitude A_{max} some-where around 4 Hz. But what is the exact frequency and amplitude of the peak? Clearly, the answer lies in understanding how the denominator of equation (6.62) interacts with frequency ω. Let's just focus on the denominator of equation (6.62) for a moment and, for convenience, define the denom-inator as a function of frequency: $\delta(\omega) = \sqrt{(k - \omega^2 m)^2 + \omega^2 c^2}$. Now let's plot $\delta(\omega)$ while vary-ing ω from 0 to 15 Hz (figure 6.17a). Notice that the minimum of this function corresponds to the resonant frequency in figure 6.16 (because the quotient will always be greatest where the denom-inator is least). By inspection, the minimum appears to be just a little more than 4 Hz. But how can we pin down *exactly* where the minimum of this function lies?

If we take the first derivative of $\delta(\omega)$, then the minimum in figure 6.17a will turn into a zero-crossing, as shown in figure 6.17b. The roots of this derivative will give us the exact location of the zero crossing.

We can use standard algebraic techniques to solve this equation for its roots. Solving the equation

$$\frac{d}{d\omega}\delta(\omega) = \frac{d}{d\omega}(k - \omega^2 m)^2 + \omega^2 c^2 = 0$$

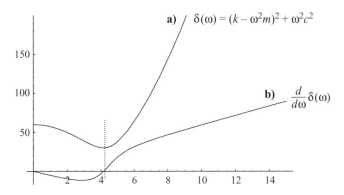

Figure 6.17
Plot of denominator of equation (6.62).

for ω yields two valid answers:

$$\omega = \pm\frac{\sqrt{2km - c^2}}{\sqrt{2}m}. \tag{6.63}$$

(Though I've dropped the square root from the definition of $\delta(\omega)$, the derivative of the positive root is still valid.) Since ω is the frequency of a phasor, the positive and negative solutions correspond to angular velocity of the clockwise and counterclockwise rotations of the phasor, respectively. We can take the positive version to indicate the positive resonant frequency, and the other to indicate the corresponding negative resonant frequency, for some combination of the parameters m, c, and k.

A slight rearrangement of equation (6.63) reveals its significance more clearly. First we square it, then expand it and simplify,

$$\omega^2 = \frac{2km - c^2}{2m^2}$$

$$= \frac{k}{m} - \frac{c^2}{2m^2},$$

then restore it by taking its square root again:

$$\omega = \pm\sqrt{\frac{k}{m} - \frac{c^2}{2m^2}}. \tag{6.64}$$

As we've honored the importance of the resonant frequency parameter in the other harmonic oscillators, let's do so here as well:

$$\omega_{rdd} = \omega = \pm\sqrt{\frac{k}{m} - \frac{c^2}{2m^2}} \qquad \begin{array}{l} \textit{Resonant Frequency of the Driven Harmonic} \\ \textit{Oscillator with Dissipation} \end{array} \tag{6.65}$$

is the *resonant frequency of the driven harmonic oscillator with dissipation*.

Notice that ω_{rdd} depends only on m, c, and k. This demonstrates, as we'd expect, that the resonant frequency varies only with the parameters of the resonator, not with the forcing function. Also, the resonant frequency is lowered by increasing dissipation.

Finding the Amplitude at Resonance   What is the amplitude of the oscillator at its resonant frequency? One way to find out would be to drive the harmonic oscillator with its resonant frequency and observe the resulting amplitude. Mathematically, we can do the same thing if we plug equation (6.65) for the resonant frequency into equation (6.62) for the magnitude of the oscillator. After some tedious but elementary fiddling around with algebra, we get

$$A_{max} = \frac{2F}{\sqrt{\dfrac{-c^4 + 4c^2km}{m^2}}}, \qquad\qquad \textit{Maximum Amplitude at Resonance} \ (6.66)$$

which is the maximum amplitude at resonance. The denominator of equation (6.66) will be zero if c, k, or m is zero. In that case, the value of A_{max} goes to infinity.

An Example   Let's take the same parameters as before, and plug $m = 3$ kg, $c = 7$ Ns/m, $k = 60$ N/m, and $F = 1$ into equation (6.66). This yields a maximum amplitude at resonance of

$$A_{max} = \frac{2F}{\sqrt{\dfrac{-c^4 + 4c^2km}{m^2}}} = \frac{6}{7\sqrt{671}} = 0.033.$$

Plugging the same parameters into equation (6.65) for the resonant frequency yields two solutions:

$$\omega_{rdd} = \pm\sqrt{\frac{k}{m} - \frac{c^2}{2m^2}} = \pm\frac{\sqrt{311/2}}{3} = \pm 4.157 \text{ Hz}.$$

From figure 6.16, it looks like we've got it about right: the amplitude is about 0.033 at a frequency just above 4 Hz.

The Effect of Damping on Resonance   What is the effect of the damping parameter c on the behavior of the resonator? Using the same parameters, $m = 3$ kg, $k = 60$ N/m, and $F = 1$, let's vary the damping factor c and see what happens to the resonance.

Figure 6.18 shows the responses of a driven harmonic oscillator when the damping factor c is set to 4, 6, 8, and 10. Note two influences of damping on resonance: the peak becomes less pronounced (which corresponds to a lower quality factor, Q), and the resonant frequency drops as c increases. As shown by the parametric line drawn through the peaks, the resonant frequency drops from 4.38 Hz to 3.8 Hz as c goes from 4 to 10.

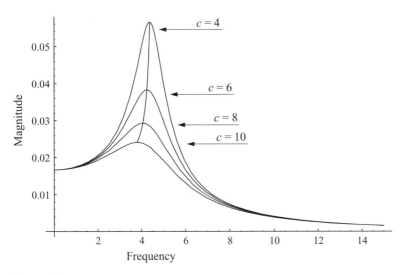

Figure 6.18
Response spectrum of a driven harmonic oscillator.

Summary

Resonance can be described with just three characteristics of vibration: displacement, velocity, and acceleration. These qualities can be seen as derivative aspects of displacement. The basic English meaning of derivative is "a fact that can be gleaned from other more basic facts." The derivative of a function $y(x)$ is the slope of a line tangent to the function at x.

The phrase "rate of change" holds the key to understanding the derivative: velocity measures the time rate of change of an object's position; acceleration measures the time rate of change of an object's velocity, and so forth. In general, the *rate of change* is defined as the slope that is tangent to the function at the point of interest. Velocity is the first derivative of displacement with respect to time. Acceleration is the first derivative of velocity with respect to time.

We can determine the derivative of a function by measuring the slope of a line tangent to the function at the point of interest. We begin with a line that intersects the curve at two points, then gradually bring one point toward the other until, in the limit, the line is tangent to the curve.

The derivative of a sine wave is a cosine wave. The derivative of a cosine wave is a negated sine wave. The derivative of e, the base of the natural logarithms, is identically e. The second derivative of a sine wave is the sine wave negated. Similarly, the second derivative of a cosine wave is the cosine wave negated.

Scaling the frequency of a sinusoid proportionately scales both the frequency and the amplitude of its derivative. Scaling the frequency of a sinusoid proportionately scales the frequency and negates and squares the amplitude of its second derivative (acceleration).

Differential equations relate a primary quantity such as displacement to derived quantities such as velocity and acceleration, allowing us to relate physical properties of vibrating systems such as musical instruments to mathematical representations of their vibration.

First-order differential equations cover physical systems where displacement is proportional to velocity. This characterizes, for example, the way energy from a plucked string dissipates through time.

Second-order differential equations cover systems that include acceleration. This describes, for example, instruments with a driven vibrating system, producing a sustained tone, such as the bowed string, woodwinds, and brasses.

Using second-order equations, we can describe transient vibrations—what happens when a driving force is applied to an instrument, and what happens when it is taken away. The harmonic equation describes simple harmonic motion in the most elegant form. This corresponds to undamped vibration. The damped harmonic equation elegantly characterizes this large class of resonators.

Differential calculus allows us to identify otherwise hard-to-explain facts about resonant systems, such as the circumstances under which they will not vibrate, depending upon values for mass, elasticity, and dissipation. There are three typical cases: underdamping, critical damping, and overdamping. We can also determine the resonant frequency and peak amplitude of a resonator, track how amplitude varies with frequency, and observe the effect of damping on resonance.

7 The Wave Equation

I do not know what I may appear to the world; but to myself I seem to have been only like a boy playing on the seashore, and diverting myself now and then by finding a smoother pebble or a prettier shell than ordinary, whilst the great ocean of Truth lay all undiscovered before me.
—Sir Isaac Newton

The characteristic movement of simple harmonic motion produces the shape of a sinusoid when its displacement is plotted against time (volume 1, chapter 8). What about the vibrating patterns of higher-dimensional shapes, such strings, membranes, and air? While it may seem daunting to consider more complex systems, it needn't be so. The *wave equation* provides a unified perspective for all forms of physical vibration in terms not much more complex than simple harmonic motion. It describes, for some point on an object, how its displacement from equilibrium changes from moment to moment based on the forces in its immediate neighborhood. This is the same approach taken for simple harmonic motion. The wave equation provides a way to describe the characteristic vibrating patterns of musical instruments of all types and to understand exactly how sound propagates in air.

It makes sense that the forces of action and reaction act locally from one point on a string to the next. The wave equation simply generalizes this to all points on a line, surface, or volume. Since we're dealing with forces and masses, Newton's second law of motion figures prominently in the analysis (see volume 1, equation (4.24)).

7.1   One-Dimensional Wave Equation and String Motion

In its simplest form, the wave equation characterizes the motion in any medium where the restoring force on every particle is proportional to its displacement from its equilibrium position. We can model a one-dimensional medium as an ideal string (one having uniform density, with no dissipation, dispersion, or stiffness). With two dimensions, we can model a membrane such as an ideal drum head with similar properties. In three dimensions, we can model vibrations in air.

Suppose we have two consecutive frames from a movie of a vibrating string separated in time by Δt (figure 7.1). How could we predict that the shape of the string in the first frame would lead to that in the second? How could we predict that the point a distance x from one end would displace from position u_1' to u_2 during the elapsed time Δt? Figure 7.2 shows these two movie frames superimposed.

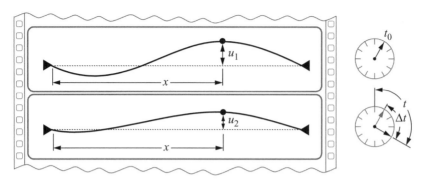

Figure 7.1
Deflection of a point on a string from equilibrium.

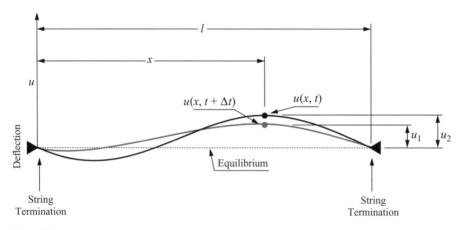

Figure 7.2
Superimposed adjacent movie frames of a vibrating string.

In general, to model the behavior of the string as a whole, we want to specify the deflection u from the string's equilibrium for *every* position x along the string and *every* time t. Mathematically, we're looking for a function that gives the correct deflection u for any point x along the string at any time t. Taken together, the deflection of all points along the string at all moments in time is a full description of the motion of the string. Let's call this function $u(x, t)$.

We restrict point x to lie between zero and the length of the string l, and restrict time to any $t \geq 0$. The function $u(x, t)$ is known as a *field function* because it specifies the deflection of the string in a two-dimensional field consisting of the one-dimensional space parameter x on one axis and time t on the other. We want to figure out what the field function $u(x, t)$ should be for a string.

Imagine a single piano string, and focus on a single point on the string. Imagine the string vibrating, and notice that nearby points on the string tug on this point and on each other as the string moves. Intuition suggests that the evolution of the string's deflection at the chosen point through time should be related to the values of the field function for nearby points.

7.1.1 Ideal String

To keep the problem manageable, let us idealize the kinds of strings we will consider by applying the following simplifying assumptions.

- The mass of the string per unit of length is constant; that is, the string is homogeneous.

- The string is perfectly elastic and offers no resistance to bending. This allows us to ignore non-linear effects due to stiffness.

- We ignore the distorting effects of gravitational force on the string because we assume its tension is much greater than gravity.

- We restrict our interest to small vibrations of the string. The deflection and slope of the string is never very large. This allows us to avoid nonlinear effects due to the fact that real strings are not infinitely elastic.

- Whereas a real string can exhibit transverse vibrations in two dimensions (up/down and backward/forward) as well as longitudinal and torsional vibrations, we restrict ourselves to examining only one-dimensional, vertical vibration. This allows us to avoid the nonlinear interactions that can arise between these modes when they are all active simultaneously.

While these assumptions simplify the analysis, the results are still useful when applied to real strings.

7.1.2 Vibration of the Ideal String

Consider a section of an idealized string between points P and Q (figure 7.3). Let vectors T_P and T_Q represent the force on the ends of the string section demarcated by points P and Q. In the moment represented in the figure, point P is experiencing a force due to string tension (x_P) pulling to the left and also a force downward (y_P). The vector sum of x_P and y_P is T_P. Point Q is experiencing a force due to string tension (x_Q) pulling to the right and also a force upward (y_Q). The vector sum of x_Q and y_Q is T_Q. Because the string has no stiffness (no resistance to bending) the tension on the string will be tangent to (that is, in line with) the curve of the string at each point, including points P and Q. The slopes of vectors T_P and T_Q vary depending upon the deflection of the string.

We can observe the slope of the string at points P and Q by breaking down the vectors T_P and T_Q into their horizontal and vertical components, as shown in figure 7.3. Angles α and β show the slope of the vectors at the two points.

Since we are excluding longitudinal wave motion, the horizontal components of the tension (x_P and x_Q in the figure) are the same and are also constant. With this in mind, we can express the horizontal components of the tension using elementary trigonometry:

$$x_P = x_Q = T_P \cos \alpha = T_Q \cos \beta = T, \tag{7.1}$$

where T is the constant horizontal tension on the string.

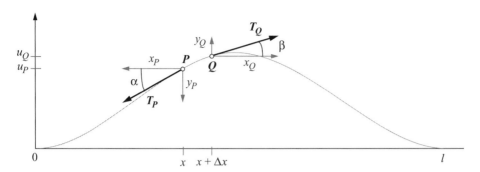

Figure 7.3
Idealized vibrating string.

We can express the forces in the vertical direction similarly. Note that since T_P happens to be directed downward at this moment, its vertical component y_P is negative. It can be expressed as $y_P = -T_P \sin \alpha$. The vertical component of T_Q can be written $y_Q = T_Q \sin \beta$. The net restoring force acting on the string segment in the vertical direction is the combination of these forces:

$$T_Q \sin \beta - T_P \sin \alpha. \hspace{4cm} \textit{String Restoring Force} \quad (7.2)$$

Because of the tension on the string, the net restoring force will be negative if the segment is above the equilibrium point (because the string is generally being pulled downward toward equilibrium) and positive when it is below.

Since the string is homogeneous, the average mass of the string segment is just its mass density ρ times the length of the segment, Δx. Since we are trying to understand the motion of this string segment, remember that Newton's second law of motion characterizes acceleration a in terms of the mass m of an object and the force f applied to it: $f = ma$ (see volume 1, section 4.12). We've already identified the vertical force and the mass, so we can represent the vertical movement of the string segment using this law as follows:

Force = Mass · Acceleration,

$$f = m \cdot a \hspace{10cm} (7.3)$$

$$T_Q \sin \beta - T_P \sin \alpha = \rho \Delta x \cdot a_x,$$

where a_x is the acceleration of some point x on the string. Divide both sides of equation (7.3) by string tension T, yielding

$$\frac{T_Q \sin \beta}{T} - \frac{T_P \sin \alpha}{T} = \frac{\rho \Delta x}{T} a_x. \hspace{5cm} (7.4)$$

But since equation (7.1) indicates that all the horizontal force vectors are equal to each other and equal to T (in particular, $T_P \cos \alpha = T_Q \cos \beta = T$), we can substitute any of these terms we like for T in equation (7.4) and obtain

$$\frac{T_Q \sin \beta}{T_Q \cos \beta} - \frac{T_P \sin \alpha}{T_P \cos \alpha} = \frac{\rho \Delta x}{T} a_x. \tag{7.5}$$

From elementary trigonometry we know that $\tan \theta = \sin \theta / \cos \theta$ (see appendix section A.2). So we can rewrite equation (7.5) as

$$\tan \beta - \tan \alpha = \frac{\rho \Delta x}{T} a_x. \tag{7.6}$$

Dividing both sides by Δx, we have

$$\frac{\tan \beta - \tan \alpha}{\Delta x} = \frac{\rho}{T} a_x. \tag{7.7}$$

Now, $\tan \alpha$ and $\tan \beta$ are the *slopes of the string* where it touches points P and Q, respectively. The *difference* between these two slopes, $\tan \beta - \tan \alpha$, is the *amount of bending* that the string segment undergoes between points P and Q. Remember from volume 1, section 4.10.1, that the *acceleration of a curve* and the *bend of a curve* amount to the same thing. With this in mind, we can define the average bend $\bar{b}$ of a string over an interval using the same reasoning we used for average acceleration:

$$\bar{b} = \frac{\tan \beta - \tan \alpha}{\Delta x}. \qquad\qquad \textit{Average Bend of a Curve} \tag{7.8}$$

This says that the average bend of a curve is the difference between its beginning and ending slope, divided by its length.

We can use this to define the bend of a curve in an infinitesimal region around a point x by letting the size of the segment Δx shrink to a vanishingly small size, just as we did for acceleration:

$$b_x = \lim_{\Delta x \to 0} \frac{\tan \beta - \tan \alpha}{\Delta x}, \qquad\qquad \textit{Bend of a Curve} \tag{7.9}$$

where b_x is the bend of the string at point x.

There is only one more detail before we can put all the elements of the wave equation together. Recall that the average mass of the string segment is $\bar{m} = \rho \cdot \Delta x$, which is the mass density times the length of the segment. Since we're now talking about an infinitesimal point on the string, we must figure out what the infinitesimal mass of the string is. Perhaps it comes as no great surprise that it's just the mass density itself, so the mass of the string in an infinitesimal region around a point x is simply

$$m_x = \lim_{\Delta x \to 0} \frac{\Delta x \rho}{\Delta x} = \rho. \qquad\qquad \textit{Infinitesimal Mass of a String} \tag{7.10}$$

7.1.3 Assembling the Wave Equation

Now we can put the wave equation together. Using the definition of bend b_x in equation (7.9) as the force component, and the definition m_x in equation (7.10) as the mass component, we can rewrite Newton's second law of motion as expressed in equation (7.7) as:

$$f = ma,$$
$$b_x = \frac{\rho}{T} a_x, \tag{7.11}$$

where a_x is the acceleration of the string at point x.

The bend at a point on a string is proportional to the acceleration at that point, scaled by the mass density and inversely scaled by the tension.

Greater mass density will tend to increase the bend, and greater tension will tend to decrease the bend. A particularly revealing form of the wave equation comes when we multiply both sides by T:

$$Tb_x = \rho a_x. \tag{7.12}$$

The restoring force, which is the string tension times the curvature of the string, is always balanced by the inertial force, which is the mass density times acceleration.

Thus, the wave equation balances the restoring force and the inertial force, just as any other system vibrating in simple harmonic motion must do.

Recall that tension T and mass density ρ are the two mechanical properties that determine the speed of wave motion in a string. The speed of sound c in a string is $c = \sqrt{T/\rho}$ (see volume 1, equation (8.15)). From this, we can rewrite equation (7.11) to read:

$$b_x = \frac{a_x}{c^2}. \tag{7.13}$$

The curvature at a point on a string is related to the acceleration of the string at that point and inversely related to the square of the speed of wave propagation in the string.

7.1.4 A Derivation Using Partial Derivatives

I have excluded the use of the calculus in the preceding derivation so as to make it more accessible. But we'll get a much clearer picture if we recast these results using it.

So far we've only considered derivatives of functions of one variable of the form

$$z = f(x). \tag{7.14}$$

But suppose we must differentiate a real function of *two* independent real variables. That is in fact what we must do with the wave equation because the displacement u of a point on the string depends not only on its position x along the string but also on time t. So we want to differentiate $u = f(x, t)$, which describes the string's displacement with two variables.

But we cannot do this. What we *can* do is to hold one of the variables steady while we differentiate the other. For instance, if we hold x constant by setting it to a particular value, $x = x_1$, then the displacement at x_1 depends on time alone, and *that* we can differentiate. Conversely, if we set time to a constant, $t = t_1$, then the displacement at t_1 depends upon position alone, and we can differentiate that.

If the derivative of $u = f(x,t)$ with respect to x for a value x_1 exists, then the *partial derivative* of $f(x,t)$ with respect to x at the point (x_1, t_1) is denoted

$$\left.\frac{\partial f}{\partial x}\right|_{(x_1, t_1)} \quad \text{or} \quad \left.\frac{\partial u}{\partial x}\right|_{(x_1, t_1)}. \tag{7.15}$$

By the definition of the derivative, we have

$$\left.\frac{\partial f}{\partial x}\right|_{(x_1, t_1)} = \lim_{\delta x \to 0} \frac{f(x + \Delta x, t_1) - f(x_1, t_1)}{\Delta x}. \tag{7.16}$$

We can define the partial derivative of $u = f(x, t)$ with respect to t in the same way by holding x constant, say, equal to x_1, and differentiating with respect to t:

$$\left.\frac{\partial f}{\partial t}\right|_{(x_1, t_1)} = \left.\frac{\partial u}{\partial t}\right|_{(x_1, t_1)} = \lim_{\delta t \to 0} \frac{f(x_1, t_1 + \Delta t) - f(x_1, t_1)}{\Delta t}.$$

This has a very simple geometrical interpretation. Suppose we take a series of transparent movie frames of a vibrating string and stack them one behind the other so that the image of the string forms a surface, as in figure 7.7. Then the partial derivative of time t with respect to some particular point on the string x_1 corresponds to a vertical plane intersecting the surface at x_1 *in the time direction*. (In terms of figure 7.7, the plane would be receding into the page.) The partial derivative of position x with respect to some particular time t_1 corresponds to a vertical plane intersecting the surface lengthwise at t_1 *across the string*.

So the partial derivative is a "divide and conquer" strategy: we hold one variable at a time steady and operate on the other. Of course, we can differentiate these derivatives once more to obtain four *second partial derivatives*: x with x, x with t, t with x, and t with t:

$$\frac{\partial^2 f}{\partial x^2} = f(x, t). \qquad\qquad \textit{Second Partial Derivative} \tag{7.17}$$

If all the derivatives are continuous (we generally can't take the derivative of a function containing discontinuities), it can be shown that the order of differentiation does not matter. And, of course, we can keep going, creating third partial derivatives, and so on, by holding all but one of the variables constant.

7.1.5 Wave Equation with Partial Derivatives

There are two places in the wave equation where second derivatives come in handy (Kreyszig 1968). The first is the acceleration term, a_x, meaning the acceleration at point x, for instance, in equation (7.13). Recalling that acceleration is the second derivative of displacement u, we can write

$$a_x = \frac{\partial^2 u}{\partial t^2}. \qquad (7.18)$$

Again, the reason for partial differential notation is that displacement of a string depends upon both position and time.

The second place a derivative can be found in equation (7.13) is in the bend parameter b_x. Think about acceleration as a bending line.

Recall from equation (7.6) that $\tan \alpha$ and $\tan \beta$ are the *slopes* of the string where it touches points P and Q, respectively. Each slope corresponds to the *derivative* of displacement at its respective location. We can write

$$\tan \alpha = \left.\frac{\partial u}{\partial x}\right|_x \quad \text{and} \quad \tan \beta = \left.\frac{\partial u}{\partial x}\right|_{x+\Delta x}.$$

The *difference* between these two slopes is the amount by which the slope changes from point x to point $x + \Delta x$. In the limit, this difference is the second derivative of displacement u. Thus, we can write

$$b_x = \lim_{\Delta x \to 0} \frac{\tan \beta - \tan \alpha}{\Delta x} = \left.\frac{\partial u}{\partial x}\right|_{x+\Delta x} - \left.\frac{\partial u}{\partial x}\right|_x = \frac{\partial^2 u}{\partial x^2}. \qquad \textit{Force Due to Tension} \quad (7.19)$$

Now we're poised to assemble the wave equation using these differentials. Substitute equations (7.18) and (7.19) into equation (7.11):

$$f = ma \qquad \text{Newton's second law of motion}$$

$$b_x = \frac{\rho}{T} a_x \qquad \text{Equation (7.11)} \qquad (7.20)$$

$$\frac{\partial^2 u}{\partial x^2} = \frac{\rho}{T}\frac{\partial^2 u}{\partial t^2}. \qquad \text{Wave equation}$$

It is customary to define $c^2 = T/\rho$, which when substituted into (7.20), rearranges it into the following canonical form:

$$\frac{\partial^2 u}{\partial t^2} = c^2 \frac{\partial^2 u}{\partial x^2}. \qquad \textit{One-Dimensional Wave Equation} \quad (7.21)$$

The inverse of ρ/T is equal to the square of the speed of propagation.[1] We write c^2 for these physical constants to remind us that it must be positive. But if we take the square root of c, we obtain

$$c = \sqrt{\frac{T}{\rho}}, \qquad (7.22)$$

which has appeared numerous times in this book in various contexts (for example, see volume 1, equation (8.15)).

Equation (7.21) is called a *field equation* because such equations govern the local motion at each point in the vibrating body.

Where the string is most bent, it is accelerating the most, and the rate of acceleration is governed by the mass, tension, and displacement of the string.

7.2   An Example

Consider the sinusoidal string vibration in figure 7.4. Where it is most bent is where it is accelerating most rapidly. The string is not bent at the point where it crosses the x-axis, so the string segment experiences no acceleration there. The actual rate of acceleration achieved by the string at any point will be governed by the mass and tension of the string and the magnitude of the string's displacement.

Equation (7.13) and equation (7.21)—its incarnation in the calculus—provide an extremely general description of the possible states of a string. We can use these equations to estimate how a string's shape will evolve, moment by moment. We can model the vibration of a real string if we first establish

- The string's *physical conditions* of mass density, length, and tension
- The *initial conditions* of the string's deflection, velocity, and acceleration at time $t = 0$
- The *boundary conditions* describing what happens at the edges of the string

With these, the wave equation can be used to describe the motion of an ideal string in detail.

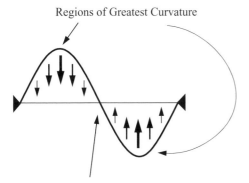

Regions of Greatest Curvature

Region of Least Curvature

Figure 7.4
Acceleration of a sinusoidal string vibration.

7.2.1 Boundary Conditions

If the ends of a string are at $x = 0$ and $x = l$, then the boundary conditions for a string tethered at both ends to an unmovable mass can be described as

$$u(0, t) = 0 \quad \text{and} \quad u(l, t) = 0 \quad \text{for all } t, \tag{7.23}$$

that is, the ends are stationary for all time. If the string is tethered to a movable mass, such as the body of a musical instrument, then some energy will be dissipated from the ends of the string into the instrument and radiated away as sound (and heat). For now, assume there is no movement at the string ends, and ignore the effects of the drag of the air and internal friction on the string.

7.2.2 Initial Conditions

Let's call $f(x)$ the initial deflection of the string at point x and the initial velocity of the string at the same point $g(x)$. Then these initial conditions will be

$$u(x, 0) = f(x) \quad \text{and} \quad v(x) = g(x), \tag{7.24}$$

where $v(x)$ is the instantaneous velocity of the string at point x.

For example, suppose we distend one of the idealized strings from its center and then release it, as shown at the top of figure 7.5. The initial conditions are $f(x) = \Lambda(x)$, where $\Lambda(x)$ is a triangular shape, and $g(x) = 0$, because the string initially has no velocity. Equation (7.13) predicts that the string will accelerate most where it is most bent, and it does: the center section accelerates fastest. Interestingly, the kink splits, then propagates from the center in both directions. The wave then reflects from the string ends and converges again.

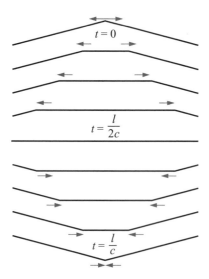

Figure 7.5
String plucked from the center.

7.2.3 Normal Modes

Suppose we were able to bend the string so that its initial shape is one half period of a sinusoid. When we released it, the string would demonstrate sinusoidal vibration (figure 7.6a). Figure 7.6b shows what would happen if we initially bent it in the shape of a whole period of a sinusoid. Since one full period of a sinusoid is exactly twice as curved as one half period, we'd expect its acceleration, velocity, and (consequently) its frequency to be doubled, which it is.

In fact, if the initial conditions are any sinusoidal shape of the form

$$u_n(x, 0) = f(x) = A_n \sin \lambda_n t \tag{7.25}$$

where A_n is the amplitude, n is an integer $1, 2, 3, \ldots$, and $\lambda_n = \pi n x / l$, where l is the string length, then the resulting shape $u_n(x, 0)$ is defined as the nth *normal mode* of the string. The functions $u_n(x, 0)$ are called the *characteristic functions* or *eigenfunctions* of the vibrating string, and the values λ_n are called the *characteristic values* or *eigenvalues* of the vibrating string. The eigenvalues constitute the frequencies of the normal mode vibrations of the string (see volume 1, section 8.5).

7.2.4 String Spectrum

The set of all eigenvalues $\lambda_1, \lambda_2, \ldots$ is called the *spectrum* of the vibration. Since

$$\sin \frac{\pi n x}{l} = 0, \qquad x = \frac{l}{n}, 2\frac{l}{n}, \ldots, n - 1\frac{l}{n},$$

the nth normal mode has $n - 1$ *nodes,* or points on the string that do not move (excluding the ends).

Perhaps most important, even if the initial condition for $u(x, 0) = f(x)$ is a sum of the eigenfunctions—even an infinite sum—the movement of the string will still be characterized properly by the wave equation, (7.13) and (7.21). Thus, not only are the eigenfunctions solutions to the wave equation but so are the sums of the eigenfunctions,

$$u(x, 0) = f(x) = A_1 \sin \lambda_1 t + A_2 \sin \lambda_2 t + \cdots$$

$$= \sum_{n=1}^{\infty} A_n \sin \lambda_n t. \tag{7.26}$$

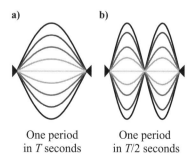

a)

One period
in T seconds

b)

One period
in $T/2$ seconds

Figure 7.6
Doubling the curvature doubles the frequency.

Such a sum is called a *harmonic spectrum*. In fact, for *any arbitrary string shape* which meets the boundary conditions $u(0, t) = 0$ and $u(l, t) = 0$ for all t, there is a set of amplitudes A_n that can be chosen such that the shape of a (possibly infinite) sum of characteristic functions matches that arbitrary shape.

In 1746, Jean le Rond d'Alembert demonstrated that in fact there are many shapes a vibrating string can assume that are not simple normal modes. He proved that (for an ideal string with no friction) one can start with *any shape,* and the shape will eventually repeat periodically in time. However, in between, it can change in very complicated ways. In response to d'Alembert's work, Euler developed the wave equation for a string, proving that arbitrary string shapes are simply (possibly infinite) sums of normal modes in suitable proportions. The upshot is that any vibration pattern of an ideal string (limited by the boundary conditions) can be created by superposing normal modes in suitable proportions. The normal modes are the basic components; all possible string vibration patterns are sums of multiples of (possibly infinitely) many normal modes.

7.3   Modeling Vibration with Finite Difference Equations

Perhaps the easiest way to see the wave equation in action is to imagine how a sampling of evenly spaced points on the string would behave if we observed them at evenly spaced moments in time. We could do this physically, for example, by painting glowing phosphorescent dots on the tips of a Shive wave machine (see volume 1, section 7.8.3), then taking a movie of its vibration in the dark.

When we observe a system only at discrete points in space and at discrete moments in time, we *discretize* the continuous dimensions of time and space. We start by fixing a string of length l horizontally along the x-axis and apply tension T. We name the point where the string is tethered at the left end x_0, then mark the string at N successive points, each a distance $\Delta x = l/N$ apart. Thus, we create a series of points on the string $x_j = x_0 + j\Delta x$, $j = 0, 1, 2, ..., N$. We can represent this as a row of points:

$$\begin{bmatrix} x_0 & x_1 & x_2 & \ldots & x_N \end{bmatrix}.$$

Similarly, we discretize time, and call the first moment of observation t_0. If moments of observation occur at a constant time interval Δt, we can name the moments of observation:

$$t_n = t_0 + n\Delta t, \qquad n = 0, 1, 2, \ldots, M,$$

where M is the number of observations.

We can represent these points as a column of times:

$$\begin{bmatrix} t_M \\ \vdots \\ t_1 \\ t_0 \end{bmatrix}$$

We are seeking to model how the string displacement at the points x_j change through the times t_n. That is, we want to find a function $u(x_j, t_n)$ that specifies the displacement u of the string at the jth position and the nth moment. But first, a more compact representation for $u(x_j, t_n)$ would be useful, so let's take j (the position index) and n (the time index) and invent the following notation:

$$u_j^n = u(x_j, t_n),$$

which says that the displacement u_j^n of the string at position j and time n equals $u(x_j, t_n)$.

Now, if we combine the row notation for the N positions with the column notation for the M times, we end up with an $N + 1 \times M + 1$ matrix U, where each entry corresponds to the displacement u_j^n of a point $x_j = j \cdot \Delta x$ on the string at a moment in time $t_n = n \cdot \Delta t$.

$$U = \begin{bmatrix} u_0^M & u_1^M & \cdots & u_N^M \\ u_0^{M-1} & u_1^{M-1} & \ddots & u_N^{M-1} \\ \vdots & \vdots & & \vdots \\ u_0^1 & u_1^1 & & u_N^1 \\ u_0^0 & u_1^0 & \cdots & u_N^0 \end{bmatrix} \quad \uparrow t$$

$$\xrightarrow{\quad\quad\quad} x$$

The bottom row of the matrix represents the displacement of the string at the start: u_j^0. The first and last columns (u_0^n and u_N^n) are determined by the boundary conditions. The other rows represent the displacement of the string while it is vibrating. We can use this matrix representation to show how the string's displacement evolves under the influence of nearby points in time and space.

But how shall we model the displacement of the string through time? If we're talking about a sequence we've already recorded, for instance, on film, we are free to investigate any particular time. But if we are watching an actual string vibrate, we have a problem: in this causal universe we don't have a priori access to events that will happen in the future. Thus we can only predict future displacements from current and past displacements. In the discrete matrix representation of the string, predicting what shape a string will take in the next moment is equivalent to knowing how to generate the next row in the matrix from the displacements of the current and previous rows. In order to make this prediction, we must understand how the wave equation relates displacement to time.

Suppose we're watching a movie of a string vibrating and stop the projector at frame n. For some point x_j on the string, the displacement will be u_j^n. How can we predict what the displacement of that point in the next frame will be? In other words, how can we predict u_j^{n+1}?

The second-order central difference approximation of acceleration (see volume 1, equation (4.16)) is just the tool we want here because it allows us to estimate acceleration from three adjacent observations. Adapting that equation to the matrix notation, the estimated acceleration a_j^n of a point $x_j = j \cdot \Delta x$ at time $t_n = n \cdot \Delta t$ is

$$a_j^n \approx \frac{u_j^{n+1} - 2u_j^n + u_j^{n-1}}{\Delta t^2}. \qquad\qquad \textit{Estimated Acceleration} \quad (7.27)$$

Recalling that acceleration and the amount of bend in a curve can be regarded as mathematically equivalent, we can adapt volume 1, equation (4.16), to estimate the bend of the string between three adjacent points. This is the spatial equivalent of acceleration. The estimated bend b_j^n of a point $x_j = j \cdot \Delta x$ at time $t_n = n \cdot \Delta t$ is

$$b_j^n \approx \frac{u_{j+1}^n - 2u_j^n + u_{j+1}^n}{\Delta x^2}. \qquad\qquad \textit{Estimated Bend} \quad (7.28)$$

Recall that equation (7.12) gives the wave equation in terms of instantaneous bend and acceleration: $Tb_x = \rho a_x$. Now, equation (7.27) expresses the estimated acceleration, and equation (7.28) expresses the estimated bend, based on finite differences. If we substitute these finite approximations into equation (7.12) it becomes

$$Tb_j^n = \rho a_j^n. \qquad\qquad\qquad\qquad (7.29)$$

Expanding equation (7.29) by substituting the definitions for finite bend and finite acceleration from (7.27) and (7.28), we have

$$T\frac{u_{j+1}^n - 2u_j^n + u_{j+1}^n}{\Delta x^2} = \rho\frac{u_j^{n+1} - 2u_j^n + u_j^{n-1}}{\Delta t^2}. \qquad\qquad (7.30)$$

Solving equation (7.30) for u_j^{n+1}, which is the displacement of point x_i at time t_{n+1}—that is, *the next moment in the future*—yields

$$u_j^{n+1} = \frac{T\Delta t^2}{\rho\,\Delta x^2}(u_{j+1}^n - 2u_j^n + u_{j-1}^n) + 2u_j^n - u_j^{n-1}. \qquad\qquad (7.31)$$

This can be simplified in practical situations by setting $\Delta t = 1$ s, and setting $\Delta x = \Delta t\sqrt{T/\rho}$, whereupon equation (7.31) is reduced to

$$u_j^{n+1} = u_{j+1}^n + u_{j-1}^n - u_j^{n-1}. \qquad \textit{Finite Difference Approximation of the Wave Equation} \quad (7.32)$$

Thus, if we know the displacement u at all the points x_j along the string at the times corresponding to t_n and t_{n-1}, we can get a numerical estimate of the displacement of the string at its *next moment*, $u(x_j, t_{n+1})$. This is known as the finite difference approximation of the wave equation. The approximation becomes exact in the limit as Δt and Δx approach zero.

Figure 7.7 shows a simulation of an ideal string with sinusoidal initial displacement. Two periods are shown, but the vibration would continue forever because there is no frictional force.

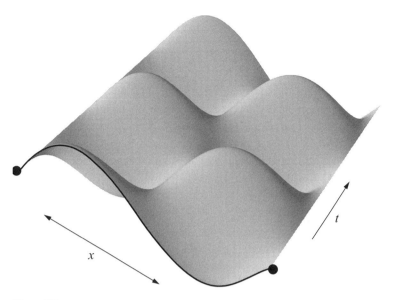

Figure 7.7
Ideal string vibration pattern.

7.3.1 Modeling Strings with Dissipation and Radiation

Any real vibrating string will show energy loss through time because of the nonconservative forces involved, such as the viscosity of the surrounding air. The internal molecular frictional forces within the string also convert energy to heat, the more so, the greater the stiffness of the string. Friction not only damps out free oscillations but also changes the frequencies of vibration (see figure 6.18). Also, the energy in the string propels the slightly yielding string anchoring points at the bridge and nut of the instrument to vibrate. Propelling forces are nonconservative because the energy is radiated away. The body of a stringed instrument receives most of the energy from the string. Its nonconservative energy is coupled to the air, which transmits it to our ears.

Dissipation   If the frictional force is linear, then the amount of friction a body experiences is proportional to the velocity of the body. We can model this by adding a term to the wave equation that drains energy in proportion to its velocity. Since velocity is the first derivative of displacement, the velocity term is

$$\frac{\partial u}{\partial t}.$$

We want to scale this by a coefficient $b_1 > 0$ that represents the amount of damping in order to control the strength of the frictional force. Therefore, we can expand the ideal wave equation (7.21) to account for energy lost by friction in the form of heat:

$$\frac{\partial^2 u}{\partial t^2} = c^2 \frac{\partial^2 u}{\partial x^2} - b_1 \frac{\partial u}{\partial t}. \qquad\qquad \textit{Wave Equation with Dissipation} \quad (7.33)$$

Radiation   The linear frictional force damps all frequencies by the same amount. But we know that higher-frequency components are typically attenuated more rapidly than lower-frequency components in musical instruments. For example, a harpsichord tone is bright at first but darkens as it dies away. (See volume 1, section 8.10.2, for an explanation of why high-frequency components die out faster.) So we must provide an additional rule to account for sound radiation.

In searching for such a second term, Ruiz (1969) and Hiller and Ruiz (1971) examined higher-order derivatives of displacement such as

$$b_2 \frac{\partial^2 u}{\partial t^2}, \; b_3 \frac{\partial^3 u}{\partial t^3}, \; \ldots, \; b_n \frac{\partial^n u}{\partial t^n}.$$

To evaluate the impact of these terms on the wave equation, they equated the motion of a given point on a string with a simple sinusoid $y = A \sin \omega t$. By the rules of differentiation (see section 6.1.4), the first four derivatives are

$$\frac{du}{dt} = A\omega \cos \omega t, \qquad \frac{d^2 u}{dt^2} = -A\omega^2 \sin \omega t, \qquad \frac{d^3 u}{dt^3} = -A\omega^3 \cos \omega t, \qquad \frac{d^4 u}{dt^4} = A\omega^4 \sin \omega t.$$

There is a subtle repeating pattern here:

$$\frac{d^3 u}{dt^3} = -\omega^2 \frac{du}{dt} \quad \text{and} \quad \frac{d^4 u}{dt^4} = -\omega^2 \frac{d^2 u}{dt^2},$$

so in general, for $n \geq 1$,

$$\frac{d^{2n+1} u}{dt^{2n+1}} = (-1)^n \omega^{2n} \frac{du}{dt} \qquad\qquad\qquad\qquad (7.34)$$

and

$$\frac{d^{2n+2} u}{dt^{2n+2}} = (-1)^n \omega^{2n} \frac{d^2 u}{dt^2}. \qquad\qquad\qquad\qquad (7.35)$$

Equations (7.34) and (7.35) thus represent classes of derivatives of integer orders. Odd-ordered derivatives are all in phase with the first derivative, velocity. All even-ordered derivatives are in phase with the second derivative, acceleration, and are in quadrature phase with the first derivative. Ruiz associates the odd-ordered derivatives with damping, which affects all frequencies alike. The even-ordered derivatives discriminate with frequency because higher modes show increased acceleration at higher vibration modes, corresponding to a decrease in density of the string material in proportion to ω^{2n}.

From this, Ruiz chooses the lowest-order damping term that discriminates among frequencies,

$$-\frac{d^3 u}{dt^3},$$

which represents a damping force proportional to ω^2. Adding the two damping terms to the wave equation with dissipation (7.33) yields

$$\frac{\partial^2 u}{\partial t^2} = c^2 \frac{\partial^2 u}{\partial x^2} - b_1 \frac{\partial u}{\partial t} - b_3 \frac{\partial^3 u}{\partial t^3}, \qquad \textit{Wave Equation with Dissipation and Radiation} \quad (7.36)$$

where $b_3 > 0$ is a coefficient determining the loss due to sound radiation.

7.3.2 Stiffness

Stiffness provides an additional restoring force that is proportional to the amount of bend in a string or membrane. The total restoring force of a string is the sum of its stiffness and tension. Because higher-order normal modes bend more in comparison to their wavelength, stiffness is a relatively stronger force for higher-frequency modes. Since the overall restoring force is greater for higher modes, but the string mass is unchanged, the higher modes have stretched (that is, higher) frequencies. This produces the octave stretching seen, for example, in pianos (see volume 1, section 8.7.4).

As the thickness of a string increases in proportion to its length, it becomes more barlike (see volume 1, section 8.7.3). We must complicate the expression for restoring force to include the ratio of the moment of inertia to the cross-sectional area of the string. Citing Morse (1960), Hiller and Ruiz give the restoring force due to stiffness of a bar with length dx as

$$F_s = -ESR^2 \frac{\partial^4 u}{\partial t^4} dx$$

for small amplitude oscillations such as commonly occur with stringed instruments, where S is the cross-sectional area of the string, E is Young's modulus, I is the moment of inertia, and $R = I/S$, the radius of gyration of the cross-section.

We must now factor this additional force into the wave equation.

By equations (7.12) and (7.19), the force due to tension T alone is

$$F_y = T \frac{\partial^2}{\partial x^2}(u),$$

and so the total restoring force is

$$F = T \frac{\partial^2 u}{\partial x^2} - ESR^2 \frac{\partial^4 u}{\partial t^4}.$$

In addition to being a factor in stiffness, the mass per unit length also depends upon the cross-sectional area S of the string. In the case of the ideal string, the infinitesimal mass is simply the mass density of the string, ρ, so for a string with thickness, the infinitesimal mass is now ρS.

Combining these elements according to Newton's second law of motion, we have

$$f = ma,$$

$$T\frac{\partial^2 u}{\partial x^2} - ESR^2\frac{\partial^4 u}{\partial t^4} = \rho S\frac{\partial^2 u}{\partial t^2},$$

or solving for acceleration to put it back into standard wave equation form,

$$\frac{\partial^2 u}{\partial t^2} = \frac{T}{\rho S}\cdot\frac{\partial^2 u}{\partial x^2} - \frac{ER^2}{\rho}\cdot\frac{\partial^4 u}{\partial t^4}. \qquad \textit{Wave Equation with Stiffness} \quad (7.37)$$

7.3.3 Combining

Combining equations (7.36) and (7.37), we have a version of the wave equation that models dissipation, radiation, and stiffness:

$$\underbrace{\frac{\partial^2 u}{\partial t^2} = \frac{T}{\rho S}\cdot\frac{\partial^2 u}{\partial x^2}}_{\text{Ideal string}} \underbrace{- \frac{ER^2}{\rho}\cdot\frac{\partial^4 u}{\partial t^4}}_{\text{Stiffness}} \underbrace{- b_1\frac{\partial u}{\partial t} - b_3\frac{\partial^3 u}{\partial t^3}}_{\text{Damping}}. \qquad \begin{array}{l}\textit{Wave Equation with Stiffness}\\ \textit{and Damping}\end{array} \quad (7.38)$$

7.3.4 Examples

The plots in figure 7.8 utilize a difference equation based on equation (7.38) to model a string with dissipation and radiation. This effect of radiation appears most clearly in figure 7.8d: since high frequencies appear choppier than low frequencies, the fact that the plot smooths out over time indicates the loss of high-frequency energy.

Each of these examples is one particular solution out of the infinite number of solutions to the wave equation. Each particular solution is a complex harmonic spectrum that can be broken down into a (possibly infinite) sum of sinusoids at integer multiples of the fundamental frequency f such that

$$f = \frac{c}{2L} = \frac{1}{2L}\cdot\frac{T}{\rho S},$$

where L is the string length.

Each plot in figure 7.8 shows position x across the page, time t receding into the page, and displacement u vertically. Figure 7.8a shows what happens if the string is struck from below at the center. The impulse splits into two impulses, which spread out left and right along the string until they hit the string terminations, whereupon they invert and reflect back toward the center and meet again, this time inverted. This demonstrates that the analytical solution to the wave equation for the ideal string can also be expressed as

$$u(x, t) = f(x - ct) + g(x + ct), \qquad \textit{Analytical Solution of the Wave Equation} \quad (7.39)$$

where functions f and g correspond to the wave shape traveling in the positive and negative directions, respectively. For any time t, they are periodic functions of x with period of $2L$.

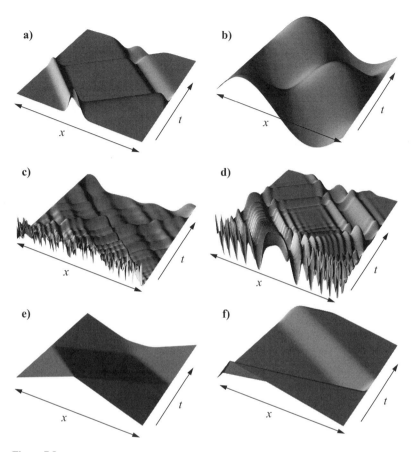

Figure 7.8
String vibrations resulting from various initial conditions.

In figure 7.8b the initial displacement of the string is one period of a sine wave (corresponding to the first overtone). Through time, all points progress smoothly with sinusoidal motion.

In figure 7.8c the initial displacement is an arbitrary random pattern. At the beginning of time, the string is jaggedly bent, indicating the presence of a great deal of high-frequency energy. Air resistance and internal string friction damp all frequencies on the string uniformly. Dissipation absorbs high-frequency energy faster than low-frequency energy, so as time passes we see the jaggedness give way to smoother (low-frequency) motion.

In figure 7.8d the string's initial displacement is an inharmonic function $u_x = \cos 10.5\,\pi x^2$. It too contains a great deal of high-frequency energy that is absorbed through time. These plots demonstrate that it is possible to put an ideal string into any arbitrary initial condition, and it will behave according to the dictates of the wave equation.

The string in figure 7.8e was plucked from the center, and the string in figure 7.8f was plucked near one end.

7.3.5 Three-Dimensional Wave Equation

It would be useful to generalize the wave equation to two and three dimensions so that we could consider, for example, the movement of waves in air. This is easy to do at this point. The ideal function of displacement in a three-dimensional medium with dimensions x, y, and z is simply

$$\frac{\partial^2 u}{\partial t^2} = c^2 \left(\frac{\partial^2 u}{\partial x^2} + \frac{\partial^2 u}{\partial y^2} + \frac{\partial^2 u}{\partial z^2} \right). \qquad \textit{Three-Dimensional Ideal Wave Equation} \quad (7.40)$$

The three-dimensional wave equation appears so commonly in the physics literature that it has been given its own conventional symbol, defined as follows:

$$\nabla^2 u = \frac{\partial^2 u}{\partial x^2} + \frac{\partial^2 u}{\partial y^2} + \frac{\partial^2 u}{\partial z^2},$$

called *del-squared* or the *Laplacian*. This reduces equation (7.40) to just

$$\frac{\partial^2 u}{\partial t^2} = c^2 \nabla^2 u.$$

An immediately useful solution to equation (7.40) is for the plane and spherical waves:

$$u = \frac{A}{r} \sin(kr - \omega t),$$

where k is a fixed vector, x is a relative vector, and radius $r = x$.

7.3.6 Applications of the Wave Equation

The one-dimensional wave equation describes motion in an ideal medium that is displaced along only one dimension. It can be adapted, for example, to model the vibration of a flute or organ pipe by substituting air pressure for string displacement, and longitudinal volume velocity for transverse string velocity. As we've seen, the equation can also be extended to two dimensions to simulate drum heads, and three dimensions to simulate dissipation of sound in air.

But where there are coupled interactions between the dimensions, the result is not necessarily a linear combination of wave equations. For example, a violin string moves in several dimensions: vertical and horizontal transverse motion as well as torsional (rotational) motion, which is caused by the bow twisting the string as it scrapes across it. The string also vibrates longitudinally. Since the elasticity in a real string is limited, large excursions in one dimension affect the tension of the string, changing its vibrating behavior in the other dimensions.

Discrete models of the wave equation can only approximate the underlying continuous wave equation. An initial disturbance can diffuse indefinitely in the continuous equation. But discretization breaks it up into individual sample points. Eventually, a discrete system will begin to repeat the same sequence because it is necessarily finite, whereas a continuous system might or might not repeat, depending on many factors.

7.4 Striking Points, Plucking Points, and Spectra

For plucked or struck strings and membranes, the position of the disturbing force, its shape, and its elastic properties have a strong effect on the resulting spectrum. A string's subsequent evolution is a function of the stiffness and frictional forces acting on it. If the force is applied continuously, such as by a bow, this force will further define the vibration.

If a drum head or string is struck near its edge, more energy is injected into the higher vibration modes, and the sound will be brighter than if the instrument had been struck near the center. Small hard mallets typically produce a brighter sound than large soft mallets. For plucked strings, sharp hard plectra typically produce a brighter sound than broad soft plectra. What all these factors share is that they affect the initial conditions of the vibration.

Consider the vibration shown in figure 7.8b. The disturbing force exactly matches the pattern of the second string mode; hence mode 2 receives essentially all the energy of the initial disturbance.

Mode Excitation Rule 1   *A mode will be excited in proportion to how closely the shape of the disturbing force matches the shape of the mode.*

However, in practice it is unlikely that a disturbing force will exactly mimic the shape of any one vibrating mode; instead, the energy in the disturbing force will often be distributed over many modes.

7.4.1 Plucking Position

We know that no matter what disturbs the equilibrium of a string or drum head, the resulting vibration will be a weighted sum of its normal modes. But how can we determine which modes will be excited and what their relative weights will be?

The plucked string provides perhaps the simplest case to analyze. Consider the vibration shown in figure 7.8e. In this figure, the string was initially stretched upward from its center and held still until it was stationary. The shape of the string in the moment before it was released forms two arms of an isosceles triangle. Therefore, the string's initial conditions consist of a nonzero initial displacement and a zero velocity. When released, the string evolves as shown in the figure: the initial kink splits in two, and these kinks propagate in opposite directions, reflecting and inverting at the string's terminations. See figure 7.5 for comparison. Now, which modes can be excited by these initial conditions? Which modes cannot?

Mode excitation rule 1 suggests that mode 1 should be excited strongly because the disturbing force (the force of the plectrum pushing the string up at its center) resembles the shape of mode 1 more than any others (see figure 7.9). That's because no other mode begins with the string all on one side of the equilibrium point. Points marked in the figure on opposite sides of the plucking point will travel in the same direction at the same time. The string moves up and down symmetrically.

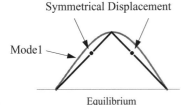

Figure 7.9
Mode 1 symmetry.

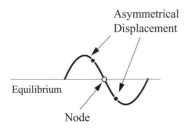

Figure 7.10
Mode 2 asymmetry.

Since the string is plucked exactly in the center, mode 2 cannot possibly receive any energy at all. That's because mode 2 vibration can only arise when equal energy is injected in opposite directions on each side of the string, creating a node in the middle (see figure 7.10). Striking the string exactly on the node of mode 2 cannot cause an upward motion on one side of the node and a downward motion on the other side. So there can be no energy in mode 2 if the string is struck or plucked exactly in the center and the striking force is perpendicular to the string. By extension, when a string is plucked in the center, odd modes $n = 1, 3, 5, \ldots$ have energy, but even modes $n = 2, 4, 6, \ldots$ have none.

We can generalize this as follows. Recall that a *node* is a point along a modal vibration where contrary vibrating forces cancel. The energy in a mode is proportional to the differential displacement on either side of the node of a mode. The energy in a mode cannot be increased by displacing the string at its nodal point because that's the one place where the differential displacement of the mode cannot be affected. Striking directly on any node provides no way for the energy to go into opposing motion on either side of the node (figure 7.11). These same ideas apply to drum head modes and acoustical modes of rooms.

Mode Excitation Rule 2   *Modes with nodes lying near or directly on a striking point receive, respectively, little or no energy from a disturbing force.*

So now we know that there are modes that won't receive energy, depending upon where we pluck or strike. But of the remaining modes, how is the energy from a disturbing force distributed among

Mode with Little Energy

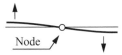

Mode with More Energy

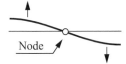

Figure 7.11
Relative mode energy.

them? For instance, as we pluck a string closer and closer to one edge, the tone of the string brightens. How do we account for this? Neither mode excitation rule sheds much light.

It can be shown that the amplitude of the nth odd mode of an isosceles triangle wave is inversely proportional to the square of the mode number (see chapter 3 and section 9.2.6). So, for a string plucked in the center, the first six modes are energized in the following proportion:

$$\frac{1}{1^2}, 0, \frac{1}{3^2}, 0, \frac{1}{5^2}, 0, \ldots.$$

If we pluck the string at the 1/8 point, we can predict by mode excitation rule 2 that multiples of mode 8 will receive no energy. But the rule describing how the energy from a pluck at that point is distributed among the remainder of the modes is not as straightforward. Figure 7.12 compares magnitude spectra for a string plucked at the midpoint and plucked at the 1/8 point. The spectrum for the midpoint pluck has most of its energy in mode 1 (corresponding to the fundamental pitch), energy drops off rapidly in higher modes, and the even-numbered modes are missing, as we'd expect from mode excitation rule 2. The spectrum for the 1/8 point pluck has less of its energy in mode 1 and more in its higher modes, and every eighth mode is missing, as we'd expect. Since the string was plucked both times with the same force, the modes received the same total energy, but plucking closer to the end of the string causes higher-order modes to receive more of the energy, resulting in a brighter sound. The mathematics governing the spectra in figure 7.12 are covered in chapter 3.

7.4.2   Breadth of the Striking Force

Whereas plucking a string at equilibrium implies nonzero initial displacement but zero initial string velocity, striking a string at equilibrium implies zero initial displacement but nonzero initial

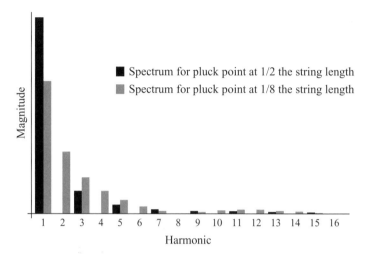

Figure 7.12
Comparison of spectra: a string plucked at midpoint vs. 1/8 point.

string velocity. A hammer blow (figure 7.13) transfers an impulsive momentum to the string along the breadth of the hammer. The broader the face, the broader the impulsive force. Thus, the shape of the striking force influences the resulting spectrum.

Figure 7.13a, for example, characterizes the displacement of an ideal string from a broad piano hammer blow, and figure 7.13b characterizes a narrow hammer blow of equal displacement. Let $f(x)$ be a function describing the distribution of the piano hammer's force along the string. For real strings, $f(x)$ is influenced by many factors, but I model it as an exponential function for simplicity. Figure 7.13c compares the normalized spectra when the distribution of force is the exponential function

$$f(x) = e^{-[2x/(wl)]^2},$$

where x is the position along the string, l is the length of the string, and w is proportional to the width of the hammer face. For figure 7.13a, $w = 1/4$ and for figure 7.13b, $w = 1/16$. Note that the narrower the displacing force, the broader the resulting spectrum, and vice versa.

7.4.3   Hardness of the Striking Force and Contact Duration

Other properties of the striking force affect the spectrum as well, such as the hardness of the hammer and precisely how that hardness yields during the strike of the hammer against the string. Soft hammers give a muffled sound, whereas very hard ones produce a bright sound.

It turns out that piano hammers and many kinds of percussion mallets are designed so that the stronger the blow, the harder the hammer becomes, because the hammer head compresses according to a power law. This means that piano hammers and some percussion mallets do not strictly obey Hooke's law, which describes a linear relation between an applied force F and resulting compression x.

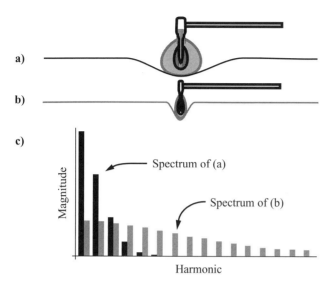

Figure 7.13
Shape of striking force and resulting spectrum.

Recall that Hooke's law is $F = kx$, where k is a constant depending upon the stiffness of the spring (see volume 1, section 8.1.2).

Research suggests that the nonlinear relation between force and compression can be approximated by $F = kx^p$, where p characterizes how the stiffness varies with the applied force (see Suzuki 1987). For instance, if $p = 1$, we have the linear Hooke's law relation, but if $p = 2$, the compression (and hence the hardness of the hammer) increases with the square of the applied force. The significance of this is that when $p > 1$, louder tones are also brighter.

Here's an intuitive explanation. Because of the nonlinear relation between force and compression, greater hammer striking force produces much greater stiffness in the felt covering of the hammer. Increased hardness in turn causes the hammer to rebound from the string more quickly, thereby reducing the contact duration. The less time the hammer is in contact with the string, the less high-frequency energy can dissipate back out of the string and into the hammer. Therefore, hammer contact duration is inversely proportional to string brightness.

Coupling spectral brightness and intensity in this way is a musically desirable feature. If there were no brightening with intensity, then striking a key with greater force would simply produce a greater sound intensity, and loud notes would sound like soft notes, only louder. This could lead to confusion as to whether the tone came from a louder piano or a closer piano. Since many other musical instruments couple increased loudness with increased spectral brightness, having this capability allows pianos and percussion instruments to blend musical dynamics effectively with, for example, an orchestra. Experiments have suggested that pianos tend to have p values in the range $1.5 < p < 3$. For p values greater than 3, there is too much contrast between the timbre of loud and soft tones; if $p < 1.5$, there is too little contrast.

Summary

The wave equation describes, for some point on an object, how its displacement from equilibrium changes from moment to moment, based on the forces in its immediate neighborhood. It is just an extension of the approach we took to describe simple harmonic motion.

The wave equation characterizes motion in any medium where the restoring force on every particle is proportional to its displacement from its equilibrium position.

To simplify the problem, we begin by modeling ideal media that have uniform density and no dissipation, dispersion, or stiffness. Geometrical analysis of an ideal string displaced from equilibrium shows that the forces acting on a segment of it obey Newton's laws of motion. Force on the segment comes from the tension on the string and the tension caused by its displacement; mass is the mass density of the string times the length of the segment; acceleration is related to the amount of bending experienced by the segment.

We then defined the bend of an infinitesimal region around a point on the string by reducing the segment length to the limit. We observed the following:

- The bend at a point on a string is proportional to the acceleration at that point, scaled by the mass density and inversely scaled by the tension.

- The restoring force (string tension times the curvature of the string) is balanced by the inertial force (mass density times acceleration).

- The curvature at a point on a string is related to the acceleration of the string at that point by the square of the speed of wave propagation in the string.

We repeated the derivation using partial derivatives. This was necessary because the displacement of a string depends upon both position and time, and partial derivatives are a way of taking the derivative of functions having multiple arguments. Partial derivatives can be thought of as a "divide and conquer" strategy: we hold one argument at a time steady and operate on the other. We used partial derivatives to relate acceleration of the string to the amount of bend in the string. We found that

- Where the string is most bent, it is accelerating the most, and the rate of acceleration is governed by the mass, tension, and displacement of the string.

We observed that wave motion is determined by initial conditions: initial displacement and initial velocity. We found that the motion of a string is completely described by its normal modes. The spectrum of a string is all the modes the string can produce.

We investigated a method of simulating wave motion using finite difference equations, which are discrete equivalents of the continuous wave equation. This allowed us to model various initial conditions and observe the results graphically.

We then looked at more realistic strings, including dissipation, radiation, and stiffness. We looked at higher-dimensional versions of the wave equation.

Finally, we considered what happens when a string is plucked, or struck with different sized plectra, or with mallets of different hardness and breadth.

8 Acoustical Systems

Why do rhythms and melodies, which are composed of sound, resemble the feelings, while this is not the case for tastes, colors or smells? Can it be because they are motions, as actions are also motions? Energy itself belongs to feeling and creates feeling. But tastes and colors do not act in the same way.
—Aristotle

We have seen how conservative forces sustain vibration within strings, pipes, membranes, and bars, holding energy in a vibrating system (volume 1, chapter 8). This chapter focuses on the way energy flows between coupled acoustical systems. The study of acoustics is greatly simplified by understanding the circumstances governing the flow of sound energy because instruments, ears, and rooms can all be viewed as networks of interconnected vibrating elements.

For simplicity, the focus here is on one-dimensional systems such as musical instrument tubes, the subject of classical hydrodynamics. With this approach we can look at wave motion in the bore of a flute but can't, for example, study the turbulent air flow from the player's lips over the fipple that drives its vibration. Fluid mechanics does not investigate the molecular basis of fluid media but assumes that the medium is homogeneous and continuously distributed in space. This approach allows us to model any vibrating medium with the concept of infinitesimal limits in a way that would not work with actual particulate air, and so it is called *continuum mechanics*. Nonetheless, the model succeeds because the molecular particles of air, being so much smaller than the volumes we consider, are effectively infinitesimal.

8.1 Dissipation and Radiation

We play a musical instrument by injecting energy into it. The instrument then dissipates the energy over time by sound radiation and heat until it returns to its original energy state. Conservative forces store energy via elasticity and inertia. Nonconservative (dissipative) forces transfer and transform energy as heat and sound. These two classes of force give us music as we know it.

Vibration arises by the interplay of the conservative forces of elasticity and inertia. We are able to hear a vibration because dissipative forces transfer the sound to our ears.

A vibrating tuning fork held in the air makes relatively little sound and rings relatively longer than a fork with its butt end held against a table top. When the fork is held in the air, energy is radiated away as sound from the tines, and converted to heat by the internal frictional forces of the tines and the viscosity of the surrounding air. Held against a table, the rate of sound radiation increases because the table couples the energy in the fork to the surrounding air more efficiently. If the same striking force is used in both cases, we hear a louder tone for a shorter duration when the fork touches the table, because the same unit of energy experiences increased dissipation.

Likewise, an electric guitar string will vibrate longer for the same amount of injected energy than a banjo string because of the greater coupling to the surrounding air that the banjo head provides.

We can achieve a very fundamental perspective on the acoustics of musical instruments, rooms, and the ear by understanding how their conservative and dissipative forces interact. The qualities of these forces and their interconnections determine how instruments make the sounds they do. This chapter investigates the properties of acoustical elements and how they can be linked together.

8.2   Acoustical Current

Acoustical current is the velocity or the rate of flow I of a volume V of air.

Suppose I had a long tube with a plunger inside, open at both ends. As I push the plunger towards the right, the positive pressure of the plunger on the air creates a current I of air which also moves to the right (figure 8.1).

We can express acoustical current as the rate at which a volume of air flows:

$$I = \frac{V}{t}, \qquad\qquad \textit{Acoustical Current, or Volume Velocity} \quad (8.1)$$

where V is volume in cubic meters and t is seconds. So we measure current of air in cubic meters per second.

Another common name for acoustical current I is *volume velocity*. This term only makes sense when applied to enclosed spaces, where we can easily measure volume. In the open air, we can equate volume velocity to *average particle velocity,* in meters per second. While a gentle breeze blowing through a window might have an acoustical current of 1 m^3/s, the air current corresponding even to a very loud sound is vanishingly small in comparison. (See volume 1, section 4.24.1.)

Figure 8.1
Acoustical current.

8.2.1 Acoustical Charge

Suppose a pump, filling a balloon, applies a current of air for some time t. The total quantity of air moved, called the *acoustical charge,* depends upon the strength of the current and how long it is applied. If we run the pump for a longer time, it will move more air; likewise if we increase the force of the pump. Solving equation (8.1) for volume, we have

$$V = It. \hspace{3cm} \textit{Acoustical Charge} \hspace{0.3cm} (8.2)$$

Total charge is equal to the current (rate of flow) times the duration of the current's flow. This sense of the term V does not mean the size of a static container filled with air. It means rather a quantity of air that flows.

8.2.2 Pressure

Pressure is force per unit of area:

$$P = \frac{F}{A}. \hspace{2cm} \textit{Pressure as Force per Unit Area} \hspace{0.3cm} (8.3)$$

The unit of measure of pressure is the pascal (abbreviated Pa), which is the force of 1 newton per square meter. For historical reasons, meteorologists usually work with a slightly different unit, a millibar (mb), defined as 100 Pa.

8.2.3 Pressure and Energy

Pressure can also be seen as the energy $W = Fs$ of a current of air $V = As$:

$$P = \frac{F}{A} = \frac{F \cdot s}{A \cdot s} = \frac{W}{V} \hspace{3cm} (8.4)$$

Eq. (8.3) $\nearrow$ $\qquad\qquad$ $\searrow$ Vol. 1, eq. (4.26)

where s is distance. This is also called *energy density*.

The kinetic energy of moving air can usefully be expressed as kinetic energy per unit volume:

$$\frac{\frac{1}{2}mv^2}{V} = \frac{1}{2}\rho v^2, \hspace{1.5cm} \textit{Kinetic Energy per Unit Volume} \hspace{0.3cm} (8.5)$$

where ρ is the density of the medium, expressed as the ratio of mass to volume:

$$\rho = \frac{m}{V}. \hspace{2.5cm} \textit{Density of a Medium} \hspace{0.3cm} (8.6)$$

The potential energy of moving air (presumed to be near the earth's surface so gravity is relatively constant) can usefully be expressed as potential energy per unit volume:

$$\frac{mgh}{V} = \rho gh, \hspace{1.5cm} \textit{Potential Energy per Unit Volume} \hspace{0.3cm} (8.7)$$

Figure 8.2
Acoustical resistor.

where g is gravity, h is height, and ρ is mass density. Recall that mass is a fundamental property of an object, measured by its inertia in terms of Newton's laws of motion $m = f/a$.

It's worth noting that air flowing within a tube tends to move in layers (*laminar flow*) with successively higher speeds toward the center of the flow, away from the walls of the tube because the outermost layer of air experiences friction with the walls of the tube. This layer in turn slows down the next most interior layer, and so on. Air may also move turbulently in a tube, such as air injected across a flute's fipple by a flutist's lips.

Resistance   Restricting air flow at the end of a tube makes it harder to push a plunger because the viscosity of the air passing through the restriction increases the *resistance R* to the flow of current. As it is driven through an acoustical resistor (figure 8.2), acoustical energy is changed into heat and/or sound. Heat is dissipated away, and sound is radiated away. Classical physics describes resistance using a dashpot (see figure 6.13). The automotive shock absorber is also a resistor that converts unwanted vibrational energy in a car's suspension system into heat. These hydrodynamic resistors all use viscosity to dissipate and radiate energy away from the system.

Ohm's Law for Acoustics   Resistance can be measured in terms of how pressure affects volume velocity. For instance, if we increase resistance R by making the hole at the end of the tube smaller but still wish to preserve the same rate of flow I, we must correspondingly increase the pressure P on the plunger.

Acoustical current is proportional to air pressure and inversely proportional to air resistance.

This is an acoustical formulation of *Ohm's law,*[1] which we can express as

$$I = \frac{P}{R}.$$ *Ohm's Law of Electricity, Adapted to Acoustics* (8.8)

Solved for R, it becomes

$$R = \frac{P}{I},$$ *Acoustical Resistance* (8.9)

which says that air resistance is the ratio of the pressure we apply to the volume velocity we obtain.

Resistance is measured in SI units of *acoustical ohms*. Unfortunately, there has been a bit of confusion about the exact units of measure for the acoustical ohm, so it is usually necessary to say explicitly which measurement is intended. My policy in this book is to use the SI units (see volume 1, chapter 4), and the SI unit of pressure is the pascal, which is 1 newton per square meter.

Acoustical current is measured as cubic meters per second, $I = V/t$. Putting the pieces together, 1 acoustical ohm Ω_a is therefore equal to 1 pascal second per cubic meter:

$$\Omega_a = \text{Pa} \cdot \text{s}/\text{m}^3. \hspace{3cm} Acoustical\,Ohm \hspace{0.3cm} (8.10)$$

8.3 Linearity of Frictional Force

In gases and fluids, resistance is generally a linear force, so that, for example, doubling the resistance halves the flow of air. However, in mechanical systems, resistance is often a nonlinear function. The brakes in an automobile typically have disproportionately more friction at low velocities than high, explaining why the brakes tend to grab at the last moment before coming to a stop.

8.3.1 Helmholtz Motion

The nonlinearity of mechanical friction is responsible for the characteristic sound of the violin. The frictional force is strong between the bow and its string when they are traveling at the same relative velocity (figure 8.3). The string motion is entrained by the bow, and friction drags the string along (a). This is the *stick phase*. Eventually, the difference between the restoring force of the string and the forward force on the bow exceeds the frictional threshold, and the string becomes unstuck (b). The restoring force of the string moves it back toward its equilibrium point. The relative velocity of bow and string is now high, so friction is low, and the bow and string slide past one another easily. This is the *slip phase*. When the string's retrograde momentum is spent, relative velocity is again small, friction increases, and the string is captured again by the bow (c). This characteristic motion of the bowed violin string, discovered by Hermann von Helmholtz, is called *Helmholtz motion* (figure 8.4).

The relative motion of the string is slow (and equal to the bow velocity) during the stick phase and rapid during the slip phase. Plotting the displacement of a point on the string under the bow, we would see a function like the one in figure 8.5. Figure 8.7 shows a plot of the force of the string against the bridge.

The period of this motion corresponds to the fundamental frequency of the violin tone we hear. But the frictional force is only part of the story.

Slip Phase   The moment before the string escapes the bow it is stretched in a V-shape, not unlike the shape of a string displaced by a plectrum about to be plucked. The movement is similar to that shown

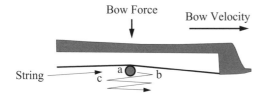

Figure 8.3
Stick/slip movement of a violin string.

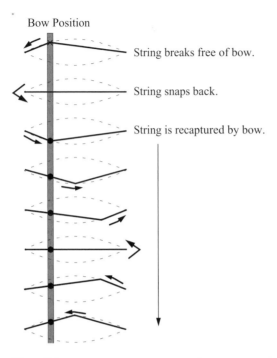

Figure 8.4
Helmholtz motion.

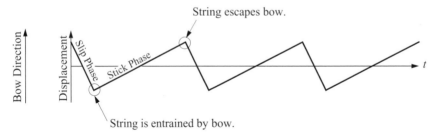

Figure 8.5
String displacement at bow position (seen from above).

for a plucked string in figure 7.8f. A single kink travels circularly around the string, reflecting from the ends.

Stick Phase   When it reaches the end of its retrograde excursion, the string begins to return to its original position and so becomes entrained by the movement of the bow. But the kink continues traveling around the string. It will reach the point where the original slip took place just in time to jar the string loose from the bow again, repeating the process.

Thus, to a first approximation, we can model the excitation force of a bow on a string as a series of small plucks, one per period, corresponding to where the nonlinear frictional force is overcome by the restoring force of the string. For harmonics, the story is basically the same. The triangular shape of the Helmholtz motion determines the spectrum of the string. Each time a kink returns to the bow position after reflecting off the far end of the string, it initiates a slip phase; when it arrives from the closer end of the string, it initiates a stick phase.

The recirculating kink acts like a governor, controlling the stick/slip motion of the string. Since its round-trip time is a function only of the mass, tension, and length of the string, the vibrating frequency remains very stable under varying bowing conditions. If only friction determined the stick/slip points, the vibrating frequency would be much less stable.

However, there are limits to the bow pressure, bow position, and bow velocity that can sustain a stable tone. When pressure or velocity exceed certain limits, they interfere with the stick/slip regime, making the tone unstable. Schelleng (1974) found that for each position of the bow, there is a maximum and minimum bowing force. Figure 8.6 (after Schelleng) shows the bow pressure

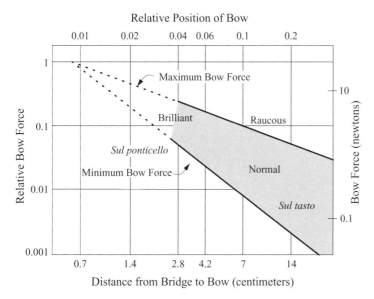

Figure 8.6
Range of bowing force as a function of distance from bridge.

Figure 8.7
Corresponding string force at bridge (seen from above).

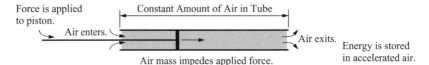

Figure 8.8
Acoustical inductor.

range for various bow positions of a cello. The closer the bow is to the bridge, the less margin the performer has between minimum and maximum. Bowing closer to the bridge (*sul ponticello*) produces a brighter but less controllable tone; bowing farther from the bridge (over the fingerboard, *sul tasto*) produces a softer tone. Only in the shaded region can a performer sustain a clear tone. Violinists typically use a bow position between 1/7 and 1/9 of the distance of the string.

In the case of an ideal string with rigid end supports, the force of the string against the bridge is a sawtooth waveform because of the impact of the circulating kinks. Figure 8.7 shows the string force at the bridge corresponding to the displacement shown in figure 8.5. The characteristic spectrum of this waveform is a sawtooth wave (see section 9.2.6). In practice, the waveform is also affected by string stiffness, mechanical leverage of the bridge, and other factors.

8.4 Inertance, Inductive Reactance

Consider what happens when we first apply a force to move a volume of air in a tube (figure 8.8). The air volume enclosed by both ends of the tube constitutes a mass—perhaps not a very massive one, but a mass nonetheless, which has some measurable inertia. The inertia of the air mass, called its *inertance,* opposes change in volume velocity. Olson (1952) gives the inertance of air in a circular tube as

$$M = \frac{\rho l}{\pi r^2}, \qquad\qquad Inertance \;\; (8.11)$$

where l is the effective length of the tube, r is its radius, and ρ is the mass density of the air.[2]

We can look at a force applied to a mass from the perspective of either the force or the mass. From the perspective of the force, the mass's inertance *reacts* against the force. From the perspective of the mass, the force's energy *induces* movement.

Because the inertia of the air dynamically reacts against the applied force, the inertial counter-force is classified as a *reactance*. Because the applied force induces energy into the accelerating air mass, the air mass is called an *inductor*. An inductor is a system that stores energy when it is accelerated. The air mass in the tube in figure 8.8 is an *acoustical inductor. Inductance* is a property of systems that store energy by acceleration. In electricity, energy is stored in the form of an electromagnetic force induced in a coil of wire because of changing current; in acoustics, energy is stored in the velocity of the air mass.

As long as the force continues, the air mass accelerates according to $f = ma$. Since the acceleration of a body is the derivative of its velocity $(a = dv/dt)$, we can write

$$f = m\frac{dv}{dt}. \tag{8.12}$$

Equation (8.12) expresses force on an object in terms of mass and acceleration. We can quickly adapt it to air. First, the equivalent of force f in air is pressure P. Second, the equivalent of acceleration a in air is the rate of change of the acoustical current dI/dt. Finally, mass m corresponds to the inertance M of the air (equation 8.11). Making these substitutions, we have

$$P = M\frac{dI}{dt}, \tag{8.13}$$

which says that the pressure driving a volume of air is proportional to the inertance of the air and its acceleration.

8.5   Compliance, Capacitive Reactance

Suppose that the air in figure 8.9a is moving at a steady velocity through the tube, driven by the piston. As the piston continues to move, suppose I suddenly close the end of the tube completely, as in figure 8.9b, trapping a volume of air inside.

As the piston continues to move to the right, the increasing air pressure within the cavity begins to impede the piston's forward movement because the stiffness of the trapped air grows as it is compressed. Eventually, the piston is forced to a stop when the counterforce of the pressure in the cavity matches the driving force on the piston.

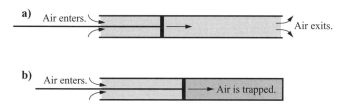

Figure 8.9
Acoustical capacitor.

The air in the tube is acting as an *acoustical capacitor,* storing energy in the relative compression of the air's particles. When the piston force is released, the stored energy pushes the piston back. The further we push the piston into the enclosed tube, the more energy is stored in the compressed air.

Capacitive reactance is proportional to displacement.

The counterforce is due partly to heat stored in the air's particles and partly to the inherent stiffness of the air. Both these counterforces are classified as reactances because both are in reaction to the applied force.

If the cavity is small, a unit change in volume results in a larger change in pressure than if the volume of air in the capacitor is large. So we can define acoustical capacitance as

$$C = \frac{dV}{dP} = \frac{\text{Change in volume}}{\text{Change in pressure}}.$$ *Acoustical Capacitance* (8.14)

Volume is measured in m^3 and pressure is measured in pascal = newton/m^2, so acoustical capacitance is measured in m^5/newton.

8.6   Reactance and Alternating Current

The acoustical capacitor *blocks the flow of direct current* because its reactance grows in direct proportion to the applied force, eventually matching the applied force so as to halt the flow (figure 8.10). By contrast, an alternating current, which by definition reverses itself periodically, will not be blocked by an acoustical capacitor because the mean displacement of an alternating current is zero (figure 8.11).

(This is only true so long as the mean displacement is small in comparison to the capacitive reactance it generates, and does not itself gradually move toward or away from the closed end

Ever-Increasing Capacitive Reactance

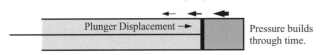

Figure 8.10
Acoustical capacitor blocks the flow of direct current.

Maximum Displacement

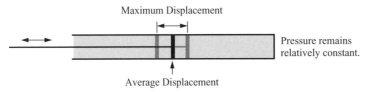

Average Displacement

Figure 8.11
Alternating current is not blocked by capacitor.

through time—that is, so long as there is no direct current component mixed in with the alternating current.)

8.7 Capacitive Reactance and Frequency

The equation for rotational energy (volume 1, equation (5.27)) shows that the energy in a wave is proportional to both its amplitude and frequency: $E = m(2\pi Af)^2$, where A is the amplitude, f the frequency, and m the mass of the medium. Solving for amplitude yields

$$A^2 = \frac{E}{4m\pi^2 f^2}.$$

If frequency f is raised while preserving E and m unchanged, the amplitude A must decrease proportionately. For example, in a simple driven spring/mass system, if the driving frequency goes up, inertial reactance of the mass will increasingly impede the mass's back-and-forth motion. If the driving force remains constant as the frequency increases, then the amplitude will have to decrease. Figure 8.12 is an example of an alternating current that goes from 0 Hz to 8 Hz while preserving E and m unchanged. Since the size of the mass does not change and the amount of driving energy does not change, the extent of the mass's excursion diminishes as frequency rises.When a signal with constant energy drives a constant mass, the amplitude of the mass's motion goes down as the frequency of its motion goes up.

Notice that as the frequency of a constant-energy/constant-mass signal increases, its average displacement around zero decreases. Since capacitive reactance is proportional to displacement, it is inversely proportional to increasing frequency.

It can be shown that the *capacitive reactance* X_c for some frequency f and capacitance C is

$$X_c = \frac{1}{2\pi fC}, \qquad\qquad\qquad\qquad \textit{Capacitive Reactance} \quad (8.15)$$

measured in acoustic ohms. A plot of this function is shown in figure 8.13.

By varying C while f is held constant, we see that

Capacitive reactance is inversely proportional to capacity.

In other words, the bigger the air cavity, the more elastic is the volume of air captured within it and the less counterforce it produces in response to an applied force. By varying f while C is held constant,

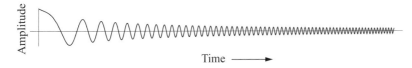

Figure 8.12
Constant-energy signal with increasing frequency.

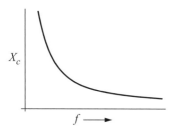

Figure 8.13
Capacitive reactance.

we see that capacitive reactance is (theoretically) infinite at 0 Hz and is (theoretically) zero at ∞ Hz. This agrees with experience, since 0 Hz corresponds to direct current (DC), and capacitors absolutely block DC. As frequency of a constant-energy signal goes higher, displacement is reduced; hence there is less reactance, so in the limit as frequency goes to infinity, capacitive reactance vanishes.

8.8   Inductive Reactance and Frequency

If we drive the piston of the acoustical inductor in figure 8.8 with simple harmonic motion, an alternating current is created. The mass of air driven by the piston reacts to its changing velocity with a counterforce proportional to its mass. As we increase the frequency of the driving force, we subject the air mass to greater acceleration, which the air mass increasingly impedes. Look again at figure 8.12, which shows that as the frequency of the driving force rises, the inductive reactance grows and the amplitude correspondingly drops. Amplitude of the alternating current is proportionately diminished for increasing frequency. In the limit as frequency goes to infinity, amplitude is diminished to zero. Thus, inductive reactance is directly proportional to frequency. Inductive reactance is a counterforce that is proportional to acceleration of a mass.

It can be shown that the *inductive reactance* X_l for some frequency f and inductance L is

$$X_l = 2\pi f L. \hspace{4cm} \textit{Inductive Reactance} \hspace{0.3cm} (8.16)$$

A plot of inductive reactance is shown in figure 8.14.

8.9   Combining Resistance, Reactance, and Alternating Current

Seldom do the forces of resistance and reactance operate singly; most real acoustical systems combine them in a variety of ways. We can study the interactions of pressure and current moment by moment in an alternating current using the mechanical systems in the next sections for thought experiments. These systems are not ideal but can be helpful to think about.

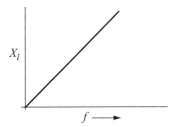

Figure 8.14
Inductive reactance.

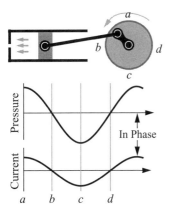

Figure 8.15
Motor and acoustical resistor.

8.10   Resistance and Alternating Current

If we attach an acoustical resistor to a motor-driven arm, as shown in figure 8.15, and measure pressure and current inside the opening for one full revolution, the graphs of rotation versus pressure and current would show that alternating current and pressure are proportional at all times. For instance, as the motor rotates from *a* to *b*, the plunger decelerates toward the hole in the tube, and both the pressure and current through the hole decrease together. So, in a purely resistive system, current and pressure are always in phase.

8.11   Capacitance and Alternating Current

Systems with capacitance do not have an in-phase relation between current and pressure. Figure 8.16 shows an acoustical capacitor attached to a motor. We must assume that virtually no air escapes from the enclosed cavity. Also assume that air pressure is balanced inside and outside of the cavity at the equilibrium point shown. Thus, the piston has no reactive force when the crank is at positions

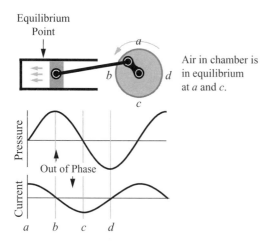

Figure 8.16
Motor and acoustical capacitor.

a and *c*. As the piston rotates from *a* to *b*, the air in the cavity is compressed. The stiffness of the trapped air increases and reaches its maximum value at *b*. The current—or rate of air flow—is highest at the start of the compression phase at *a* because at that point there is no extra stiffness in the trapped air in the cavity and hence no heightened stiffness to oppose the motion of the motor. At *b* the reactive counterpressure exactly matches the force applied by the motor. At this moment (*b*) the current in the capacitor goes to zero.

While the pressure decreases from *b* to *c*, the current flows in the reverse direction, indicated by the current's becoming negative. We see from this that the current and pressure are 90° out of phase, corresponding to one quarter cycle.

Say that the instantaneous pressure *P* fluctuates sinusoidally according to $P = r \sin 2\pi f t$, where *r* is the radius of the motor's crank, *f* is its frequency, and *t* is time. Then since $\pi/2$ corresponds to a quarter cycle, the instantaneous current *I* fluctuates sinusoidally one quarter cycle ahead, or $I = r \sin(2\pi f t + \pi/2)$, which equals $I = r \cos 2\pi f t$. Since the current flow reaches its maximum ahead of the pressure, we say that the *current leads pressure* by a phase angle of a quarter cycle, or 90°.

8.11.1   Inductance and Alternating Current

Figure 8.17 shows an acoustical inductor attached to a motor. The inductor consists of the mass of air captured in the tube on both sides of the plunger, plus an end correction consisting of some amount of the surrounding air that is influenced by the movement of the air in the cylinder. (In this example, we must ignore the inertia of the piston and its linkages, which would otherwise swamp the small effect of the air mass.)

The trapped air's velocity is greatest at *a* and *c*, but its acceleration at these points is zero. Since there is no acceleration at these points, there is no inertial reactance. The acceleration (rate of current change) is greatest at *b* and *d*. Since inertia is in proportion to acceleration, the inertial reactance is

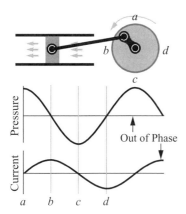

Figure 8.17
Motor and acoustical inductor.

greatest here. Just as any mass opposes change in velocity, so a volume of air opposes any change in current. When induced pressure increases, it *pushes against* the current; when induced pressure decreases, it *pushes with* the current.

As with capacitance, the current and pressure in an inductor are out of phase. However, for the inductor, the current flow reaches its maximum *after* the pressure does, and so we say that the *current lags pressure* by a phase angle of a quarter cycle, or 90°.

If we say that the instantaneous pressure P fluctuates according to $P = r \sin 2\pi ft$, where r is the radius of the motor's crank, f is its frequency, and t is time, and since $-\pi/2$ corresponds to a lag of a quarter cycle, then the instantaneous current I fluctuates according to $I = r \sin(2\pi ft - \pi/2)$, which equals $I = -r \cos 2\pi ft$.

We can also appeal to calculus for an explanation. The acoustical inductor opposes change in current but does not oppose direct current (except for resistance due to friction in a non-ideal inductor). The time-varying pressure $P(t)$ across an inductor with inductance L equals the derivative of the time-varying current $I(t)$:

$$P(t) = L\frac{dI}{dt}.$$

Suppose we have an alternating current defined as $I(t) = A \sin 2\pi ft$. Its derivative is:

$$\frac{dI}{dt} = 2\pi fA \cos 2\pi ft,$$

but because of the definition of $P(t)$ above, we have

$$P(t) = 2\pi fLA \cos 2\pi ft.$$

Current $I(t)$ and pressure $P(t)$ are off by a quarter cycle in an ideal inductor.

8.11.2   Quadrature

In summary, resistance gives rise to a pressure that is in phase with the current, whereas inductance and capacitance give rise to pressures that are *in quadrature* with the current because they are out of phase by $\pi/2$.

Recall that in the complex plane, multiplying by i corresponds to rotating counterclockwise by $\pi/2$, and multiplying by $-i$ corresponds to rotating clockwise by $\pi/2$. This suggests that a complex representation of resistance and reactance would be worth considering.

8.12   Acoustical Impedance

We have identified three kinds of opposition to current: resistance R, inductive reactance X_l, and capacitive reactance X_c. The term *impedance* combines all three of these forces. We can generalize Ohm's law given in equation (8.8) to cover impedance by rewriting it as follows:

$$I = \frac{P}{Z}, \qquad\qquad \text{Ohm's Law of Impedance} \quad (8.17)$$

where the impedance Z is defined as

$$Z = \sqrt{R^2 + (X_l - X_c)^2}. \qquad\qquad \text{Impedance} \quad (8.18)$$

In general terms, impedance is how hard one must push for the momentum obtained. We can generalize impedance as follows:

- In the air, Z is the ratio of air pressure to volume velocity.
- In a string, Z is the ratio of force to string velocity.
- In an electrical circuit, Z is the ratio of pressure (volts) to current (amperes).

In all these cases, the force can be direct or alternating. If there is no reactance (either because X_l and X_c are both zero or because they cancel), then equation (8.18) reduces to just $Z = R$. But if there is nonzero reactance, then X_l and X_c will impede an alternating current flow to an extent determined by their reactances and by the frequency of the alternating current. Resistance R impedes the current regardless of its frequency.

We can put Z into the form of the Pythagorean theorem, where the length of the hypotenuse c of a right-angled triangle is defined by the sides a and b, by $c = \sqrt{a^2 + b^2}$. All we need to do is define $a = R$ and $b = X_l - X_c$; then

$$Z = \sqrt{a^2 + b^2}.$$

This suggests that we can graph impedance in two dimensions, showing *resistance,* which impedes current regardless of frequency on the x-axis, and *reactance,* which impedes current depending on frequency on the y-axis. Associating Z with the hypotenuse and R and X with the horizontal and vertical sides, respectively, we have the triangle shown in figure 8.18.

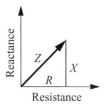

Figure 8.18
Impedance and resistance.

In terms of the complex plane, this definition of impedance suggests that R can be thought of as the real component of opposition to current, X_l and X_c are the imaginary components, and Z is the magnitude of these combined forces.

When there is no reactance, impedance reduces to resistance, represented as a vector on the x-axis. When reactance is present, we can represent impedance as the vector sum (the length of the hypotenuse) of the resistance and reactance.

Substituting for Z in equation (8.17) we can rewrite Ohm's law of impedance as

$$I = \frac{P}{\sqrt{R^2 + (X_l - X_c)^2}}. \tag{8.19}$$

Notice that, for any value of pressure P or resistance R, the current I will reach its maximum value when $X_l = X_c$ because then the term $X_l - X_c$ vanishes from the denominator. Since X_l goes up with increasing frequency f while X_c goes down, all we need to do to make them equal is to find the frequency $f = f_r$ where their curves intersect. Then, for that particular value of f_r, and by the earlier definitions of X_l and X_c, we can write

$$2\pi f L = \frac{1}{2\pi f C}.$$

Solving for f, we obtain

$$f_r \equiv f = \frac{1}{2\pi}\sqrt{\frac{1}{LC}}, \qquad \textit{Resonant Frequency in Terms of Reactance} \quad (8.20)$$

where f_r is the *resonant frequency,* the frequency where the current flow is at a maximum and the amplitude is highest.

Let's go back to the full definition of Ohm's law, showing the components of impedance explicitly by combining equations (8.15), (8.16), and (8.19):

$$I = \frac{P}{\sqrt{R^2 + \left(2\pi f L - \dfrac{1}{2\pi f C}\right)^2}}. \tag{8.21}$$

This shows that current depends upon pressure, resistance, frequency, inductive reactance (L), and capacitive reactance (C). *Impedance is frequency-selective when it contains reactance of any kind.* Since R, the expression for resistance in equation (8.21), does not have frequency as a term, pure resistive impedance is *not* frequency-selective. Thus, purely dissipative forces apply with equal force to all frequencies. If a system contains only reactance, that is, $R = 0$, and $X_l - X_c \neq 0$, it is said to be *lossless* because reactance conserves energy. If the system contains only resistance, $X_l - X_c = 0$, and $R \neq 0$, it is called *memoryless* because systems containing only resistance do not sustain vibration but dissipate energy rapidly. Memoryless systems depend only upon the current state of the system, not on any preexisting condition such as a prior amount of momentum. For example, the current in an acoustical resistor depends only on the instantaneous pressure differential across it, not on how the pressure came to have that value.

8.12.1 Dissipation

The reactance of an acoustical capacitor is determined by its stiffness. *No energy is dissipated by a pure capacitor.* Similarly, pure inductors store energy by accelerating a mass, then return that energy during deceleration. Their reactance is based on inertia. *No energy is dissipated by a pure inductor.*

Resistance R is therefore the only component of impedance that dissipates energy (see equation (8.18)). The energy dissipation of a musical instrument is the heat energy and sound energy it gives off to its environment as a consequence of being played. This suggests a neat division of the complex plane, where conservative forces are associated with the imaginary axis and nonconservative forces are associated with the real axis.

8.12.2 Resonance and Impedance

If we look at any vibration mode of a musical instrument, the frequencies below its resonant point are those for which $X_l < X_c$, and we say the mode is *stiffness-limited* because X_c predominates, and capacitance is based on stiffness. The frequencies *above* its resonance point are those for which $X_l > X_c$, and we say the mode is *inertia-limited* because X_l predominates, and inductance is based on inertia.

At precisely the frequency where $X_l = X_c$, the system is *dissipation-limited* because it is at this frequency and this frequency only that the reactance term $X_l - X_c$ vanishes, leaving only resistance, and only at this frequency is the current I therefore at a maximum, as shown in equation (8.19).

8.12.3 Quality Factor

I described Q as a measure of the steepness of a resonant peak (see volume 1, section 8.9.6). A more penetrating interpretation is available to us now.

Quality factor relates conservative and nonconservative forces in a resonant system.

Energy can be dissipated out of a resonant system as sound or as heat loss. The higher the Q, the more slowly the system dissipates energy. More precisely, Q is the ratio of the energy stored in a resonator because of conservative forces E_c (reactance) to the energy dissipated because of nonconservative forces E_{nc} (resistance) per radian of oscillation:

$$Q = \frac{E_c}{E_{nc}} = \frac{\text{Energy stored per radian}}{\text{Energy dissipated per radian}}. \tag{8.22}$$

If resistance is small compared to reactance ($E_{nc} \ll E_c$), the resonant peak is very sharp (high Q). Energy is dissipated slowly because resistance is relatively small. When such a system receives an impulsive input (such as when a string on a piano is struck), it rings for a relatively long time.

If resistance is large compared to reactance ($E_{nc} \gg E_c$), the resonance is broad (low Q). Energy is dissipated quickly because resistance is large. When such a system receives an impulsive input (such as when a string on a banjo is struck), it rings for a relatively short time.

8.12.4 Acoustical Power

The preceding paragraphs informally characterize how resonances dissipate energy through time. *Average power* $\bar{p}$ is the rate at which energy is transferred or transformed, measured as work W expended per unit of time t:

$$\bar{p} = \frac{\text{Work}}{\text{Time}} = \frac{W}{t}.$$

For more about the definition of power, see the equation for average power in volume 1, equation (4.32).

From equation (8.4), pressure P = energy/volume = W/V. And from equation (8.1), current I = volume/time = V/t. Combining, we have

$$\bar{p} = \frac{W}{t} = \frac{W}{V} \cdot \frac{V}{t} = \text{Pressure} \cdot \text{Current.} \qquad \begin{array}{c} \textit{Acoustical Power in Terms} \\ \textit{of Pressure and Current} \end{array} \tag{8.23}$$

So *acoustical power* can be viewed as the *product of pressure and current:*

$$\bar{p} = PI. \tag{8.24}$$

But remember that Ohm's law given in equation (8.17) supplies an alternative definition of current: $I = P/Z$. Plugging this definition for current into equation (8.24) gives

$$\bar{p} = P \cdot I = P \cdot \frac{P}{Z} = \frac{P^2}{Z}, \qquad \textit{Acoustical Power in Terms of Pressure and Impedance} \tag{8.25}$$

which expresses *acoustical power* in terms of *pressure and impedance.*

Finally, we can rearrange Ohm's law to define pressure as $P = IZ$. Substituting this definition of pressure into equation (8.24) yields

$$\bar{p} = P \cdot I = IZ \cdot I = I^2 Z, \qquad \textit{Acoustical Power in Terms of Current and Impedance} \tag{8.26}$$

which expresses *acoustical power* in terms of *current and impedance.*

Consider (8.25) and (8.26). These equations define power dissipation entirely in terms of current, pressure, and impedance. So, to double the power transmitted, one could either square the current or square the pressure; but one needs only to halve the impedance to achieve the same effect. Therefore, impedance is a much more potent determinant of the behavior of acoustical systems than pressure or current.

8.12.5   Impedance as the Ratio of Pressure and Current

Still more insights can come from Ohm's law. Let's rearrange equation (8.17), which defines Ohm's law as $I = P/Z$, to define impedance as

$$Z = \frac{P}{I}. \tag{8.27}$$

This means impedance can also be viewed as the *ratio of pressure to current*. For instance, if impedance is doubled, the current flow is halved for the same pressure or remains the same if the pressure is doubled.

Since impedance is the combination of resistance and reactance, this definition holds for all currents, alternating or direct.

Recall that equation (8.1) defines current as $I = V/t$, where V is volume and t is time. Thus current can also be thought of as volume velocity, measured in cubic meters per second. Volume velocity is the volume of air that flows through a cross-sectional area per second. Substituting equation (8.1) into equation (8.27) yields

$$Z = \frac{P}{V/t} = P \cdot \frac{t}{V}. \tag{8.28}$$

The following summarizes equations (8.27) and (8.28).

Impedance is the ratio of force (pressure) to volume velocity.

Impedance is the constant of proportionality between the pressure that must be applied and the flow that results. If pressure P is in units of pascal (abbreviated Pa), t is seconds (s), and V is cubic meters (m^3), then by equation (8.28), impedance is measured in pascal seconds per cubic meter:

$$Z = \frac{Pa \cdot s}{m^3}. \qquad\qquad Rayl \tag{8.29}$$

Impedance Z is called an *acoustical ohm* in some texts and a *rayl* in others.[3] The inverse of impedance is *admittance*.

8.13   Sound Propagation and Sound Transmission

If unimpeded, sound waves spread out spherically as they propagate. As the wavefront expands, its curvature per unit area diminishes so that a listener far from a sound source experiences the sound wave front as virtually plane.[4]

Another way to create plane waves is to present an impedance barrier to their spherical expansion. For example, the walls of a tube form an impedance barrier that prevents spherical spreading of sound waves inside it. If the cross-sectional area of a tube is constant, the sound's intensity does not diminish with the square of distance, as with spherical radiation, because the energy in the wave cannot expand. (Of course, frictional forces still dissipate the wave's energy.)

When there is no impediment to sound's spreading in a medium, we say it *propagates* its energy. If it is constrained to one dimension, such as by a tube, we say it *transmits* its energy. If the tube is bent, the direction of transmission follows the bend. Even though the tube itself occupies two or three dimensions, the wave trapped inside remains essentially a plane wave in one dimension. Acoustical systems such as tubes that guide waveforms in this manner are *acoustical waveguides*. Wind and brass instruments are essentially acoustical waveguides. In fact, anything that impedes the spherical expansion of a wave can be construed as a waveguide, including rooms, strings, or a piano sounding board.

One-dimensional waveguides are bidirectional transmission lines that can carry signals from a transmitter to a receiver as well as vice versa. For example, a megaphone can also act as an ear trumpet, depending on the direction of the signal. For more about waveguides, see section 9.7.

8.13.1 Propagation Delay

When sound is initially propagated in a uniform medium, it will appear later at some predictable distance from the source regardless of the nature of the disturbance that caused it. The reason propagation is not instantaneous is because the inertial and elastic properties of air delay the buildup of pressure near the disturbing force, limiting the rate at which sound can travel through the medium (see volume 1, section 7.3). Restated, the rate of sound propagation is a function of the medium's *characteristic impedance,* also called *wave impedance,* which is the characteristic ratio of pressure to volume velocity *of the medium itself*. If the capacitive reactance and inductive reactance of a medium are known, the time required for a waveform to travel a unit distance can be determined easily.

Consider a tube with a plunger to push a direct current of air along inside it. Recall equation (8.2): $V = It$, which says that total charge V is equal to the current flow I times the duration t. Solving equation (8.14) for V, we have $V = CP$, which states that the total charge is equal to the capacitance C multiplied by the pressure on the capacitor P. We can relate these two definitions of charge to study propagation delay.

Let's think about the ideal (frictionless) one-dimensional coupled spring/mass system shown in figure 8.19 as a model for packets of air in a tube. To begin with, the air packets in the tube are in static equilibrium. Now suppose the left end is suddenly compressed by a disturbing force such as a piston's entering the tube, charging the air in the tube. Energy enters the system according to $V = It$. In the first instant, the air packets just inside the tube become slightly more compressed according to $V = CP$. Since no energy is lost in this ideal system, the total energy leaving the piston is equal to the total energy entering the tube, and we can write

$$V = CP = It. \tag{8.30}$$

Now let's switch perspective from the air packets back to the spring/mass model. The charge CP on the first spring creates a pressure difference across the first mass because the

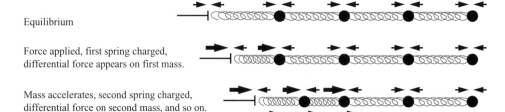

Equilibrium

Force applied, first spring charged,
differential force appears on first mass.

Mass accelerates, second spring charged,
differential force on second mass, and so on.

Figure 8.19
Propagation of disturbance on a spring/mass system.

spring to its right is less charged. Therefore, the mass experiences a net force to the right and accelerates in that direction. Its displacement starts charging the second spring, which in turn causes the second mass to experience a pressure difference and begin to accelerate. We see a continuing movement of the initial pressure disturbance moving across the springs that, assuming no dissipation, would continue indefinitely.

By equation (8.13), the pressure across each mass is directly proportional to inertance and the acceleration of the volume velocity:

$$P = M\frac{dI}{dt} = M\frac{\Delta I}{\Delta t}. \tag{8.31}$$

If we start with current and time both set to zero, then the change in time (Δt) and the change in acoustical current (ΔI) are equal to the final time t and final current I, so we can simplify and write $P = MI/t$. Rearranging,

$$Pt = MI,$$

which says that if pressure P is applied for time t upon the inertance M, the final current will be I.

Let's put equations (8.30) and (8.31) side by side to show how the terms are related:

$$It = CP,$$
$$Pt = MI. \tag{8.32}$$

Notice that both equations share terms t, I, and P.

8.13.2   Finding the Propagation Delay of a Medium

We can solve the two equations in (8.32) for t in terms of M and C:

$$It \cdot Pt = CP \cdot MI,$$
$$PIt^2 = MCPI,$$
$$t^2 = MC,$$
$$t = \sqrt{MC}. \qquad\qquad \textit{Propagation Delay} \;\; (8.33)$$

Equation (8.33) gives the time it takes for a pressure change to travel a unit length, since M and C are defined in terms of length. The velocity of a wave over some distance D may be found with

$$v = \frac{D}{t} = \frac{D}{\sqrt{MC}}. \qquad\qquad\qquad \textit{Propagation Velocity} \quad (8.34)$$

This is the distance a wave travels over a unit length D.

8.13.3 Finding the Characteristic Impedance of a Medium

To determine a medium's characteristic impedance, divide the two equations in (8.32):

$$\frac{Pt}{It} = \frac{MI}{CP}.$$

Then multiply both sides by P/I:

$$\frac{P^2 t}{I^2 t} = \frac{MIP}{CPI},$$

$$\frac{P^2}{I^2} = \frac{M}{C},$$

$$Z_0 \equiv \frac{P}{I} = \sqrt{\frac{M}{C}}. \qquad\qquad\qquad \textit{Characteristic Impedance} \quad (8.35)$$

Characteristic impedance Z_0 is the ratio of pressure P to current I in a medium.

From (8.35) we see that Z_0 is a function of the density and compressibility of a medium, which is a measure of the speed of sound in the medium. As the density increases, Z_0 increases. As the compressibility increases, Z_0 decreases. Characteristic impedance measures the extent to which a medium impedes the progress of a signal moving through it.

8.13.4 Characteristic Impedance of Air

The characteristic impedance of a medium such as air is the product of its density and the speed of sound. In terms of equation (8.35), we let

$$Z = \sqrt{\rho \gamma P} = \rho c, \qquad\qquad \textit{Characteristic Impedance in Open Air} \quad (8.36)$$

where ρ is the density of air (its mass per unit volume), P is barometric pressure, and $\gamma = 1.4$ is the ratio of the specific heat of air at constant pressure to constant volume. In a tube the characteristic impedance is

$$Z \equiv \frac{\rho c}{A}, \qquad\qquad \textit{Characteristic Impedance of Air in a Tube} \quad (8.37)$$

where A is the cross-sectional area of the tube. Impedance goes up with decreasing cross-sectional area because of the increased influence of the boundary layer of air next to the tube surface.

According to Beranek (1986), the density of air ρ is approximately

$$\rho = 1.29\left(\frac{273}{T+273}\right)\left(\frac{P}{0.76}\right) \text{kg/m}^3, \qquad \text{Density of Air} \quad (8.38)$$

where T is temperature in degrees Celsius, and P is barometric pressure in meters of mercury. In the equation for the speed of sound (volume 1, equation 7.14), we calculated the speed of sound in air to be

$$20.0345\sqrt{T} = 331.116 \text{m/s} \quad \text{at STP.}$$

So the characteristic impedance of air is approximately 427.14 SI rayl at STP, and the characteristic impedance of air in a tube is $Z = 427.14/A$ SI rayl at STP.

The characteristic impedance of a vibrating string is

$$Z = \frac{K}{c} = \frac{K}{\sqrt{K/\rho}} = \sqrt{K\rho} = \rho c, \qquad \text{Characteristic Impedance of a String} \quad (8.39)$$

where K is the string tension, ρ is the linear mass density (mass per unit length), and c is the speed of sound in the string. Together, equations (8.36) and (8.39) show that characteristic impedance can be interpreted as the geometric mean of the two forces that resist displacement: elasticity (or equivalently, tension) and inertia (Smith 2004).

8.13.5 Transmission and Reflection at Impedance Barriers

We've seen that a transmission line of any kind exhibits a characteristic impedance. Think about a single point on a transmission line of some sort, such as an instrument string, brass instrument tube, or electrical wire. If the characteristic impedance at the next point along the line is the same as the current point, the wave (ideally) continues to propagate all its energy down the line. But if the characteristic impedance changes, some energy is transmitted, some is reflected back, and the energy that continues is subject to distortion by refraction and diffraction (and of course, some energy is converted to heat, as usual). *Return loss* is the ratio of signal power transmitted into a system to the power reflected (returned, and therefore lost to the destination).

Suppose we have a transmission line with a barrier of some kind at the end. If we wish to transmit the maximum amount of power across the barrier, we want the barrier's impedance to match the characteristic impedance of the transmission line. If the characteristic impedance is the same across the barrier, then it is impossible to tell from the sending end that the transmission line is not infinitely long because all the signal that is fed into it is propagated and none is reflected.

If, on the other hand, we wish to reflect the maximum amount of power back into the transmission line, then we want the barrier's impedance to be as contrasting as possible to that of the transmission line.

Note that reflected energy may form resonances if the system is underdamped.

8.13.6   Examples of Channeling Acoustical Energy

One can view the analysis of musical instrument design, listening environments, and even the mechanical components of hearing as the *channeling of acoustical energy through media of various impedances*. The materials from which musical instruments are constructed—air, wood, steel, and brass—each has its own characteristic impedance. In the case of concert halls, the media also include building materials, curtains, and even the impedance characteristics of the human bodies in the audience. In some cases, such as a violin string or the bore of a clarinet, the aim is to reflect a great deal of energy back into the vibrating system. In other cases, such as a piano sounding board, the aim is to dissipate the energy, not only to maximize sound radiation, but because a highly damped system will have a broader, flatter frequency response, so that all frequencies are transmitted with similar efficiency. The production of music hinges on the transmission and reflection of energy in acoustical systems. Consider the following examples.

The Guitar   To create a guitar tone, we must generate a sustained frequency from energy injected into the string by plucking, and we must radiate enough energy to hear it comfortably some distance away.

1. The string receives impulsive energy from the performer.

2. The string seeks to dissipate the energy as fast as possible to return to its equilibrium. But the substantial impedance mismatch at the nut and bridge reflects most energy back into the string.

3. Being underdamped, the string conserves the energy by vibrating.

4. Some energy is transmitted through the bridge to the top plate of the guitar, forcing it to vibrate sympathetically.

5. The top plate is highly damped by the surrounding air, inside and outside the guitar's cavity. The top plate radiates broadband energy outward into the air.

6. The top plate also vibrates the air inside the cavity, which acts as a Helmholtz resonator. The resonator radiates some of its bandpass-filtered energy out through the sound hole.

7. The vibrating air inside the cavity also vibrates the back and sides, which radiate some sound. The edges of the top also transmit vibration from the top to the sides and back of the body.

8. As the string loses energy, its tone decays because it is not being replenished.

9. The rate at which the tone dies corresponds to the rate at which sound is radiated from the guitar string, plus heat dissipation from internal frictional forces in the string and the guitar top, plus losses due to air viscosity.

The energy injected into the string by the pluck can be used in two ways, depending upon how the instrument designer sets up the impedance mismatch between the strings and the body of the guitar. For example, the impedance mismatch at the bridge and nut of an electric guitar is very large; little acoustical energy escapes from the strings per unit time, so the tone is quite sustained, but the instrument does not generate much acoustical power (without amplification). The impedance mismatch of a nylon-stringed acoustical guitar is much less; the tone is less sustained, but the acoustical power is much greater. It is hard to hear an unamplified electric guitar across the room, whereas a good-quality acoustical guitar can fill a concert hall.

The reason the electric guitar is made from solid wood also has to do with impedance matching: the drastic impedance mismatch between the air and the solid body reduces the sensitivity of the electric guitar to the acoustical energy field of its own amplified loudspeakers. This helps prevent an acoustical feedback loop from forming that would drive the guitar and amplifier system into oscillation.

Hearing   In the discussion of how hearing works (volume 1, section 6.2), I described the impedance mismatch between the air surrounding the tympanum and the much denser fluid within the inner ear. Whereas the impedance mismatch is desirable in a guitar to sustain its tone, it is undesirable in the ear because the basilar membrane must be provided with as much signal strength as possible from the tympanum.

The tympanum and the bones of the middle ear act as a transformer, matching the characteristic impedances of the air and the inner-ear fluid. The pinnae also act as impedance-matching systems. Similarly, the bell of a trumpet matches the impedance of its tube to the surrounding air, increasing its radiation efficiency.

Sound Absorption   Structural materials such as brick and concrete have much higher impedance than air. But the impedance mismatch at wall boundaries is greater for high frequencies, so these reflect more, whereas lower frequencies tend to transmit through the wall. So when sound radiating from a stage reaches the walls, high frequencies tend to be reflected, low frequencies tend to be transmitted, and both high and low frequencies are absorbed.

Sound absorption in a room can be enhanced by adding materials that are close to the characteristic impedance of air. For maximum absorption, the wavelength of sounds to be absorbed must be on the same scale as the pores of the absorptive material.

A plane wave incident upon a hard wall creates a particle velocity node at the wall surface. Since there is no wave movement at the wall surface, there can be no friction, hence no absorption. But one quarter wavelength away from the wall, there is a particle velocity antinode; placing absorptive materials at that distance from the wall produces maximum friction for that wavelength and provides the best absorption. Therefore, to attenuate a particular frequency, place a thin absorbent material at one quarter of its wavelength from the wall. To absorb all higher frequencies as well, place an absorbing material against the wall that is as thick as one quarter wavelength of the lowest frequency to be absorbed. This explains why a thin layer of acoustical tile works much better suspended from the ceiling than glued to it.

Acoustical Insulation   Acoustical insulation prevents sound from penetrating a barrier, such as between apartments in a building, or from a dance hall to the surrounding neighborhood. It also can prevent unwanted noise from entering a recording studio, anechoic chamber, or concert hall. For maximum acoustical insulation, we wish to have the walls provide the greatest possible impedance mismatch to the surrounding air. Heavy, inert building materials like concrete, filled cinder block, and stone are most effective. Doubling the mass per unit area can generally improve existing insulation by 6 dB SPL. A practical way to achieve good insulation is to erect a separate wall system that is decoupled from the first—to build, in essence, a room within a room in such a way that the two sets of walls do not couple vibration through each other at any point. Another approach is to use sheets of foam impregnated with metal pellets. These approaches essentially require the impinging sound to experience multiple closely spaced impedance barriers.

As the neighbors of garage rock bands can surely attest, sound absorbers do not make good sound insulators. Sound absorbers work either by converting acoustical energy to heat or by passing it through to where it will not return (an open window is a perfect sound absorber by this definition). Sound insulators provide very little absorption and a great deal of reflection. Sound absorbers are typically best used to reduce unwanted room reflections; sound insulators are best used to protect yourself or your neighbors from intrusive sounds.

8.14   Input Impedance: Fingerprinting a Resonant System

The acoustical properties and playability of woodwinds and brasses can be effectively studied by watching how the impedance at the input end of the instrument varies as a function of frequency. If we think of the player's mouth and lips as a pressure-controlled valve, then we can see that resonance occurs when pressure waves from the performer's mouth combine constructively with waves reflected from the open end of the instrument.

Benade (1976) has provided a most penetrating discussion of these effects. He credits Wilhelm Eduard Weber for discovering (in 1830) that the player's lips act as a yielding termination of the tube. He cites Helmholtz (1863) for the first brief yet complete analysis of the basic mechanism of the interaction between a reed, the player's lips, and an acoustic tube. The following is adapted from Benade's exposition.

8.14.1   Incident and Reflected Waves

Suppose we have an open air pipe with a loudspeaker mounted at its left end. If we connect the loudspeaker to a battery and close the switch, the loudspeaker diaphragm sends a high-pressure wave p_0^+ down the tube.

By equation (8.37), the characteristic impedance of air in the tube changes where the cross-sectional area changes, causing a low-pressure wave p_0^- to be reflected back up the tube from its mouth (figure 8.20). The round-trip time is $2cl$, where c is speed of sound, and l is the length

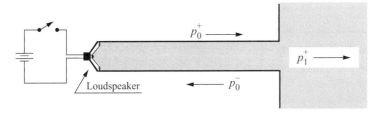

Figure 8.20
Electrically driven resonator.

of the tube. If we open the switch the moment the low-pressure wave arrives back at the loud-speaker, the diaphragm moves out of the tube, which further lowers the low-pressure reflected wave arriving from the mouth. This heightened low-pressure area draws in higher-pressure air from the right, causing a low-pressure wave to propagate back toward the mouth. When this low-pressure area arrives at the mouth, it is again reflected with a phase reversal, returning up the tube as a high-pressure wave.

8.14.2   Sustaining Oscillation by Feedback

If we adjust the rate at which we connect and disconnect the loudspeaker to match the phase of the reflected wave, then the impulses from the loudspeaker can be made to cooperate with the wave energy stored in the tube from previous impulses to sustain vibration. The energy injected by the loudspeaker makes up for losses due to dissipation and radiation of sound.

We can automatically time when the loudspeaker injects energy into the tube by connecting it to a pressure-actuated switch (figure 8.21). In this arrangement, a high-pressure wave will push the flexible diaphragm out of the tube, thereby closing the switch. It is opened again when a low-pressure area pulls the diaphragm back into the tube. This is not fundamentally different from how the lips work on the mouthpiece of a brass instrument, or how single- or double-reed instruments create sustained oscillations.

Sustained vibration is created in woodwind and brass instruments by the lips in conjunction with a reed or mouthpiece. The lips provide a yielding termination to the tube, which the pressure switch and loudspeaker arrangement emulates. We must momentarily force the switch closed the first time to begin oscillation, but after that the rate at which the loudspeaker injects energy is governed by the propagation velocity of pressure waves through the tube, just as it is for a musical instrument.

8.14.3   Regime of Oscillation

The spectrum of a woodwind or brass instrument typically contains many harmonics because, as in the example of the switch-activated loudspeaker, the volume velocity generated in the tube by the lips is a nonlinear function of pressure. The exciting function injects broad spectral content into the tube, activating many modes of vibration.

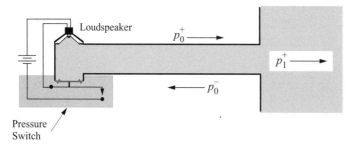

Figure 8.21
Sustaining driven resonator.

If the phases of the pressure waves of these modes align at the mouthpiece, they cooperatively reinforce each other, helping to stabilize the oscillation. But if they arrive out of phase with each other, they either make no contribution to the stability of the oscillation or possibly even detract from it. The frequency of oscillation chosen by the air column will be the one that maximizes total energy generation, which is then shared among the various components in a well-defined way.

The choice of fundamental frequency will be driven by the fundamental mode and whichever of the higher-order modes have frequencies that are closest to whole-number multiples of the fundamental, whether or not they are related to the "natural" harmonic components. We could say that these cooperative components set up a kind of hegemony—a *regime of oscillation,* in Benade's terms—that determine the oscillating frequency. The more modes that cooperate in a regime, the more quickly the instrument speaks and the more focused is its tone. So oscillation is a subtle function of how the phases of the vibrating modes interact with each other, and how these vibrations interact with resonances in the mouthpiece or reed.

8.14.4 How Does Input Impedance Change with Frequency?

In addition to determining the fundamental frequency, the tube contributes to shaping the harmonic structure of the instrument and even the playability of certain pitches. To understand impedance at the input to the tube, let's look at the ratio of pressure to volume velocity—the impedance—at different frequencies. We drive a loudspeaker with a sine wave oscillator, and insert a microphone in the tube next to the loudspeaker (figure 8.22).

Pressure   To sustain resonance, the loudspeaker must push in phase against a returning high-pressure wave or pull in phase with a returning low-pressure wave. The magnitude of the pressure fluctuation near the microphone is greatest at resonance. At other frequencies, the returning wave moves sometimes in opposition to the loudspeaker and sometimes with it, so the magnitude of pressure fluctuation near the microphone is less, on average, than it is at resonance. The pressure magnitude in the tube varies as a function of frequency, reaching a maximum near each resonant mode and dropping in between.

Volume Velocity   Since the amount of power available to the loudspeaker from the oscillator is relatively constant regardless of frequency, the volume velocity remains relatively unchanged. (In

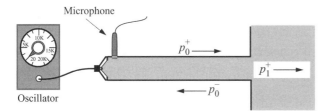

Figure 8.22
Impedance detector.

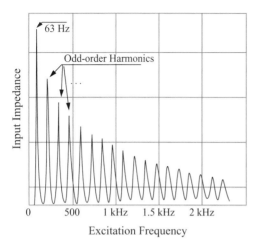

Figure 8.23
Impedance of a 140 cm cylindrical pipe. Adapted from Benade (1976).

practice, we'd have to go to some lengths to ensure that the loudspeaker produces constant volume velocity at all frequencies under test. It can be done, and the interested reader is referred to Benade's excellent solution.)

Measuring Impedance   Since volume velocity is relatively constant in our setup (figure 8.22), and impedance is the ratio of pressure to volume velocity, the impedance at the input to the tube is directly proportional only to the pressure at the input.

Since the diaphragm of the microphone is sensitive only to pressure change, maintaining constant volume velocity in the tube means the microphone's signal can be read directly as relative impedance.

If we plot input impedance against frequency of a cylindrical pipe (figure 8.23), we observe that its maxima correspond exactly to the "natural" resonant mode frequencies because this is precisely where the most pressure must be applied to maintain volume velocity (Benade 1976). Since this is effectively a pipe closed at one end, the vibration modes are odd multiples of the fundamental, as governed by the equation for frequency modes of a pipe closed at one end (volume 1, equation 8.21). Energy drops off with increasing mode number because of frictional and thermal losses within the tube.

Functions of Bell and Mouthpiece   Because of the sharp impedance mismatch at its opening, the small sound intensity that can be emitted from a cylindrical length of tube is not loud enough to be musically useful. This can be experimentally validated as follows. Insert a French horn mouthpiece into one end of a garden hose about the same length as a French horn, and have a horn player perform on it. The tone will be only about as loud as ordinary speech. Adding a large funnel to the other end acts as a primitive but effective flared bell, increasing the obtainable loudness.[5] The flared bell allows more efficient radiation of sound by impedance matching the high acoustical

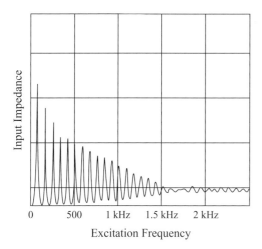

Figure 8.24
Impedance of a cylindrical pipe with trumpet bell. Adapted from Benade (1976).

pressure inside the tube to the lower pressure outside. The flared bell also makes the radiation pattern of the instrument more directional, that is, most of the sound intensity is propagated directly out the front of the bell.

The major drawback of a flared bell is that it progressively lowers the frequencies of higher-order harmonics because the effective length of a flared tube increases with frequency. This causes the travel time of the higher harmonics in the tube to be out of phase with lower ones, so that they no longer cooperatively reinforce oscillation. Additionally, because the horn progressively radiates higher frequencies more efficiently, there is less high-frequency energy in the reflected waves within the tube, which also decreases the stability of oscillation. Figure 8.24 shows that above 1500 Hz effectively no energy is returned into the tube from the bell whereas for the pipe without flared bell, substantial energy is returned even above 2000 Hz.

Figure 8.25 shows Benade's measurements of a cornet made by the British craftsman Henry Distin in 1865. Principally added here is the mouthpipe and mouthpiece. The frequencies are shifted somewhat by the addition of the mouthpipe, and impedance peaks in the range of 500 to 1000 Hz are enhanced, corresponding to the resonance of the mouthpiece. Benade noted that the third and fourth impedance peaks do not follow the smoothly rising sequence of surrounding peaks because of constrictions and misalignments of the tubing where it joins the valves, bell, and mouthpipe.

Figure 8.26 shows the input impedance of a B♭ trumpet. Again, peaks in the vicinity of the mouthpiece resonance are enhanced. The figure marks the impedance peaks that are implicated in sounding two tones, B♭4 (written C4) and F4 (written G4). Peaks 2, 4, 6, and 8 correspond to the partials of B♭4. When played softly, the forcing function from the player's lips are nearly sinusoidal, and only impedance peak 2 is excited. As the level of loudness increases, nonlinearity of the forcing function increases, increasing the participation of peaks 4, 6, and 8. A player of this

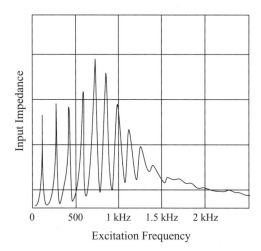

Figure 8.25
Impedance of a cornet. Adapted from Benade (1976).

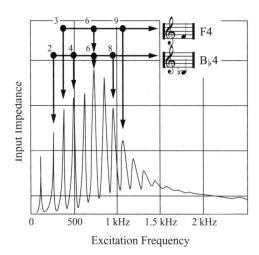

Figure 8.26
Impedance of a B♭ trumpet. Adapted from Benade (1976).

instrument would notice that louder tones at this pitch are much easier to control than soft tones because the addition of the higher modes creates a more stable regime of oscillation. In the case of the pitch F4, the peaks involved are 3, 6, and to a lesser degree, 9. Since the peak at 3 is higher than the one at 2, the fundamental at low volume for pitch F4 is more stable than the one for B♭4. With increasing loudness, the much stronger peak at 6 is excited, making this a very easy note to play on the instrument.

8.15 Scattering Junctions

Solutions of the one-dimensional wave equation (chapter 7) always have the form of a positive-going (right-traveling) wave $f^+(t)$ and negative-going (left-traveling) wave $f^-(t)$ (see, for example, figure 7.8). In terms of the definition of characteristic impedance, $Z = P/V$, which governs wave propagation speed in a one-dimensional medium, we can define these traveling pressure waves as

$$f^+(t) = ZV^+(t),$$

$$f^-(t) = -ZV^-(t),$$

where total pressure $f(t) = f^+(t) + f^-(t)$.

The analysis in this section follows Smith (2005). Equation (8.37) shows that impedance changes inversely with cross-sectional area of a tube, and equation (8.39) shows that impedance in a string varies directly with tension and mass density. So, wherever an acoustic tube changes cross-sectional area, or where a string changes mass density (it's essentially impossible for a string's tension to change along its length), there will be an impedance mismatch, and scattering will occur at the boundary. A traveling wave impinging on the boundary will be split into a reflected wave and a transmitted wave. The total force and volume velocity must be conserved across this boundary, so the amount of reflected energy and transmitted energy will equal the total energy. However, the proportion of reflected versus transmitted energy will be determined as a function of the impedances of the medium and termination (string and bridge, or tube and opening or cap).

We can derive the coefficients of reflection and transmission as follows. Let the pressure waves on the left side of the impedance boundary be $f_0 = f_0^+ + f_0^-$, and the pressure waves on the right be $f_1 = f_1^+ + f_1^-$, and in general, $f_n = f_n^+ + f_n^-$. Also define the corresponding velocity waves on the left $v_0 = v_0^+ + v_0^-$, and on the right $v_1 = v_1^+ + v_1^-$, and in general, $v_n = v_n^+ + v_n^-$. In the case of a single impedance boundary, total force $f = f_0 + f_1$ and total velocity $v = v_0 + v_1$ at the junction. In the general case, at an N-way junction we have

$$f = \sum f_n \quad \text{and} \quad v = \sum v_n.$$

Suppose we have a vibrating string with an impedance discontinuity at some position $x = 0$. Let us say that a positive string velocity v corresponds to a vertical movement of the string at that point. String velocity must be unchanged across the impedance junction (otherwise the string would break), so $v = v_0 = v_1$, and the vertical force at the junction must balance: $f = f_0 = -f_1$.

For an acoustic tube, the pressures across an impedance junction must be unchanged $f = f_0 = f_1$, and the volume velocities must balance $v = v_0 = -v_1$. A positive volume velocity is one that flows toward the junction, so for the tube, having $v_0 = -v_1$ means that a right-going wave on the left is still a right-going wave on the right of the junction.

Let's continue deriving reflection and transmission coefficients by examining the ideal string. String velocities must match across the junction: $v = v_0 = v_1$. By earlier definitions, we can expand the velocity terms $v = v_0^+ + v_0^- = v_1^+ + v_1^-$, and in general, $v = v_n^+ + v_n^-$. Rearranging,

$$v_n^- = v - v_n^+. \tag{8.40}$$

Also, string forces must balance: $f = f_0 + f_1 = 0$. Expanding,

$$f = (f_0^+ + f_0^-) + (f_1^+ + f_1^-) = 0.$$

Recalling the definition of impedance, we can write

$$f = Z_0(v_0^+ - v_0^-) + Z_1(v_1^+ - v_1^-) = 0,$$

where Z_0 and Z_1 are the characteristic impedances of the left and right sides of the junction, respectively. Now, by equation (8.40),

$$f = Z_0[v_0^+ - (v - v_0^+)] + Z_1[v_1^+ - (v - v_1^+)]$$

$$= Z_0(2v_0^+ - v) + Z_1(2v_1^+ - v).$$

Recalling that $f = 0$, and solving for v, we have

$$v = 2\frac{Z_0 v_0^+ + Z_1 v_1^+}{Z_0 + Z_1}, \tag{8.41}$$

which says that the string velocity at point $x = 0$ is a function of the arriving velocity waves and the impedances of the two sides of the junction. We can use this to define the departing velocity waves, which will reveal what happens when these arriving waves reflect and transmit.

Again, by equation (8.40),

$$v_0^- = v - v_0^+ = \frac{2Z_1}{Z_0 + Z_1}v_1^+ - \frac{Z_1 - Z_0}{Z_0 + Z_1}v_0^+, \tag{8.42}$$

$$v_1^- = v - v_1^+ = \frac{2Z_0}{Z_0 + Z_1}v_0^+ + \frac{Z_1 - Z_0}{Z_0 + Z_1}v_1^+. \tag{8.43}$$

The derivation of equations (8.42) and (8.43) is given in the appendix section A.11. These equations say that the departing waves are combinations of the transmitted and the reflected waves, scaled by the two impedances. If we define

$$R = \frac{Z_1 - Z_0}{Z_0 + Z_1}, \qquad\qquad\qquad\qquad\qquad\qquad \textit{Reflection Coefficient} \;\; (8.44)$$

then we can simplify:

$$v_0^- = (1 + R)v_1^+ - Rv_0^+. \quad \text{Left departing wave}$$

$$v_1^- = (1 - R)v_0^+ + Rv_1^+. \quad \text{Right departing wave}$$

(8.45)

So v_0^-, the left-traveling departing wave, is constructed by transmitting some of v_1^- and reflecting some of v_0^+. (Note that v_0^+ is subjected to a 180° phase shift at the junction when it is reflected.) Similarly, v_1^-, the right-traveling departing wave, is constructed by transmitting some of v_0^- and reflecting some of v_1^+. As a thought experiment, assign some value to R in these equations and note that they conserve total energy. We can memorialize equations (8.45) in a state diagram as shown in figure 8.27, the Kelly-Lochbaum scattering junction. (Markel and Gray 1976).

Equation (8.44) shows that unless $Z_0 = Z_1$, some of the total power supplied by an arriving wave to the junction is reflected back. If this were, for example, a junction in a power transmission line, any impedance mismatch across the junction would reduce the power sent. Since the reflection coefficient R scales arriving velocity v_n^+, it can be thought of as the coefficient of velocity reflection with respect to the total incoming velocity.

We can apply the Kelly-Lochbaum scattering junction to construct a model of scattering within a musical instrument tube or string, as shown in figure 8.28. (Rabiner and Schafer 1978). (Note that the order of the junction operators in figure 8.28 has been permuted with respect to

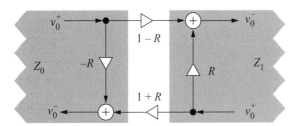

Figure 8.27
Kelly-Lochbaum scattering junction.

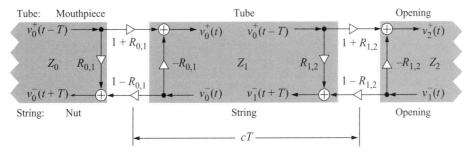

Figure 8.28
Modeling waves traveling across two scattering junctions.

their order in figure 8.27 so that all right-traveling waves are on the top and left-traveling waves are on the bottom.)

We have three regions of varying impedance, Z_0, Z_1, and Z_2, which we can define as the impedance changes between a mouthpiece, tube, and opening or between a nut, string, and bridge. Now define a time interval T corresponding to the propagation time along the medium with speed of sound c. If we let $v_n^+(t)$ and $v_n^-(t)$ be the wave velocities at the extreme left of a segment n, then the traveling waves at the extreme right will correspondingly be $v_n^+(t-T)$ and $v_n^-(t+T)$.

Because the vocal tract is essentially a tube that varies in diameter over its length, we can model its filtering properties as the concatenation of a set of cylindrical tubes of varying diameters. The impedance changes at the junctions between these piecewise tube sections reflect and transmit energy among the sections. Resonances arise as a consequence of energy being trapped in each section because of reflection. The Kelly-Lochbaum junction was originally developed to study vocal filtering. This analysis is developed further in the discussion of linear prediction (section 9.5.2). This analysis has also been the basis for a wide range of techniques for musical instrument physical modeling known as waveguide synthesis, discussed in section 9.7.2.

Summary

This chapter discussed how conservative forces operate within strings, pipes, membranes, and bars to sustain vibration, and how energy flows are transferred between coupled acoustical systems. We play a musical instrument by injecting energy into it. The instrument stores and dissipates the energy over time by sound radiation and heat until it returns to its original energy state. Conservative forces store energy via elasticity and inertia. Dissipative forces transfer and transform energy as heat and sound. These two classes of force give us music as we know it.

Acoustical current is the velocity or the rate of flow of a volume of air. The total quantity of air moved, called the acoustical charge, depends upon the strength of the current and how long it is applied. Pressure is force per unit of area. Pressure can also be seen as the energy of a volume of air, called the energy density.

An acoustical resistor impedes the flow of air by dissipating energy as heat or radiating it away as sound. In gases and fluids, resistance is generally a linear force. However, instruments such as the violin depend upon a nonlinear frictional force to make their characteristic vibration.

The inertia of the air mass, called inertance, opposes change in volume velocity. Opposition to applied force because of inertia is called inertial reactance. It impedes the applied force because of the inertia of the air. Because an air mass stores energy by accelerating, it is called an acoustical inductor.

An acoustical capacitor is a mass of air that stores energy in the molecular motion of the air's particles. Capacitive reactance is proportional to displacement.

Capacitive reactance and inductive reactance are so named because they are forces in reaction to the applied force. Capacitive reactance is inversely proportional to increasing frequency, and inductive reactance is directly proportional to frequency.

In a purely resistive system, current and pressure are always in phase. Resistance gives rise to a pressure that is in phase with the current, whereas inductance and capacitance give rise to pressures that are said to be in quadrature with the current because they are out of phase by $\pi/2$.

Impedance is the vector sum of resistance, capacitive reactance, and inductive reactance. In general, impedance is how hard one has to push for the momentum obtained. In air, it is the ratio of air pressure to volume velocity.

The resonant frequency of a system is where the current flow is at a maximum and the amplitude is highest. It corresponds to where capacitive reactance and inductive reactance of a system are balanced. Where capacitive reactance dominates, the system is stiffness-limited; where inductive reactance dominates, the system is inertia-limited. At precisely the frequency where the two reactances balance, the system is resistance-limited.

Impedance is frequency-selective when it contains reactance of any kind. Resistance is the only component of impedance that dissipates energy. No energy is dissipated by a pure capacitor or a pure inductor.

Quality factor relates conservative and nonconservative forces in a resonant system. Quality factor is the ratio of the energy stored in a resonator because of conservative forces (reactance) to the energy dissipated because of nonconservative forces (resistance) per radian of oscillation.

From the basic definitions of power as work per unit of time, and current as volume per unit of time, we can find acoustical power in terms of pressure and current, pressure and impedance, or current and impedance. From these relations, impedance can be viewed as the ratio of pressure to volume velocity. Impedance is the constant of proportionality between the pressure that must be applied and the flow that results.

When there is no impediment to sound's spreading in a medium, we say it propagates its energy. If unimpeded, sound waves spread out spherically as they propagate. If a sound is constrained to one dimension, such as by a tube, we say it transmits its energy. The rate of sound propagation or transmission is a function of the medium's characteristic impedance, also called wave impedance, which is the ratio of pressure to volume velocity of the medium itself. The characteristic impedance in a string can be interpreted as the geometric mean of the tension and inertia.

If the characteristic impedance of a medium changes, some energy is transmitted, some is reflected back, and what continues is distorted by refraction and diffraction (and as usual, some energy is converted to heat). Return loss is the ratio of signal power transmitted into a system to the power reflected (returned). If we wish to transmit the maximum amount of power across a barrier, we want the barrier's impedance to match the characteristic impedance of the transmission line. If we wish to reflect the maximum amount of power back into the transmission line, we want the barrier's impedance to be as contrasting as possible.

One can view the analysis of musical instrument design, listening environments, and even the mechanical components of hearing as the channeling of acoustical energy through media of various impedances. The acoustical properties and playability of woodwinds and brasses can be effectively studied by watching how the impedance at the input end of the instrument varies as a function of frequency.

Impedance changes inversely with cross-sectional area in a tube. Impedance in a string varies directly with tension and mass density, so wherever an acoustic tube changes cross-sectional area, or where a string changes mass density, there will be an impedance mismatch, and scattering will occur at the boundary. A traveling wave impinging on the boundary will be split into a reflected wave and a transmitted wave. The total force and volume velocity must be conserved across this boundary, so the amount of reflected energy and transmitted energy will equal the total energy. However, the proportion of reflected versus transmitted energy will be determined as a function of the two impedances. The Kelly-Lochbaum scattering junction provides a simulation of these effects that can be used to synthesize physical models of musical instruments.

Suggested Reading

Fahy, Frank J. 1985. *Sound and Structural Vibration: Radiation, Transmission, and Response*. New York: Elsevier. San Diego: Academic Press.

Fletcher, Neville H., and Thomas D. Rossing. 1999. *The Physics of Musical Instruments*. New York: Springer-Verlag.

Kinsler, Lawrence E., Austin R. Frey, Alan B. Coppens, and James V. Sanders. 1999. *Fundamentals of Acoustics*. 4th ed. New York: Wiley.

Main, Iain G. 1993. *Vibrations and Waves in Physics*. 3d ed. New York: Cambridge University Press.

Morse, Philip M., and K. Uno Ingard. 1986. *Theoretical Acoustics*. Princeton, N.J.: Princeton University Press.

Pierce, Allan D. 1989. *Acoustics: An Introduction to Its Physical Principles and Applications,* esp. 22, 107–108, 320–321. New York: Springer-Verlag.

9 Sound Synthesis

Mathematical science . . . has these divisions: arithmetic, music, geometry, astronomy. Arithmetic is the discipline of absolute numerable quantity. Music is the discipline which treats of numbers in their relation to those things which are found in sound.
—Cassiodorus

9.1 Forms of Synthesis

Fourier synthesis can be used to create any periodic vibration (see chapter 3). But Fourier methods are only one of an essentially limitless number of techniques that can be used to synthesize sounds.

9.1.1 Linearity and Synthesis

Linear synthesis techniques can generally be used to reproduce a sound that is identical to the original. The Fourier transform, for example, can be used to reproduce any periodic waveform. This is part of the reason we call it a transform: we can analyze a signal into its transformed state and then use its inverse transform to reproduce the original. Nonlinear techniques generally provide no way to reproduce a sound that is identical to an original, but they may have other compelling advantages, such as being economical to calculate or intuitive to use.

9.1.2 Linear Synthesis

The criteria of linear systems are superposition and proportionality, described here with an emphasis on synthesis.

Superposition   Air and water are linear media (at least for the strength of signals we are discussing) because waves superimpose without distortion. For two signals x_1 and x_2, if $F(x_1) = y_1$, and $F(x_2) = y_2$, and if $F(x_1 + x_2) = y_1 + y_2$, then function F is linear (if it also meets the proportionality criterion); otherwise it is nonlinear.

Proportionality   In air and water, little waves pass through big ones, and vice versa, without being modified by the encounter. Linear systems are independent of amplitude. If $F(x_1 + x_2) = y_1 + y_2$, and $F(ax_1 + bx_2) = ay_1 + by_2$, then function F is linear (if it also meets the superposition criterion); otherwise it is not. Examples of nonlinear media include photographic film and magnetic tape, both of which can saturate if the amplitudes of signals being recorded on them are cumulatively too strong.

9.1.3 Overview of Linear Synthesis Types

Linear transforms are characterized by adding or subtracting weighted basis functions of some kind. In the case of Fourier analysis and synthesis, the basis functions are the family of sinusiods. A non-Fourier linear synthesis technique called *Walsh-Hadamard synthesis* has basis functions that are square waves. The wavelet transform (see chapter 10) has a variety of nonsinusoidal basis functions.

We can synthesize a periodic function corresponding to an arbitrary spectrum via the inverse Fourier transform. Cycling the resulting waveform repeatedly is equivalent to the original infinite periodic function. This is the basic idea of *waveform synthesis* (see section 9.2.5).

Although the Fourier transform is mathematically valid only when applied over all time, it can be adapted to analyze and reproduce sounds that change through time. *Additive synthesis* is an extension of the Fourier transform to signals that can change over time (see figure 9.9). The technique can be implemented, for example, by driving a bank of oscillators with time domain functions representing the frequencies and amplitudes of the sound's components. Since in this case the frequencies need not be harmonics, this technique can also generate inharmonic spectra. The general case of additive synthesis is discussed in chapter 10. Even a simple mixing console can be thought of as a kind of additive synthesizer, in the sense that it superposes and scales its input signals through time.

As the name suggests, *subtractive synthesis* removes energy from a spectrum by filtering. Like additive synthesis, it can be applied to signals that change through time. Subtractive synthesis is discussed under *linear predictive coding* (LPC) (see section 9.5.2).

The advantage of linear synthesis systems is that, since they are based on a transform, we can use them to both analyze and reproduce a sound. But this is generally not a musically interesting thing to do—why not just use the original sound? For musical applications, we generally wish to create effects not otherwise obtainable. We can manipulate the analysis data of linear systems to achieve some very interesting effects, and doing so is usually fairly intuitive because linear techniques involve only superposition and proportionality.

The liability of many linear synthesis systems is that the amount of analysis data can be dauntingly large, requiring huge amounts of calculation or data storage or both in order to produce realistic-sounding synthesis. Nonlinear techniques typically are much more economical, but they are also less general.

9.1.4 Nonlinear Synthesis Types

Any technique that does not meet the superposition and proportionality criteria is a nonlinear technique. This book covers only a small fraction of the amazing variety of such techniques, but it hits many of the high points.

The primary limitation of nonlinear synthesis techniques is that they have no inverse transform, so there is no way to use them to exactly reproduce an original sound. Although no direct analysis is possible, that doesn't mean we can't predict the kinds of sounds we'll get from a nonlinear technique; we just won't necessarily be able to produce a particular sound exactly. It may not be possible for a nonlinear technique to produce a particular type of sound regardless of the parameters used.

Nonlinear systems are generally more computationally efficient than linear synthesis forms. Some, such as frequency modulation, model instrumental timbres very well with few parameters and little computation.

A central element of many nonlinear techniques is *modulation,* which is just a latinate word meaning *change.* In sound synthesis, it has the particular meaning of applying a time-varying change to a signal.

The aspects of signals that are exploited by nonlinear synthesis include

- *Amplitude,* where, for example, a signal is saturated or clipped, or where there is uneven amplitude response to different frequencies (filtering).

- *Frequency,* where new frequencies are produced in response to an input signal.

- *Phase,* where the phase of the output signal is not a linear function of the input phase.

Musically, many nonlinear synthesis systems have a quixotic character that can sometimes be counterintuitive, whereas linear systems tend to offer fewer surprises. Since composition is the art of controlled expectation, nonlinear techniques are very useful in the composer's tool kit. However, they are not a panacea. If overused, the inner structure of nonlinear techniques can become horribly clichéd.

9.2   A Graphical Patch Language for Synthesis

It is worthwhile to have a way to construct sound-generating modules to illustrate the synthesis techniques. We need three elements: a way to keep time, a way to pass signals around, and a way to transform signals.

Here's a clock that produces a monotonically increasing value corresponding to the flow of time. We need a way to observe the output of the clock. The function plotter shown here takes two input signals: the *x* input moves a pen horizontally, and the *y* input moves the pen vertically.

Connecting the clock to both inputs of the plotter produces a diagonal line. Since the clock's values are applied to both inputs, it moves the pen an equal distance vertically and horizontally through time, producing a ramp at a 45° angle (the identity function).

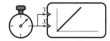

We need a way to transform inputs to create outputs. The sine wave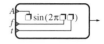
oscillator shown here implements the function $A \sin 2\pi f t$. The inputs
on the left side, A, f, and t, are connected to the indicated terms of the
formula. The output appears at the right side and can be the input to
another module.

If we wire the sine wave oscillator with the clock and plotter, and supply parameters for A, f, and
t, we have the setup shown in figure 9.1. Note that the amplitude input is associated with variable
A and frequency with variable f. If left unspecified, these variables indicate parameters we can set
any way we like, to obtain, in this case, any desired amplitude and frequency. If an input should
receive a particular value, it will be shown associated with a constant.

Let's call a configuration of such modules a *patch*. This patch scales time t by $2\pi f$. The sine of
the result is then scaled by A. The time value is also applied to the plotter and moves the pen horizontally. The value of the sine result drives the pen vertically.

In order to represent increasingly complex modules, we can use the standard rules of mathematical equality to define new functions to encapsulate existing definitions. Taking the oscillator,
for example, if we let

$$m \cdot g(t) = A \sin 2\pi f t,$$

we could interpret this graphically as shown in figure 9.2.

These diagrams can at times get cluttered with many lines connecting modules, so I occasionally use variables to temporarily hold a value produced in one part of a patch for reuse elsewhere. For example, the oscillator patch could have been drawn as in figure 9.3. The clock sets
the value of the variable t, and t is referenced by other module inputs. Here, the variable t holds
the output of the clock, and the inputs to the oscillator and plotter reference its value whenever

Figure 9.1
Simple oscillator patch.

Figure 9.2
Encapsulating patches.

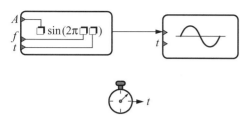

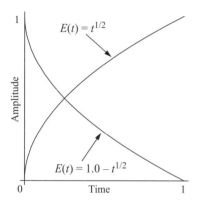

Figure 9.3
Patch using variables instead of interconnections.

Figure 9.4
Envelope curves.

they need an input. This comes in so handy at times that I may only indicate the variable t when I want to imply the clock signal.

9.2.1 Constructing a Simple Synthesis Instrument

This simple graphical patch language is flexible enough to allow representation of a wide range of synthesis and processing techniques. Let's create a tone that has a natural attack and decay, control over amplitude and frequency, and control over timbre. The oscillator provides a way to generate a particular frequency, but we need a way to control its amplitude through time.

Simple Envelope Generator   Let's invent an envelope generator that has a simple exponential attack, variable-length steady state, and exponential decay. We want to be able to control the overall duration, the attack time, and the decay time individually. The function $y = E(t) = t^{1/n}$ produces an exponential curve that goes from 0 to 1 over the range $0 < t < 1$ seconds. The function $y = 1.0 - E(t)$ necessarily goes from 1 to 0 over the same interval. If $n = 1$, $E(t)$ is linear. The variable n specifies the time constant of the function; values of n farther away from 1.0 are more sharply curved (figure 9.4).

We can scale the function to an arbitrary time length by redefining the envelope function to be

$$y = E(t, d, n) = \left(\frac{t}{d}\right)^{1/n}, \qquad \qquad \textit{Envelope Kernel} \quad (9.1)$$

where d is the duration in seconds.

If we use one envelope function $E_a(t, d_1, n_1)$ for the attack and concatenate another, $E_d(t, d_2, n_2)$, to its end for the decay, we can use it, for example, to vary the instantaneous amplitude of an oscillator, creating the amplitude envelope shown in equation (9.2):

$$\mathrm{env}(a, n_1, d, n_2, t) = \begin{cases} t < a, & (t/a)^{1/n_1}, \\ a \le t \le a+d, & 1.0 - [(t-a)/d]^{1/n_2}, \end{cases} \quad \textit{Simple Envelope Generator} \quad (9.2)$$

where t is time, a is the attack time, d is the release time, and each n determines the steepness of its part of the envelope curve, with n_1 controlling the curve of the attack and n_2 controlling the curve of the decay. The higher the value of n, the more quickly the function changes. For example, figure 9.5 shows a function with attack time $a = 0.1$ s, release time $d = 0.6$ s, $n_1 = 2$, and $n_2 = 1.5$. Putting it all together, we can construct a simple tone generator with an amplitude envelope, as shown in figure 9.6 where the module env( ) is as defined in (9.2).

In order to realize a tone with this setup, we assign values to the variables and start the clock running at $t = 0$. For example, setting envelope values to those in the previous paragraph, and setting $f = 32$ Hz and $A = 1$ produces the waveform shown in the figure 9.7.[1]

Controlling Frequency   If we make the system of representing pitch flexible enough, we can use this simple instrument to study the scale systems described in volume 1, chapter 3.

For equal-tempered pitches, recall volume 1, equation (3.4), repeated here:

$$f_{k, v} = f_R \cdot 2^{(v-4)+k/12},$$

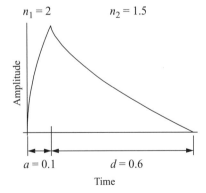

Figure 9.5
Simple envelope generator function.

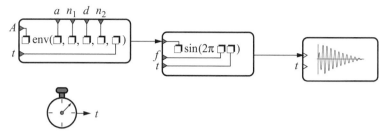

Figure 9.6
Simple synthesis instrument with amplitude envelope.

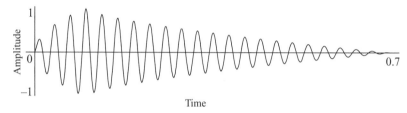

Figure 9.7
Tone produced by the synthesis instrument.

where k is an integer signifying one of the 12 pitch classes numbered 0 to 11, R is the reference frequency, such as A440, and v is the desired octave. If we construct symbols for each chromatic pitch, such as C♯4 $= f(1,4)$, we can set the frequency parameter $f = $ C♯4 to achieve that pitch in the simple instrument shown in figure 9.6.

We can approach just intonation along the lines worked out at the end of volume 1, section 3.8. Recall volume 1, equation (3.11), reproduced here:

$$C_\pi(v) = R \cdot \frac{16}{27} \cdot 2^{v-4}.$$

We constructed a set of symbols such as $F_\pi(v) = C_\pi(v) \cdot 4/3$, which provides the "πthagorean" pitch F_π in any octave v. We can set the frequency parameter $f = F_\pi(v)$ to achieve Pythagorean pitch F_π in octave v using the simple synthesis instrument. We can also use the more elaborate constructions for pitch in MUSIMAT (see volume 1, appendix sections B.1 and B.2).

Adding Vibrato   *Vibrato* is a periodic pitch modulation around a target pitch. Depending upon the instrument, the *vibrato rate* for musical instruments typically ranges from 1 to 7 Hz, and *vibrato depth* ranges from about one tenth of a semitone (100 cents) up to a maximum of about a minor third (think: Wagnerian soprano). Perhaps the simplest way to model vibrato is just to add the output of a slowly time-varying oscillator to the frequency parameter, and use the result as the frequency input of the oscillator generating the waveform. Call the oscillator that generates the

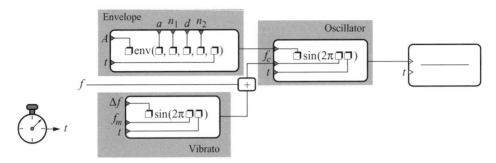

Figure 9.8
Synthesis instrument with vibrato and amplitude envelope.

waveform the *carrier oscillator,* and the oscillator that modulates the target frequency the *modulating oscillator.* The calculation for the frequency f_c of the carrier oscillator is

$$f_c = f + \Delta f \cdot \sin 2\pi f_m t, \hspace{4cm} \textit{Vibrato} \hspace{0.3cm} (9.3)$$

where f is the target frequency, Δf is the vibrato depth, and f_m is the vibrato rate. Clearly, if $\Delta f = 0$, there is no vibrato, and the carrier frequency is just the target frequency f. But if $\Delta f \neq 0$, the carrier frequency will rise and fall at the rate of f_m. The patch with vibrato added is shown in figure 9.8.

What value do we assign to Δf in order to achieve a desired depth of modulation? Recalling that pitch is logarithmic in frequency, if we want the vibrato depth to be a constant interval, such as a semitone, then Δf must grow with increasing pitch. Recall from volume 1, section 3.2.2, that the size of a tempered semitone ratio is about 1.06. That is, for some frequency f_n, the pitch f_{n+1} a semitone above is $f_{n+1} \cong 1.06 \cdot f_n$. If we set $\Delta f = 1.06 \cdot f$, where f is the target frequency, we actually get a vibrato depth of a whole tone, because the range of the sine function is ± 1.0, which provides a range of a semitone above and below the target pitch. So instead we set $\Delta f = 0.53 \cdot f$ to get a semitone of vibrato depth.

Of course, performers don't produce exactly sinusoidal vibrato. An improvement is to introduce some randomness into the vibrato to achieve a more naturalistic effect. This subject begins to verge into modeling of musical instrument performance. It's an enormous subject; I investigate just the basics.

9.2.2 Static Control over Timbre

With figure 9.8 we have achieved a synthesis instrument that has control over pitch, duration, amplitude, amplitude envelope, and vibrato. But its timbre is a simple sinusoid. In order to improve this, we turn now to ways of synthesizing timbre.

Timbre is actually the subject of key interest in synthesis because it largely determines the perceived quality of the synthetic tone. Musicians often have contradictory aims for timbre. On the

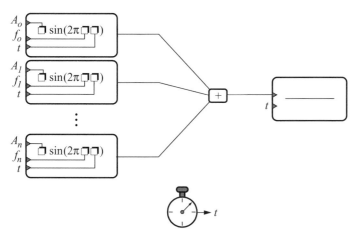

Figure 9.9
Additive synthesis.

one hand, having the most naturalistic possible sound is the goal of those who want to model standard acoustic instruments like pianos and clarinets. For orchestral composers, for example, this provides an inexpensive means to test out musical ideas. Unfortunately, realism in music synthesis is a kind of holy grail, often sought, seldom achieved.

On the other hand, sound synthesis can be used to extend the palette of timbres available to musicians beyond what can be produced by conventional instruments. Synthesis can create sounds that metamorphose from the familiar to the extraordinary. For example, linear predictive coding (LPC) synthesis can create hybrid sounds such as a talking flute. Synthesis can create unearthly sounds, such as the Shepard scale illusion (see volume 1, section 6.4.7).

A straightforward way to control timbre is to specify the spectrum of a sound by controlling the amplitudes and frequencies of a bank of oscillators, as in figure 9.9.

We could express this mathematically as follows:

$$f(t) = \frac{1}{\rho}\sum_{n=1}^{N} A_n \sin 2\pi f_n t, \qquad\qquad \textit{Oscillator Bank Synthesis} \quad (9.4)$$

where N is the number of components, starting with the fundamental. Their amplitudes and frequencies are given by vectors A_n and f_n, respectively, each of length N.[2] The sum of the amplitudes A_n is usually normalized so that the amplitude of $f(t)$ does not vary with the number and strengths of the components. This is done by setting

$$\rho = \sum A_n.$$

Oscillator bank synthesis is a generalization of Fourier synthesis (see section 3.1.1) because the amplitudes and frequencies of each oscillator can be any value whatever: we are not constrained to harmonic spectra and periodic waveforms. However, the spectra specified by A_n and f_n are static. Not many musical instruments produce entirely static spectra, so let's extend this to handle dynamic amplitudes and frequencies that can change through time.

9.2.3   Dynamic Oscillator Bank Synthesis

Dynamic frequency control allows realistic synthesis of music instrument tones, glissandos, and vibrato. Dynamic amplitude allows us to trace the spectral evolution of tones. All we have to do is make the amplitudes and frequencies of the spectral components be functions of time as well so that we have an array of amplitude functions $A_n(t)$ and frequency functions $f_n(t)$, as shown in equation (9.5).[3]

$$f(t) = \sum_{n=1}^{N} A_n(t) \sin[2\pi f_n(t) \cdot t]. \qquad\qquad \textit{Dynamic Oscillator Bank Synthesis} \quad (9.5)$$

Normalization, as in equation 9.4, can be added if desired.

9.2.4   Advantages and Disadvantages

There are two primary disadvantages of oscillator synthesis. First, many oscillators are typically required to synthesize a realistic sound. A single piano tone might have 30 harmonics containing significant energy. Multiply that by five fingers on both hands, and the number of oscillators required jumps to 300. If the damper pedal is held down, the number of oscillators would be in the thousands.

Second, this approach requires that the amplitudes and frequencies of each spectral component be specified in detail. Furthermore, if the frequencies and amplitudes of the components change through time, as is true with natural musical instruments, the amount of data required to control the oscillators can be many times greater than the resulting signal.[4]

However, this is the most general way of synthesizing sound. Since summation is linear, oscillator bank synthesis can be driven by Fourier analysis, producing sounds that have a dazzlingly realistic quality. Not only can we synthesize any sound we can analyze, we can also synthesize *any sound at all,* limited only by our ability to dream it up. In fact, the combination of a fast enough computer, a loudspeaker, and additive synthesis is perhaps the most general-purpose musical instrument ever created.

But too much generality can be paralyzing. If, for instance, a composer must make every decision at every level, the task can become overwhelming. Therefore, generally, the goal of sound synthesis is to generate sounds that are expressive without requiring extreme micromanagement to produce interesting results.

9.2.5   Waveform Synthesis

Waveform synthesis preserves some of the generality of oscillator bank synthesis at significantly less cost in complexity because the amplitudes and frequencies of the components are constants instead of functions of time. It can be expressed as follows:

$$y(t) = \frac{1}{\rho} \sum_{n=1}^{N} A_n \sin 2\pi n f t, \qquad\qquad \textit{Waveform Synthesis} \quad (9.6)$$

where A_n is an array of component amplitudes of length N. This technique is essentially identical to Fourier synthesis (see section 3.1.1) using oscillators, but it is less general than dynamic oscillator bank synthesis (section 9.2.3) because it is limited to harmonic spectra and periodic signals. The sum of the amplitudes A_n should be normalized if the amplitude of the output should not vary with the number and strengths of the components by setting

$$\rho = \sum A_n.$$

A discrete version of equation (9.6) known as wavetable synthesis works well within the hardware and software constraints of typical modern computer systems, and it is widely used to synthesize sound (see section 9.2.8).

9.2.6 Fourier Series Waveforms

Generalizing equation (9.6) to include all possible real harmonic waveforms, we have

$$f(t) = \sum_{k=0}^{\infty} a_k \cos(2\pi k t) + b_k \sin(2\pi k t), \qquad\qquad \textit{Real Fourier Series} \quad (9.7)$$

where coefficients a_k and b_k scale the cosinusoidal and sinusoidal contributions, respectively, of harmonic k. Equation (9.7) does not have the frequency parameter f appearing in equation (9.6) because here we want to focus only on wave shape without concerning ourselves about frequency. We can always add frequency back in to realize the specified wave shape at a particular desired frequency. Term a_0 scales any energy at 0 Hz because $\cos k = 1$ when $k = 0$. (b_0 makes no contribution at 0 Hz because $\sin k = 0$ when $k = 0$.) The complex form of the Fourier series is

$$f(t) = \sum_{k=0}^{\infty} (a_k + i b_k) e^{i 2\pi k t}. \qquad\qquad \textit{Complex Fourier Series} \quad (9.8)$$

The behavior of this equation might be a little surprising. For instance, if we set all a_k and b_k to zero except $a_1 = 1$, equation (9.8) reduces to

$$e^{i 2\pi t} = \cos 2\pi t + i \sin 2\pi t,$$

while equation (9.7) reduces to just $(\cos 2\pi t) + 0$.

9.2.7 Geometrical Waveforms

Some interesting specimens among the family of wave shapes are implied by equations (9.7) and (9.8). Following are some well-known shapes.

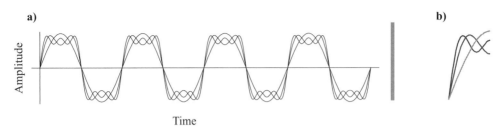

Figure 9.10
Square wave, Fourier series.

Square Waves   If we only sum odd-numbered sine-phase harmonics and arrange their amplitudes to be odd reciprocals,

$$f(t) = \sum_{k=0}^{N-1} \frac{1}{2k+1} \sin[2\pi(2k+1)\cdot t], \qquad N > 0, \qquad \textit{Square Wave, Fourier Series} \quad (9.9)$$

then the series converges as $N \to \infty$ to a *square wave*. The series expansion of the first few terms of (9.9) is

$$f(t) = \sin 2\pi t + \frac{1}{3} \sin 2\pi 3t + \frac{1}{5} \sin 2\pi 5t + \frac{1}{7} \sin 2\pi 7t + \cdots.$$

The first few waveforms in this sequence ($k = 0$, 1, and 2) are demonstrated in figure 9.10a. When $N = 1$, equation (9.9) produces a sine wave, and for $N > 1$, it produces the sum of a sine wave and odd harmonics. The following table shows the harmonics produced and their amplitudes for the first few values of k.

k	0	1	2	3	4	5	6	7	8
Amplitude	1	1/3	1/5	1/7	1/9	1/11	1/13	1/15	1/17
Frequency	1	3	5	7	9	11	13	15	17

The spectrum of the square wave contains components that are odd harmonics of the fundamental. The amplitudes diminish with increasing harmonic number. In figure 9.10, note the slight overshoot and ringing that occurs at the ends of the vertical excursion of the waveforms. The effect, called the *Gibbs phenomenon* (or more colorfully, *Gibbs' horns*), is shown magnified in figure 9.10b. The horns indicate that Fourier series functions only approximate the discontinuous points of the non-band-limited square wave function defined in equation (9.10).[5]

The geometric square wave, also called the *non-band-limited square wave,* can be expressed as

$$f(t) = \begin{cases} 1, & 0 \le t < \pi, \\ -1, & \pi \le t < 2\pi. \end{cases} \qquad \textit{Square Wave, Non-Band-Limited} \quad (9.10)$$

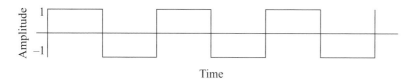

Figure 9.11
Non-band-limited square wave.

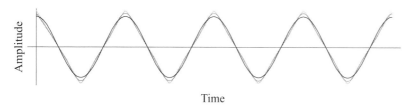

Figure 9.12
Triangular wave Fourier series.

as shown in figure 9.11. This is the same as equation (9.9) with $N = \infty$, so it has an infinite number of odd harmonics.

The non-band-limited square wave takes on only two states through time: -1 and $+1$; it is binary and discontinuous. This function has two discontinuities per period, where it transits instantaneously from above to below and vice versa.

The *Walsh-Hadamard transform* uses rectangular waves as its basis functions instead of using sinusoids or phasors (see appendix section A.7).

Triangular Waves   By summing odd-numbered cosine harmonics and arranging their amplitudes to be squared odd reciprocals,

$$f(t) = \sum_{k=0}^{N-1} \frac{1}{(2k+1)^2} \cos[2\pi(2k+1 \cdot t)], \qquad N > 0, \quad \textit{Triangular Wave, Fourier Series} \quad (9.11)$$

the series converges as $N \to \infty$ to a *triangular wave*, as demonstrated in figure 9.12. The series expansion of the first few terms of equation (9.11) is

$$f(t) = \cos 2\pi t + \frac{1}{9}\cos 2\pi 3t + \frac{1}{25}\cos 2\pi 5t + \frac{1}{49}\cos 2\pi 7t + \cdots.$$

The following table shows the harmonics and amplitudes for the first few values of k.

k	0	1	2	3	4	5	6	7	8
Amplitude	1	1/9	1/25	1/49	1/81	1/121	1/169	1/225	1/289
Frequency	1	3	5	7	9	11	13	15	17

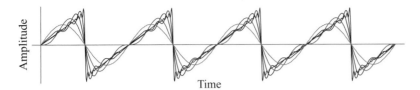

Figure 9.13
Sawtooth wave Fourier series.

The spectrum of the triangular wave contains components that are odd harmonics of the fundamental. The amplitudes diminish with the square of increasing harmonic number. Most of the energy in this signal is in the lowest harmonics.

The geometric triangular wave can be expressed as

$$f(t) = 2|2\lfloor t/(2\pi)\rfloor - 1| - 1, \qquad\qquad\qquad \text{\textit{Triangular Wave, Non-Band-Limited}} \quad (9.12)$$

where $\lfloor \dots \rfloor$ is the floor function. The Fourier series convergence of the triangular wave is shown in figure 9.12.

Sawtooth Waves   The sawtooth wave is created by summing all odd harmonics and subtracting all even harmonics, with amplitudes as the reciprocal of the harmonic number:

$$y(t) = \sum_{k=1}^{N} \frac{(-1)^{k+1}}{k} \sin 2\pi kt, \qquad N > 0, \qquad\qquad \text{\textit{Sawtooth Wave, Fourier Series}} \quad (9.13)$$

The series converges as $N \to \infty$ to a *sawtooth wave,* as demonstrated in figure 9.13. The series expansion of the first few terms of equation (9.13) is

$$f(t) = \sin 2\pi t - \frac{1}{2}\sin 2\pi 2t + \frac{1}{3}\sin 2\pi 3t - \frac{1}{4}\sin 2\pi 4t + \cdots.$$

The geometric sawtooth wave can be expressed as

$$y(t) = \begin{cases} t, & 0 \le t < \pi, \\ t - 2\pi, & \pi \le t < 2\pi. \end{cases} \qquad\qquad \text{\textit{Sawtooth Wave, Non-Band-Limited}} \quad (9.14)$$

Sum of Cosines Waves   Summing equal-amplitude cosine harmonics,

$$t(t) = \frac{1}{N}\sum_{k=1}^{N} \cos 2\pi kt, \qquad N > 0, \qquad\qquad\qquad\qquad \text{\textit{Sum of Cosines}} \quad (9.15)$$

the series converges as $N \to \infty$ to an impulse train signal. The series expansion of the first few terms of equation (9.15) is

$$f(t) = \frac{1}{N}(\cos(2\pi t) + \cos(2\pi 2t) + \cos(2\pi 3t) + \cdots).$$

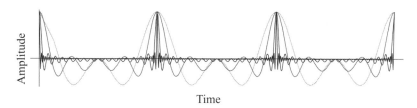

Figure 9.14
Sum of cosines converges to an impulse train signal.

Figure 9.14 shows the sum of cosines for $N = \{2, 4, 16, 32\}$.

In the limit, as $N \rightarrow \infty$, the sum of cosines converges to an impulse train function:

$$f(t) = \frac{1}{N} \cdot \frac{1 - \cos 2\pi k}{1 - \cos(2\pi k/N)}. \qquad \textit{Sum of Cosines, Closed Form} \quad (9.16)$$

The complex form of the sum of cosines—I suppose we should call it the *sum of phasors*—is the same as equation (9.15) with a phasor instead of a cosine:

$$f(t) = \frac{1}{N} \sum_{k=1}^{N} e^{i2\pi kt}. \qquad \textit{Sum of Phasors} \quad (9.17)$$

Here is its closed form:

$$f(t) = \frac{1}{N} \cdot \frac{1 - e^{i2\pi k}}{1 - e^{i2\pi k/N}}. \qquad \textit{Sum of Phasors, Closed Form} \quad (9.18)$$

As N grows, the sum of phasors becomes the complex helix form shown in figure 9.15.

A Note about Non-Band-Limited Signals   The geometric forms of the equations for all these waveforms exist only in the limit when $N \rightarrow \infty$; therefore they have infinite bandwidth. All the preceding geometrical waveforms (except for the sum of cosines) decrease in amplitude (the triangular wave the most quickly) with increasing harmonic number k and eventually become insignificant. However, when synthesizing geometric waveforms in a sampled system such as a computer, one should use the Fourier summations, being careful to adjust the limit of summation N to prevent the highest harmonic from exceeding the Nyquist frequency if aliasing is not desired.

9.2.8   Wavetable Synthesis

The limitations of waveform synthesis are the same as for the Fourier transform: the waveform must be periodic, and the resulting spectrum must be harmonic. But a variant of this approach provides a very efficient way to synthesize an arbitrary harmonic spectrum on a computer.

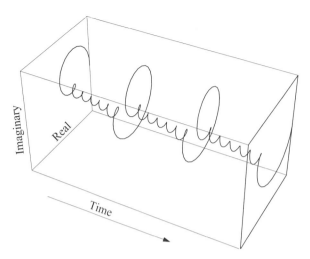

Figure 9.15
Sum of phasors.

Equation (9.6) is written in terms of continuous time t. To synthesize sound using this equation would require either an analog electronic synthesizer or an analog computer because they operate in continuous time. To adapt this approach to a digital computer coupled to a DAC, we must sample this continuous-time function by defining a sampling interval T (see section 1.4). Then defining $t = xT$, the discrete version of equation (9.6) is

$$y(x) = \sum_{n=1}^{N} A_n \sin 2\pi nfxT,$$ *Discrete Waveform Synthesis* (9.19)

where x is the sample index, f is the fundamental frequency, T is sample period and n is the harmonic number.

Creating a Wavetable   Since every period of equation (9.19) is the same, we can drastically reduce the amount of computation required to generate it by precomputing one period of the desired waveform and storing it in a table. We then iteratively read out the precomputed sample values from the table in real time to generate the sound. The only real-time computation needed is amplitude scaling and a small amount of arithmetic to determine the order in which to extract the samples from the table. To capture just one period of the waveform requires a modification of equation (9.19):

$$\tau_s = \sum_{n=1}^{N} A_n \sin 2\pi n \frac{s}{L}.$$ *Wavetable Synthesis* (9.20)

Figure 9.16
Sampled wave table.

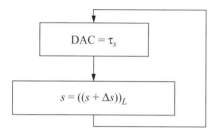

Figure 9.17
Table-lookup oscillator.

Here, τ_s represents a table of length L, indexed by s, that holds one period of the sampled waveform. The amplitudes of the desired harmonic components are given by the array A_n of length N. Figure 9.16 shows a sampled wavetable, given $A_n = \{1, 1/3, 1/5\}$, $N = 3$, and $L = 32$.

Indexing a Wavetable   Having constructed a wavetable, we must now develop a means to extract values from it at a rate that will produce a desired synthesis frequency. Suppose the table stores one waveform period composed of $L = 1024$ samples. We must develop a computer program that will read the samples out of the table in any order we specify, one sample at a time. The samples are then converted through a DAC and sent to a loudspeaker. If we read out samples at a rate of $R = 8192$ samples per second, looping back to the beginning of the table when we run off the end, the frequency we'd output is $f = R/L = 8$ Hz. Figure 9.17 shows a simple two-step procedure that performs this operation.

This procedure first outputs the sample of table τ indexed by s to the DAC, then increments s by an amount Δs, then repeats these steps. The variable s is the *index*, and Δs is the *increment*. The index corresponds to the instantaneous phase of the oscillator, and Δs corresponds to the instantaneous frequency. In the example under discussion, $\Delta s = 1$. When s is incremented past the end of the table such that $s \geq L$, then the modulus operator applies, and s is reset back into the range of the table. (The expression $((q))_p$ means the remainder after integer division of q by p. See appendix section A.5.)

If we set $\Delta s = 2$ so that we skip every other sample, we'd run through the table twice as fast, and the frequency would double. Thus, for some sample rate R and table length L, the formula for the frequency f is

$$f = \Delta s \frac{R}{L}. \tag{9.21}$$

In this example we're limited to frequencies that are multiples of 8 Hz so long as Δs must be an integer. But if Δs is a real variable, we could theoretically generate any frequency. For example, if we set $\Delta s = 1.5$, equation (9.21) indicates we'd be able to generate a frequency of 12 Hz. But remember that the wavetable τ is just a list of samples, and there is nothing between them to index, unless we make up rules to define how to interpret the space between sample values. Here are some common choices for how to ascribe meaning to the space between samples.

Truncation   Choose the nearest sampled value toward 0. For example, if $s = 3.7$, we throw away the 0.7 and choose sample 3. Truncation is performed with the floor operator, $\lfloor x \rfloor$, which returns the largest integer not larger than x. So, for example, $\lfloor 3.7 \rfloor = 3$. Since truncation simply discards the fractional part of the index, the index it produces can be off by nearly an entire sample. For example, if $s = 3.999$, truncation still indexes sample 3, although sample 4 might be a better choice in this case.

Rounding   We can improve on truncation by performing rounding on the index before selecting the sample. For example, if $\Delta s = 3.7$, then because its fractional part $0.7 \geq 0.5$ we round up, choosing sample 4. We can express rounding as follows:

$$
x = \begin{cases} x - \lfloor x \rfloor < 0.5, & \lfloor x \rfloor, \\ x - \lfloor x \rfloor \geq 0.5, & \lfloor x \rfloor + 1, \end{cases}
$$

which sets x to $\lfloor x \rfloor$ if its fractional part is less than 0.5 and otherwise to $\lfloor x \rfloor + 1$.

Linear Interpolation   Another approach is to assume that a straight line joins adjacent sample points and to select a point on this line that is proportional to the fractional part of the index.

For example, suppose we arbitrarily select two samples from a series: $y_{14} = 3.2$, and $y_{15} = 4.5$. Say the real index is $s = 14.7$. Draw a ramp connecting adjacent samples, as shown in figure 9.18, and measure off a distance between the samples corresponding to the size of the fractional part of the index, then measure the distance from the ramp to the x-axis at that point. In this example, the interpolated result would be $y = 0.7y_{15} + (1 - 0.7)y_{14} = 4.11$. In general, we proceed in two steps:

Step 1. $\sigma = s - \lfloor s \rfloor$.

Step 2. $y = \sigma y_{\lfloor s+1 \rfloor} + 1 - \sigma y_{\lfloor s \rfloor}$.

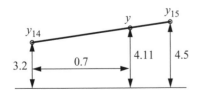

Figure 9.18
Linear interpolation.

Either truncation or rounding can be used effectively to traverse continuous and noncontinuous functions alike because only the actual samples are indexed. Truncation is computationally the simplest approach, rounding is slightly more complicated, and linear interpolation is most complex. Truncation can produce results that err by almost an entire sample, whereas rounding can err by at most half a sample. The error introduced by linear interpolation depends upon the precision of the available arithmetic, but it generally produces an error far less than the other two if indeed the underlying function is at least approximately linear between samples.

In digital audio, truncation and rounding errors appear as a kind of *phase jitter* because the signal is misindexed slightly by the error. It's as though the indexed region of the waveform is shoved forward or backward in time by the size of the truncation or rounding error. Phase jitter is a kind of frequency modulation, so it introduces distortion into the signal. For the truncating oscillator, truncation error can be reduced by increasing the number of samples of the underlying waveform. Linear interpolation also introduces some error because rarely will a linearly interpolated value agree exactly with the actual value of the underlying function; however, the error will generally be less than for truncation or rounding.

Spectrally, linear interpolation is similar to a second-order lowpass filter with a triangular impulse response. Take another look at figure 4.4, which shows that the convolution of two rectangular functions is a triangular function. The Fourier transform of the triangular impulse response is the sinc-squared function (see figure 4.36).

The main lobe of the sinc-squared function contains the spectral bandwidth of the original signal. It progressively filters out high-frequency components. The first null in the function occurs at half the sampling rate, so it behaves like a rather poor-quality anti-aliasing filter. The side lobes of the sinc-squared function contain attenuated copies of the original spectral bandwidth and cause aliasing (Smith 2004). The frequency response of linear interpolation is not ideal for at least two reasons. The spectrum is progressively lowpass-filtered near half the sampling rate and is nowhere flat, and aliasing contributed by the first side lobe is down only about 26 dB.

But that's not all. Linear interpolation is commonly used to obtain a fractional delay copy of a signal, for example, to stretch or compress a segment of audio in time. If the point of interpolation sits right on top of a source sample, no spectral change is introduced by the linear interpolation, and the frequency response is allpass. But if the interpolation point is halfway between two samples, the source is lowpass-filtered (by averaging) and suffers distortion from aliasing. If the interpolation point changes through time, which it typically does, linear interpolation introduces an objectionable variable filtering effect.

Additionally, linear interpolation only works if the underlying function from which the samples were derived was smoothly continuous before it was sampled; then interpolation may produce a reasonable approximation of what the function might have been between samples. Otherwise, the result is just a guess.

None of these techniques is ideal, and each introduces some distortion. But they are computationally quite efficient and are widely used. The method of band-limited interpolation (see section 10.2.7) generally yields the highest-quality results, though at greater computational cost.

9.2.9   Table Lookup Oscillator

To construct a table lookup oscillator, we want to choose a sample from table τ as defined in equation (9.20) indexed by s and then advance by Δs samples and repeat, starting over when we fall off the end of the table. Both the index s and increment Δs are real-valued variables, allowing us to progress through τ_s at an arbitrary rate. For simplicity, we'll truncate s to index each sample we output.

Oscillator with Constant Amplitude and Frequency   The result will be a waveform with a constant amplitude and frequency. The wave produced depends on the contents of the wavetable τ. For each sample n, we perform the following steps:

Step 1. $y_n = A\,\tau_{\lfloor s \rfloor}$.

Step 2. $s = ((s + \Delta s))_L$.   *Static Table Lookup Oscillator Procedure* (9.22)

Step 3. $n = n + 1$.

Step 4. Repeat.

y_n is the array of output samples, A is amplitude, L is the length of the table, the notation $\lfloor x \rfloor$ is the floor function, and the notation $((x))_L$ means the value of x modulo L.

1. We start by outputting the currently indexed value: $y_n = A\,\tau_{\lfloor s \rfloor}$. The output sample y_n is the table value τ indexed by the floor of the integer table lookup index s and scaled by the amplitude A.

2. Next, we determine which table value to index next time around: $s = ((s + \Delta s))_L$. The next table index s is obtained by adding the current table index and the increment Δs, modulo the length of the table L.

3. Finally, we advance to the next sample time, $n = n + 1$, then repeat the calculation for as many output samples as required.

The increment Δs determines the rate at which we progress through the table, and hence the frequency f of the oscillator. Suppose we have some fixed value of Δs. If we increase the sampling rate R holding Δs constant, the frequency goes up. And if we increase the length of the table L holding Δs constant, the frequency goes down. Since τ holds exactly one period of the waveform, the length L of the table corresponds to 2π radians. Thinking along these lines, we see that the increment Δs corresponds to the frequency f:

$$\Delta s = f\frac{L}{R}.$$ *Oscillator Increment and Frequency* (9.23)

Thus, to achieve a frequency $f = 8$ with a table length $L = 1024$ and sample rate $R = 8192$, we set $\Delta s = 1$.

Oscillator with Variable Amplitude and Frequency   In order to be musically useful, we must allow the oscillator's instantaneous frequency to change through time, for instance, to let the pitch glide up and down over time. The simplest way to accommodate this is to have an array of increments, Δs_n, one for every output sample n, representing a sequence of instantaneous frequencies we wish to synthesize. For every sample we output, we determine the next table value τ_s to pick by taking the current table index s_n and adding the current increment Δs_n to it. The array of increments Δs_n allows us to vary frequency through time. Let's also have an array of amplitudes A_n so that amplitude can vary per sample as well. A procedure that takes these requirements into account is as follows (Mathews 1969):

Step 1. $y_n = A\tau_{\lfloor s_n \rfloor}$.

Step 2. $s_{n+1} = ((s_n + \Delta s_n))_L$.   *Table Lookup Oscillator Procedure* (9.24)

Step 3. $n = n + 1$.

Step 4. Repeat.

The steps are as before except that potentially we have a different table index s_n and increment Δs_n on each sample n, and a different amplitude A_n on each sample. (I say "potentially" because, of course, we could set all the A_n and Δs_n the same, in which case this would revert to the static oscillator.)

1. We start by outputting the currently indexed value: $y_n = A_n\tau_{\lfloor s_n \rfloor}$. The nth output of the oscillator is formed from the product of the nth amplitude value times the sample indexed by the floor of the nth index value.

2. Next we compute the index of the subsequent output. $s_{n+1} = ((s_n + \Delta s_n))_L$. The next index is obtained by adding the current index and increment, modulo the length of the table.

3. We advance to the next moment, then repeat the calculation: $n = n + 1$.

The instantaneous increment Δs_n corresponds to the instantaneous frequency f_n:

$$\Delta s_n = f_n \frac{L}{R}.$$ *Instantaneous Oscillator Increment and Frequency* (9.25)

We can model the operation of the truncating table lookup oscillator as shown in figure 9.19. For each sample n, the current frequency f_n is multiplied by L/R to create Δs_n. Then Δs_n is added to the previous value of s_n to make the next real index value. We then take the modulus of the real index, producing the new s_n. We take the floor of s_n to give us an integer index that we can use to

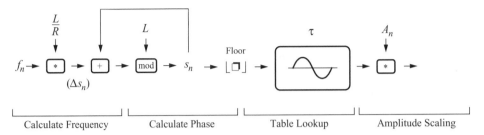

| Calculate Frequency | Calculate Phase | Table Lookup | Amplitude Scaling |

Figure 9.19
Truncating lookup oscillator patch.

select one sample from the wavetable τ. Finally, the result from the table is scaled by the current amplitude A_n.

9.3 Amplitude Modulation

Amplitude modulation (AM) varies the instantaneous amplitude of a signal, usually in a periodic manner. Figure 9.20 shows a sinusoid with a periodically varying instantaneous amplitude and a constant frequency.

Consider the patch shown in figure 9.21. The output of the first cosine wave oscillator, the *modulating oscillator*, is connected to the amplitude input of the second, the *carrier oscillator*.[6] In this way, the amplitude of the carrier oscillator is dynamically controlled by the modulating oscillator. I've chosen to use cosines here because it will help later with the mathematics, but sine waves would have done just as well in practice. In this example, the frequency of the modulating oscillator is f_m, and the frequency of the carrier is f_c. The amplitude input of the modulating oscillator is I, which stands for *modulation index*. Note that a constant value of 1.0 is added to the output of the modulating oscillator before it is fed into the carrier oscillator's amplitude input.

Writing out the equation for this patch, we have

$$f(t) = (1.0 + I\cos 2\pi f_m t)\cos 2\pi f_c t .$$

Simplify by letting $\omega_c = 2\pi f_c$, and $\omega_m = 2\pi f_m$:

$$f(t) = \underbrace{(1.0 + I\cos\omega_m t)}_{M}\underbrace{\cos\omega_c t.}_{C}$$

Amplitude Modulation with Carrier in Output   (9.26)

The terms labeled M correspond to the modulator, and the terms labeled C to the carrier. The formula essentially multiplies M and C terms to create the amplitude-modulated signal $f(t)$.

Note that when $I = 0$, the M term equals 1.0 and drops out, so the equation reduces to $f(t) = \cos\omega_c t$, a simple sinusoid at the carrier frequency.

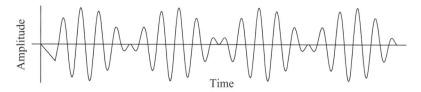

Figure 9.20
Amplitude modulation, carrier present in the output.

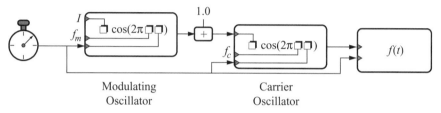

Modulating
Oscillator

Carrier
Oscillator

Figure 9.21
Amplitude modulation.

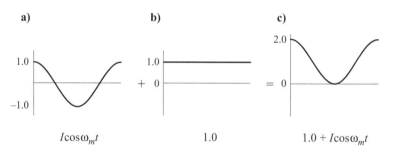

Figure 9.22
Graphical view of the M term.

    When $I \neq 0$, the M term essentially plots a cosine wave at frequency ω_m and adds 1.0 to every point. The effect is shown in figure 9.22 for a single period of the modulating cosine wave with $I = 1.0$. Figure 9.22a shows the term $I \cos \omega_m t$ and figure 9.22b shows the constant function 1.0. When summed in 9.22c, they form the M term.

    Note that since $I = 1$, in this example the value of the function in figure 9.22c ranges from a maximum of 2 to a minimum of 0. This function is then multiplied by the carrier term C in equation (9.26). When the M term's value is momentarily 0, it nullifies the carrier signal, and when it is at its maximum value of 2, it doubles the amplitude of the carrier signal. In between, the amplitude of the carrier follows the contour of the cosine wave as determined by the M term. If we restrict the range of the index to $0 \leq I \leq 1.0$, the result of the M term is always greater than or equal to 0, as shown in figure 9.22c,

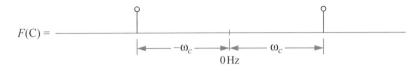

Figure 9.23
Spectrum of C terms.

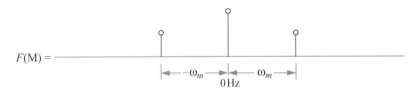

Figure 9.24
Spectrum of M terms.

and so it is an unsigned quantity. When $I = 1$, we say that the *depth of modulation* is 100 percent. If $I = 0$, the depth of modulation is zero. The waveform in figure 9.20 shows a modulation index of 100 percent. (I've also set the carrier and modulator frequencies in that example so that $\omega_m << \omega_c$. Thus many periods of the carrier oscillator are affected by one period of the modulator, making it easier to see the pattern.) The result we see is that the modulating signal dynamically controls the *amplitude envelope* of the carrier signal.

What is happening spectrally? Recall from section 2.6.8 that a real cosine waveform with unity amplitude is the vector sum of two half-amplitude phasors of equal sign and opposite frequency (see especially equation (2.52)). Since the C term in (9.26) is a cosine waveform, its spectrum is a cosine spectrum with half-amplitude frequency components $\pm\omega_c$ (figure 9.23).

The M term in (9.26) is the sum of the constant function 1.0 and a cosine waveform. The spectrum of a constant is just its magnitude at 0 Hz (see section 4.7.1). Thus the spectrum of the M term is a component with magnitude 1.0 at 0 Hz and a cosine spectrum with half-amplitude frequency components $\pm\omega_m$ (figure 9.24).

We can tie these two spectral plots together by recalling that since the M and C terms are multiplied in the time domain, their spectra are convolved in the frequency domain. The consequence is that copies of the M components are placed around each of the components of C (figure 9.25). Thus amplitude modulation can be thought of as frequency-shifting the spectrum of the modulating signal by the frequency of the carrier ω_c.

Since the entire spectrum of the modulating signal is shifted up and down by ω_c and $-\omega_c$, this form of amplitude modulation is sometimes called *double sideband modulation*. In broadcasting applications, it is an unnecessary redundancy to have identical upper and lower sidebands, and some broadcast systems filter out one sideband or the other, a technique known as *single sideband* modulation, in order to reduce the amount of radio frequency spectrum required by the transmitter.

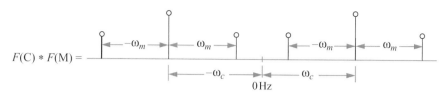

Figure 9.25
Convolution of M and C.

Here is the equation for amplitude modulation again for reference:

$$f(t) = (\underset{\uparrow}{1.0} + \underset{\uparrow}{I} \cos \omega_m t) \cos \omega_c t.$$

Duplicate of the carrier   Amplitude of the sidebands

Comparing figure 9.24 and 9.25, note that the spectral component corresponding to the modulation constant 1.0 is convolved to reside at the position of the carrier frequency $\pm \omega_c$. The amplitude of the sidebands that surround the carrier are determined by I.

9.3.1 Finding the Spectrum

We can derive the spectrum for amplitude modulation as follows. First, notice that the equation for amplitude modulation has the general form: $(a + b)c$, where $a = 1.0$, $b = I \cos \omega_m t$, and $c = \cos \omega_c t$. If we expand $(a + b)\,c$ to $ac + bc$ and substitute, we have

$$f(t) = 1.0 \cdot \cos \omega_c t + \underbrace{I \cos \omega_m t \cdot \cos \omega_c t}_{}.$$

$\qquad\qquad\qquad\qquad$ Product of two cosines

(9.27)

Note that the right-hand terms in (9.27) represent the product of two cosines. There is a handy trigonometric identity that shows what we can do with the product of two cosines (see appendix section A.4):

$$\cos x \cos y = \frac{1}{2}[\cos(x + y) + \cos(x - y)].$$

Substituting $\cos x = I \cos \omega_m t$, and $\cos y = \cos \omega_c t$ from this trigonometric identity into (9.27), we have

$$f(t) = \cos \omega_c t + \frac{I}{2}[\cos(\omega_c t + \omega_m t) + \cos(\omega_c t - \omega_m t)]$$

$$= \underbrace{\cos \omega_c t}_{} + \underbrace{\frac{I}{2}\cos(\omega_c + \omega_m)t}_{} + \underbrace{\frac{I}{2}\cos(\omega_c - \omega_m)t}_{}.$$

(9.28)

$\qquad\quad$ Carrier $\qquad\qquad$ Upper sideband $\qquad\quad$ Lower sideband

Equation (9.28) shows the spectral form of amplitude modulation. The first term represents the carrier frequency, and the second and third represent the upper and lower sidebands, respectively. Note that the carrier frequency does not depend in any way upon ω_m; all the dynamic properties of amplitude modulation are controlled by the frequencies and amplitudes of the sidebands, which are each one half of the value of I.

The discussion has assumed that the modulating function is a cosine oscillator, and while that is a good assumption for analysis purposes, in practice any signal can act as the modulating function. For instance, if ω_c is a radio frequency, and $M(t)$ is an announcer's microphone signal, the result could be an AM radio broadcast signal according to

$$f(t) = [1.0 + IM(t)]\cos \omega_c t. \qquad \textit{Amplitude Modulation, General Case} \quad (9.29)$$

9.3.2 Ring Modulation

Equation (9.26) for amplitude modulation is a two-quadrant multiply, a signed quantity times an unsigned quantity (see section 3.1.7). Since the M term is always greater than or equal to 0 (because we require $0 \leq I \leq 1.0$), it is an unsigned quantity. And because the C term is an acoustical waveform, it goes positive and negative, so it is signed.

Now, if we left out the $+1.0$ from the M term in equation (9.26) it would simply be $I\cos \omega_m t$, which is a signed term. The resulting equation,

$$f(t) = (I\cos \omega_m t)\cos \omega_c t, \qquad \textit{Ring Modulation} \quad (9.30)$$

is a four-quadrant multiply because the M and C terms are both signed. The spectrum corresponding to equation (9.30) is shown in figure 9.26.

Ring modulation is the same as amplitude modulation but without the carrier frequency present in the output spectrum. For a more rigorous proof of the spectral consequences of ring modulation, see appendix section A.10.

9.3.3 Musical Uses of Amplitude Modulation

If the frequency of the modulating oscillator is subaudio, amplitude modulation is called *tremolo*. This is the effect, for instance, of the tremolo control on a Fender electric guitar amplifier.

Probably the most important use of amplitude modulation is the central role it plays in spectral analysis and synthesis (see chapter 3). The Fourier transform essentially ring-modulates the probe phasor and the input signal as the first step to determine the spectrum of the sound.

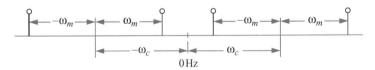

Figure 9.26
Spectrum of ring modulation.

Since there is not necessarily a connection between the spectrum of the carrier and the modulating signal, amplitude modulation and ring modulation are really good ways to create inharmonic spectra. Depending upon the materials and treatment, the effect can still retain enough of the original timbre of the modulating signal to identify its source, though it sounds mutated in a strangely dissonant but still coherent way. Some Hollywood science fiction movies from the 1950s used ring modulation to give the aliens spooky-sounding voices.

9.4   Frequency Modulation

Frequency modulation (FM) varies the instantaneous frequency of a signal in a periodic manner. Figure 9.27 shows a frequency-modulated sinusoid with time-varying frequency and a constant amplitude.

The frequency of a carrier sinusoid is varied by a modulator sinusoid, and the strength of the modulator's effect on the carrier frequency is called the depth of modulation. In figure 9.27 the carrier frequency is about ten times higher than the modulating frequency, and the depth of modulation is quite strong.

The frequency of an oscillator can be varied in time by patching the output of the modulating oscillator to the frequency input of the carrier oscillator (figure 9.28). Writing out the equation for this patch, we have

$$f(t) = A\sin(\omega_c t + \Delta f \sin \omega_m t), \qquad\qquad \textit{Frequency Modulation}\ \ (9.31)$$

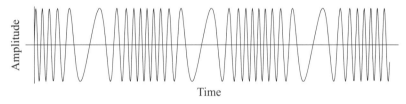

Figure 9.27
Frequency-modulated sinusoid.

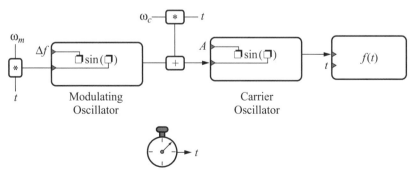

Figure 9.28
Frequency modulation patch.

where A is amplitude, $\omega_c = 2\pi f_c$ is the carrier frequency, and $\omega_m = 2\pi f_m$ is the modulating frequency. The term Δf, called *peak frequency deviation*, determines the amplitude of the modulating oscillator's output, which controls the swing of the carrier oscillator's frequency. If $\Delta f = 0$, equation (9.31) reduces to $f(t) = A \sin \omega_c t$, which is a sine wave at a constant frequency. However, if $\Delta f \neq 0$, a multitude of sidebands appear, positioned above and below the carrier at multiples of $\pm f_m$.

To get a feel for this, let's use equation (9.31) to create some sample spectra with different values of Δf. Let's set $f_c = 1000$ Hz, and $f_m = 100$ Hz, and increase Δf gradually, starting at 0. When $\Delta f = 0$, we have a sine wave at constant frequency with a magnitude spectrum as shown in figure 9.29a.

As Δf grows, we see the carrier decline in amplitude, and additional sidebands at fixed frequencies $f_c \pm n f_m$ enter the spectrum, where n is the integer order of the sidebands. It appears that as Δf increases, energy is stolen from the carrier frequency and distributed over an ever wider number of sidebands. However, in figure 29d, where $\Delta f = 3.6$, the carrier makes a forceful reappearance, even as the number of sidebands continues to grow.

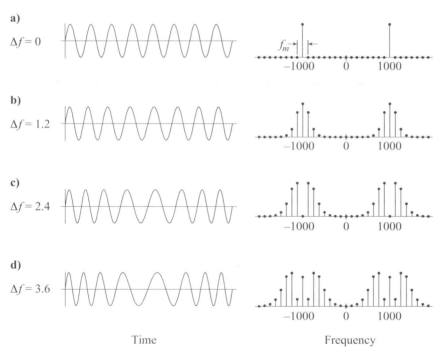

Figure 9.29
Sample FM spectra, various values of frequency deviation.

9.4.1 FM as the Combination of Fixed Frequency Components

How can it be that if we sweep a frequency continuously across a certain limited frequency range it suddenly looks like we have a collection of fixed frequency components spread at discrete points across a wide range of the spectrum? And what accounts for the comings and goings of the sidebands?

It turns out that if fixed frequency sinusoids are combined at exactly the right frequencies, phases, and amplitudes, the result is identical to a frequency-modulated sinusoid. For this to work out, the sidebands of the same order above and below the carrier must be equal in amplitude, and the odd-ordered lower sidebands must be 180° out of phase from the even-ordered sidebands. Additionally, the amplitudes of the various orders of sidebands must be carefully chosen. An example will give a flavor of this.

The three sine waves shown in figure 9.30a are defined as follows:

$$x(t) = 0.98 \sin 2\pi f t, \qquad\qquad \text{Carrier}$$

$$y(t) = 0.24 \sin 2\pi (f + \Delta f) t, \qquad \text{Upper sideband}$$

$$z(t) = -0.24 \sin 2\pi (f - \Delta f) t, \qquad \text{Lower sideband}$$

where $f = 1$ and $\Delta f = 1/16$. Figure 9.30a shows 16 periods of $x(t)$, just enough time for the upper and lower sidebands to precess against each other once. These three waves are summed, $s(t) = x(t) + y(t) + z(t)$, to create the wave labeled $s(t)$ in figure 9.30b. The lighter sinusoid in

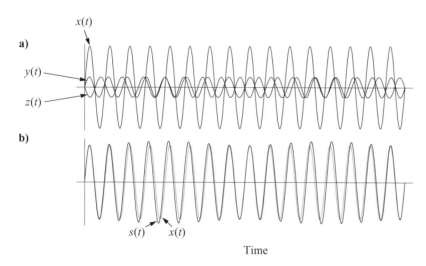

Figure 9.30
Sum of three sine waves creates frequency modulation.

line figure 9.30b is just $x(t)$ shown again for reference. The curves look almost the same, but looking carefully, it's evident that they are not.

Notice that $s(t)$ first lags behind, then advances past $x(t)$. It looks as though the frequency of $s(t)$ changes over time. This is not an optical illusion: in fact, the frequency of $s(t)$ is changing. Summed sinusoids with carefully chosen amplitudes and frequencies are equivalent to a single waveform that changes frequency over time.

Notice in figure 9.30a that when the sidebands are in phase with each other, they are 90° out of phase with respect to the carrier, and when they are out of phase with each other, they are −90° out of phase with the carrier. The sum of the three components appears to have nearly constant amplitude, but it advances and lags in phase relative to the carrier. Figure 9.27 shows this to be exactly the behavior of a frequency-modulated sinusoid: its phase velocity advances and retreats, alternately increasing and decreasing its frequency slightly.

Although the amplitude of $s(t)$ changes slightly, this is because we're summing only three sinusoids together. In order to eliminate the amplitude change, we'd need to sum an infinite number of sinusoids at ever wider frequencies and ever smaller amplitudes according to a particular formula.

What is the formula that determines how to mix the carrier and sidebands together so that a constant-amplitude frequency-modulated sinusoid results?

9.4.2   Return of the Bessel Functions

The amplitudes of the FM-induced sidebands are determined by Bessel functions of the first kind and nth order. Interestingly, they are the same Bessel functions that characterize the motion of vibrating membranes (volume 1, figure 8.21). $J_0(I)$ determines the amplitude of the carrier, $J_1(I)$ determines the amplitude of the first upper and lower sidebands, and in general, $J_n(I)$ determines the amplitude of the nth upper and lower sidebands.

The frequency-modulated sinusoid given in equation (9.31) can be seen to be equivalent to a set of fixed-frequency sinusoids whose amplitudes and phases are determined by the Bessel functions, according to the following trigonometric identity:

$$
\begin{aligned}
f(t) &= A\sin(\omega_c t + I\sin\omega_m t) \\
&= A\sin(\theta + I\sin\beta) \\
&= J_0(I)\sin\theta + \sum_{n=1}^{\infty} J_n(I)[\sin(\theta + n\beta) + (-1)^n \sin(\theta - n\beta)].
\end{aligned}
\tag{9.32}
$$

If we expand the terms of the infinite summation in equation (9.32), we obtain

$$
\begin{aligned}
f(t) = {}& J_0(I)\sin\theta \\
& + J_1(I)[\sin(\theta + \beta) - \sin(\theta - \beta)] \\
& + J_2(I)[\sin(\theta + 2\beta) + \sin(\theta - 2\beta)] \\
& + J_3(I)[\sin(\theta + 3\beta) - \sin(\theta - 3\beta)] \\
& + \cdots.
\end{aligned}
\tag{9.33}
$$

The sidebands are at multiples of the modulating frequency, odd lower sideband components are negative, and the amplitudes of successive orders of sidebands are determined by the corresponding orders of Bessel functions.

9.4.3 FM and Bessel Functions

We can make the following observations about equation (9.32) and about Bessel functions to help explain what makes FM synthesis sound the way it does.

- The summation of sidebands goes on to infinity. Only when $I = 0$ can we say that there is absolutely no energy in any sidebands. Therefore, we can say that a frequency-modulated sinusoid has an infinite spectrum, but depending upon the value of I, most of the sidebands have insignificant energy.

- The odd-ordered lower sidebands alternate sign. The term $(-1)^n$ accomplishes this sign reversal for the odd lower sidebands.

  Looking at the characteristic shape of the Bessel functions themselves,

- The Bessel functions look somewhat like damped sinusoids, since their peak amplitudes diminish gradually as the modulation index I increases.

- Note that J_0 is different from all the other Bessel orders because $J_0(0) = 1$, whereas for all other orders $n > 0$, $J_n(0) = 0$. So when $I = 0$, $J_0(I)$, which controls the amplitude of the carrier, has all the available energy.

- Higher-order Bessel functions start up from zero more and more gradually. This means that higher-order sidebands have significant energy only when the modulation index I is large.

- Depending upon the index, some Bessel functions will be strongly positive, and others—even nearby components—might be near zero or at zero, or negative. For example, in volume 1, figure 8.22, $J_2(5)$ is almost zero while $J_3(5)$ is near its maximum.

- Also notice in that figure and in figure 9.29d that when $I > 2.5$, some Bessel functions produce a negative scaling coefficient for some values of I.

- As I increases, the amplitudes of the sidebands increase and decrease in strength in a characteristic damped sinusoidal way.

  Thinking about all these points, and especially the last one, if we have a way to change the modulation index dynamically, the spectrum will change in complex ways, tracking the ups and downs of the Bessel function curves. Altogether, the spectrum of the sound will exhibit a lively character as we sweep I from one value to another during the time a note is sounding. This is the particular charm of FM synthesized sounds, that they contain an interesting built-in spectral evolution, determined by the Bessel function curves, as the modulation index varies. If an important aim of good synthesis techniques is to elicit complex yet predictable behavior from simple controls, then FM synthesis is certainly a very good technique indeed.

9.4.4   Modulation Index

At this point, we have two equations for frequency modulation. First there is equation (9.31):

$$f(t) = A \sin(\omega_c t + \Delta f \sin \omega_m t),$$

which expresses the strength of the modulating oscillator in terms of peak frequency deviation Δf. Then there is equation (9.32):

$$f(t) = A \sin(\omega_c t + I \sin \omega_m t),$$

which expresses spectra in terms of the Bessel function index I. If I could be related to Δf, we could use the Bessel functions to predict the spectrum we would get if we employ a particular peak frequency deviation. The rest of this section searches out this connection.

The expression for instantaneous phase in equation (9.32) is $\omega_c t + I \sin \omega_m t$. In graph theory, this is an expression of the form $ax + b$, defining a slope, where $a = \omega_c$ determines the rate at which the slope grows, $x = t$ marks units on the x-axis, and $b = I \sin \omega_m t$ is the offset. This is not a linear slope because b is not a constant; rather it is itself a slope that changes steepness through time sinusoidally. But it's still a slope—just a rather bumpy one.

Here is an example. For convenience, let $\theta(t) = \omega_c t + \Delta f \sin \omega_m t$. If we set $\Delta f = 0$, then $\theta(t) = \omega_c t + 0$, which is just a linear ramp function that intersects the y-axis at zero. Curve (a) in figure 9.31 shows the slope of $\theta(t) = 2\pi \cdot 2 + 0$ over a time interval of 4 seconds. The slope of $\theta(t)$ determines the frequency of the sinusoid in equation (9.32), which will be 2 Hz. If, as with function (b), we set $\omega_c = 2\pi \cdot 8 + 0$, it would have a steeper slope, and the frequency of the sinusoid would go up to 8 Hz.

Now let's make this more interesting. Setting $\omega_c = 4$ Hz, $\omega_m = 2$ Hz, and $\Delta f = 2$, we get curving function (c) in figure 9.31. Here, the instantaneous frequency varies, or *modulates,* according to the instantaneous slope of $\theta(t)$. Where the slope rises more sharply, the frequency goes up; where it rises less sharply, the frequency goes down. The most positive part of the slope of this function will approximately double the frequency when t is in that region; where the slope momentarily goes to zero (where it becomes horizontal), the frequency momentarily goes to zero.

Last, consider the slope of (d) in figure 9.31. Here, $\omega_c = 6$ Hz, $\omega_m = 2$ Hz, and $\Delta f = 4.8$. Notice that sometimes the slope of this curve goes negative. That means the value of $\theta(t)$ decreases through time in that region, and the phase velocity of the oscillator is negative. We're producing *negative frequencies* in that region of time. We can leverage these ruminations about $\theta(t)$ into a precise understanding of how modulation index I relates to peak frequency deviation Δf, as follows.

If the ramp function $\theta(t)$ defines frequency, then its derivative defines instantaneous frequency. According to the rules laid out in section 6.1, the derivative of $\theta(t)$ with respect to time can be expressed in terms of equation (9.32):

$$\frac{d\theta}{dt} = \frac{d}{dt}\omega_c t + \frac{d}{dt} I \sin \omega_m t$$

$$= \omega_c + I \omega_m \cos \omega_m t.$$

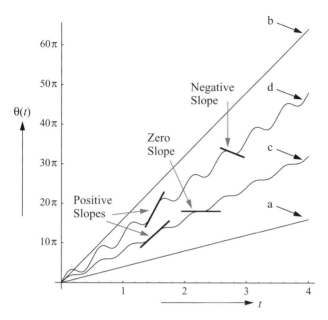

Figure 9.31
Linear and sinusoidally driven slope functions.

Similarly, the derivative of $\theta(t)$ in terms of equation (9.31) is

$$\frac{d\theta}{dt} = \omega_c + \Delta f \cos \omega_m t \,.$$

Since both these expressions define the derivative of $\theta(t)$ with respect to time t, we may equate them:

$$\omega_c + \Delta f \cos \omega_m t = \omega_c + I \omega_m \cos \omega_m t \,.$$

Solving for Δf yields

$$\Delta f = I \omega_m, \qquad\qquad\qquad\qquad \textit{Peak Frequency Deviation} \quad (9.34)$$

and solving for I produces

$$I = \frac{\Delta f}{\omega_m} \,. \qquad\qquad\qquad\qquad \textit{Modulation Index} \quad (9.35)$$

Equations (9.34) and (9.35) allow us to relate depth of modulation, the modulation frequency, and the index of the Bessel functions. In practical terms, if we want to use FM to create a spectrum that has the strengths of the Bessel functions at some particular index I, and we want the components to be separated in frequency by ω_m, then we must choose a depth of modulation Δf according to these formulas (F. R. Moore 1990, 318).

9.4.5   Calculating FM Spectra

As we've seen, for some positive carrier frequency, as index I grows,

- The spectral bandwidth grows.
- The upper sidebands grow toward higher frequencies.
- The lower sidebands grow toward lower frequencies.

If the index I grows sufficiently, sidebands below 0 Hz eventually become active. What happens to the lower sidebands that venture below 0 Hz? They turn into negative frequencies, or equivalently, we hear a phase-reversed positive frequency. Both interpretations are valid mathematically: a negative frequency can be thought of simply as a phasor spinning in the opposite direction of a corresponding positive frequency. Alternatively, 0 Hz can be thought of as a mirror that reflects negative frequencies back into the positive domain, but with a 180° phase shift. In a nutshell,

$$\sin(\theta) + \sin(-\theta) = \sin(\theta) - \sin(\theta) = 0. \tag{9.36}$$

In other words, reversing the phase of a negative frequency makes it equivalent to a corresponding positive frequency with inverted sign (figure 9.32).

Let's look at a realistic example FM spectrum that includes sidebands that have strayed into negative frequency territory. Let's set $f_c = f_m = 100\,\text{Hz}$, and $I = 4.9$. The upper sidebands will be at 200, 300, 400, . . . Hz, and the lower sidebands will be at 0, –100, –200, . . . Hz. We can calculate the spectrum from the Bessel functions by looking up the strengths of the n sidebands for $J_n(4.9)$, as shown in figure 9.33.

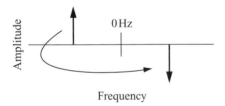

Figure 9.32
Reversing the phase of a negative frequency.

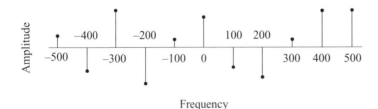

Figure 9.33
Sample FM spectrum.

The Bessel functions determine the relative strengths and signs of the components. The component at 0 Hz corresponds to a positive constant offset in the waveform. If we wrap the negative frequencies around to positive frequencies with a change of sign (or a 180° phase shift), they land on top of and sum with any components at the same frequency. They either add energy to or subtract energy from their positive frequency counterparts, depending upon whether their signs agree. This process is shown in figure 9.34. The result of combining the wrapped negative frequencies is the positive-frequency spectrum shown in figure 9.35. Our ears don't distinguish negative amplitude components from positive components. If we take the magnitude of this spectrum, we obtain what we'd actually hear (figure 9.36).

The wrapped negative components landed on top of positive frequency components because the ratio of carrier to modulating frequency was unity. Whenever $f_m/f_c = 1$, the resulting spectrum will always be harmonic. What about other possible ratios?

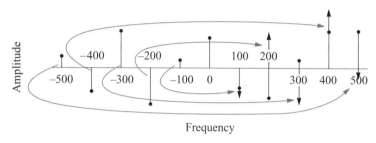

Figure 9.34
Components reflecting around 0 Hz.

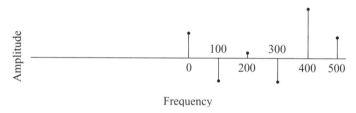

Figure 9.35
Resulting positive spectrum.

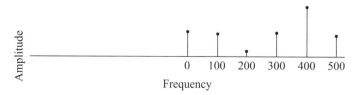

Figure 9.36
Magnitude FM spectrum.

Truax (1977) suggested a *harmonicity ratio,* defined as the ratio of modulating frequency to carrier frequency, such that $H = f_m/f_c$, where H can be either an integer or a real number.

There are two major classes of FM spectra to consider:

- If H is rational, the spectrum is harmonic
- If H is irrational, the spectrum is inharmonic

Rational Harmonicity   The preceding example spectrum belongs to the class of rational harmonicity, $H = 1$. As we've seen, this spectrum is harmonic, and the carrier frequency is also the fundamental.

An interesting subclass of spectra consists of $H = 1/m$, where m is a positive integer. For example, if $H = 1/2$, the carrier frequency is twice the modulating frequency. The spectrum is still harmonic, but the carrier is the second component, not the first. In fact, all spectra $H = 1/m$ are harmonic, and the carrier is positioned at component m.

The other interesting subclass is spectra where $H = n$, and $n > 1$. These spectra are also harmonic. Here, however, some components are missing, depending upon the value of n. An important example is $H = 2$, where the modulating frequency is twice the carrier frequency ($f_m = 2f_c$). Sidebands occur at $f_c \pm 2kf_m$ (where $k = 0, 1, 2, \ldots$) leading to a spectrum that omits all even harmonics, making this a convenient way of modeling the clarinet timbre. The first upper sideband component appears at $f_c + 2f_c = 3f_c$, and the next appears at $f_c + 4f_c = 5f_c$ and so on. Going down, the first lower sideband component appears at $f_c - 2f_c = -f_c$, the second at $f_c - 4f_c = -3f_c$, and so on. The negative frequencies wrap around on top of the positive ones, resulting in a spectrum with no even components. In general, for $H = n$, and $n > 1$, only components $kn + 1$ will be present.

Irrational Harmonicity   In general, if $H = n/m$, n and m are positive integers, the spectrum is harmonic. But if H is irrational, the spectrum is inharmonic. In this case, the components in the spectrum do not fall at integer multiples of the carrier frequency. Second, the negative frequencies that wrap around 0 Hz typically land between, not on, the positive frequency components, making the spectrum denser. Consider, for example, $H = \sqrt{2}$, which has components spreading out from the carrier at $f_c \pm kf_c\sqrt{2}$, $k = 0, 1, 2, \ldots$.

Again, there are two subclasses of interest. For $H = 1/m$, where m is a positive irrational number, components cluster increasingly densely around the carrier as m increases, mimicking the spectra of drums and gongs. There tends to be no clear fundamental for these timbres; hence they have no distinct pitch.

For $H = n$, $n > 1$, and n is a positive irrational number, the components spread out increasingly widely as n increases. This is a useful class of spectra for mimicking metal bar percussion, where components tend to stretch wider than the integer harmonic sequence. Even though the components are inharmonic, their relative sparseness can still contribute a sense of pitch under the right conditions.

9.4.6 A Historical Note

Prior to the 1970s frequency modulation was used primarily in radio theory to characterize FM broadcast transmission. It was the inspiration of John Chowning to wonder what such spectra would sound like translated into the audio domain. Initially led to the phenomenon by a mistake in a sound synthesis program he had written, Chowning (1973) went on to develop and patent an astonishingly powerful application of FM synthesis of audio spectra that included realistic simulations of every major family of Western musical instrument tones. His work was extended in many directions, both auditory (Schottstaedt 1977) and compositional (Truax 1977; Dashow 1980). Chowning worked with the Yamaha corporation to develop the DX7 synthesizer, the first mass-produced all-digital synthesizer that used audio band FM synthesis.[7] F. R. Moore (1990) asserts that in terms of the numbers of units sold, the DX7 qualified at the time as the most popular keyboard instrument.

9.4.7 Angle Modulation

There is some controversy as to whether what Chowning invented was an application of frequency modulation or phase modulation (PM). In the language of communications theory, both frequency modulation and phase modulation are kinds of *angle modulation*, which is defined as a sinusoid with a time-varying frequency: $f(t) = A \sin \theta(t)$. In frequency modulation, the instantaneous frequency of the carrier is varied from its center frequency by an amount proportional to the instantaneous value of the modulating signal. In phase modulation, the phase of the carrier is controlled by the modulating waveform. Clearly, varying the phase shift of a carrier also modulates its frequency, so FM and PM are very closely related.

We know that instantaneous frequency is the derivative of phase. The instantaneous phase is given by $\theta(t) = \omega_c t + \phi(t)$, and the instantaneous frequency is given by

$$\omega(t) = \frac{d}{dt}\theta(t) = \omega_c + \frac{d}{dt}\phi(t).$$

If the frequency of $f(t)$ is proportional to phase deviation $\phi(t)$, we have *phase modulation*. If the frequency of $f(t)$ is proportional to the derivative of phase deviation, $\frac{d}{dt}\phi(t)$, we have *frequency modulation*. By these criteria, Chowning's method is clearly frequency modulation. Not that it matters much. Schottstaedt (2003) writes, "Of course, you can't tell which is in use either from the waveform or the mathematical expression of the waveform—you have to know what the modulating signal was. That is a roundabout way of saying that in computer music applications there is no essential difference between frequency and phase modulation."

9.4.8 A Few of Chowning's Examples

According to the foregoing theory of FM, we can characterize Chowning's synthesis technique as follows. His method was based upon a frequency modulation patch (see figure 9.28), which by itself produces only a static FM spectrum. To this, Chowning added dynamic amplitude control using an envelope generator (see figure 9.6), and dynamic control over the modulation index I, also using an envelope generator.

One of Chowning's best ideas was that by varying the modulation index I through time, the Bessel functions would cause the spectrum to evolve in interesting ways, and this could be controlled to provide a sense of aliveness that marked the sound of traditional acoustical instruments. Another idea was to exploit particular carrier/modulator frequency ratios to model the spectrum of particular instrument families, for example, using $H = 2$ for clarinets and irrational ratios for percussion.

His method consisted of artistic but reasoned choices for carrier-to-modulator frequency ratios, based on the spectrum of the instrument he was modeling, and suitable choice of amplitude envelope and modulation envelope parameters to make a convincing demonstration. He had available a great deal of practical information about the spectral ballistics of musical instruments from the work of Risset (1969) and Mathews at Bell Telephone Laboratories to help him.

A version of Chowning's classic FM instrument patch appears in figure 9.37. The principal parameters are

A, the peak amplitude

$\omega_c = 2\pi f_c$, the phase velocity of the carrier oscillator, calculated from the carrier frequency f_c

$\omega_m = 2\pi f_m$, the phase velocity of the modulating oscillator, calculated from the modulating frequency f_m

$\Delta f = I / f_m$, the peak frequency deviation, expressed as the ratio of modulation index and modulating frequency

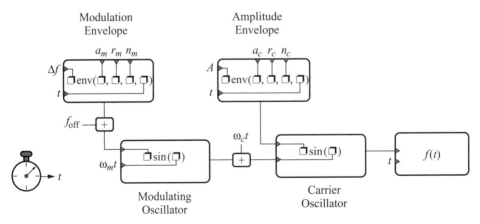

Figure 9.37
Dynamic frequency modulation patch.

$f_{\text{off}} = \alpha I/f_m$, where α is the initial or baseline frequency deviation (The idea is that f_{off} stipulates a minimum spectral brightness for a tone, and its maximum spectral brightness corresponds to $\Delta f + f_{\text{off}}$.)

a_m, r_m and n_m, attack time, release time, and sharpness of attack and release for the frequency deviation envelope

a_c, r_c and n_c, attack time, release time, and sharpness of attack and release for the amplitude envelope

Here are a few of the simulations Chowning created with this setup.

FM Trumpet Risset (1969) had shown that the spectral brightness of trumpets is proportional to their amplitude. Chowning simulated this by having the modulation envelope track the same envelope contour as the amplitude envelope. The spectrum of a trumpet is harmonic, with all components present, suggesting $H = 1$ would make a successful simulation. He made some artistic guesses about the shape of a trumpet amplitude envelope and the value for the peak modulation index that would sound trumpetlike, based on information from Risset, the acoustics literature, and his own ear. The result was his first successful simulation.

FM Bell Chowning knew that the amplitude and spectral brightness of bell tones diminishes exponentially over time, so he coupled the amplitude and modulation envelope trajectories. The spectrum is inharmonic, suggesting an irrational ratio for H such as $\sqrt{2}$ or e. He found various choices for the value for the peak modulation index and synthesized realistic simulations of gongs, vibraphones, and bells.

FM Bassoon Unlike the previous examples, the spectral evolution of the bassoon attack starts off mostly with high frequencies, then fills in lower frequencies as the tone becomes more solid. Chowning realized he should employ $H = 1/n$ for this timbre and make the carrier frequency high, so that as the modulation envelope grows, the effect the ear hears is of lower-frequency components entering after high-frequency components are already present.

FM Clarinet For the clarinet, Chowning used $H = 2$, resulting in a spectrum containing only odd harmonics. He adjusted the amplitude envelope to match a family of clarinet tones and adjusted the modulation index to match its overall spectrum.

FM Voice Chowning (1989) subsequently explored the use of FM to synthesize the singing voice. The most appropriate configuration he found was to have multiple carrier oscillators (typically three) driven by a common modulating oscillator, enabling him to construct rather arbitrary spectra. The frequency of the common modulator is set to the fundamental of the vocal tone to be synthesized. Each carrier frequency is set to whatever integer multiple of the modulating frequency places it nearest the center of each of the three primary vocal formants. Because of this approach, it is impossible to vary continuously between vowel sounds on the same pitch, or to vary the pitch keeping the same vowel. In spite of this limitation, the system was capable in Chowning's masterful hands of delivering compelling vocal synthesis.

Importance of Vibrato for Spectral Fusion   Perhaps the most interesting result of Chowning's vocal synthesis system was to show the importance of vibrato (correlated perturbation of harmonic frequency) for the psychoacoustic fusion of vocal timbres. I heard him demonstrate this once. First he played the sound of a rich male voice synthesized with his system, singing an "ah" timbre, with vibrato. Then he played the same timbre without vibrato. I still recognized it as a male voice but no longer as rich. Then he applied a bell-like exponentially decaying envelope with a sharp attack to the timbre, again without vibrato. I thought I was hearing a bell tone; it did not sound like a voice at all.

Finally, he repeated the bell-like envelope with the same timbre, and again I heard a bell, but this time, half-way through, he switched the vibrato back on. I experienced a severe case of cognitive dissonance. I observed my perception do a surprising and quick readjustment from thinking I was hearing bells back to hearing the rich male vocal timbre. For every listener hearing the demonstration, the cognitive interpretation of a vocal timbre won out over the bell the moment the vibrato started.

He did another demonstration to a similar effect with two vocal synthesis instruments set to "sing" the interval of a major third, but without vibrato and using the bell envelope. The timbre started out sounding richly bell-like. The vibrato he subsequently turned on was correlated within each voice but uncorrelated between voices. The complex bell timbre vanished, replaced by two male voices singing. We are hard-wired, evidently, to fuse harmonics that vary according to a correlated time function of frequency. This is not too surprising, given the survival value of sound source separation and system identification that our ancestors must have evolved long ago.

9.4.9   Critique of FM Synthesis

Chowning pioneered the application of FM to audio band spectra. The technique is extremely efficient for digital sound synthesis and is flexible enough for a wide variety of compositional applications. Handled well, it can achieve astonishing effects inexpensively. But otherwise, FM synthesis sounds like . . . well, FM synthesis. The ear is a wonderfully adaptable organ, and our contemporary sonic culture now includes the characteristic burble of an FM synthesizer sweeping through the Bessel functions.

Since FM synthesis is a nonlinear synthesis technique, there is no formal analysis method that would allow us to exactly resynthesize an arbitrary timbre with FM. Using FM to model instrumental timbres is strictly an art.

9.4.10   Composing with FM

Some interesting uses of FM synthesis have a purely compositional motive. Just as composers organize pitch space into musical scales and chords, Barry Truax suggested compositional methods for organizing the carrier-to-modulating frequency ratio (the $c:m$ ratio). Besides the degree of harmonicity defined as $H = f_m/f_c$, Truax (1977) developed other ways to organize the $c:m$ ratio that would help project structure in music composed using FM sounds. His methods included "predicting the precise interval between the carrier and the actual fundamental and relating that interval to just or tempered scales, and predicting sets of $c:m$ ratios producing unique spectra and those producing exactly the same spectrum (i.e., the same set of sidebands)"(70).

The composer James Dashow (1980) has sought ways to treat spectra as musical chords, to convincingly combine inharmonic spectra with traditional performed musical instruments. One of his methods was to pick out, for instance, a pair of chromatic pitches (a diad) and specify them as anchor points in an inharmonic spectrum that would include them as components. If, for instance, the pitches were E♭ and A, he would additionally specify the position of these components in the resulting spectrum; they might be, for example, the 5th and 7th components of a spectrum created using FM synthesis. When sounded together with, for instance, a piano playing E♭ and A, the ear senses the spectral correspondence, and a pleasing kind of musical unity-in-diversity results.

9.4.11 Waveshaping

Waveshaping synthesis is an improvement over FM synthesis in that arbitrary spectra can be directly specified instead of being mandated by the shape of the Bessel function curves. In its simplest form, it only creates harmonic spectra, but it can be extended to create inharmonic spectra.

Waveshaping is also sometimes called *nonlinear distortion synthesis*. A flat mirror reflects light linearly, but a curved mirror distorts the image; waveshaping synthesis is like reflecting a sinusoid off a curved mirror.

We are accustomed to interpreting $\sin \omega t$ as a linear ramp function ωt used to index a sine function, as in figure 9.38a. Another perspective on waveshaping is to say that it turns the tables (so to speak) on the standard sine wave oscillator. Instead of using a ramp function to index the sine function, it uses the sine function to index a ramp function. As shown in figure 9.38b, function W is the identity function that linearly reflects the sinusoid indexing it. The equation for the output waveform is

$$s(t) = W(\sin \omega t). \tag{9.37}$$

What other possible functions are there for W? We can try out any arbitrary reflecting function to see what happens. Figure 9.39 shows a cosine function indexing several kinds of reflecting functions. In figure 9.39a, a cosine reflected by the identity function generates a cosine. In 9.39b, a bump in the identity function creates two corresponding bumps in the cosine wave. In 9.39c, two complementary ramps create two complementary cosines with pointy ends. In 9.39d, a parabola

a) Sine Wave Generation

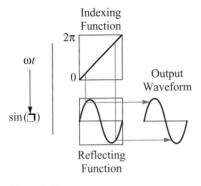

b) Waveshaping

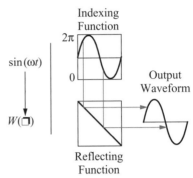

Figure 9.38
Sine wave generation and waveshaping.

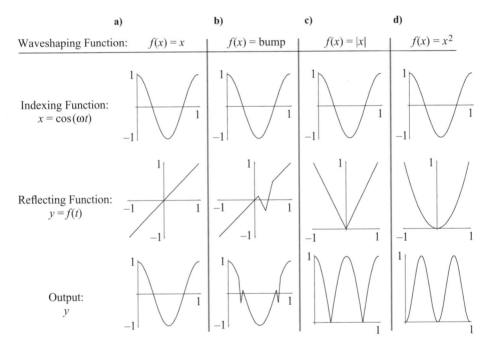

Figure 9.39
Arbitrary waveshaping functions.

warps the cosine wave into two periods of an inverted cosine wave. If it is not the identity function, then the reflecting function distorts the input waveform in a characteristic way. Any such distortion will introduce new spectral components.

Waveshaping as Function Composition   Nonlinear waveshaping is an application of function composition. A *composable function* is one that can be another function's argument. For instance, if $y = f(x)$, and $z = g(y)$, then

$$z = g(f(x)) \qquad\qquad\qquad Function\ Composition \quad (9.38)$$

is the composition of g with f because f is the argument to g.[8] For example, if $y = f(x) = x + 1$, and $z = g(y) = y^2$, then

$$z = g(f(x)) = g(x+1) = x^2 + 2x + 1.$$

We can see that waveshaping as defined by equation (9.37) is clearly based on function composition as follows. If in equation (9.38) we let $W = g$, and $f(x) = \sin \omega t$, then $g(f(x)) = W(\sin \omega t)$.

But frequency modulation can also be viewed as function composition. If $y = f(x) = cx + I \sin mx$, and $z = g(y) = a \sin y$, then

$$z = g(f(x)) = g(cx + I \sin mx)$$

$$\qquad\qquad\qquad FM\ as\ the\ Composition\ of\ Functions \quad (9.39)$$

$$= a \sin(cx + I \sin mx).$$

Compare equation (9.39) to equation (9.31). So FM and waveshaping are closely related synthesis techniques.

Chebyshev Polynomial Reflecting Functions   The Chebyshev[9] polynomials are a very useful class of reflecting functions. Their use in sound synthesis was discovered, virtually simultaneously, by LeBrun (1979) in the United States and Arfib (1979) in France. These functions have the following two remarkable properties when used as reflecting functions. By suitable addition of Chebyshev polynomial functions, a reflecting function can be constructed that produces any mixture of harmonics at any combination of strengths. Varying the amplitude of the indexing sinusoid varies the spectral brightness of the output of the Chebyshev polynomial reflecting function.

When the indexing sinusoid has low amplitude, its excursion covers only a small region at the center of the reflecting function and produces relatively simple harmonic spectra. But when its excursion covers the entire range of the reflecting function, the full harmonic spectrum coded in the function is produced. Thus, waveshaping has a capacity (like FM) to generate sounds with dynamic spectral evolution. But waveshaping (unlike FM) allows one to specify an arbitrary harmonic spectrum not limited to those provided by the Bessel functions.

In order to construct a set of known harmonic spectra, we want a set T_n of shaping functions that, when combined, produce the desired spectrum. (We choose T because one romanized version of Chebyshev's name begins with that letter.) Function T_0 should produce 0 Hz, T_1 should produce the fundamental, T_2 should produce the second partial, T_3 the third partial, and so on. We can define the first two functions trivially as follows:

$$T_0(x) = 1.$$

No matter what x is, the output is 1, so this produces 0 Hz.

$$T_1(x) = x.$$

This is the identity function. If x is sinusoidal, the output is identically sinusoidal.

We have T_0 and T_1, but what's the next function in the series? What we're looking for is a function that takes a sinusoidal index function like $\cos\theta$ and gives back the nth harmonic of θ. That is, we want a function T that works like this:

$$T_n(x) = T_n(\cos\theta) = \cos n\theta. \tag{9.40}$$

The solution arises in the context previously considered in ring modulation, which is the product of two sinusoids. Recall the handy trigonometric identity that defines the product of two cosines (see appendix section A.4.1):

$$\cos u \cdot \cos v = \frac{\cos(u+v) + \cos(u-v)}{2}.$$

If we set $u = n\theta$, and $v = \theta$, then this becomes

$$\cos n\theta \cdot \cos\theta = \frac{\cos(n\theta + \theta) + \cos(n\theta - \theta)}{2}$$

$$= \frac{\cos[(n+1)\theta] + \cos[(n-1)\theta]}{2}.$$

There are several interesting things in this equation. Notice on the left-hand side the term $\cos n\theta$. By equation (9.40) we can rewrite this as $T_n(\cos\theta)$. Notice on the right-hand side the terms $\cos[(n+1)\theta]$ and $\cos[(n-1)\theta]$. We can rewrite these as $T_{n+1}(\cos\theta)$ and $T_{n-1}(\cos\theta)$, respectively, yielding

$$T_n(\cos\theta) \cdot \cos\theta = \frac{T_{n+1}(\cos\theta) + T_{n-1}(\cos\theta)}{2}.$$

We can simplify this by letting $\cos\theta = x$, yielding

$$T_n(x) \cdot x = \frac{T_{n+1}(x) + T_{n-1}(x)}{2}.$$

This equation defines T_n in terms of the next function in the series, T_{n+1}, and the previous function in the series, T_{n-1}. If we solve it for T_{n+1}, we have

$$T_{n+1}(x) = 2xT_n(x) - T_{n-1}(x), \tag{9.41}$$

which is a recursive formula that defines the next function in terms of the current function and the previous function. This is just what we need because we have exactly two solutions at hand: T_0 and T_1. Setting $T_0 = T_{n-1}$ and $T_1 = T_n$, we can use equation (9.41) to find $T_{n+1} = T_2$. Then, having found T_2, we can continue this process recursively to produce T_3, T_4, and so on. The recursive solutions to equation (9.41) are called Chebyshev polynomials of the first kind.

Here's how we derive T_2 using equation (9.41):

$$T_0(x) = 1,$$
$$T_1(x) = x,$$
$$T_2(x) = 2x \cdot x - 1 = 2x^2 - 1.$$

Similarly we derive T_3 from T_1 and T_2:

$$T_1(x) = x,$$
$$T_2(x) = 2x^2 - 1,$$
$$T_3(x) = 2x \cdot (2x^2 - 1) - x = 4x^3 - 3x.$$

The reader can derive T_4, ending up with $8x^4 - 8x^2 + 1$. This series of polynomials gets large and complex rather quickly. What does it have to do with sound synthesis? When we substitute $x = \cos\theta$ in these equations,

$$T_0(\cos\theta) = 1,$$
$$T_1(\cos\theta) = \cos\theta,$$
$$T_2(\cos\theta) = 2\cos^2(\theta) - 1$$
$$= 2\left[\frac{\cos(\theta + \theta) + \cos(\theta - \theta)}{2}\right] - 1$$
$$= \cos 2\theta + \cos 0 - 1$$
$$= \cos 2\theta.$$

Continuing this series, we obtain

$$T_3\cos\theta = \cos 3\theta,$$
$$T_4\cos\theta = \cos 4\theta,$$
$$T_5\cos\theta = \cos 5\theta.$$

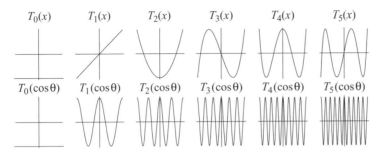

Figure 9.40
Chebyshev polynomial functions.

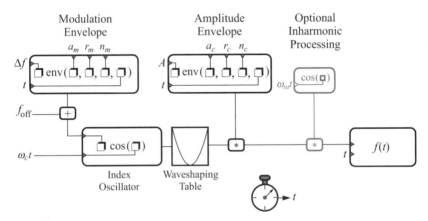

Figure 9.41
Dynamic waveshaping instrument patch.

Remarkably, the Chebyshev polynomial of order n produces the nth harmonic of the cosine index function. Figure 9.40 shows the first four Chebyshev polynomial functions and the corresponding output they produce when they are driven by $\cos\theta$. To achieve a composite spectrum, all we must do is to make a weighted sum of the desired T_n:

$$f(x) = \frac{h_0}{2} + \sum_{k=1}^{N} h_n T_n(x),\qquad(9.42)$$

where h_n are the weights of the n components.

A synthesis patch for waveshaping comparable to the patch for FM is shown in figure 9.41. Like FM, the instrument has an amplitude envelope and a modulation envelope allowing amplitude and spectral content to evolve through the duration of a tone, providing a heightened sense of realism. Unlike FM, however, the spectral components are strictly harmonic multiples of the frequency of the index oscillator. If inharmonic spectra are desired, an optional processing step can be added, drawn with light lines in figure 9.41, that multiplies the spectrum of the output by another sinusoid, creating ring modulation (see section 9.3.2).

9.5   Vocal Synthesis

Synthesis is all about characterization and modeling: how to make something like something else. We synthesize either because the synthesized model provides some expressive power we don't otherwise have, or to emulate a sound for less than the cost of the original. For musical applications, we are typically more concerned with using synthesis to enhance expressive power.

Perhaps because the voice is such a rich, expressive sound source, there are many approaches to vocal synthesis. Most models begin by observing that the vocal tract resembles a tube about 17 cm long, made up of sections with varying diameters, connected together in series, and branching at the end toward the nasal tract (figure 9.42). At the head end of the tube, energy can escape through both nose and mouth.

The glottis provides a driving function at the lower end of the vocal tract that produces a broad-spectrum periodic impulse train during voiced speech. Because the glottis can adjust its frequency and amplitude, we can speak with inflection and sing. Figure 9.43 shows two periods of the waveform of the author singing the sound "ah." The impulsive nature of the glottal source is clearly visible.

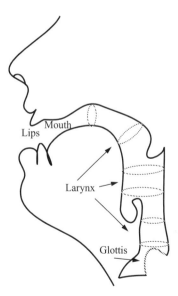

Figure 9.42
Vocal tract.

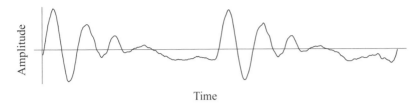

Figure 9.43
Two periods of a vocal waveform.

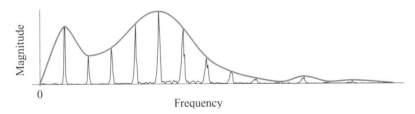

Figure 9.44
Spectrum of voice with formant regions.

Because some reflection occurs where the diameter of each tube section changes, each behaves like a Helmholtz resonator (see volume 1, section 8.3.3). The maxima of this resonant structure, called *formants,* are regions of the vocal frequency response that transmit energy very efficiently. Therefore, components that fall within formant regions tend to be emphasized in the resulting spectrum. The mouth can change its volume, and the lips can change their shape and aperture, allowing some formants to change their frequency and sharpness (see volume 1, equation 8.26). Other formants, corresponding to the inflexible segments, remain relatively fixed. Our hearing uses the formants to help identify speech. Figure 9.44 shows the spectrum of the voice signal in figure 9.43. The spectral envelope of the formants is outlined by the top curve. Three or four formants can be observed in this spectrum.

9.5.1 Waveform Synthesis

Perhaps the easiest way to model the spectrum of a vocal timbre is simply to extract one period of the waveform in figure 9.43 and make it into the wavetable of an oscillator. For example, start with the simple synthesis instrument shown in figure 9.8, but instead of a sine wave oscillator, substitute a table lookup oscillator (section 9.2.9), where the table is one period of figure 9.43. Care must be taken to ensure that there is no discontinuity at the ends of the table, or a click is heard. This can be accomplished by windowing the waveform with a smoothing function. The envelope and vibrato parameters can be adjusted to taste.

This synthesis technique is used by many commercial wavetable synthesizers. Although this simple technique can be refined a great deal and can provide realistic and interesting vocal timbres, the results are limited (see, for example, McNabb, 1981). First, the technique only produces vowels. Second, big changes to the oscillator's fundamental frequency appear to make the physical size of the speaker change. The reason is that as the frequency shifts, the entire spectrum of the sound is shifted linearly up and down in frequency, so the formants change as well. But our hearing equates low-pitched formants with a big person and high-pitched formants with a little person. To be effective over a wide range, many tables must be sampled from the vocalist over his or her entire range, and the nearest one must be selected for the desired fundamental frequency. Individual tables can only be transposed by a few semitones before the size-shifting psychoacoustic effect kicks in. We can do better.

9.5.2 Linear Prediction

Because vocal production depends upon current input (the glottal function) plus current and past outputs (caused by reflections at the mouth and between sections of the vocal tract), the effect of the vocal tract on the spectrum of speech can be modeled reasonably well as an IIR filtering process (see chapter 5).

To model the vocal formants as an IIR filter, we must discover the coefficients of a filter that matches their frequency response. We begin with a difference equation, a *linear predictor*, that expresses each new sample of the speech as a linear combination of previous samples. The coefficients of the linear predictor, the *prediction coefficients*, can be made to provide a highly accurate model of the formants. Estimating the coefficients requires us to solve a set of linear equations and then possibly to adjust the resulting data in order to obtain convergence to a unique and stable solution, but this will prove easier to do than it might sound. The result is a filter that matches the frequency response of the vocal tract. By taking such measurements periodically, the dynamic frequency response of the vocal tract can be characterized over time.

Having characterized the resonances of the vocal tract, the next step is to isolate the pure unfiltered sound of the glottis. To do so, we invert the frequency response of the vocal tract filter and drive it with the original speech sound that we are analyzing, thereby neutralizing the filtering effect of the vocal tract on the glottis. The unfiltered sound source is the *residue signal*. For vowels, the unfiltered glottal function looks like an impulse train and sounds like a buzzer. For consonants, the source sounds like a broadband noise.

We can resynthesize the original sound by driving the residue signal through the (uninverted) vocal tract filter. This analysis/synthesis method, called *linear predictive coding* (LPC), can perform very high-quality speech analysis and synthesis.

Why Do Linear Prediction?   LPC's first application was in telecommunications, where the fundamental interest is to reproduce speech at a distance using a representation requiring the least bandwidth to transmit. The original goal was to reduce as much as possible the amount of data transmitted while maintaining intelligibility, and indeed one can achieve dramatic data reduction with LPC.[10]

However, we typically can't afford to suffer any loss of quality for musical applications.[11] For music, there is generally no point in directly resynthesizing the original sound. Why not just use the original? Therefore, LPC's musical usefulness lies in its ability to modify the sound, preferably in ways that are either difficult or impossible to do otherwise.

Musically, LPC can be used to model the speaking and singing voice, woodwind instruments, birds, whales, violins, and any resonant sound source. Since the LPC model separates the speech into a formant filter and a residue signal, we can alter the timing and pace of the analyzed speech or music or modify its spectral content. We can also cross the formant filters from one signal with any other sound to create hybrid sounds, called *cross-synthesis*. For instance, filtering the sound of a flute with the prediction coefficients of a speech segment creates the illusion of a talking flute.[12]

What Does Prediction Have to Do with Filtering?   If we know the position and orientation of an airplane a moment ago, and also know how its controls are set, we can predict with fair certainty where it will be a moment from now. Similarly, the response of an IIR filter depends upon its current inputs and past outputs. If we know the coefficients of a filter and its initial conditions, we can predict its value in the future. But prediction implies uncertainty and therefore estimation. For every prediction, there will possibly be an error of some magnitude against the actual outcome.

Premise of LPC   If we can predict a signal's behavior in the time domain, we can characterize its behavior in the frequency domain. To predict a signal's time domain behavior, we construct a filter with a response that closely matches the signal's time domain behavior. Since the Fourier transform of the filter's impulse response is its frequency response, the resulting filter characterizes the resonant properties of the signal.

   To make this system work, we must have a method of constructing a filter that matches the time domain behavior of the analysis signal. With this in mind, let's consider just the IIR portion of equation (5.41):

$$y_n = x_n - \sum_{s=1}^{N} b_s y_{n-s}. \qquad\qquad \textit{IIR Filter as a Linear Predictor} \quad (9.43)$$

(The notation y_{n-s} is equivalent to $y(n-s)$.) The standard interpretation of this is that the filter reads input sample x_n and subtracts from it a weighted combination of past outputs to create the current output y_n. But we can also use this equation as a way to predict the next value of y_n from the previous N samples y_{n-s} for all integers $s = 1, ..., N$. In this interpretation, the term x_n is the *difference* between the *true value* y_n and the *predicted value* as indicated by the summation of past outputs in equation (9.43). Isolating x_n in equation (9.43),

$$x_n = y_n + \sum_{s=1}^{N} b_s y_{n-s},$$

we see that x_n is the amount by which the value y_n differs from what would be predicted by recent past outputs alone. A successful predictor would have $|x_n| \ll |y_n|$ for all n because this would indicate that when the *linear prediction coefficients* b_s are applied to the past outputs, they accurately predict future outputs. Overall, we seek the smallest error ε such that the mean squared error

$$\varepsilon = \sum_{n=0}^{N-1} x(n)^2$$

is as small as possible, given N, the length of the signal being analyzed.

Entropy, Redundancy, and Information   If an outcome is *predictable* based on available information, then any additional information of the same kind is *redundant*. If the samples of $y(n)$ can be perfectly predicted from a weighted sum of the previous N samples, then any additional samples would supply no additional information. This redundancy in the signal is what the prediction coefficients are characterizing. If $y(n)$ can be perfectly predicted, that is, if $\varepsilon = 0$, then all we need in order to regenerate it are its prediction coefficients and its initial conditions.

For example, consider the transfer function of the two-pole filter, equation (5.96). Setting $z^{-x} = y(n - x)$, we can directly convert this into the filter equation:

$$y(n) = \delta(n) + 2R(\cos\theta)y(n - 1) - R^2 y(n - 2)$$

where $\delta(n) = 1$, $n = 0$, else 0.

In terms of equation (9.43), the coefficients of this filter are $b_1 = 2R\cos\theta$ and $b_2 = R^2$. If we set the radius $R = 1$, then the effect is to move the poles of this filter onto the unit circle. If we supply this filter with initial conditions for $y(n - 1)$ and $y(n - 2)$ such as the values 1.0 and 0, the impulse response of the filter will ring forever, producing a pure sinusoid at frequency θ.

These two filter coefficients can be looked upon as perfect predictors of $y(n)$ because only the coefficients and the initial conditions are required to predict all possible values of $y(n)$. If we set $R < 1$, then $y(n)$ will describe a sinusoid at frequency θ with an exponentially decreasing amplitude. If $R > 1$, $y(n)$ will be a sinusoid at frequency θ but with exponentially increasing amplitude. All these functions can be perfectly predicted from the initial conditions and the two prediction coefficients. Since they perfectly characterize $y(n)$, we can replace $y(n)$ with the initial conditions plus the two prediction coefficients without any loss of information, thereby considerably compressing the amount of data required to characterize the signal.

Now consider a random sampled signal $u(n)$. If it is purely random, there is no way to predict its next value from any number of previous values—in fact, this is a good definition of randomness. There is no way we could compress this signal as we did $y(n)$ unless we constructed a filter that was as long as the signal. But since that would result in no data compression, what would be point? Using the terminology of information theory, we say that $u(n)$ has a high degree of entropy. Similarly, since we wouldn't be able to squeeze any additional redundancy out of the prediction coefficients for $y(n)$, they also have a high degree of entropy.

We can define *information* as the degree of entropy in a signal. We can define *entropy* as the number of states required to characterize a system.[13] We want the resonant properties of speech to have *minimum entropy* because then the states required to characterize the resonance will be merely a handful of prediction coefficients. We want the residue properties of speech to have *maximum entropy* because then we have made sure that everything that can be predicted is accounted for in the prediction coefficients.

In the case of the filter generating a sinusoid, there is no residue signal because there is no information in the signal that is not characterized perfectly in the prediction coefficients. But for real signals, there will always be a residue. The better we are at extracting redundancy into the prediction coefficients, the less predictable the residue becomes, and its entropy increases. The greater its entropy, the broader and more uniform (whiter) its spectrum becomes. So the result of successful linear prediction is that the residue has the flattest spectrum possible.

LPC is attractive in telecommunications because most acoustical signals have high redundancy and low entropy, allowing for a great deal of compression, thereby lowering communication costs. Since the spectrum of speech usually does not change much in 30 ms, we can reduce that 30 ms of sound down to a handful of prediction coefficients plus the residue signal. The magnitude of the residue signal is generally rather small, and for typical speech applications it can be made to fit into 8 bits per sample. This way, intelligible speech can be transmitted at a rate of about 5000 bits per second or less.

The reason LPC is attractive to musicians has more to do with how it divides a signal into an all-pole filter and a residual signal. For cross-synthesis, the residual is not used and needn't be computed because the filter will be applied to an entirely different sound.

Estimating the Prediction Coefficients   To make this scheme work, we must have a means of identifying prediction coefficients that squeezes the maximum amount of entropy out of the signal. Let's think about what equation (9.43) is describing. According to the discussion of all-pole filters in chapter 5, equation (9.43) specifies some sort of complex spectral function in the z plane whose frequency response is defined by its N conjugate pole pairs.

We can observe what kind of spectrum a set of coefficients will generate by looking at their Z transform, as defined by equation (5.49). The spectral estimate for a sampled function s_k can be written

$$H(z) = \sum_{k=0}^{N-1} s_k z^k. \tag{9.44}$$

We know from equation (5.99) that the Z transform of equation (9.43) is

$$P(z) = \frac{a}{1 + \displaystyle\sum_{n=1}^{N} b_n z^{-n}}. \qquad\qquad \textit{All-Poles Model (9.45)}$$

The difference between equations (9.44) and (9.45) is that whereas $H(z)$ can have only zeros of transmission, $P(z)$ can have only poles, corresponding to infinite spectral density at their centers. Filters with poles are good at modeling spectra that have sharp, discrete lines (Dirac delta functions), such as the voice. It is much harder to model the spectral signatures of the voice with filters containing only zeros, so equation (9.45) is generally used. It's called the *all-poles model*.

The Wiener-Khinchin theorem (equation 4.26) states that the Fourier transform of the autocorrelation of a function f is equal to the power spectrum of that function: $F\{\mathrm{corr}(f,f)\} = |F(\omega)|^2$.

That means we have two ways in which to describe spectra: equation (9.45) describes the spectrum of an N-pole filter in terms of its coefficients in the z plane, and equation (4.26) describes the power spectrum of an arbitrary signal in terms of the autocorrelation of a sampled time function. Let's relate these two views.

First, let's convert the all-poles model, equation (9.45), into a power spectrum by taking the absolute value of the square of both sides:

$$|P(z)|^2 = \left| \frac{a}{1 + \sum\limits_{n=1}^{N} b_n z^{-n}} \right|^2 .$$

Next, remember from equation (4.25) that autocorrelation is written $\phi_n = \text{corr}(f, f)(n)$, and by the Wiener-Khinchin theorem, we know that the Fourier transform of ϕ_n is equal to the power spectrum $|P(z)|^2$. But the Fourier transform is just the Z transform evaluated on the unit circle with $z = e^{i\omega}$. That means we can express the power spectrum of ϕ_n as a Z transform:

$$\Phi_n = \sum_{n=0}^{N-1} \phi_n z^{-n} .$$

Now we have a power spectrum $|P(z)|^2$ specified in terms of the Z transform of the prediction coefficients that we want to discover, and a power spectrum Φ_n specified in terms of the Z transform of the autocorrelation of the signal we want to analyze. When we relate them, $|P(z)|^2 \approx \Phi_n$, we have

$$\left| \frac{a}{1 + \sum\limits_{n=1}^{N} b_n z^{-n}} \right|^2 \approx \sum_{n=-N}^{N} \phi_n z^{-n} . \tag{9.46}$$

The left-hand side is the Z transform of the filter we want to design, and the right-hand side is the Z transform of the autocorrelation of the waveform we are trying to match. The value N is both the number of coefficients we want to discover and the extent of the autocorrelation function we will use to find them. Thus, N corresponds to the order of the filter we are seeking. Though we can set N to as high as the total number of autocorrelations available, in practice we want it to be much smaller.

The solution to equation (9.46) can be shown to have the maximum possible entropy of all possible solutions; it is therefore called the *maximum entropy method* (MEM). This equation implies that there is a linear set of relations between the autocorrelation function and the prediction coefficients. To actually solve it requires solving a set of simultaneous equations based on the series expansions of both sides. It can be shown that the filter coefficients satisfy the matrix equation

$$
\begin{bmatrix}
\phi_0 & \phi_1 & \phi_2 & \cdots & \phi_N \\
\phi_1 & \phi_2 & \phi_3 & \ddots & \phi_{N-1} \\
\phi_2 & \phi_3 & \phi_4 & & \phi_{N-2} \\
\vdots & & & & \vdots \\
\phi_N & \phi_{N-1} & \phi_{N-2} & \cdots & \phi_0
\end{bmatrix}
\cdot
\begin{bmatrix}
1 \\
b_1 \\
\vdots \\
b_N
\end{bmatrix}
=
\begin{bmatrix}
a_0 \\
0 \\
0 \\
0
\end{bmatrix}
. \tag{9.47}
$$

The matrix form is just a notational convenience. To write out the equations to be solved simultaneously in standard polynomial form would require much more ink. We'd have to write

$$(\phi_0 \cdot 1) + (\phi_1 \cdot b_1) + (\phi_2 \cdot b_2) + \cdots + (\phi_N \cdot b_N) = a_0,$$
$$(\phi_1 \cdot b_1) + (\phi_2 \cdot b_2) + (\phi_3 \cdot b_3) + \cdots + (\phi_{N-1} \cdot b_{N-1}) = 0,$$
$$\vdots \qquad\qquad\qquad \vdots$$
$$(\phi_N \cdot b_N) + (\phi_{N-1} \cdot b_{N-1}) + (\phi_{N-2} \cdot b_{N-2}) + \cdots + (\phi_0 \cdot b_0) = 0,$$

and so on. Fortunately, because equation (9.47) is constant along its diagonals, it is a so-called *Toeplitz matrix,* and some very efficient algorithms have been developed to solve them on a computer (Press et al. 1988; Makhoul 1975; Rabiner and Schafer 1978).

Using LPC   Generally, we want to set N, the order of the predictor, to a value somewhat higher than the number of sharp spectral features we wish to discover. Though we can set it as high as the number of autocorrelations available, numerical instabilities can arise. Depending on the nature of the signal, a surplus of poles can lead to false spectral peaks. By limiting the number of poles, the spectrum may be smoothed out somewhat, but this is generally a good thing.

The difference between the LPC analysis of a signal and that provided by the DFT is that the LPC spectrum is continuous in frequency whereas the DFT is sampled in frequency. One consequence of this is that the estimated LPC spectrum may have very sharp spectral features that can be overlooked if the spectrum is not evaluated finely enough. One possible way to properly direct the MEM algorithm is to take the DFT of the signal first, then compare it to the spectral features derived by the MEM, using the DFT as a guide as to which spectral features may be spurious.

Another problem with LPC has to do with the stability of the filters derived by the MEM. If, for instance, a component is increasing in amplitude during an analysis frame, MEM will turn that into an unstable recursive filter in order to correctly model its increase in gain. The instability, if mishandled by the LPC analysis/synthesis system, can result in some very loud, very unpleasant sounds that are bad for loudspeakers and listeners alike. The MEM algorithm can be quirky, and under the right conditions it is perfectly capable of putting poles of the filter far outside the unit circle all on its own. If the arithmetic precision of the computer implementing the filter is not sufficient, or if there are other subtle numerical problems, dreadful, deafening howling noises can ensue.

To avoid these problems, we have basically two choices (F. R. Moore 1990):

- Ignore the problems, and hope for the best. This becomes a losing strategy as the number of samples in the analysis frame grows because we'll be running potentially unstable filters for a longer time, increasing the risk of insufficient internal arithmetic precision in the filter. Our chances get slimmer as we either lengthen the analysis frame size or increase the sampling rate (or both).

• Reduce the magnitude of the poles. If we reduce an unstable pole's magnitude so it lies on the unit circle, it goes from an exponentially increasing sinusoid to a constant-amplitude sinusoid. If we bring it inside the unit circle along its radius, it becomes an exponentially decreasing sinusoid. The choice we make depends upon what we want. If it seems that the sound is relatively steady-state, maybe putting the unstable poles on the unit circle is the right thing. If we think the unstable poles are the result of quirky MEM behavior, we can move them inside the unit circle. Or we can kill them off altogether by moving them all the way inside to a magnitude of zero.

LPC as a Data Reduction System   A drastic way to economize on transmitted information is actually to discard the residue signal altogether. For vowels, the wave shape of the residue buzz signal remains relatively constant, varying mostly in frequency and amplitude. For consonants, the residue noise signal is spectrally very bright and mostly only varies in amplitude. If the source is a vowel, we can transmit just its frequency and amplitude parameters; if it is a consonant, we just send its amplitude. The parameter that indicates whether the synthesizer should produce noise instead of buzz is called the voicing parameter. These parameters are transmitted along with the prediction coefficients to the synthesizer, which first reconstructs the residue signal, then filters it. The result sounds quite mechanical, but is surprisingly intelligible and highly efficient to implement. This is the basis of the LPC-10e algorithm described in U.S. Federal Standard 1015.[14]

Better-quality speech that does not require much additional transmission bandwidth uses a *codebook,* which is a table of frequently encountered residue signals set up in advance. The analyzer compares the character of the residue signal to the signals in the codebook, choosing the best match using a least-squares fit, then transmits just the index code for this signal. The synthesizer receives the code index, looks up the corresponding residue signal in its identical codebook, and regenerates the signal. This is called Code Excited Linear Prediction (CELP).

The bigger the codebook, the better this approach can model the original speech. But the bigger the codebook, the longer it takes to search for the best match. If the codebook must also take frequency and amplitude information into account, it would have to be quite large. One common way to address this is to have two codebooks, the first of which contains prototype residue signals set up in advance. The second starts off empty and is used during operation as a kind of scratch-pad memory, containing copies of the previous residue signals delayed by an amount that matches the frequency of the signal being encoded. This is the CELP algorithm described in U.S. Federal Standard 1016.[15]

A Caveat   LPC succeeds best with sounds that are well modeled as all-pole resonators. The voice is not a pure IIR process because it includes side branches—principally the nasal passages. Side branches introduce zeros in the transfer function of the vocal tract that can't be easily modeled with an all-pole IIR filter. Thus LPC does not model nasal sounds well. More precisely, the spectrum of nasal sounds is not well captured in the prediction coefficients. As a consequence, the nasal components of the spectrum remain in the residue signal. A more complicated approach (based on equation 5.41) is required to model both poles and zeros, the pursuit of which is left to that admirable personage, the interested reader.

9.5.3   FOF Synthesis

FOF synthesis demonstrates a clever use of time/frequency domain symmetry to simulate spectral features of the singing voice and other sounds. It was developed by Rodet (1984).

Each *formant wave function* (*Forme d'Onde Formantique,* or FOF) consists of a windowed sinusoid with a frequency corresponding to the center of one of the vocal formants. By adjusting the shape of the envelope, the spectral bandwidth of each FOF can be adjusted to match the spectral shape of a vocal formant. Although the technique can model many timbres, it excels at modeling vocal resonances.

As shown in figure 9.45, a FOF is a single momentary pulse. To create a continuous tone, a train of FOFs must be generated at a rate corresponding to the desired fundamental frequency. To simulate the principal vocal formants of the voice, between three and five trains of FOFs are generated and summed (figure 9.46).

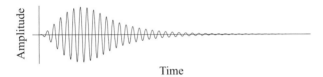

Figure 9.45
Single FOF.

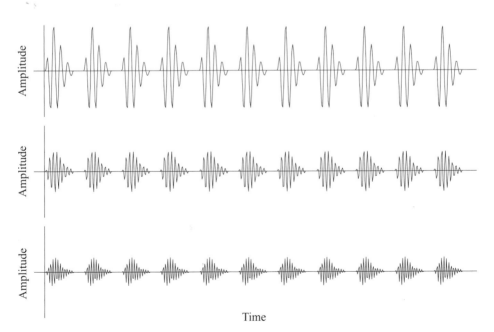

Figure 9.46
Train of FOFs triggered at fundamental frequency.

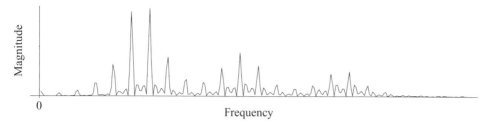

Figure 9.47
Spectrum of figure 9.46.

Figure 9.47 shows the positive frequency magnitude spectrum of the sum of the FOFs in figure 9.46. It has a voicelike spectrum with three formants.

FOF Generation A FOF generator consists of an amplitude envelope generator similar to equation (9.2), controlling the amplitude of a sine wave generator, similar to figure 9.6. The FOF amplitude envelope generator determines the shape of the window applied to the sinusoid.

We have seen the effects of multiplying a sinusoid and a window function in numerous contexts. Multiplying in the time domain convolves in the spectral domain, and so the shape of the window determines the shape of the spectrum of the product. We've seen that windows with sharp edges splatter energy over a broad range of frequencies, whereas smoother windows only slightly broaden the bandwidth of the signal they are multiplied by. FOF synthesis exploits this insight.

Amplitude Envelope Generation The amplitude envelope for a single FOF generator is

$$A(\beta, \alpha, n) = \begin{cases} n < 0, & 0 \\ n < \pi/\beta, & \frac{1}{2}(1 - \cos\beta n)e^{-\alpha n} \quad \text{Attack} \\ n \geq \pi/\beta, & e^{-\alpha n} \quad\quad\quad\ \text{Decay} \end{cases} \tag{9.48}$$

where n is time in samples. The basic envelope is the exponential function $e^{-\alpha n}$ with a decay rate controlled by α. The attack duration lasts π/β samples. During this time, the function $0.5(1 - \cos\beta n)$ smooths the sharp discontinuity at the beginning of the exponential.

The cosine expression for the attack during $n < \pi/\beta$ is shown in figure 9.48a. The exponential component is shown in 9.48b. Their product is shown in 9.48c. The final decay of the exponent is shown in 9.48d, and the composite envelope is shown in 9.48e.

Figure 9.49 shows how the shape of the envelope sharpens as π/β becomes smaller with α fixed. There is no first- or second-order discontinuity at the end of the attack. Since β determines the impulsiveness of the attack, it primarily determines how broadly energy is distributed in the spectrum.

Figure 9.50 shows the effect of α for a fixed setting of β. As α increases, the decay becomes more rapid. Thus, α also has an impact on spectrum, but not as dramatic as β.

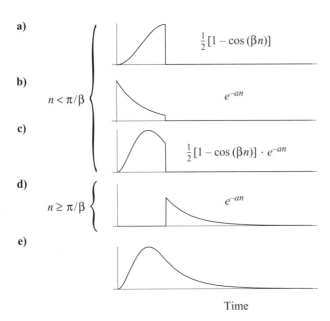

Figure 9.48
FOF envelope components.

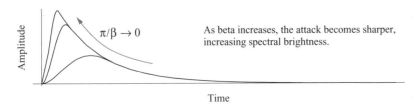

Figure 9.49
FOF envelope, increasing β.

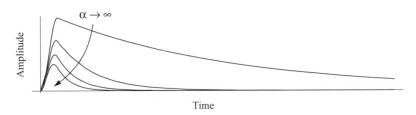

Figure 9.50
FOF envelope, increasing α.

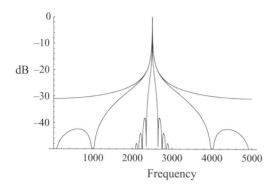

Figure 9.51
FOF spectral brightness as a function of β.

Note also in both figures 9.49 and 9.50 that peak amplitude changes as a function of α and β. In a later step, we will normalize the amplitudes.

FOF Sine Wave Generator   Combining the envelope and the sinusoid generator, we have the complete FOF tone generator:

$$s_n(\beta, \alpha, \omega, \phi) = \begin{cases} n < 0, & 0 \\ n < \pi/\beta, & \frac{1}{2}(1 - \cos\beta n)e^{-an}\sin(\omega n + \phi) \\ n \geq \pi/\beta, & e^{-an}\sin(\omega n + \phi) \end{cases} \qquad \begin{array}{l} \textit{Forme d'Onde} \\ \textit{Formantique (FOF)} \end{array} \; (9.49)$$

The FOF envelope function scales the sinusoid $\sin(\omega n + \phi)$, where $\omega = 2\pi f$, and f is the center frequency of the formant being synthesized. No filtering occurs here, yet each FOF generates a single impulse with a spectrum that can mimic the frequency response of one vocal formant.

Figure 9.51 shows the magnitude spectrum for various values of β. Large values of β set the narrowest skirts because the attack time is slowest. From narrowest to widest, the attack times correspond to $\pi/\beta = 10$ ms, 1 ms, and 0.1 ms, respectively, with a formant center frequency of 2500 Hz and a sampling rate of 10,000 Hz.

Cascading FOFs   To simulate a voice containing three to five formants, we must run three to five trains of FOF synthesizers in parallel (figure 9.52). Each FOF is triggered at a rate corresponding to the fundamental frequency at times τ_n. The bandwidth of each formant range is controlled by β_n, and the band center is controlled by ω_n.

Summary of FOF Synthesis   FOF synthesis is computationally less challenging than filter-based approaches to speech synthesis such as LPC. This provides a number of advantages, including simplified arithmetic, no fear that unstable filters will make ballistic projectiles out of the loudspeaker cones, and relatively simple calculation of control parameters. It has been used to synthesize

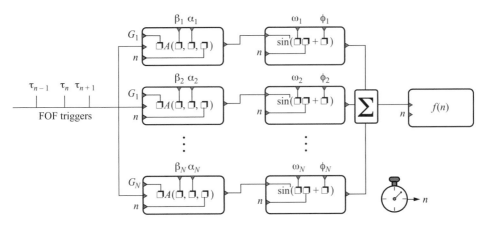

Figure 9.52
FOF synthesis of multiple formants.

a great number of vocal, instrumental, and other sounds, including violins and nonresonant instruments such as cymbals.

Its principal disadvantages over LPC are twofold. The FOF model does not include nonvoiced speech, so it can't easily be made to talk. Also, the parameter adjustments required to produce realistic-sounding vocal timbres are quite complex, depending on the timbre of the voice being synthesized, its vocal register, volume, the quality of vibrato, and so on. For analysis-based synthesis such as LPC, the model of the vocal timbre is implicit in the analyzed voice. For FOF synthesis, the rules governing vocal timbre must be made explicit. The rules of speech and singing have been studied extensively by Sundberg (1991), among others.

Rodet, Potard, and Barrière (1984) constructed a program called CHANT (French for *song*), which embeds a set of rules for modeling general vocal synthesis with FOFs. A user of this system supplies high-level parameters, and CHANT controls the FOF synthesis. A more ambitious program called FORMES, based on the Lisp programming language, was written to extend CHANT into a completely general compositional programming environment for FOF synthesis.

9.5.4 Granular Synthesis

FOF synthesis creates complex timbres by combining a sequence of many simple individual sound impulses, each having a characteristic envelope and frequency (see section 9.5.3). If we think of these individual sound impulses as grains of sound, we can then describe FOF synthesis is a kind of *granular synthesis*. This in turn suggests that there could be other forms of granular synthesis, and indeed there are many.

The idea of granular synthesis stems from papers published by Dennis Gabor in the late 1940s. Chapter 10 examines Gabor's ideas in more detail, but for now suffice it to say that in general a grain can be literally any sound, although by convention grains are usually brief, typically on the order of a few milliseconds in duration.

grain can be literally any sound, although by convention grains are usually brief, typically on the order of a few milliseconds in duration.

As with FOF synthesis, grains can be windowed sinusoids, or they can be prerecorded sounds, or sounds of any sort whatsoever. Each grain can be thought of as a kind of musical note in the technical sense of that term developed in the discussion of tones, notes, and scores (see volume 1, section 2.2), except that the grains are so brief that they are more like note fragments: hundreds or thousands must be strung together to last long enough to make an audible sound. (Gabor 1946).[16]

Gabor's Kinematical Frequency Converter   In addition to setting out the theory of granular synthesis, Gabor (1946) invented a device based on his theory that was capable of changing the time scale of a sound (called time dilation) without changing its pitch, or vice versa. He called this invention the Kinematical Frequency Converter.[17] It was based on a modification of the optical audio track of a film projector.

Movie cameras of his day encoded audio waveform fluctuations as a continuous track on the edge of the film. The most positive amplitude was encoded as transparent, the negative amplitude as black, and the intermediate amplitudes as various shades of grey. The audio track of the film was passed over a narrow slit between an exciter lamp and a photocell. As the audio encoded on the film moved past the slit, light intensity from the bulb was modulated by the relative transparency of the audio track on the film. Variations in light striking the photocell produced an electrical analog of the audio signal encoded on the film, which was then amplified so it could drive a loudspeaker in the theater.

Gabor adapted this as shown in figure 9.53. He added a slotted drum that could rotate at an independent velocity. Each slot would sweep light across the film in the window, thereby projecting onto the photocell the segment of the audio captured in the window. To blend successive sweeps together, Gabor progressively shaded the window so that it was transparent in the center and opaque at its edges, using a Gaussian density distribution. Gabor reported best results when the slots were separated by approximately one half of the window's width.

To get a feel for how it worked, suppose the film is stationary while the wheel rotates. Each slot sweeps over the same area of film, so we hear pulses of sound whose character is determined by the audio encoded on the film that lies within the window. If the window is positioned over a vowel sound such as "a", we would hear "aaaa. . .".

Now suppose the film travels at its normal rate while the wheel rotates at a fixed rate so that we hear the audio as it was originally recorded. But if we now slow the film speed a bit, the pitch of the audio remains the same, but the *rate of spectral evolution* of the sound that we hear is slowed. This is because the rate at which the slots sweep across the audio in the window determines the frequency of audio pulses, but the rate of film movement determines the rate at which new audio material moves into the window. Hence, film speed now only determines the rate of spectral evolution, not frequency. Holding the wheel radial velocity constant while the film moves at a varying rate accomplishes change in tempo without change in frequency.

We can use Gabor's device together with a variable-speed audio recorder to change frequency without changing tempo. Suppose we set the film speed to half the speed of the original

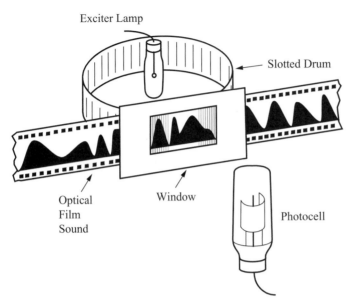

Figure 9.53
Gabor's kinematical frequency converter. From Gabor (1946, 446).

recording. Thus it will take twice as long to play the whole film, though we hear the audio at its original frequency. If we record the output to a machine running at half speed, then play it back at full speed, we will hear the audio transposed up an octave but lasting the length of the original recording. In this way, Gabor was able to demonstrate for the first time independent control of tempo and pitch.

The performance of Gabor's machine was not ideal. The apparatus could produce strong beats depending on the number of slots, their rate, the size of the window, and other factors. He eventually developed an improved version that used magnetic tape. A modern implementation of Gabor's technique, called the phase vocoder, is discussed in chapter 10.

Composers of Granular Synthesis The pioneer of granular synthesis techniques in music was Iannis Xenakis. From exposure to Gabor's work, Xenakis (1971) discovered that "all sound is conceived as an assemblage of a large number of elementary grains adequately disposed in time." He used analog oscillators and tape-splicing techniques to assemble his composition *Analogique A-B* for string orchestra and tape in 1959. Whereas Gabor used grains to modify the time/rate information of existing audio signals, Xenakis did not start with an existing sound. Instead, he proposed a sound grid consisting of successive *screens* of amplitude frequency functions describing the distribution of sound grains at discrete moments through time. He explored musical strategies for filling the screens with clouds of grains. He primarily used stochastic techniques (see volume 1, section 9.13) for his compositional strategies.

Given the small size of the sound grains, the biggest challenge to their use is organizing their distribution in time, frequency, and intensity. The possible organizing principles are seemingly limitless. Rigorous exploration of the possible compositional methodologies had to wait until the common availability of digital computers. Many composers have explored granular synthesis. The composer Curtis Roads (1988) has developed the theory and practice of granular synthesis since 1974 and has continued to research techniques and compose music utilizing granular synthesis. Truax (1988) implemented a real-time granular synthesis system in Canada in the late 1970s and early 1980s. Many implementations of granular synthesis are now commonly available, for example, on the World Wide Web.

Granular synthesis is reminiscent of the post-Impressionist pointillistic painting technique exemplified by the painter Georges Seurat. Other painterly analogies include air-brushing or stenciling. Granular synthesis can also be compared to collage art, in the sense that it starts with ready-made objects (the grains) and composes them into a pattern or skein. This suggests that the principal dimensions of the process are the morphology of grains and their disposition in time and frequency. The interested reader is referred to Roads (2001), whose writings on the subject are definitive and highly recommended.

9.6 Synthesizing Concert Hall Acoustics

Since sound travels at a constant rate, a sound introduced at one end of a tube will propagate to the other end after a constant delay (figure 9.54).

A *delay line* is a simple functional unit that can be used to simulate a sampled traveling wave in an acoustical tube (figure 9.55). We use the notation z^{-N} to indicate a delay of N samples through a delay line (see section 5.11).

It is natural and convenient to think of delay lines as building blocks for acoustic spaces such as concert halls and amphitheaters because the effects of propagation delay in these contexts can be understood as echoes and reverberation. They can also be used as building blocks for digital simulation of physical models of musical instruments. The treatment here prepares the way for a discussion of waveguide models of musical instruments.

$x(n) \longrightarrow$  $\longrightarrow y(n)$

Figure 9.54
Wave propagating in a tube.

$x(n) \longrightarrow$ 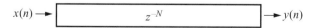 $\longrightarrow y(n)$

Figure 9.55
N-sample delay line.

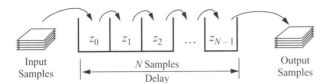

Figure 9.56
Bucket brigade.

9.6.1   Delay Lines

Here are two ways to implement a delay line.

Bucket Brigade   Imagine an array of N empty storage bins with a courier who can access them sequentially. As the courier walks past them right to left, he advances each previous sample to each subsequent bin.

Suppose there are N bins labeled z_0, z_2, ..., z_{N-1} (figure 9.56). First, the courier removes the sample from bin z_{N-1} and puts it on the Output stack. Next, he removes the sample in z_{N-2} and moves it to bin z_{N-1}. He continues, advancing each previous sample to reside in its subsequent bin. When he advances the contents of bin z_0, he takes a sample from the Input stack and puts it in z_0. Then he returns immediately to the last bin (z_{N-1}) and repeats the whole procedure until all samples have been transferred to the Output stack. The data move through a sequence of bins like a bucket brigade moves a series of buckets from person to person.

The total delay time is the time it takes the courier to move one sample multiplied by N, the number of bins. If the sample interval is T seconds per sample, the total delay time is NT seconds.

In-place Method   A simpler approach leaves the data in place. As before, we have an Input stack, an Output stack, and an array of storage bins. For each bin, the courier removes the oldest sample from its bin and puts it on the Output stack, then replaces it with a sample from the Input stack, then moves on to the next oldest bin. The courier will have visited all the other bins before returning to the one he started with, by which time the sample just placed there will have been delayed by N sample times. As before, if the sample interval is T seconds per sample, the total delay time is NT seconds.

Let D be a delay line of length N, i the integer index of the current bin, x the next sample from the Input stack, and y the delayed sample headed for the Output stack. Then we can write the procedure for an in-place delay line as follows:

$y = D(i)$	Remove sample from bin indexed by i; send it to Output stack.
$D(i) = x$	Insert next value from Input stack into the now empty bin.
$i = ((i+1))_N$	Increment i, wrapping around to beginning if it goes off the end.
Return to step 1.	Repeat until there is no more input.

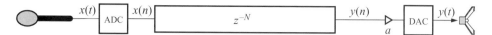

Figure 9.57
Digital delay line.

9.6.2 Modeling Echoes

If the delay line inputs x arrive sequentially from a microphone via an analog-to-digital converter (ADC), and the outputs y are passed sequentially to a digital-to-analog converter (DAC) and then to an amplifier and loudspeaker, the sound from the loudspeaker will be a copy of the input, delayed by N samples (figure 9.57).

Acoustically, this is similar to the echo we might hear standing some distance from a large wall in an open space. So long as the acoustical delay time is great enough to exceed the threshold of the precedence effect (see volume 1, section 6.13.3), if we clap our hands, we will hear a single sharp reflection called a *slap echo* from the wall. Figure 9.57 is an essential building block for commercial delay line audio effects. Its operation is similar to the simple tape delay shown at the beginning of chapter 5.

9.6.3 Modeling Traveling Waves in Air

Note the term a just before the DAC in figure 9.57. We can use it to attenuate the signal, modeling the inverse square law of distance for a spherical wave, so that the delay line can model sound coming from a distance. The proper attenuation value for a depends upon knowing the total delay time d. If a delay line is N samples long and the sample period is T seconds, then the delay time is $d = NT$ seconds. Recalling that the pressure of a spherical waveform drops off as $1/d$, we can model the attenuation of a pressure wave by setting $a = 1/(NT)$. Since intensity drops off as $1/d^2$, we can model the attenuation of intensity by setting $a = 1/(NT)^2$.

Air absorbs varying amounts of high-frequency energy as sound travels a unit distance, depending mostly on humidity but also on temperature and pressure. We can simulate this effect of humidity by adding a lowpass filter to the output of the delay line, and calibrate its attenuation of high frequencies due to humidity conditions from tables in acoustics texts. In practice, the exact attenuation values are less important than the fact that increasingly distant sounds become progressively more muffled.

9.6.4 Multipath Wave Propagation—Comb Filtering

Suppose we have a flutist and a microphone separated by some distance in an open space (figure 9.58). There are two paths sound can take to the microphone: the direct signal path and a reflected signal path off the floor. Suppose further that the two paths together make an isosceles right triangle, as shown. If the length of the reflected path is 2 m, then the length of the direct signal path will be $\sqrt{2} \cong 1.414$m, and the reflected path will arrive at the microphone delayed by $(2 - 1.414)\text{m}/c_s = 0.586/331.1 = 1.76$ ms. The pressure waves of the two paths meet at the

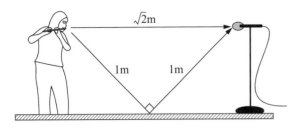

Figure 9.58
Multipath wave propagation.

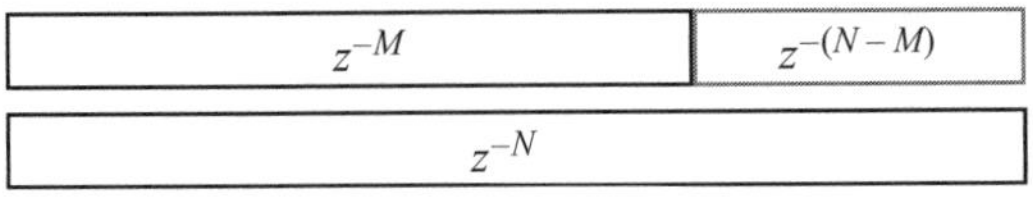

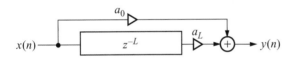

Figure 9.59
Composite delay lines.

![Comb filter diagram]
$$x(n) \longrightarrow \boxed{z^{-L}} \; a_L \triangleright \oplus \longrightarrow y(n)$$

with $a_0 \triangleright$ path.

Figure 9.60
Comb filter.

microphone diaphragm and sum together. At some frequencies, the wave fronts reinforce at the diaphragm, while at others they cancel. Thus, the interaction of these two waves at the microphone causes a kind of filtering that colors the tone quality of the instrument being recorded. Most of the surfaces of a good recording studio are made to be absorbent so as to minimize all signal paths but the direct one, thereby reducing spectral coloration from multipath propagation.

We could simulate the preceding effect with two delay lines, one for the direct signal path, another for the reflected path (figure 9.59). But since we are generally more interested in the relative delay between paths than in the absolute delay, we simply delay the reflected path by the difference between the delay lengths, $L = N - M$.

Taking this approach, we can model a dual-path acoustical system as shown in figure 9.60, where a_0 is the signal attenuation experienced by the direct signal path, and a_L is the attenuation through the delay of length L. The summation point $\oplus$ corresponds to the point where the acoustical paths mix together at the microphone's diaphragm. The filter structure is a *comb filter*.

Here's how to determine the frequencies at which the two paths reinforce and cancel their signals at the microphone's diaphragm. Set a_0 and a_1 to positive nonzero values, and then drive the system with a sine wave of amplitude ± 1. The key is to note that *the phase of the sine wave entering the delay line is also the phase of the sine wave entering the summation point $\oplus$ via the direct signal* *path, since the direct path has zero delay.* The gain is maximum and equal to $a_0 + a_L$ when a whole number of periods fits in the L samples of the delay line because then both ends of the delay line are exactly in phase. This occurs at frequencies $\omega_k T = k2\pi/L$, $k = 0, 1, 2, \ldots$. Note that these are harmonics of the fundamental frequency $\omega_1 T = 2\pi/L$ of the delay line. Thinking along the same lines, gain is minimum and equal to $|a_0 - a_L|$, where an odd number of half periods fits in the L samples of the delay line because then the ends of the delay line are exactly out of phase. Thus, minima occur at frequencies $\omega_k T = (2k+1)\pi/L$.

Returning to the flute example, the frequency corresponding to a delay of 1.76 ms is 568 Hz. So one period of a 568 Hz sine wave fits the difference between the direct and reflected path, and the microphone diaphragm experiences maxima at that frequency and its harmonics. It experiences minima at one half that frequency, 284 Hz and its harmonics.

We can derive the frequency response of the comb filter as follows. By inspection, we can express the difference equation as

$$y(n) = a_0 x(n) + a_L x(n - L). \tag{9.50}$$

Compare equation (9.50) to the difference equation of the simple lowpass filter given in equation (5.3). They are basically the same except that here the second term is delayed by z^{-L} instead of just z^{-1}. Therefore, the transfer function is

$$H(z) = a_0 + a_L z^{-L}.$$

Evaluating this on the unit circle by setting $z = e^{i\omega T}$, we obtain the complex frequency response,

$$H(e^{i\omega T}) = a_0 + a_L e^{-iL\omega T}. \tag{9.51}$$

Look again at figure 5.10, which shows the transfer function of a simple lowpass filter. Essentially, all that has changed here is that the exponent of z now has an extra coefficient L. Whereas the lowpass transfer function goes counterclockwise 180° as $\omega \to \pi$, this transfer function goes through $L/2$ rotations. We end up with a frequency response like the one in figure 5.13, except repeated $L/2$ times.

Converting equation (9.51), the complex frequency response, into the magnitude response, we have

$$G(\omega) = |H(e^{i\omega T})| = |1 + e^{-iL\omega T}| = 2\left|\cos\frac{L\omega T}{2}\right|. \tag{9.52}$$

Compare this to equation (5.28), the frequency response of a lowpass filter. Figure 9.61 shows the comb filter spectrum for $L = 7$. It should now be clear how the comb filter got its name.

The comb filter spectrum has a series of $L - 1$ *notches* or *nulls* in the Nyquist interval where the energy goes to zero, so we could call it a kind of multiple band reject filter. When heard, this effect is immediately recognizable as *flanging,* which is a technique where a signal is summed with a slightly delayed copy of itself. If the length of the delay varies over time, the number and position of the nulls change, superimposing a kind of hollow swish sound over whatever material is played through it.[18]

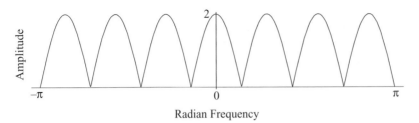

Figure 9.61
Comb filter frequency response.

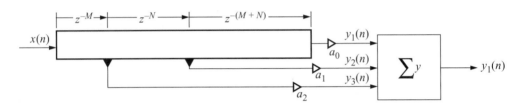

Figure 9.62
Tapped delay line.

9.6.5 Tapped Delay Line

In most natural settings, there are typically many more sources of reflection than there are sources of sound. For example, in a concert hall, sound from a single source on the stage will be reflected by the numerous acoustical features of the hall. We can adapt the delay line to this reality by placing additional outputs, called taps, at various points within the delay line. Each tap registers the signal at its location and sums it with any other taps to create the final output (figure 9.62).

We can simulate how a listener will experience the early reflections in a concert hall. For a simplified example, start with the floor plan of a concert hall, such as shown in figure 9.63. For an arbitrary listening position, trace the distance of the first several room reflections as rays between the stage and the listener. Using the speed of sound, convert these distances to delay times. Now adjust the positions of the taps in the tapped delay line to correspond to these delays. Next, calculate to the inverse square law of spherical spreading, air absorption, and sound absorption by the walls. Set the coefficients for each tap appropriately. Since the tapped delay lines are modeling the sum of reflections at each ear, it might be convenient to have two tapped delay lines, one for each ear. Driving relatively dry music through these tapped delay lines gives a spacious quality to the result.

An interesting effect is created when the taps can be adjusted in real time. With appropriate control software, it becomes possible for a listener to "walk around" in a virtual acoustic space and have the reflections track the listener's movement in the space. Such techniques are used to simulate artificial acoustic environments in virtual reality simulations of physical spaces.

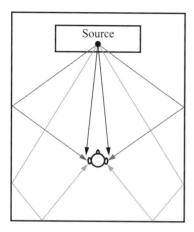

Figure 9.63
Early side reflections.

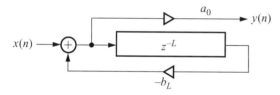

Figure 9.64
Recirculating comb filter.

9.6.6 Recirculating Delay Lines

If we feed the output of a delay line back to its input, it becomes a kind of recursive filter that, for fairly long delays, behaves like the Echoplex tape machine, described at the beginning of chapter 5.

Figure 9.64 shows the output of a recirculating delay line. Inverted copies of sound $x(n)$ delayed by L samples are added back to the delay line's input and also to the output $y(n)$. If $x(n)$ is the sound of a speaker saying "ECHO", then for $a_0 > 0$, $b_L > 0$, and $b_L < 1$, $y(n)$ will sound like "ECHO Echo echo echo echo . . .", mimicking the way a sound dies exponentially as it reflects between the walls of a room. The greater the length of the delay line, the longer it takes the echo to return, and as the feedback gain $b_L \to 1.0$, the echoes take longer and longer to die away. When $b_L = 0$, this degenerates into a simple gain control of a_0.

Using a recirculating delay line to model room reverberation produces a rather unnatural effect, in that each pass of the sound through the delay produces a copy that is spectrally identical to the original, only quieter (if $b_L < 1$). But in a natural setting, air absorption and absorption by walls is not spectrally flat; rather, high-frequency energy is absorbed at a higher rate, so the successive reflections are progressively more muffled. We can simulate this effect by substituting a simple

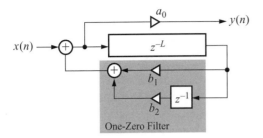

Figure 9.65
Recirculating delay with one-zero filter.

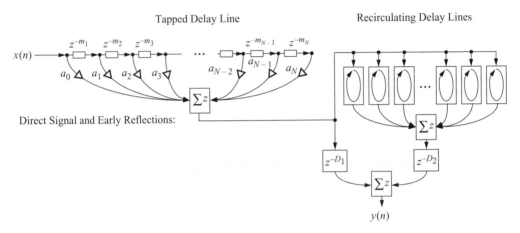

Figure 9.66
Hall simulation system using delay lines.

one-zero lowpass filter for the feedback term (figure 9.65). If we set $b_1 = b_2 = 1/2$, the filter averages adjacent samples, so that each pass through the delay line progressively removes high frequencies. Other coefficient settings accelerate or retard this effect (see section 5.13.1).

9.6.7   Hall Simulation

Schroeder (1970) experimented with combinations of allpass and comb filter elements to create realistic synthesis models of room acoustics. Recall volume 1, figure 7.36, which showed how the impulse response of a room can be analyzed into a direct signal, early reflections, and reverberation tail. The direct signal and early reflections can be modeled with a single tapped delay line. The reverberation tail can be modeled as a set of recirculating delay lines. Moorer (1979a) extended Schroeder's work, adding early reflections as well as air and wall absorption. A flow graph of Moorer's system is shown in figure 9.66.

The input signal $x(n)$ is fed both to the input of a tapped delay line and to the summed output of the tapped delay line. The coefficient a_0 scales the contribution of the direct signal. The lengths of tap delays z^{-m} are chosen to match the geometry of the early reflections of the room and are scaled respectively by coefficients a_m to account for spreading loss, air absorption, wall absorption, and so on. The sum of the direct signal and early reflections constitutes the first 40 to 80 ms of the decay.

The direct signal and early reflections are sent both to a set of recirculating delay lines and to an additional delay that is eventually summed into the final output. The lengths of the recirculating delay lines can be chosen, for example, to model the distance between the three pairs of opposing walls in a typical shoebox concert hall, and their respective gains can model accumulated losses due to spreading and absorption. The output of the recirculating delays is summed, sent through another delay, then summed with the early reflections.

The advantage of Moorer's topology is that all the density of the first N reflections is forwarded to the recirculating delay, which then recirculates the reflections many times, so that the buildup of reflections is quite rapid, as is the case with better concert halls. The delays D_1 and D_2 are set so that the first echo from the recirculating delays coincides with the end of the last echo from the early reflections. This means that either D_1 or D_2 will be zero, depending on whether the total delay of the early reflections is longer or shorter than the shortest recirculating delay line.

Unfortunately, a good deal of fiddling around is required to obtain good-sounding reverberation with this setup. The delays commonly stack up in a highly composite way, even if the delay lengths are chosen to be mutually prime, so that their prime factorizations contain no common factors. Since the delays cascade through the system, their lengths are added together, and the sum of multiple primes is by definition no longer prime. Moorer gives example settings that sound good, but that are heuristically derived.

Performance is improved greatly if lowpass filters are used in the feedback path of the recirculating delays instead of simple gains. However, this introduces another slight problem: the later parts of the reverberation tail become unnaturally tubby because only low-frequency energy can linger. In a natural concert hall, low-frequency energy is lost due to admittance through the walls.[19] This can be overcome by inserting a highpass filter just prior to the final output to simulate the admittance.

The reverberation system developed by Jot (1997) goes a step further, providing control over frequency-dependent reverberation time, allowing independent control over the coloration of a reverberator and its decay time. It is based on a scheme for interconnecting recirculating delay lines that uses a matrix interconnection between a set of recirculating delay lines. (Gerzon 1976). This approach also provides faster and denser buildup of reflections than is obtainable with comb filter networks.

9.7 Physical Modeling

Physical modeling studies the causal interactions of vibrating systems that are the basis of natural sounds. This includes the way energy travels from a performer into and through a musical instrument, how its resonances affect the resulting vibration, and how these vibrations are propagated into the surrounding air. The reverberation and ambient sound systems developed previously are also examples of physical models.

Since it is based on the physics of the vibrating system, physical modeling can also capture the idiosyncrasies of a particular instrument. Physical models capture not only the steady-state behavior of an instrument but also its transient characteristics. As energy floods through a vibrating system, the instrument responds in a characteristic way, and its response provides fertile information to the ear to help characterize the sound source.

There are many approaches to physical modeling. The conventional physical modeling approach was outlined in chapter 7. Starting with a differential equation for an ideal vibration, refinements are added to accommodate other important factors affecting the vibrating system, such as friction, radiation, driving force, and stiffness (see also volume 1, chapter 8, and in this volume, chapters 6 and 8).

Digital waveguide models have been developed that share the advantages of physical modeling but are computationally more efficient as well as intuitively more appealing than classical physical modeling (Smith 1996). Preparation for understanding this material can be found especially in chapter 8. This section focuses on waveguide models, beginning with Karplus-Strong synthesis, a simple extension of the delay line to model plucked strings.

9.7.1   Karplus-Strong Synthesis

We can create a flexible and very inexpensive physical model of a plucked string out of a delay line of length L. First remove the input $x(n)$ from the delay line. Then preload a sequence of random samples into the memory cells of the delay line, $U_0, U_1, \ldots, U_L$, so it looks like figure 9.67a. When the delay line is switched on, it outputs the preloaded random sequence one sample at a time, and if it's connected to an audio DAC with sample rate R, we hear a brief burst of noise as the delay line drains, which sounds kind of like "fffft."

The next step is to recirculate the delay line's output back to its input, as in figure 9.67b. The samples in the recirculating delay line form a periodic sample pattern that the ear detects as a

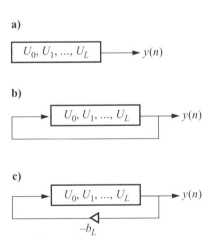

Figure 9.67
Karplus-strong delay lines.

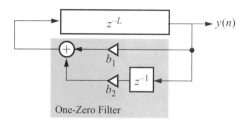

Figure 9.68
Plucked-string synthesis.

complex tone, with a frequency f corresponding to the length of the delay line and the sampling rate, $f = R/L$. Even though the pattern is random, the ear interprets it as a steady, buzzy timbre.

Let's try adding a multiplier b_L on the feedback path (figure 9.67c). Setting the multiplier coefficient b_L to 1 gives us the behavior described in the previous paragraph. Setting it to 0 gives us the one-shot "fffft" sound we started with. But if we set it somewhere in the range $0 < b_L < 1$, then the periodic noise will be attenuated exponentially to silence as it recirculates. During the first pass through the delay line, the samples are scaled by b_L. During the second pass, they are scaled by b_L^2, and on pass n, the samples are scaled by b_L^n. The ear hears this as a complex pitched tone with an exponential decay, suggestive of a plucked or struck string.

A characteristic of plucked or struck string instruments that we'd like to emulate is that the higher harmonics die away more quickly than the lower harmonics, so that the sound becomes more muffled over time. The delay line instrument could be greatly improved if we replace the multiplier b_L with a simple one-zero lowpass filter, as shown in figure 9.68.

If we set $b_1 = b_2 = 1/2$, the filter is just the simple moving average lowpass filter studied in section 5.3. As the preloaded signal recirculates through the delay line, samples pass through the lowpass filter, and on each recirculation, remaining high-frequency energy is attenuated more rapidly than low-frequency energy. The result is that the bright periodic noise we hear at the beginning doesn't just die away, it mellows through time. This lends a dynamical spectral behavior to the simulation that can be made to sound very convincingly like mandolins, banjos, harps, plucked violins, guitars, and drums, depending upon the signal preloaded into the delay line and the kind of filtering to which it is subjected (Karplus and Strong 1983).

The reach of the Karplus-Strong synthesis model was dramatically extended and related to the physics of the plucked string by Jaffe and Smith (1983). The discussion here follows their analysis and that of F. R. Moore (1990, 282).

The dynamic spectral effects introduced by the moving-average filter in the feedback path is what makes this synthesis technique interesting. But some questions must be answered to make this a really useful synthesis technique:

How long does it take for a sound to die away?

Can we control the decay rate of the harmonics?

What is the effect of the phase delay of the filter on frequency?

If we set $b_1 = b_2 = 1/2$ in figure 9.68, the filter's impulse response is

$$y(n) = \frac{x(n) - x(n-1)}{2}.$$

We studied this filter extensively in chapter 5 and know that its transfer function is

$$H(z) = \frac{1 + z^{-1}}{2}.$$

Its frequency response is obtained by taking the magnitude of the transfer function evaluated at $z = e^{i\omega} = e^{i2\pi f/R}$, which we know from chapter 5 evaluates to

$$G(f) = \cos\frac{\pi f}{R}.$$

So the frequency response of the averaging filter is shaped like the first quadrant of a cosine wave, going from 1 to 0 as frequency goes from 0 Hz to the Nyquist rate, $R/2$ Hz. It is a lowpass filter with a gentle roll-off of high frequencies. After the signal has recirculated once around the delay loop, its frequency content is attenuated according to $G(f)$. After the second time around, frequency content is attenuated by $G(f)^2$, and on the nth time around, attenuation is $G(f)^n$.

The phase delay of the filter is the phase of $H(e^{i\omega})$ divided by ω, which we found in chapter 5 evaluates to

$$\angle H(e^{i\omega}) = -\frac{1}{2}.$$

So the averaging filter introduces a one half sample delay to signals passing through. The filter is *linear-phase* because the delay is constant regardless of frequency. We can express the fundamental frequency f_1 of the recirculating delay line as the length of the delay L plus the length of the averaging filter with respect to the sampling rate: $f_1 = R/(L + 0.5)$ Hz. Define the frequency of harmonic k as $f_k = kf_1$. Then the rate at which harmonic k dies away is $G(f_k)^n$.

Jaffe and Smith (1983) determined that the time in seconds required for harmonic k to be attenuated by Q dB is given by

$$\tau_Q(f_k) = \frac{\ln 10^{Q/20}}{f_1 \ln|G(f_k)|}. \tag{9.53}$$

Equation (9.53) demonstrates that the decay time of harmonic k is related to the fundamental frequency of the delay loop f_1 and to the attenuation of the filter at that frequency, $G(f_k)$. This means that higher harmonics decay faster than lower ones, and also that high-pitched tones decay faster overall than low-pitched tones, just as we wanted. Happily, these behaviors map well onto the timbre of plucked and struck string tones. Unhappily, the range of variation from high to low pitch is actually too wide in practice and sounds unnatural. Besides, one size does not fit all: each kind of plucked string instrument has a unique trade-off between high-frequency rolloff and tone length.

To control note duration, we must be able to shorten and lengthen notes arbitrarily. To do this, we must modify the averaging filter characteristics. A simple way to shorten notes is to add an additional gain control ρ in the feedback path, so the impulse response becomes

$$y(n) = \rho\frac{x(n) - x(n-1)}{2}, \qquad 0 < \rho < 1.0,$$

and the frequency response becomes

$$G(f) = \rho\cos\frac{\pi f}{R}.$$

This shortens the overall decay time and also decreases the range of variance between low and high harmonic decay. But we need also to be able to lengthen decay time, especially for high pitches. We can do this by modifying the filter so that high frequencies are less attenuated, thus increasing overall decay time. Jaffe and Smith set the filter coefficients so that $b_1 = 1 - S$, $b_2 = S$, where S is a decay-stretching factor. Together with the decay-shortening factor ρ, the combined impulse response becomes

$$y(n) = \rho[(1 - S)x(n) + Sx(n-1)].$$

If $\rho = 1$ and $S = 0.5$, this is exactly the same filter as before. The frequency response of this filter is

$$G(f) = \rho\sqrt{(1 - S)^2 + S^2 + 2S(1 - S)\cos(2\pi f/R)}. \tag{9.54}$$

If $S \neq 0.5$, the decay time will be longer than for $S = 0.5$. (For 0 or 1, S will not cause any decay; when $S = 0$, the filter passes $x(n)$ directly, and when $S = 1$, it just passes $x(n-1)$ directly.)

The phase response of this modified filter is also more complicated:

$$\angle H(e^{i\omega}) = \frac{\tan^{-1}\dfrac{-S\sin\omega}{(1 - S) + S\cos\omega}}{\omega}.$$

If $S \neq 0.5$, the phase delay will no longer be linear, and different frequencies will be delayed by different amounts. Therefore, the value of S will affect the frequency of the harmonics. For $S < 0.5$, higher harmonics will be progressively sharper. Happily, this allows us to mimic the stretching of upper harmonics that occurs in, for example, pianos. For $S > 0.5$, harmonics will be progressively flatter.

The last practical problem has to do with fine-tuning pitch. As defined, the Karplus-Strong model has a fundamental frequency of

$$f_1 = \frac{R}{L + 0.5}\text{Hz}.$$

It would be most convenient to run the instrument at a constant sampling rate R, but since L is also an integer, we are limited to the quantized pitches dictated by this formula. As fundamental frequency rises, L shrinks. At very high frequencies, the frequency jump between successive integer values of L can be so large that we can't match them to any scale. We need a way to achieve fractional delay times through the delay loop so we can achieve arbitrary pitch at a fixed sampling rate. Jaffe and

Smith point out that the allpass filter exactly performs this function (see section 5.13.4). The allpass filter only affects phase and has no effect on frequency. Jaffe and Smith cascade an allpass filter together with the averaging filter to achieve arbitrary delay and hence arbitrary fundamental frequency. They use a simple allpass filter with impulse response

$$y(n) = Cx(n) + x(n-1) - Cy(n-1),$$

which has a transfer function of

$$H(z) = \frac{C + z^{-1}}{1 + Cz^{-1}}, \qquad 0 < C < 1.$$

The phase delay is

$$\angle H(e^{i\omega}) = -\frac{1}{\omega}\tan^{-1}\frac{-\sin\omega}{C + \cos\omega},$$

which, for ease of calculation, is approximately equal to a delay of

$$D = \frac{1-C}{1+C} \tag{9.55}$$

at low frequencies, where D is the delay through the allpass filter. The phase delay of the allpass filter is not the same for all frequencies. Upper harmonics are slightly flattened in the range $0 < C < 1$, and are increasingly sharpened in the range $-1 < C < 0$. So long as $|C| < 1$, we can change C dynamically to achieve vibrato effects.

We can achieve simple overall gain control by directly scaling the output of the instrument with an amplitude term. To emulate the piano damper pedal, for example, we could have two values for ρ, a value closer to 1.0 during note sustain, and a value closer to 0 when the damper pedal is lifted. This would accelerate the attenuation of all energy in the delay loop, as does the damper pedal. This technique can also help reduce the possibility that arithmetic quantization errors in the filters would prevent the signal from dying away completely to zero. Another way to help with this problem is to prescreen the random values prior to filling the delay line: we calculate their mean value and substract this value from all samples, so the signal has no DC bias.

It would be convenient if we could specify the time τ required for a tone at frequency f to decay by Q dB. Solving equation (9.53) for $G(f)$, we have

$$G(f) = 10^{-Q/\tau}.$$

Frequency f may be a fundamental or a harmonic frequency. We compare this value to the one produced by the unadulterated delay loop:

$$G_{\text{nom}}(f) = \cos\pi\frac{f}{R},$$

which is the nominal attenuation assuming $\rho = 1$ and $S = 0.5$. If $G_{\text{nom}} < G$, we must lengthen the decay; if $G < G_{\text{nom}}$, we must shorten the decay. Lengthening requires solving equation (9.54) for S. This can be done by arranging equation (9.54) into standard form:

$$(2 - 2\cos\omega)S^2 + (2\cos(\omega) - 2)S + 1 - G(f)^2 = 0.$$

Since it has the general form of $ax^2 + bx + c = 0$, it can be solved by the quadratic formula. Two solutions will result, ranging generally between 0 and 1, on either side of 0.5. Choosing the value less than 0.5 results in harmonic stretching, and it is generally the one we'd want to mimic string instruments.

Last, we need to tune the allpass filter parameters to achieve the correct frequency. The nominal delay introduced by the moving-average filter is S samples. In order to calculate the length of the delay loop L, we must first compute the desired delay $L_d = R/f_1$. The integer part of L_d can be used to set the length of the delay loop L, and the fractional part can be used to fine-tune the pitch with the allpass delay. We extract the integer part of L_d with the floor function, $L = \lfloor L_d \rfloor$. If $L + S$ is less than the desired length, we subtract 1 from L and make up the difference by choosing an all pass coefficient C that produces the desired fractional delay D. Solving equation (9.55) for filter coefficient C in terms of D, we obtain

$$C = \frac{1 - D}{1 + D}.$$

This value of C is applied to the allpass filter, and the delay length is set to L, thereby obtaining the correct frequency.

There are many extensions to this technique. For example, to effect a softer stroke or pluck, the noise signal can be lowpass-filtered before preloading it into the delay line. To create timbres with only even harmonics, initialize the delay line to contain two identical periods of random values. This doubles the frequency and the decay time. To create drumlike timbres, one can modulate the feedback filter coefficients according to a random variable. Karplus and Strong (1983) suggest a feedback filter with an impulse response of

$$y(n) = \begin{cases} [x(n) + x(n-1)]/2, & b, \\ -[x(n) + x(n-1)]/2, & 1 - b, \end{cases}$$

where b is a random variable in the range $0 < b < 1$.

One can also dynamically replace the contents of the delay line with successive samples of an arbitrary sound. The delay line acts rather like the resonating sympathetic strings of a sitar or viola d'amore to capture and sustain frequency content that is harmonically related to its length.

Karplus-Strong synthesis combines computational efficiency with realism and great expressive control. But Karplus-Strong synthesis is just a special case of a much broader and even more powerful synthesis model. Julius Smith (2004) generalized the Karplus-Strong model employing the theory of digital waveguides, to which we turn next.

9.7.2 Waveguide Synthesis

Waveguides can be thought of as acoustic tubes or strings, like the ones discussed in chapter 8, but they can be used to model any one-dimensional wave motion. (I've found it useful to think about waveguides as water-filled troughs, for example.) The discussion here follows Smith (2004).

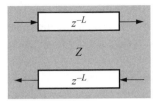

Figure 9.69
Waveguides.

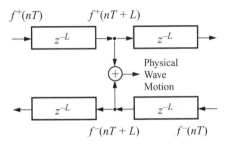

Figure 9.70
Tapping a waveguide delay line.

We observe experimentally that a disturbance in a one-dimensional medium causes waves to propagate away from the disturbing force in both directions (for example, see figure 7.5), and we've seen that solutions of the one-dimensional wave equation have the form of a positive-going (right-traveling) wave $f^+(nT)$ and negative-going (left-traveling) wave $f^-(nT)$ (see chapter 7). We can emulate this property of one-dimensional media by the simple expedient of combining two one-directional delay lines (see section 9.6) that propagate waves in opposite directions according to some characteristic impedance Z, an arrangement called a *waveguide* (figure 9.69).

Waveguides can be used to model any one-dimensional acoustical vibration. Physical force, pressure, or velocity waves can be represented as the sum of the traveling waves in the two delay lines.

We can realize a physical wave at any point along the waveguide by tapping and summing the two delay lines at the desired delay (figure 9.70).

Delay lines model wave propagation in a medium with constant characteristic impedance, but we must also be able to model wave propagation across impedance discontinuities, such as walls, string terminations, or cross-sectional area variation in tubes. The scattering junction (see section 8.15) describes the reflection and transmission of acoustical energy across a change of impedance (see especially figures 8.27 and 8.28). The complex reflection coefficient

$$R_{0,1} = \frac{Z_1 - Z_0}{Z_1 + Z_0}$$

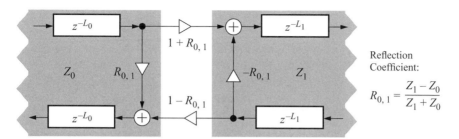

Figure 9.71
Waveguide scattering junction.

determines the way energy is reflected and transmitted by the junction as a function of the impedances to the left (Z_0) and right (Z_1) of the junction.

Combining waveguides with scattering junctions (figure 9.71) allows us to realistically emulate
the physics of many natural vibrating systems such as strings, air columns, and halls without having to numerically integrate the wave equation.

9.7.3 Plucked-String Synthesis

We observed in figure 7.8 that a traveling wave on a string reflects from its terminations with a sign
inversion. A string and its terminations can be modeled as a waveguide for the string and a pair of
scattering junctions for the terminations. By setting the impedances of the string and terminations
to appropriate values, the scattering junctions model reflection of energy back into the string at its
terminations and also model energy transmission into the body of the instrument.

For example, let the vertical displacement of a string be $u(t)$, made up of a positive-going
(right-traveling) wave $u^+(t)$ and negative-going (left-traveling) wave $u^-(t)$. Suppose it takes
$NT = cl$ sample times of duration T for the wave components to propagate the length of a string
of length l at speed c. If we sample these traveling wave components at $N+1$ equidistant points
along the string, we can preload the waveguide with these sampled values, and the samples will
be indexed in the waveguide as $u(n)$, where n is the index.

Since NT is the propagation time of the string in one direction, $2NT$ is the round-trip string loop
delay time. To model this, we must have two waveguides of length N, one to propagate forward,
the other to propagate back. Using the terminology of stringed instruments, let's call one of the terminations the bridge at position $x = 0$ along the string, and the other end the nut termination at
position $x = N$.

If the terminations are idea lly rigid and massive, and the string is ideally flexible and light-
weight, then all energy is reflected from the junctions back into the string. Note in figure 9.72
that the signs of the coefficients that feed back energy from one waveguide to the other are negative. This is appropriate for displacement waves, velocity waves or acceleration waves, which
reflect with phase inversion. Pressure waves and force waves reflect without phase inversion.

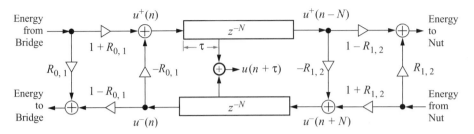

Figure 9.72
Simple waveguide plucked string synthesis.

Since we are modeling displacement of a string, the terminations are inverting, as shown in figure 9.72. If we set

$$R_{0,1} = R_{1,2} = 1 = \frac{Z_{m+1} - Z_m}{Z_{m+1} + Z_m} = \frac{\infty - 0}{\infty + 0},$$

then the scattering junctions actually don't scatter; they reflect all energy back into the string (with phase inversion). By choosing a value for R in the unit interval $0 < R < 1$, some energy is transmitted, and the remainder is reflected back into the string. If $0 < R \ll 1$, most energy is transmitted at the junction, and the string's vibration ceases quickly, like a banjo string or sound in a room with heavy curtains and carpeting. If $0 \ll R < 1$, most energy is reflected, and the vibration lasts longer, like a low-pitched piano string or sound in a room with stone surfaces.

We can usefully define the way in which energy is reflected at a scattering junction as the *reflection transfer function*. We can correspondingly define the way energy is transmitted as the *transmission transfer function*. Virtually all the energy at the nut of an unstopped guitar or violin string is reflected, so in these cases, the reflection transfer function at the nut is $R_{1,2} \approx 1$. When a violin string is stopped by the player's finger, energy is lost by friction of the string against the flesh of the finger, and so $R_{1,2}$ is quite a bit lower; consequently, a pizzicato (plucked) violin string tone dies away much more quickly if the string is stopped by the finger than if it is open (unstopped). Since a stopped guitar string is terminated by a metal fret that introduces little friction, its reflection transfer function is about the same whether it is stopped or not. At the bridge end of stringed instruments, some energy is transmitted into the body of the instrument, so $R_{0,1} < R_{1,2}$.

Because the body of a stringed instrument is essentially a Helmholtz resonator, the scattering junction at the bridge end of the string is frequency-dependent, so the value of the transmission transfer function is complex, and its magnitude frequency response has a lower $R_{0,1}$ for frequencies near resonance because at resonance the body is better able to radiate its energy.

Even if a musical instrument has multiple resonances, its overall transmission transfer function will be linear. To see this, imagine if we added an additional resonance to an instrument. For example, the Indian rudra vina (figure 9.73) has a resonating gourd at both the bridge and the nut. The effect of the two resonances on the overall spectrum of the instrument is additive, and overall its

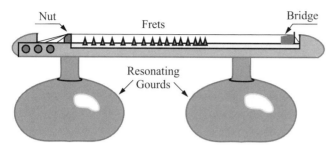

Figure 9.73
Vina.

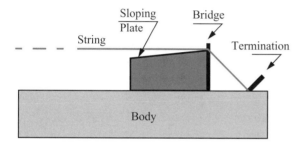

Figure 9.74
Vina bridge.

transmission transfer function is linear. In general, the resonances introduced by the bodies of musical instruments combine linearly.

9.7.4 Nonlinear Scattering Junctions

Not all scattering junctions are linear, and again the vina is a good example. An enlarged view of the bridge (figure 9.74) shows that a sloping bone plate comes up to meet the string as it approaches the bridge. As it vibrates vertically, the end of the string slaps against this plate. The impedance of this junction is nonlinear because the vertical (but not the horizontal) displacing force in the string is constrained by the plate, so that in this case force is not proportional to string displacement. The result is that high-frequency harmonic energy is injected into the string, giving it a characteristic sizzling sound of instruments that have this kind of plate next to the bridge (notably the vina and sitar), and the bandwidth increases with greater vertical excursion of the string.

9.7.5 Clarinet Synthesis

The clarinet headpiece also is a nonlinear scattering junction similar to the vina: the reed is clamped down against a plate in the clarinet headpiece in such a way that it closes against the headpiece in proportion to the pressure difference between the inside of the player's mouth and the inside

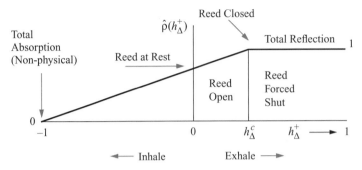

Figure 9.75
Simple clarinet reflection function.

of the clarinet bore. For waveguide synthesis, Julius Smith uses a simple signal-dependent and embouchure-dependent nonlinear reflection coefficient to terminate the bore at the headpiece. Figure 9.75 shows Smith's (2004) qualitative analysis of the reed reflection coefficient as a function of mouth pressure. The formula for this figure is:

$$\hat{\rho}(h_\Delta^+) = \begin{cases} 1 - m(h_\Delta^c - h_\Delta^+), & -1 \leq h_\Delta^+ \leq h_\Delta^c, \\ 1, & h_\Delta^c \leq h_\Delta^+ \leq 1, \end{cases} \tag{9.56}$$

for $m = 1/(h_\Delta^c + 1)$. The corner point h_Δ^c is the smallest pressure difference giving reed closure. Embouchure and reed stiffness correspond to the choice of offset h_Δ^c and slope m.

The scattering junction at the bell of the clarinet acts as a highpass filter for transmitted energy and as a lowpass filter for reflected energy, so it is a linear junction that behaves somewhat like a loudspeaker cross-over network.[20] Since the bore of a clarinet is effectively cylindrical, simple waveguides can be used to model the propagation delay in the bore. Because the main control variable is air pressure at the mouth, it is convenient to use pressure wave (noninverting) scattering junctions. The scattering junction for the bell can also serve to model the round-trip absorption losses of the bore, so we don't have to calculate these separately.

Smith's block diagram for a waveguide clarinet is given in figure 9.76. The reflection filter and output filter are complementary lowpass and highpass, filters, respectively, with cross-over frequency at around 1500 Hz for the bell. The output filter implements bell and tone-hole losses; the reflection filter passes the complementary low-frequency energy back into the bore. Smith suggests that the simplest practical implementation for the tone-hole losses is to use the bell filter settings unchanged for the tone holes, as though the clarinet were cut to the length of the fingered tone hole for each note. The cylindrical bore of the clarinet can be modeled with a simple waveguide.

The termination at the headpiece is what distinguishes the clarinet timbre from brasses that also have a cylindrical bore and flared bell. Two input controls determine oscillation: mouth air pressure $p_m(n)$ and embouchure. Ignoring the embouchure for a moment, the reflected signal $p_b^+(n)$ returning through the bore from the bell is summed with the mouth pressure to generate the total pressure at the headpiece.

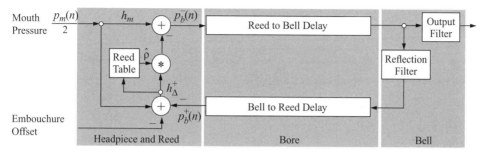

Figure 9.76
Waveguide model of a clarinet.

Total pressure h_Δ^+ is then used to determine how open or closed the reed should be. If the total pressure inside the headpiece is low relative to the mouth pressure, the reed will be sucked shut, and the reflection coefficient of the headpiece should be closer to 1.0, because most energy will be reflected. If the total pressure inside the headpiece is high relative to the mouth pressure, the reed will be blown open, and the reflection coefficient of the headpiece should be closer to zero because most energy will be transmitted into the mouth. The headpiece reflection coefficient $\hat{\rho}$ is therefore determined by using total pressure h_Δ^+ to index the function $\hat{\rho}(h_\Delta^+)$, given in figure 9.75, to establish the instantaneous reflection coefficient, depending upon the instantaneous position of the reed.

The embouchure is primarily a function of two variables: the position of the lips along the length of the reed and the clamping pressure of the lips on the reed. However, a simpler view is to note that these two parameters principally control the position of the pressure drop where the reed begins to open at the knee of the curve in figure 9.75. In light of this, the role of the embouchure can be seen as a simple offset that adjusts the position of the knee, and this is how it is implemented in figure 9.76. This one parameter can then emulate the strength and position of the bite and even the stiffness of the reed. Other effects, such as a general brightening, can be achieved by altering the slope of m in equation (9.56) or by making it an exponential instead of a linear function.

9.7.6 Bowed Strings

The bow divides the string into two sections, creating two nonlinear junctions with parts of the string on either side. A simplified block diagram for the basic waveguide bowed string according to Smith (2004) is shown in figure 9.77. When a rosined horsehair bow is drawn across a violin string, at first the frictional force of the bow drags the string along. When the restoring force of the string exceeds the bow frictional force, the string breaks loose and slides easily against the bow because the function of bow friction versus string velocity is nonlinear in such a way that as bow velocity increases, friction—which is initially quite high—drops quickly to a low point. When the string moves once again in the direction of the bow, it is entrained once again by the bow's frictional force and dragged along. Thus energy is added to the string by the bow as though by tiny plucks synchronized with the movement of the string in the direction of the bow. Since the primary control is bow velocity, the waveguide data are treated as velocity waves. (See section 8.3.)

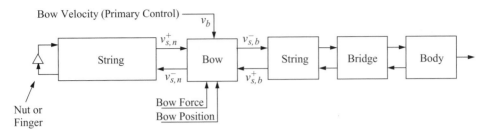

Figure 9.77
Block diagram of a waveguide bowed string.

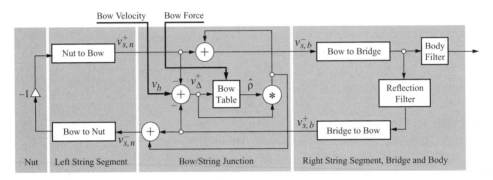

Figure 9.78
Waveguide model for a bowed string.

Figure 9.78 shows a waveguide model for a bowed string. The waveguide at the nut end of the string carries left-going $v_{s,n}^-$ and right-going $v_{s,n}^+$ velocity waves. The waveguide at the bridge end of the string carries right-going $v_{s,b}^-$ and left-going $v_{s,b}^+$ velocity waves. The + indicates waves traveling toward the bow. The fundamental pitch of the system can be adjusted by changing the absolute lengths of the waveguides, and the position of the bow on the string can be adjusted by changing the relative lengths of the bow-end and body-end waveguides. Continuous change in bow position and continuous change in pitch (glissando) can be implemented by interpolating between delay line samples.

The nut can be modeled as a unit reflection coefficient because virtually all energy is reflected from it. The bridge and nut are phase-inverting because we are interpreting the waveguide contents as velocity waves. The bridge has complex-valued losses as a consequence of the frequency response of the body, which acts like a Helmholtz resonator. Fingered notes on a violin are more highly damped and lowpass-filtered in comparison to the bright tone of the open strings. Rather than adding another filter to account for this, we can combine the reflection transfer function of the nut with that of the bridge to economize calculation, as well as the round-trip attenuation and dispersion of wave energy in the strings.

Bow velocity and bow force are the control inputs from the performer that determine the quality of vibration. On a violin, when the difference between bow velocity and string velocity is near zero, the frictional force increases, and the bow entrains the string, increasing its energy by dragging it along. We can simulate this as follows: energy from the left and right string segments arriving at the bow is passed along "underneath" the bow to the other string segment; additionally, depending on the velocities of bow and string, energy will be injected into the string or dissipated from it by the bow.

Here's how the bow/string junction works. In figure 9.78, the right string segment output $v_{s,b}^+$ and left string segment output $v_{s,n}^+$ are passed along "underneath" the bow to their respective string segments $v_{s,n}^-$ and $v_{s,b}^-$ through the adders on the outer rails of the bow/string junction. Additionally, $v_{s,b}^+$ and $v_{s,n}^+$ are sent into the bow/string junction to determine whether energy should be drained or added to the string velocity.

When bow and string velocities are similar, we inject energy; when they are opposed, we do not, or if bow pressure is heavy, we subtract energy.

In order to determine the coefficient of friction between bow and string, we must obtain the differential v_Δ^+ between the instantaneous string velocity at the bow, $v_{s,b}^+ + v_{s,n}^+$, and the bow velocity v_b. That is, we calculate $v_\Delta^+ = v_b - (v_{s,b}^+ + v_{s,n})$. When bow velocity and string velocity are opposed, v_Δ^+ is large, and we want the output of the bow table $\hat{\rho}(v_\Delta^+)$ to be close to zero so that the multiplication that follows the table injects little or no energy into the string. When bow velocity and string velocity are similar, v_Δ^+ is small, and we want the output of the bow table $\hat{\rho}(v_\Delta^+)$ to be large so that the multiplication that follows the table injects more energy into the string, thereby overcoming frictional and dissipative forces in the string and sustaining vibration.

The contents of the bow table defines the way friction and bow/string differential velocity interact. Smith gives a simple quantitative function (figure 9.79). The flat center section of the table shows the bow and string stuck together with a high coefficient of friction. The skirts of the function correspond to reduced friction when the string has broken away from the bow by exceeding v_Δ^c, the breakaway/capture differential velocity.

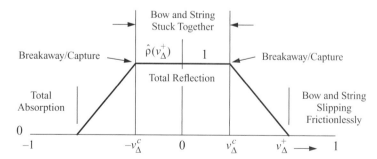

Figure 9.79
Simple string friction function.

The bow table function is defined by

$$\hat{\rho}(v_\Delta^+) = \frac{r(v_\Delta(v_\Delta^+))}{1 + r(v_\Delta(v_\Delta^+))}, \tag{9.57}$$

where the ratio of the bow impedance Z_b to the string impedance Z_s is given by $r(v_\Delta) = 0.25 Z_b(v_\Delta)/Z_s$, and $v_\Delta = v_b - v_s$ is the bow/string velocity differential.

The bow impedance function $Z_b(v_\Delta)$ is the coefficient of friction of the bow against the string; that is, by the definition of impedance, bow force $F_b = Z_b(v_\Delta) \cdot v_\Delta$. Nominally, $Z_b(v_\Delta)$ is constant and positive when the magnitude of the bow/string velocity differential is less than the breakaway/capture threshold, that is, when $|v_\Delta| < v_\Delta^c$, where v_Δ^c is both the capture and breakaway threshold. This is the static coefficient of friction. For $|v_\Delta| > v_\Delta^c$, $Z_b(v_\Delta)$ rapidly falls toward a relatively low dynamic coefficient of friction, and in an ideal system continues towards a zero coefficient (which is impossible in practice). With a real bow, the breakaway/capture parameter is not a single point but is a function of a hysteresis parameter. Hysteresis in this context means that the transition from stuck to slipping requires slightly greater force than is required to transition from slipping back to stuck. This simplified model ignores hysteresis.

9.7.7 Critique of Waveguides

The simple digital waveguide models of musical instruments share the advantages of physical modeling but are computationally more efficient. Smith suggests that waveguide models require $O(1)$ computations per sample, whereas conventional physical models require $O(N)$, where N is the number of discrete modeling cells, corresponding to the number of samples in the waveguides. For realistic tasks, Smith estimates that waveguides are on the order of 300 times more efficient than numerical integration of the wave equation. It is possible to construct complicated waveguide synthesis models that can be calculated in real time.

If the model is fitted to a suitable physical controller, it can be performed live. And that's a really good thing because, in spite of the uncanny way in which waveguide models seem to efficiently and intuitively capture the acoustical process of musical instruments, they do not capture any of the nuance of a skilled performer. Audition of samples generated by the simple models sound flat and wooden, as if they were performed by an orchestrion. We should not be surprised at this, and in all fairness, this critique must also be leveled at all the other synthesis methods discussed in this chapter. It's only that, with waveguide instruments, the gap between the sometimes quite high degree of realism of the sound and the clumsiness of algorithmic control begs for better ways to perform these models.

Fortunately, this problem is getting some attention. If a physical model is implemented in a real-time computer system and equipped with sensors to track a performer's gestures, it can be performed like a regular instrument, with very convincing results. By changing the parameters of the model, the controls can be used to create a wide variety of different instrumental timbres. We can model the response of an instrument to the articulations of a performer, and we can characterize both the ordinary and degenerate vibrational modes of an instrument. For example, a physical model of a violin can capture how excessive bow pressure produces a wolf tone, or how an overblown clarinet produces a multiphonic, or how a misplaced embouchure on a flute produces a breathy tone.

One of the aims of music synthesis articulated at the beginning of this chapter is to obtain realization of music without performers. But all of the synthesis techniques discussed in this chapter are like instruments without performers, they incorporate no model of the performer/instrument interaction. To obtain a human touch still requires a human. It is certainly an area of fruitful research to characterize and understand human performance.

9.8   Source Models and Receiver Models

Data compression is a potential application for all synthesis techniques described in this chapter. Audio compression techniques such as MPEG achieve compression by employing a psychoacoustic model that removes unhearable components. Thus, MPEG compression can be called a *receiver model,* and by this reasoning, physical modeling is a *source model.* (However, MPEG-4 incorporates both source and receiver models.) Typically, control parameters of source models require far less bandwidth than the parameters of a receiver model. The Musical Instrument Digital Interface (MIDI) also qualifies as a source model. With MIDI, a note can be started with as little as three bytes of data and ended with only two more.

For a more general example, we can interpret common music notation as the control parameters of a source model, namely, a standard symphony orchestra. The memory required to store the notated score of Beethoven's Ninth Symphony as a set of note parameters is small in comparison to the memory required to store an MPEG-encoded representation of an orchestra playing that score. Recent versions of the MPEG standard, such as MPEG-4, incorporate source modeling to allow for this kind of audio compression.

We gain something and lose something with source models. With receiver models, *any* signal the ear can hear can be encoded (with varying levels of quality, depending on the encoding process). With source coding, all one can encode are sounds for which there is a suitable model. To achieve realism, the model must either be performable in real time by a person, or one must incorporate gestural knowledge of a competent performer into the model. It would be ideal if a violin physical model could discriminate, for example, between a Stradivarius and a Guarneri violin, for example, because certainly a competent listener can do so from a high-quality MPEG encoding. Unless the model is performed in a concert space, the model must also supply an acceptable simulation of such an acoustic space.

Thus source models must explicitly embody information that receiver models can take for granted. Therefore, it is unlikely that source models will supplant receiver models, but they will supplement them where appropriate. And there are plenty of places where they are appropriate and desirable, such as to provide musicians with novel and adaptable instruments and to supply composers with realistic simulations of musical instruments that they can use to preview their works.

Summary

Linear synthesis techniques can generally be used to reproduce a sound that is identical to the original. Nonlinear techniques generally provide no way to reproduce a sound that is identical to an original but may have other compelling advantages, such as being economical to calculate or intuitive to use.

Linear transforms add or subtract weighted basis functions of some kind, such as the sinusoids used by the Fourier transform. Additive synthesis combines individual components to create complex timbres. Subtractive synthesis removes energy from a spectrum by filtering. Linear systems are fairly intuitive to use but can require a great deal of analysis data. Nonlinear techniques typically are much more economical, if less general.

We developed a patch system to specify synthesis algorithms and used it to create patches for a wide variety of synthesis techniques. We first got control over pitch, duration, amplitude, amplitude envelope, and vibrato, then investigated a set of synthesis techniques to control timbre. Some sound synthesis techniques are optimized to provide the most naturalistic sound possible. Others produce hybrid sounds, or unearthly sounds, or sounds that metamorphose.

Oscillator bank synthesis generates a weighted sum of fixed-frequency sinusoids at static harmonics of a fundamental. Adding functions of frequency and time to each oscillator provides for time-varying (dynamic) synthesis of arbitrary spectra.

We considered a set of geometrical waveforms obtainable from Fourier series, including square wave, triangular wave, sawtooth wave, and sum of cosines.

By substituting a wavetable for continuous sinusoids, we can discretize oscillator bank synthesis and provide efficient and highly general synthesis. The wavetable can be any discrete function, so the resulting synthesis is completely arbitrary. We developed a table lookup oscillator.

Amplitude modulation (AM) varies the instantaneous amplitude of a signal in a periodic manner. Its spectrum can be viewed as a combination of fixed frequency components. Amplitude modulation without the carrier present is called ring modulation.

Frequency modulation (FM) varies the instantaneous frequency of a signal in a periodic manner. Its spectrum also can be viewed as a combination of fixed frequency components. The amplitudes of the sidebands are controlled by Bessel functions. By varying depth of modulation and the frequency ratio of carrier to modulating oscillator, a wide variety of musical instrument simulations can be created.

Waveshaping synthesis is an improvement over FM synthesis in that arbitrary spectra can be directly specified instead of being mandated by the shape of the Bessel function curves. In its simplest form, it only creates harmonic spectra, but it can be extended to create inharmonic spectra.

Waveform synthesis can be used to model vowel sounds. Linear prediction can synthesize high-quality vocal utterances and is relatively economical to calculate. However, because it relies on recursive filtering techniques, is must be used carefully. If an outcome is predictable based on available information, then any additional information of the same kind is redundant. We can define information as the degree of entropy in a signal. We can define entropy as the number of discrete states required to characterize a system. We want the resonant properties of speech to have minimum entropy because then the states required to characterize the resonance will be merely a handful of prediction coefficients. We want the residue properties of speech to have maximum entropy because then we have made sure that everything that can be predicted is accounted for in the prediction coefficients. Most acoustical signals have high redundancy and low entropy, allowing for a great deal of data compression, thereby lowering communication costs. LPC is attractive to musicians because it divides a signal into an all-pole filter and a residual signal.

Formant wave functions in FOF synthesis consist of windowed sinusoids with frequencies corresponding to the center of each vocal formant. The bandwidth of each FOF can be adjusted to match the spectral shape of a vocal formant. Although the technique can model many timbres, it excels at modeling vocal resonances. FOF synthesis is computationally less challenging than filter-based approaches to speech synthesis such as LPC, but it can't produce consonants.

Granular synthesis, like FOF synthesis, builds up complex sounds out of individual grains of sound. It is an idea related to the work of Dennis Gabor that combines "frequency language" and "time language" in one sonic event.

Concert hall acoustics can be modeled using delay lines, recirculating delay lines, and allpass filters in various combinations. Acoustically, a room is essentially a multipath wave propagation system.

Physical modeling synthesis models the causal interactions of vibrating systems that are the basis of natural sounds. They capture not only the steady-state behavior of an instrument but also its transient characteristics. Karplus-Strong synthesis can be used to create inexpensive physical models of a plucked string. It can be thought of as a simplified form of waveguide synthesis, which is capable of synthesizing also winds, brasses, and bowed strings.

A potential application for all synthesis techniques described in this chapter is data compression. Audio compression techniques such as MPEG achieve compression by employing a psychoacoustic model that removes unhearable components. Thus, MPEG compression can be called a receiver model, and by this reasoning, physical modeling is a source model.

10 Dynamic Spectra

What do we hear? The answer of the standard text-books is one which few students, if any, can ever have accepted without a grain of salt. According to the theory chiefly connected with the names of Ohm and Helmholtz, the ear analyzes the sound into its spectral components, and our sensations are made up of the Fourier components, or rather of their absolute values. But Fourier analysis is a timeless description, in terms of exactly periodic waves of infinite duration. On the other hand, it is our most elementary experience that sound has a time pattern as well as a frequency pattern. This duality of our sensations finds no expression either in the description of sound as a signal $s(t)$ in function of time, or in its representation by Fourier components $S(f)$. A mathematical description is wanted which ab ovo takes account of this duality. Let us therefore consider both time and frequency as co-ordinates of sound, and see what meaning can be given to such a representation.
—Dennis Gabor

Because musical signals demonstrate highly complex interactions between time and frequency, music notation systems throughout the world are designed to convey information that is local to both time and frequency. But up to this point in this book, the time and frequency aspects of sound have been held strictly separate. In this chapter, we examine ways to treat both in one unified representation.

In the opening quotation Gabor characterizes Fourier analysis as "a timeless description." While this is true, there are shades of meaning. Gabor is reminding us that the Fourier transform (defined in equation 3.5) must be evaluated for all time from $-\infty$ to ∞. From this it is easy to suppose that the Fourier transform provides no temporal information. This is not true because the complex spectrum preserves phase information. However, taking the magnitude of the Fourier transform discards phase information. Only the intensities of the various frequencies across the entire Fourier transform window are preserved in a magnitude spectrum. By way of proof, figure 10.1 shows that the magnitude spectrum of a *sequence* of tones is the same as the magnitude spectrum of the *sum* of the tones, so long as the total energy is equal. Total energy is the same in both waveforms. In figure 10.1a, the tones are sequential, and in figure 10.1b, the tones are simultaneous. The magnitude Fourier transform is taken over the entire duration shown in the plots. The two magnitude spectra in figures 10.1a and 10.1b are substantially identical, showing that the magnitude Fourier transform disregards everything but frequency content within its analysis window.

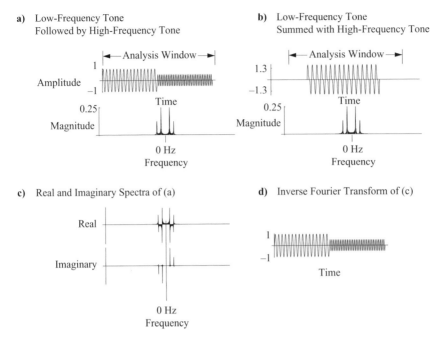

Figure 10.1
Magnitude fourier transform of two signals.

But the (complex) Fourier transform preserves all information in the signal. Figure 10.1c shows the real part and the imaginary part of the Fourier transform of figure 10.1a. Figure 10.1d shows the inverse Fourier transform of the complex spectrum in figure 10.1c. We see that we obtain the identical signal as in figure 10.1a, showing that the Fourier transform preserves both temporal and spectral content.

Since the Fourier transform does preserve temporal information, we would like to be able to separate a sound's temporal information from its frequency information, for example, to track the way the energy of a sound's harmonics changes through time. But the Fourier transform binds frequency information with temporal information in the complex spectrum, and it is not immediately obvious how to separate them. What is lacking is a way to observe how spectral information changes through time.

10.1 Gabor's Elementary Signal

Gabor (1946; 1947) suggested a new approach that reconciled time and frequency in one representation in an optimal way, achieving greater temporal and spectral precision than is available with the Fourier transform. His insight paved the way for new families of temporal-spectral analysis techniques, including the short-time Fourier transform (STFT) and phase vocoder. His thinking also portended so-called multirate analysis, including the wavelet transform.

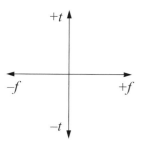

Figure 10.2
Gabor's information diagram.

a)   b)

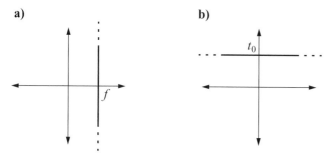

Figure 10.3
Signals and the information diagram.

10.1.1 Information Diagrams

Gabor's analysis begins with a two-dimensional representation of sound that he calls an *information diagram,* where time t and frequency f are laid down on the y-axis and x-axis, respectively (figure 10.2). The aim of the representation is to explore the symmetrical relation between time and frequency.

Clearly, any new representation must carry forward all prior understandings. Thus, for instance, a simple harmonic oscillation such as $s(t) = \sin 2\pi f t$ will be represented in Gabor's information diagram as a single vertical line of infinite extension at frequency f (figure 10.3a). This shows that the ideal sine wave exists for all time $\pm t$. Similarly, an impulse at time t_0 will be represented in Gabor's information diagram as a single horizontal line of infinite extension (figure 10.3b). This shows that the ideal impulse includes all frequencies $\pm f$.

10.1.2 Quantum Uncertainty

So far, the information diagram has replicated our understanding of ideal waveforms and impulses. However, the goal is to construct a uniform representation for nonideal signals that we can hear. Since hearable signals can't be timeless, their spectra can't be of infinite extent; they must have some

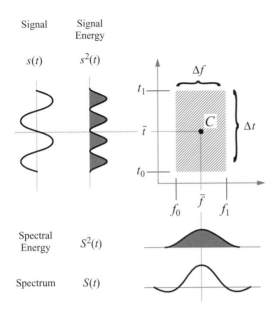

Figure 10.4
Heisenberg box.

effective bandwidth and duration. We can imagine such a hearable signal as occupying a region in the information diagram that we can call a *Heisenberg box,* with effective duration Δt over an interval from t_0 to t_1 and effective bandwidth Δf from f_0 to f_1 (figure 10.4).[1] We can then define a mean epoch $\bar{t}$ and mean frequency $\bar{f}$. This gives us a point C that is the time frequency center of the signal.

Gabor saw that this approach could be used to represent sounds as contiguous areas in time and frequency, except for one thing: time and frequency are not symmetrical in such a simple way. If some sound $s(t)$ is real, its Fourier transform $S(f)$ is in general complex and includes both positive and negative frequencies. This is unfortunate because we'd like to identify just one frequency, $\bar{f}$, as the mean frequency for a sound, just as we can easily identify one time, $\bar{t}$, as its mean epoch.

Gabor's solution is to convert real input signals into analytic signals (section 2.11) so that only positive frequencies remain. This allows us to represent finite signals as Heisenberg boxes (figure 10.5) because now we have a single mean frequency and a single mean time.

The Heisenberg box suggests graphically the physical limitations we experience about all frequency-based phenomena such as sound: we can't know an instantaneous frequency at an instantaneous time. Of course, we can know a local frequency at a local time, but we must choose whether we'd prefer to have better temporal resolution and poorer frequency resolution, or vice versa. This implies that *only the area of the Heisenberg box is fixed, not its shape*. In other words, for some effective duration Δt and some effective bandwidth Δf,

$A \le \Delta t\, \Delta f,$ *Gabor's Uncertainty Relation (after Heisenberg)* (10.1)

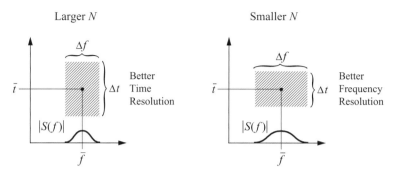

Figure 10.5
Heisenberg boxes.

where A is the area of the Heisenberg box. We can interpret equation (10.1) as saying that the area A of a Heisenberg box, which represents the quantum uncertainty between time and frequency for any signal, is constant. Therefore, any increase in Δt or Δf must be compensated by a decrease in Δf or Δt, respectively.

As we observed in chapter 3, the Fourier transform of a signal N samples long is a spectrum with N complex components—an N-to-N mapping between time and frequency. And since the ratio is always $N/N = 1$, we can say that the information diagram of the Fourier transform contains one complex datum (or two real data) per Heisenberg box, and that these boxes represent *quanta of information*. Gabor was drawing on classical quantum wave mechanics to propose a quantum view of sound energy.

10.1.3 Acoustical Quanta

Gabor (1947) researched windowing functions that could turn equation (10.1) into an equality and could therefore be used to represent elementary signals, or acoustical quanta. The essential feature of Gabor's elementary signals is to provide a local view of the time and frequency content of a signal that is optimal for the frequency and time extent covered. He proposed using a Gaussian window because it possesses a compact combination of effective duration and width, thereby minimizing the uncertainty principle inequality. This would "assure the best utilization of the information area." Gabor's elementary signal is thus a phasor with a Gaussian envelope:

$$s(t) = e^{-\alpha^2(t-t_0)^2} e^{i2\pi ft}, \qquad \text{\textit{Gabor's Elementary Signal}} \ (10.2)$$

where α is a real constant determining the width of the Gaussian window. Elementary signals are thus harmonic vibrations of some frequency f, modulated by a Gaussian "probability pulse." Figure 10.6 shows an example of equation (10.2) for $\alpha = 0.5, f = 4$, and $-5 < t < 5$.

Depending upon whether we take the imaginary or real view, these elementary signals come in sinusoidal or cosinusoidal flavors; depending upon the value of α, the Gaussian envelope is either

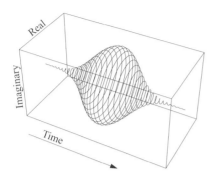

Figure 10.6
Gabor's elementary signal.

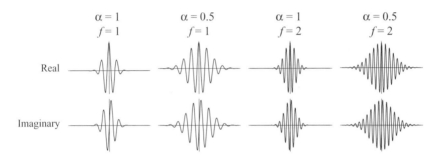

Figure 10.7
Examples of Gabor's Gaussian probability pulses.

long or short; depending upon the value of f, the frequency is high or low. Some examples are shown in figure 10.7. Since only the area of the information quantum matters, how do we assign it dimensions? Gabor suggests defining the effective duration as $\Delta t = \sqrt{\pi}/\alpha$ and the effective bandwidth as $\Delta f = \alpha/\sqrt{\pi}$. By these definitions, if $\alpha = 0$, the elementary signal becomes a simple phasor, and if $\alpha = \infty$, it becomes an impulse.

Gabor gives the Fourier transform of equation (10.2) as

$$S(f) = e^{-(\pi/\alpha)^2(f-f_0)^2} e^{i2\pi ft}. \tag{10.3}$$

Gabor (1947) notes that any arbitrary signal can be analyzed in terms of these elementary signals. Complex sounds can be thought of as a matrix, or a tiling of Heisenberg boxes, where each tile has the dimensions $\Delta t \Delta f = (\sqrt{\pi}/\alpha)(\alpha/\sqrt{\pi}) = 1$. Each tile corresponds to one quantum of sound. We can also interpret them as cells in a two-dimensional array. He states, "This method of analysis contains 'time language' and 'frequency language' as special extreme cases. If the cells are infinite in the time direction we obtain Fourier analysis; if they are infinite in the frequency direction we obtain an expansion into delta functions, that is to say, the function $s(t)$ itself."

He concludes that, "the 'quantum of sound' is a concept of considerable physiological significance." This has certainly been borne out in practice, as his work constitutes the theoretical underpinnings of a number of major revolutions in audio communications, including such popular technologies as MPEG audio and MP3, and related psychoacoustic sound coding techniques.

10.2   Short-Time Fourier Transform

Gabor (1946) suggested adapting Fourier windowing techniques (section 3.6) to create a uniform representation of time and frequency. If windowing a signal can be likened to taking a snapshot of it, then taking successive windowed snapshots can be likened to taking a movie of the signal. The snapshots, or *analysis frames,* capture successive time periods of the signal, and the Fourier transform of these analysis frames can reveal the spectral evolution of the signal.

Following Gabor's lead, we can extend the Fourier transform in order to localize a particular analysis frame in time as well as frequency. Let $X_k(n)$ be the discrete *short-time Fourier transform* (STFT) of a discrete signal $x(n)$, where n indexes sample time and k indexes frequency. If we arrange the values of $X_k(n)$ in a matrix with rows n and columns k, we see that each column corresponds to a single Fourier transform of length N positioned at sample n, and each cell corresponds to one of Gabor's quanta of information.

Discrete
frequency

$X_N(-\infty)$	$\ldots$	$X_N(-1)$	$X_N(0)$	$X_N(1)$	$\ldots$	$X_N(\infty)$
$\vdots$	$X_k(-n)$	$\vdots$	$\vdots$	$\vdots$	$X_k(n)$	$\vdots$
$X_1(-\infty)$	$\ddots$	$X_1(-1)$	$X_1(0)$	$X_1(1)$	$\ddots$	$X_1(\infty)$
$X_0(-\infty)$	$\ldots$	$X_0(-1)$	$X_0(0)$	$X_0(1)$	$\ldots$	$X_0(\infty)$

Discrete time

We obtain seemingly very different interpretations of the STFT depending upon how we traverse this matrix.

10.2.1   Windowed Fourier Transform View

If we leave n fixed at some particular time and let frequency k vary over its entire range, we move up and down within a single column. Each column represents the spectrum of a region of the input signal that is local to time n.

10.2.2   Filterbank View

Alternatively, if we leave k fixed at some frequency and let time n vary over its entire range, we move across a row, obtaining a time-varying function describing the energy in a band of frequencies local to k.

Figure 10.8 is a plot of the magnitude spectrum of a sequence of STFTs shown from the windowed Fourier transform perspective. Figure 10.18 is a plot of the same signal from the filterbank

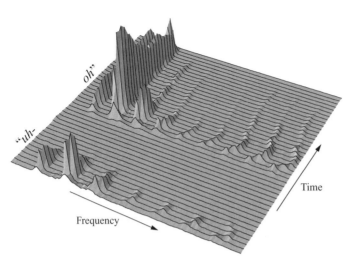

Figure 10.8
Waterfall diagram—Fourier transform view of a speech signal.

perspective. The lines in figure 10.8 correspond to the columns of the matrix, and the lines in figure 10.18 correspond to its rows. It is evident that both perspectives reveal the same underlying shape, so we see that the STFT has both "time language" and "frequency language" (in Gabor's words), thereby unifying time and frequency in one representation.

Figures 10.8 and 10.18 are called *waterfall diagrams* because of their appearance. They both show a speaker saying "uh-oh." The even spacing of the components in frequency reveals the basic harmonic structure of the human voice. We can easily observe how the frequencies and amplitudes of the harmonics change through time. Note that the pitch of the voice drops during the "oh," and the amplitude trails off, as is common for English speakers of this phrase.

Because the STFT is based on the Fourier transform, the spectral characteristics of the signal must be slowly varying with respect to the size of the analysis window so that the requirements of the Fourier transform about stationarity and periodicity of signals are not too badly violated. By suitable parameter choices, this condition can be met for most signals.

10.2.3 Discrete Short-Time Fourier Transform

The formula for the discrete short-time Fourier transform of a signal $x(n)$ is

$$X_k(n) = \sum_{r=-\infty}^{\infty} x(r)h(n-r)e^{-i2\pi kr/N}, \qquad \textit{Discrete Short-Time Fourier Transform (STFT)} \quad (10.4)$$

where n indexes discrete time for all $-\infty < n < \infty$, k indexes discrete frequency over the range $0 \le k < N$, and N is the length of the discrete Fourier transform in n, and by the principle of quantum uncertainty, N is also the number of frequency bands into which the spectrum is divided (see section 10.1.1). Input function $x(n)$ is a real-valued discrete-time signal defined for all integers n.

Equation (10.4) is nearly the same as the windowed DFT (compare this to equation 3.33) but with one extra trick: the ability of the windowing function $h(n)$ to be repositioned over different regions of the input signal by varying n.

What Is $h(n)$? The function $h(n)$ is a real-valued discrete-time analysis window defined for all integers n. In practice, it is usually equal to zero outside of some central range $h_1 \leq n \leq h_2$ so that its effective length is $N_h = h_2 - h_1 + 1$. It is generally designed to be narrow in time, or frequency, or both, and is normalized so that $h(0) = 1$. Bear in mind that N_h, the extent of the window function $h(n)$, is not the same as N, the extent of the Fourier transform window. I further describe $h(n)$ in the following sections and show how the size of the windowing function and the size of the Fourier transform window interact.

What Does k Index? The frequency index k does not vary from $-\infty$ to ∞, as does the time index n, because phasors such as $e^{-i2\pi kr/N}$ lie on the unit circle and are periodic in k.[2] Therefore, the DFT and STFT represent the entire spectral range of the input signal as one period of 2π radians, divided into N equally spaced frequency bands, indexed by k. Let's define the *frequency sampling interval* of the STFT as

$$\Omega = \frac{2\pi}{N} \text{ radians.}$$

The radian velocity corresponding to the kth frequency band is thus $k\Omega$ radians.

Figure 10.9 shows how the DFT and STFT carve up the spectrum on the unit circle into $N = 8$ pie slices. The points $\Omega_{0, 1, 2, \ldots}$ correspond to the successive frequency band centers, and Ω_c is the cutoff frequency that divides the successive bands.

Expressing k as radian velocity instead of frequency is convenient because it allows us to operate on spectra independent of actual sampling rate. But we can tie k to frequencies in Hertz if we have a specific sampling rate in mind. If the input signal $x(n)$ was sampled at f_s Hertz, then by the sampling

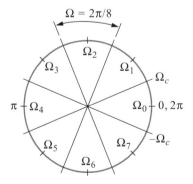

Figure 10.9
Division of the spectrum by DFT and STFT.

theorem, $f_s = 2\pi$ radians. We can determine the center of the kth frequency band f_k in Hertz by solving the following proportion for f_k:

$$k\Omega = k\frac{2\pi}{N} = \frac{f_k}{f_s}.$$

Isolating f_k yields the relation between band center frequency in Hertz and the index k:

$$k\Omega f_s = f_k \text{ Hz.} \hspace{4cm} \textit{Frequency of the kth Band} \quad (10.5)$$

10.2.4 Interpreting the STFT

We saw that the STFT can be interpreted either as a sequence of windowed Fourier transforms or as a bank of filters. These seemingly contradictory interpretations coexist in equation (10.4), rather like an optical illusion. We can shift perspectives simply by how we group the three multiplicands, $x(\cdot)$, $h(\cdot)$, and the probe phasor $e^{-i2\pi kr/N}$. Since there are three possible groupings, there are actually three different interpretations worth our consideration. Not only will these interpretations shed light on the operation of the STFT, but more important, they will help solidify the fundamental basis of signal processing mathematics presented in this book so far.

STFT as a Sequence of Windowed Fourier Transforms   It will be helpful to define the sequence

$$x(n, r) = x(r)h(n - r). \hspace{6cm} (10.6)$$

Think about equation (10.6) as a succession of frames along a line r indexed by n (Dantus 1980). For example, if we fix $n = 0$ and let r vary across its range from $-\infty$ to ∞, the signal $x(0, r)$ will be the original signal $x(n)$ multiplied by the window $h(n)$ positioned so that $h(0)$ lies over $x(0)$, as follows:

$$\ldots \quad x(-2)h(2) \quad \boxed{x(-1)h(1)} \quad x(0)h(0) \quad x(1)h(-1) \quad x(2)h(-2) \quad \ldots$$

Figure 10.10 provides a visualization for equation (10.6). Imagine two boards sliding past each other separated by a dowel. The boards are positioned at the product shown shaded in the figure above. Compare equation (10.6) to equation (4.1) to see that this is the kernel of the convolution operation.

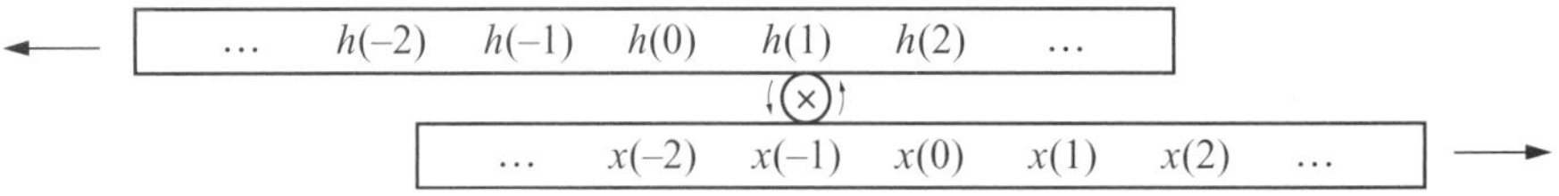

Figure 10.10
Sequence of windowed Fourier transforms.

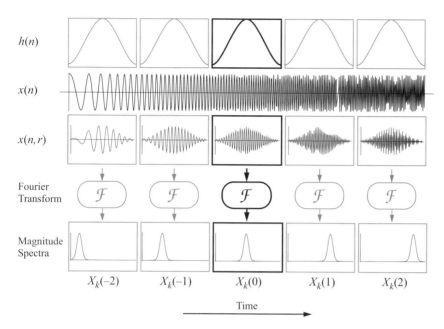

$h(n)$

$x(n)$

$x(n, r)$

Fourier
Transform

Magnitude
Spectra

$X_k(-2)$ $X_k(-1)$ $X_k(0)$ $X_k(1)$ $X_k(2)$

Time

Figure 10.11
Windowed Fourier transform interpretation of STFT.

Substituting equation (10.6) into equation (10.4), we have

$$X_k(n) = \sum_{r=-\infty}^{\infty} x(n, r)e^{-i2\pi kr/N}. \qquad \textit{Fourier Transform View of STFT} \quad (10.7)$$

Equation (10.7) appears to be a standard Fourier transform of the sequence $x(n, r)$ along the variable r with n held fixed. We can therefore interpret $X_k(n)$ as a sequence of local spectra of $x(n)$ indexed by n. We can visualize this interpretation as shown in figure 10.11. Figure 10.11 shows the signal to be analyzed $x(n)$ as a sinusoid sweeping monotically from a low frequency to a higher frequency over sampled time n. A sequence of window functions $h(n)$ selects frames of the input signal shown in the row $x(n, r)$. The Fourier transform of each $x(n, r)$ frame is taken, resulting in the sequence of magnitude spectra shown on the bottom row. As the average frequency in each frame increases, we see the energy content in each successive magnitude spectrum correspondingly rise in frequency. The sequence of magnitude spectra captures the way the frequency content of the signal changes through time, thereby providing both "time language" and "frequency language" information about the signal.

STFT as a Demodulating Lowpass Filter   It will be helpful in this section to define $x'(n, r)$ as follows:

$$x'(n, r) = x(r) \cdot e^{-i2\pi kn/N}. \qquad (10.8)$$

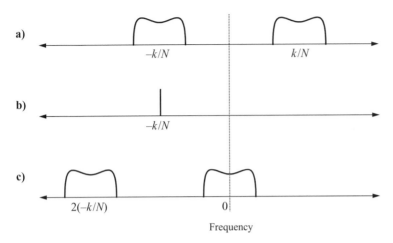

Figure 10.12
Demodulation step.

Then equation (10.4) can be read as

$$X_k(n) = \sum_{r=-\infty}^{\infty} x'(n, r)h(n - r). \tag{10.9}$$

Recalling the operation of convolution (equation 4.1), we can see that equation (10.9) is a standard convolution sum of the sequence $x'(n, r)$ along the variable r for each n held fixed. We can view the operation of equation (10.9) in two stages:

1. The first step is summarized in the definition of $x'(n, r)$ in equation (10.8). From what we learned in section 2.9, we know that multiplying $x(r)$ by the negative-frequency phasor $e^{-i2\pi kn/N}$ will shift down the frequency of all components in $x(r)$ by k/N radians. If $x(r)$ is real, its spectrum is conjugate symmetric around 0 Hz. Suppose the band center of $x(r)$ is at $\pm k/N$ radians (figure 10.12a). Multiplying it by $e^{-i2\pi kn/N}$ (figure 10.12b) shifts the spectrum so that the positive frequency band center is shifted to 0 Hz (figure 10.12c).

2. The second step performed in equation (10.9) convolves the demodulated signal created in the first step with the window function $h(n)$. Here's where we get some insight on what to do with the window function: if $h(n)$ is chosen to be the impulse response of a suitably designed narrowband lowpass filter, then the convolution step can be seen as a lowpass filtering operation that filters out the lower sideband, leaving only the baseband. This is shown in figure 10.13, where a real spectrum (10.13a) is modulated by a phasor (10.13b) to shift its frequency to a band center of 0 Hz, and then a lowpass filter (10.13c) is applied to isolate the modulated band.

In this interpretation, $X_k(n)$ is viewed as a set of demodulated and lowpass filtered signals. Figure 10.14 shows the operations performed by this interpretation of the STFT.

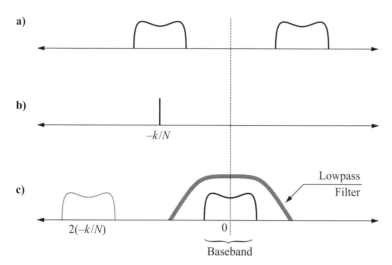

Figure 10.13
Filtering step.

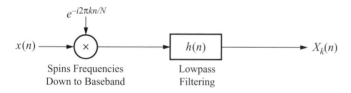

Figure 10.14
STFT by demodulation and lowpass filtering.

STFT as a Bandpass Filter Bank A third very telling interpretation of equation (10.4) reveals the STFT as a bank of bandpass filters. Here is equation (10.4) again for reference:

$$X_k(n) = \sum_{r=-\infty}^{\infty} x(r)h(n-r)e^{-i2\pi kr/N} .$$

If we substitute $r = n - r$ into (10.4), it becomes

$$X_k(n) = \sum_{r=-\infty}^{\infty} x(n-r)h(r)e^{-i2\pi k(n-r)/N} .$$

Expanding by the rule $e^{a-b} = e^a e^{-b}$,

$$X_k(n) = \sum_{r=-\infty}^{\infty} x(n-r)h(r)e^{-i2\pi kn/N} e^{i2\pi kr/N} .$$

Since $e^{-i2\pi kn/N}$ is constant for each term of the summation, we can factor it out:

$$X_k(n) = e^{-i2\pi kn/N} \sum_{r=-\infty}^{\infty} x(n-r)h(-r)e^{i2\pi kr/N} .$$

For convenience, define $h_k(r)$ as the product of the window function $h(r)$ and the phasor:

$$h_k(r) = h(r)e^{i2\pi kr/N} . \qquad (10.10)$$

Substitute this back into equation (10.4) and swap the positions of $h(n)$ and $x(n)$ to reveal the form of a standard convolution sum:

$$X_k(n) = e^{-i2\pi kn/N} \sum_{r=-\infty}^{\infty} h_k(r)x(n-r). \qquad (10.11)$$

What does equation (10.10) mean? The real impulse response $h(r)$ is modulated in the time domain by $e^{i2\pi kr/N}$. We know that multiplying signals in the time domain convolves their spectra. Therefore, the frequency response of $h(r)$ is convolved with the spectrum of the phasor. The result is that the frequency response of the lowpass filter is shifted up by k/N radians (figure 10.15). The bandwidth and shape of the resulting bandpass filter are determined by $h(r)$. Thus this operation converts the lowpass filter into a bandpass filter.

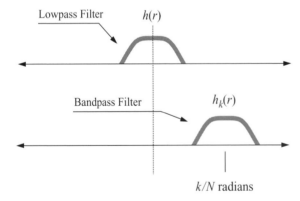

Figure 10.15
Lowpass filter shifted to become a bandpass filter.

The first thing that equation (10.11) does is to convolve the bandpass filters $h_k(n)$ with $x(n)$, thereby filtering out all but the frequencies within the kth passband. Finally, in order to create $X_k(n)$, equation (10.11) demodulates the convolution sum by the phasor $e^{-i2\pi kn/N}$, thereby shifting the filtered spectrum back to the baseband.

This $X_k(n)$ is exactly the same as in the interpretations of $X_k(n)$ in the previous two sections. But this interpretation has some advantages over the other two. To see this, we must look at just the bandpass-filtered outputs of equation (10.11). That means we want to leave out the final demodulator that creates $X_k(n)$. Therefore, let's define a new function $y_k(n)$ to be just the bandpass-filtered outputs:

$$y_k(n) = \sum_{r=-\infty}^{\infty} h_k(r)x(n-r). \qquad \textit{Bandpass-Filtered Outputs of STFT} \quad (10.12)$$

Note that equation (10.12) is the same as equation (10.11), but omits the demodulating phasor. If we fix k and vary n over the range of all input values of $x(n)$, we can interpret $h_k(n)$ as a set of bandpass filters, and $y_k(n)$ represents their distinct outputs.

Now here's the advantage: because the $y_k(n)$ directly represents all the frequency bands of $x(n)$, all we have to do to recreate the output $y(n)$ of the STFT is to directly sum all the $y_k(n)$, without needing to form $X_k(n)$. Figure 10.16 shows the bandpass filter interpretation of the STFT based on equation (10.12). So the STFT can be viewed as a bandpass filtering operation. Since filtering is a linear time-invariant process (see section 5.5), the STFT is a linear time-invariant process as well, with all the simplifying assumptions that brings.

Figure 10.17 shows a diagram of one STFT filter bank. The impulse response $h(n)$ is multiplied by the phasor $e^{i2\pi kr/N}$, convolving its spectrum into a bandpass filter with radian velocity k/N. Input signal $x(n)$ is convolved with the bandpass filter, creating the filterbank outputs $y_k(n)$.

If desired, we can create $X_k(n)$ from the bandpass-filtered outputs $y_k(n)$ simply by performing the step we left out of equation (10.12), namely, demodulating $y_k(n)$ by $e^{-i2\pi kn/N}$.

Figure 10.18 demonstrates the STFT of the "uh-oh" speech signal using the bandpass filterbank interpretation. The filterbank outputs $y_k(n)$ correspond to the lines in figure 10.18, showing variation in spectral energy through time at a particular frequency k/N radians. Compare this to figure 10.8, where the lines show variation in spectral energy through frequency at a particular time according to $X_k(n)$.

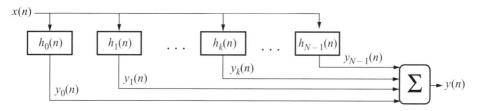

Figure 10.16
STFT as a bank of bandpass filters.

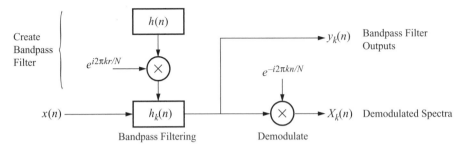

Figure 10.17
Bandpass filter interpretation of STFT.

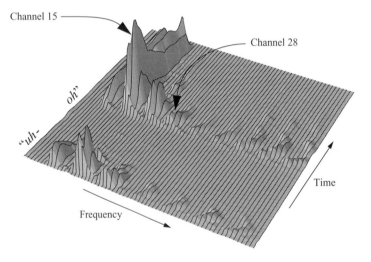

Figure 10.18
Bandpass filter interpretation of a speech signal.

Are the Filterbank and Windowed Fourier Interpretations Really the Same? We can reconcile the filterbank and windowed Fourier transform interpretations of the STFT by relating the bandpass filterbank output $y_k(n)$ to the windowed Fourier transform output $X_k(n)$ (Portnoff 1976).

Start with the definition of the filterbank output, equation (10.12).

$$y_k(n) = \sum_{r=-\infty}^{\infty} h_k(r)x(n-r).$$

Substitute $r = n - r$ and swap the positions of x and h to form a standard convolution sum.

$$y_k(n) = \sum_{r=-\infty}^{\infty} x(r)h_k(n-r).$$

Expand h_k by its definition in equation (10.10).

$$y_k(n) = \sum_{r=-\infty}^{\infty} x(r)[h(n-r)e^{i2\pi k(n-r)/N}].$$

Split the phasor in the above equation into a phasor in n and another in r. Factor out the phasor in n.

$$y_k(n) = e^{i2\pi kn/N}\left(\sum_{r=-\infty}^{\infty} x(r)h(n-r)e^{-i2\pi kr/N}\right).$$

Note that the summation is identical to equation (10.4), so

$$y_k(n) = e^{i2\pi kn/N}X_k(n).$$

Rearranging, we see that filterbank functions $y_k(n)$ are related to Fourier transform functions $X_k(n)$ by $e^{i2\pi kn/N}$:

$$\frac{y_k(n)}{X_k(n)} = e^{i2\pi kn/N}.$$

In other words, $X_k(n)$ is the demodulated equivalent of $y_k(n)$.

10.2.5 Examples

To see how this works for realistic signals, let's look at some examples.

First Example Let's set the input signal $x(n) = \sin 2\pi ft$ and set $f = k\Omega$. When analyzed, this signal will have all its energy in the center of the kth STFT frequency band. Suppose we take the STFT of this signal, resulting in outputs $X_k(n)$ and $y_k(n)$.

Varying n, examine the kth row of $X_k(n)$ (figure 10.19a). We observe that it is a constant function of time n, corresponding to the interpretation of $X_k(n)$ as a series of spectra: since the frequency of the input signal is exactly on the band center, its demodulated frequency is zero, and since its frequency doesn't vary, each $X_k(n)$ does not vary.

Still varying n, examine the kth row of $y_k(n)$ (figure 10.19b). We observe that it is a helical function of time n spinning at frequency $f = k\Omega$, corresponding to the interpretation of $y_k(n)$ as a series of bandpass-filtered output signals. Its trajectory is helical because $X_k(n)$ is complex.

Thus, for any constant k, $y_k(n)$ is just $X_k(n)$ modulated by $e^{i2\pi kn/N}$.

Second Example Suppose we set $f = k\Omega + \varepsilon$, where $\varepsilon \ll \Omega$. This means we have increased the frequency of the input signal $x(n)$ by a slight amount—much less than the width of a frequency band—so that it is not exactly on the band center frequency $k\Omega$. Taking the STFT of this modified input signal, we examine the outputs.

Varying n while examining the kth channel of $X_k(n)$, we observe that it is no longer a constant function of time (figure 10.20a). Rather, it is a helical function of n with frequency $f = \varepsilon$.

a)

b)

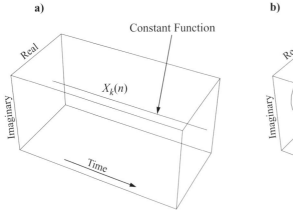
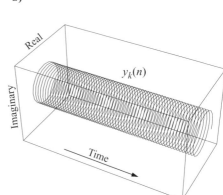

Figure 10.19
Sine wave at the kth STFT band center frequency.

a)

b)

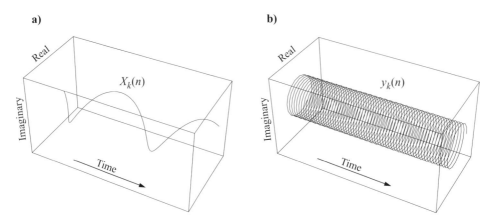

Figure 10.20
Sine wave slightly off kth STFT band center frequency.

   Still varying n but now examining the kth channel of $y_k(n)$, we observe it to also be a helical function of time n, but at frequency $f = k\Omega + \varepsilon$ (figure 10.20b).

   We see again that for any k, $y_k(n)$ is just $X_k(n)$ modulated by $e^{i2\pi kn/N}$. But this example provides another benefit: we see how the STFT represents real-world signals that are not locked on to a band center. Frequency components of the input that lie within the kth band are routed by the STFT into the kth channel of $y_k(n)$ at their original frequency. Similarly, they are routed to the kth channel of $X_k(n)$, but in this case the frequency is demodulated to the baseband.

a)   b)

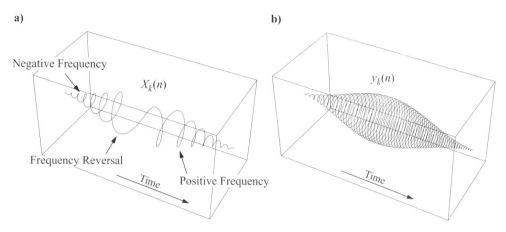

Figure 10.21
Sweeping a sine wave through a frequency band.

Third Example   This time let the input signal start at the lower band edge of the kth band $k\Omega - \Omega_c$, and then let its frequency increase gradually until it exceeds the upper band edge at radian velocity $k\Omega + \Omega_c$.

Look first at $y_k(n)$. We see in figure 10.21b that the rate of spin gradually increases through time. As the frequency approaches the band center, its magnitude increases. Its magnitude decreases as its frequency rises above the band center. Its direction of spin is the same throughout.

In figure 10.21a the amplitude contour of $X_k(n)$ is the same as $y_k(n)$, but the signal has been demodulated to the baseband. Initially, its frequency is $-\Omega_c$. As the input frequency continues to rise, the rate of spiral slows down, corresponding to the frequency's becoming less negative, until when the input frequency equals the band center frequency $k\Omega$, we see the spiraling cease momentarily. If the input frequency stopped rising at this point, we'd have a constant line, as in the first example. But it keeps rising, so as the frequency goes above the band center, $X_k(n)$ starts spinning in the opposite direction, corresponding to increasingly positive frequencies. As it passes out of the band, its magnitude decreases gradually until it disappears. As the input frequency continues to rise at the same rate, band $X_{k+1}(n)$ repeats the helical trajectory, and so on.

So, in general, we see that an input frequency causes the output of $X_k(n)$ to spin at a rate corresponding to the difference between the input frequency and the frequency of the band center. Thus frequencies below the band center are represented in $X_k(n)$ as negative frequencies (clockwise rotation); $X_k(n)$ represents frequencies above the band center as positive frequencies (counterclockwise rotation).

Fourth Example   Suppose we let $x(n) = A_1\sin(2\pi f_1 t) + A_2\sin(2\pi f_2 t)$, where f_1 and f_2 both lie in the same analysis frequency band k. There are three interesting cases:

- If $f_1 = f_2$, then $X_k(n)$ will behave as in the previous examples, with magnitude proportional to $A_1 + A_2$.

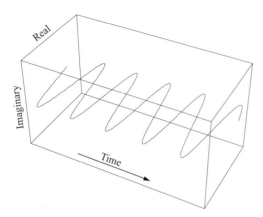

Figure 10.22
Two sine waves cancel real parts.

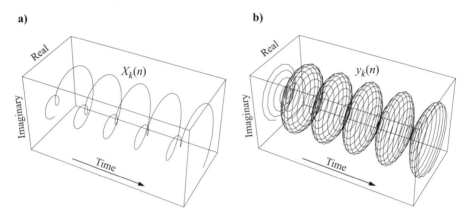

Figure 10.23
Two sine waves beat in a single STFT bandpass filter.

- If f_1 and f_2 lie an equal distance above and below the band center $k\Omega$, then (because the test signals are sinusoids) their real parts cancel within the band, and only the imaginary component of the sum shows up in $X_k(n)$ (figure 10.22). If they were cosines, the imaginary parts would cancel.

- For all other cases, the two frequencies beat against each other within the band (figure 10.23). The beat frequency equals the difference between the two input frequencies. In figure 10.23, $f_1 = k\Omega + \Omega(1/3)$ and $f_2 = k\Omega + \Omega(2/3)$. The compound helix shows the magnitude is subject to amplitude modulation.

If our aim is simply to resynthesize the original signal from $X_k(n)$, this beating is no impediment; the multiple frequencies are decoded back to the original combination by the inverse STFT. If, however, our aim is to analyze a sound to understand the behavior of individual harmonics, then we should adjust the length of the STFT analysis window to be large enough so that the width of each filter band becomes so narrow that each captures only one harmonic. To analyze a harmonic spectrum, this requires setting N large enough that so that f_s/N is less than the lowest frequency to be analyzed.

10.2.6 Inverse STFT

If we have used an appropriate windowing function $h(n)$ during analysis, we can recover the exact original sequence $x(n)$ from $X_k(n)$ by applying the inverse discrete short-time Fourier transform for each n:

$$y(n) = \frac{1}{N}\sum_{k=0}^{N-1} X_k(n)e^{i2\pi kn/N}.$$

Inverse Discrete STFT (10.13)

If we take the view of the STFT as a demodulating lowpass filter, the purpose of the modulating phasor in the inverse STFT becomes clear: each $X_k(n)$ starts out demodulated to the baseband; so, to get our original spectrum back, we must remodulate the $X_k(n)$ and then sum the result.

Figure 10.24 shows the remodulation of $X_k(n)$ back to its original frequency band. Functions $y_k(n)$ represent the remodulated spectral band centered around the kth frequency.

The final output $y(n)$ is just the sum of the remodulated spectra $y_k(n)$ for all n and k:

$$y(n) = \frac{1}{N}\sum_{k=0}^{N-1} y_k(n),$$

Provided that the analysis frame rate is sufficiently great, and that the shape of the window $h(n)$ obeys some fairly elementary requirements, the output signal $y(n)$ will be identical to the input $x(n)$. But what analysis frame rate is sufficient, and what are the requirements on $h(n)$?

Restrictions on the Window Function In order for the output of the inverse STFT to be identical to the STFT input, some requirements on $y(n)$ must be met. We've established that the STFT is a linear time-invariant system, and we know from chapter 5 that such systems are completely characterized by their impulse responses. If we interpret the STFT as a bank of bandpass filters $h_k(n)$, then it is completely characterized by the sum of its impulse responses (Portnoff 1978).

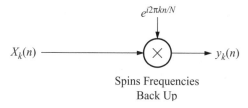

Figure 10.24
Inverse STFT by remodulation.

Letting $\tilde{h}_k(n)$ be the sum of the STFT impulse responses, we can define

$$\tilde{h}_k(n) = \sum_{k=0}^{N-1} h_k(n).$$

Recalling equation (10.10), we can rewrite this as

$$\tilde{h}_k(n) = \sum_{k=0}^{N-1} h(n)e^{i2\pi kn/N}.$$

Since $h(n)$ does not vary with k, we can factor it out of the summation, which now contains just the sum of phasors at harmonics of the analysis frequency:

$$\tilde{h}_k(n) = h(n)\sum_{k=0}^{N-1} e^{i2\pi kn/N}.$$

Like the sum of cosines, equation (9.15), this can be viewed as an impulse train function with impulses spaced every N samples.

$$\tilde{h}_k(n) = h(n)\frac{1 - e^{i2\pi n}}{1 - e^{i2\pi n/N}}.$$

In the limit as $N \to \infty$, the sequence of impulses becomes a sequence of infinitely narrow delta functions.

$$\tilde{h}_k(n) = h(n)\delta((n))_N.$$

Perhaps $\delta((n))_N$ requires a little explanation. The expression $((n))_N$ means n reduced modulo N (see appendix section A.5). The delta function $\delta(n)$ equals 1 if $n=0$ and zero otherwise. Therefore, $\delta((n))_N$ denotes an impulse train with unit impulses spaced every N samples. That means $\tilde{h}_k(n)$, the product of $h(n)$ and $\delta((n))_N$, is just $h(n)$, the lowpass filter impulse response of the STFT, sampled every N samples (see section 4.8). So we can describe $\tilde{h}_k(n)$, the STFT impulse response, as follows:

$$\tilde{h}_k(n) = \begin{cases} h(n), & n = 0, \pm N, \pm 2N, \ldots, \\ 0 & \text{otherwise.} \end{cases} \tag{10.14}$$

Now here's the problem: because the STFT convolves the input signal with this impulse response, $\tilde{h}_k(n)$ characterizes the filtering properties of the STFT itself. We'd like to prevent the STFT from introducing any spectral modifications of its own because otherwise the spectrum of the output signal could never be identical to its input. Fortunately, by clever design of the lowpass filter, we can avoid this problem: if we design $h(n)$ to be a lowpass filter such that $h(0) = 1$, and $h(n) = 0, n = \pm N, \pm 2N, \ldots$, then, by its definition, $\tilde{h}_k(n)$ in equation (10.14) will be zero everywhere except for a single unit impulse at $n = 0$. Since the convolution of any function with a single unit impulse at $n = 0$ is identically the same function (see section 4.3.1), the STFT introduces no spectral coloration if and only if the two preceding conditions are met.

Fortunately, these two conditions are easy to meet in practice. But what about the rest of the points of the function $h(n)$? What requirements do they have to meet?

Designing the Window Function We've seen that the bandpass filter interpretation of the STFT depends upon $h(n)$ being a lowpass filter. Ideally, we'd use the impulse response of the ideal lowpass filter (see appendix section A.8):

$$\text{sinc}(n) = \frac{\sin \pi n}{\pi n}. \qquad \textit{Impulse Response of the Ideal Lowpass Filter} \quad (10.15)$$

The ideal lowpass filter is sometimes called a "brick wall" filter because its frequency response is flat up to the cutoff frequency $\pm \Omega_c$, and zero outside this range (figure 10.25). The frequency range between zero and the cutoff is the passband, and the range outside these limits that is blocked by the filter is the stopband.

Modulating this ideal filter to create the appropriate bandpass filters $h_k(n)$ would allow us neatly to carve up the input spectrum into adjacent nonoverlapping frequency bands (figure 10.26).

The problem with using the ideal impulse response is that its length is infinite because it is defined over all n. Apart from the difficulties associated with implementing an infinite-length impulse response, such an exercise would be totally misguided: the whole point of the STFT is to obtain *local spectra,* which we will not get from a window of infinite time extent. So for a practical system, we must settle for a nonideal lowpass filter.

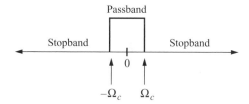

Figure 10.25
Ideal lowpass filter.

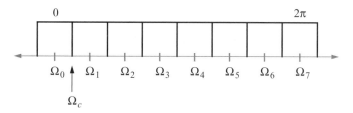

Figure 10.26
Ideal bandpass filter.

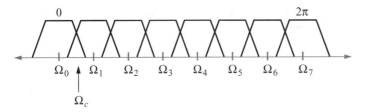

Figure 10.27
Nonideal bandpass filter.

The primary nonideal characteristic of nonideal lowpass filters is that the passband is not uniformly flat, and the roll-off from passband to stopband is not abrupt. The result is that components lying in the areas between frequency bands contribute some energy to more than one band (figure 10.27).

Nonetheless, it is sufficient if the frequency responses of the bandpass filters created from this lowpass filter combine to produce a relatively uniform frequency response.

In order to meet the criteria, the impulse response of this filter must cross through 0 every N samples. This can be arranged by substituting $n = n/N$ into equation (10.15), yielding

$$\text{sinc}\left(\frac{n}{N}\right) = N\frac{\sin(\pi n/N)}{\pi n}. \tag{10.16}$$

A straightforward way to make a nonideal finite-length lowpass filter is simply to multiply the ideal lowpass filter impulse response with a window, such as a Hamming window. The N bandpass filters can then be constructed by modulating the windowed lowpass filter with phasors of the appropriate frequencies.

Sampling in Time and Frequency   When sampling continuous functions of time, we know from the Nyquist sampling theorem that the sampling rate needed to avoid aliasing is $f_s \geq 2f_{max}$, where f_{max} is the highest frequency in the function being sampled, and f_s is the sampling rate. Since the STFT is defined for both discrete time and discrete frequency, we must be careful to avoid aliasing in the time domain as well as the frequency domain.[3] In practical terms, this means we must understand the interaction of the windowing function $h(n)$, its length, N_h, and the length of the Fourier transform, N.

Time Domain Requirements   Consider the frequency response of some nonideal lowpass filter $h(n)$ (figure 10.28). Its *effective bandwidth* spans the range of its main lobe around 0 Hz, bounded by the first zero of its spectrum.

If its effective bandwidth is β Hz, then by the sampling theorem, we must sample it at a rate of at least 2β Hz to avoid aliasing. For example, if $h(n)$ is a Hamming window, the effective bandwidth can be shown to be (Oppenheim and Schafer 1975)

$$\beta = \frac{4\pi}{N_h}\frac{\text{rad}}{\text{sec}} = \frac{2f_s}{N_h}\text{ Hz}. \tag{10.17}$$

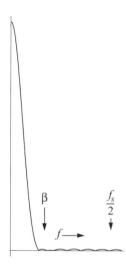

β

$\dfrac{f_s}{2}$

$f \longrightarrow$

Figure 10.28
Nonideal lowpass filter.

The positive frequency response of the Hamming window is shown in figure 10.28.

For example, if $N_h = 1024$ and $f_s = 44,100$ Hz, then the bandwidth of the Hamming window is $\beta \cong 86$ Hz. By the sampling theorem, the analysis frame rate (call it f_X) of the Hamming window must be $f_X \geq 2\beta$ or 172 Hz. Said another way, we must apply a new analysis frame every $\lambda_X = f_s/\beta$ samples. In this example, we can move the analysis window forward by at most 512 samples, skipping the window forward by at most one half the length of the window.

Frequency Domain Requirements Because of quantum uncertainty, a Fourier transform of length N samples the spectrum at $2\pi/N$ points around the unit circle. The STFT performs an N-point Fourier transform at every time n.

Quantum Sampling Rate We can combine these considerations about sampling in the time domain and frequency domain by recalling Gabor's quantum of information. Define the *quantum sampling rate f_Q* as the minimum STFT sampling rate required to sample information quanta in time and frequency without aliasing. If we set $N = N_h$, so that the length of the Fourier transform equals the length of the window function, and use a Hamming window for $h(n)$, then the minimum quantum sampling rate needed to avoid both frequency domain and time domain aliasing is

$$f_Q \geq 2\beta N \text{ quanta per second.}[4]$$

How does f_Q compare to the time domain sampling rate f_s? If we form their ratio, f_Q/f_s, we see how the data rate for $X_k(n)$ compares to the data rate for the original signal $x(n)$. For the Hamming window $\beta = 4$, so $f_Q/f_s \geq 8N/1$, meaning that STFT representation of the signal will require at least $8N$ times as much data as $x(n)$. This is an enormous explosion of data compared to the size of the original signal. Fortunately, there are ways to optimize this by a factor of almost $2N$, so the data explosion can be reduced to a factor of 4 in practice.

We can get a factor of 2 improvement right away if we limit ourselves to real input signals. Because such signals are conjugate symmetric, the negative frequencies are mirror images of the positive frequencies, and we only need to store one wing of the spectrum, cutting the data explosion in half. More economies are described in the next section.

Summary of the STFT   The STFT provides a very general representation of the input sequence $x(n)$ in terms of the parameters $X_k(n)$, allowing us to model any signal's dynamic spectral characteristics. But it causes an explosion of data, at least in the form considered here. Also, the calculation of successive STFTs taken one sample apart involves $N - 1$ of the same input samples in the computation, so there's a great deal of redundancy in the calculations from one frame to the next. The next section looks at ways to eliminate these and other inefficiencies. Along the way, we discover how the STFT can modify the time domain and frequency domain separately.

10.2.7   Toward a More Efficient STFT

As we've seen, if we hold k constant and vary n, we can view $X_k(n)$ as a demodulated and lowpass-filtered spectral band of the input signal $x(n)$. Demodulation causes components that were centered on frequency $k\Omega$ to be centered on 0 Hz. The lowpass filter then removes components outside the passband of $h(n)$, thereby limiting the bandwidth of each channel to $\pm\Omega_c$.

The limited bandwidth of $X_k(n)$ for constant k can be exploited to drastically reduce the amount of redundant computations and storage. Up to this point, we've been sampling $X_k(n)$ at the input sampling rate f_s. But since the bandpass filters carve up the input spectrum into N frequency bands, each frequency band requires just $1/N$ of the total spectrum provided by the input sampling rate. Therefore, we can lower the sampling rate for each $X_k(n)$ by up to a factor of N without significant loss of information.

Down-Sampling and Decimation   Because a continuous function is observable at every point, it can be sampled at whatever rate we like (so long as the requirements of the sampling theorem are met). And we can resample a continuous function at a new rate just by choosing other points on the function to sample corresponding to the new rate. But if we have discarded the original continuous function and have only preserved samples taken at a particular rate, then we can only lower that sampling rate by omitting samples by a process called *decimation*.[5]

We can decimate a sample sequence such as $\{0, 1, 2, 3, 4, 5, 6, 7, \ldots\}$ by applying a rule to select only some elements, such as to take only even elements, $\{0, 2, 4, 6, \ldots\}$. The result is a smaller set of data points.

Suppose we have a sequence of integers $q(n) = n$ for all integers n. Then decimating its odd values corresponds to applying the rule $n = sR$, where $R = 2$ and s is any integer. For example, if we set $s = \{1, 2, 3, \ldots\}$ and $R = 2$, we obtain the decimated sequence $q_D(n)$ as follows:

$$q_D(n) = q(sR) = \{2, 4, 6, \ldots\}.$$

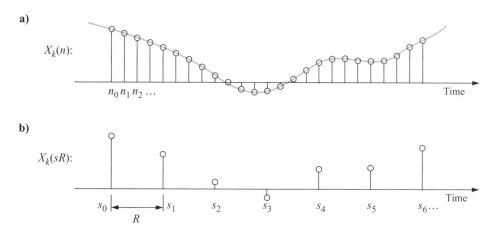

Figure 10.29
Decimation of the input signal by subsampling.

R is used to determine the skip size, and s indexes the decimated sequence. Decimation shortens the length of the sequence by a factor of R.

Decimating the STFT   We can downsample $X_k(n)$ by decimating it in n. Let R be the integer skip size between STFT frames in order to down-sample $X_k(n)$. Substituting $n = sR$ in equation (10.4) we have:

$$X_k(sR) = \sum_{r=-\infty}^{\infty} x(r)h(sR - r)e^{-i2\pi kr/N}, \qquad \textit{Decimated STFT} \quad (10.18)$$

where s denotes the index of the STFT analysis frame, and R is the analysis frame skip size.

Figure 10.29a shows the output of a hypothetical STFT bandpass filter at the input sample rate f_s. Figure 10.29b shows this signal decimated by $R = 4$.

Stride   We can conveniently express the number of samples skipped in terms of a percentage of the STFT window size, called the *stride,* defined as (R/N) percent. If the stride is 100 percent, we skip the analysis window forward by exactly the size of the analysis frame; a 50 percent stride overlaps $N/2$ samples between adjacent frames, and so on.

If the aim of the analysis is just to make a graphical display, then the stride can be chosen arbitrarily, depending upon the signal. If the signal changes rapidly, the stride should be short, possibly just a few milliseconds between frames. A high degree of overlap results in a more precise spectral display but requires more redundant computations and more data. If the signal is mostly stationary, we can generally get away with just one frame per note. If, however, the aim is to resynthesize the input signal, possibly with spectral alterations, the stride and the window function must be chosen carefully so that the combined output has as little ripple as possible due to the overlapping of successive windows. Any ripple will introduce some amplitude modulation distortion into the output.

We saw in section 10.2.6 that if $h(n)$ is a Hamming window function, then the stride must be no greater than 50 percent if we want accurate reconstruction.

Actually, decimation in this case is just a conceptual operation; what it means in practice is that we only need to compute the STFT every R samples—a substantial savings in computation. So we have reduced by decimation the number of STFTs from one every sample to one every R samples. This is fine, but how can we reconstruct the whole input signal $x(n)$ if all we have are STFTs taken every R samples? How do we get the missing information back?

Band-Limited Interpolation   There are many ways to reconstruct the missing samples from a decimated filter channel—in fact, infinitely many ways. Possibilities include simple averaging, linear interpolation, or any other method we care to make up. The reason this doesn't matter is that we threw this information away, and we don't have it any more, so unless we can come up with the original somehow, all we can do is make an educated guess as to what's missing. In this particular case, since we're working strictly with signals that are band-limited, the best educated guess will be provided by *band-limited interpolation,* which we perform in two stages.

Step 1: Expand the Decimated Sequence   Pad the sequence with zero-valued samples according to the rule

$$q_e(n) = \begin{cases} q_D(n/R), & n = 0, \pm R, \pm 2R, \ldots, \\ 0 & \text{otherwise.} \end{cases}$$

for all integers n. This creates the space needed to hold the values to be interpolated. Thus, for $R = 2$, the sequence of even integers $\{0, 2, 4, 6, 8, \ldots\}$ becomes

$$q_e(n) = \{0, 0, 2, 0, 4, 0, 6, 0, 8, \ldots\}. \tag{10.19}$$

Expanding the decimated sequence in figure 10.29b with $R = 4$ produces the sequence shown in figure 10.30.

This expanded signal is an impulse train function at the original sampling rate. Recall from section 4.8 that the spectrum of such an impulse train signal is not band-limited (in fact, it is infinitely periodic in frequency).

Step 2: Lowpass-Filter the Expanded Sequence   To extract just the baseband of the expanded sequence shown in figure 10.28, we must apply an anti-aliasing filter to remove all periodic copies

Figure 10.30
Expanded sequence from figure 10.29b.

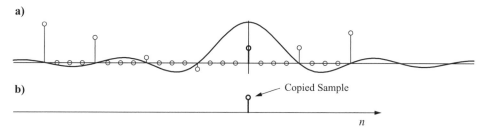

a)

b)

Copied Sample

n

Figure 10.31
Signal and sinc function aligned on an original sample.

of its spectrum (see figure 10.36). We do this by convolving the expanded sequence x with the impulse response of a lowpass filter f (Shafer and Rabiner 1973):

$$x(n) = \sum_{n=L^-}^{L^+} f(n - sR)x(sR), \qquad \qquad \textit{Band-Limited Interpolation} \quad (10.20)$$

We know that convolving a signal with a sinc function lowpass-filters it (see appendix section A.8). The sinc function has a main lobe with a value of 1.0 at location 0, and trails off in ever decreasing waves to the left and right (figure 10.31).

Let's position the peak of the main lobe of the sinc function f directly over one of the original samples of the expanded sequence x. We must also scale the sinc function along the x-axis so that all other original samples line up with where the sinc function crosses through zero (figure 10.31a). With the sinc function positioned this way, if we sum the product of the sinc function and the expanded sample series, we simply isolate that sample because the rest are either zero to begin with (the interpolated zero-valued samples) or are scaled by 0.0 because they sit on zero crossings of the sinc function. We copy out the isolated sample to tally a new function (figure 10.31b).

But that's not all: we can use the same operation to reconstruct the value of the samples we threw away during decimation. We do so by positioning the peak of the main lobe of the sinc function directly over the sample we wish to reconstruct (figure 10.32a). We multiply each sample in the expanded series by the corresponding value of the sinc function and sum the products. The result is the reconstructed value for that sample. We copy out the reconstructed sample to tally a new function (figure 10.32b). If we tally out these steps for each sample of the expanded sequence, we obtain the original function (figure 10.33). Compare this to figure 10.29a.

Thus, the shape of the sinc function provides a "recipe" for how to mix together the adjacent sample values to perfectly reconstruct each missing sample. This restores the original sampled sequence. Interpolation lengthens the sequence by a factor of R. Of course, this only works "perfectly" if we take the sum of the products over the entire extent of the sinc function, from negative to positive infinity. If we window the sinc function to step in from infinity, we lose the corrective influence of the remotest samples (if they exist), and the result may have some distortion.

a)

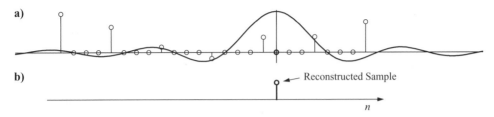

b)   ⬡ ← Reconstructed Sample

n

Figure 10.32
Signal and sinc function aligned on an interpolated sample.

n

Figure 10.33
Sum of the product of signal and sinc function.

Because the sinc function is of infinite extent, we must decide how to window it to make the computation finite. If we take too little, the interpolated function may be distorted; if we take too much, we needlessly waste computation. Regarding equation (10.20), Portnoff (1978) has shown that if $f(n)$ is a 1-to-R interpolating lowpass filter of order Q, the summation limits should be $L^+ = \lfloor n/R \rfloor + Q$ and $L^- = \lfloor n/R \rfloor - Q - 1$, where $\lfloor x \rfloor$ expresses the largest integer less than or equal to x (also called the floor function). Dantus (1980) recommends a value of $Q \approx 12$ for speech work.

Restoring an STFT Filter Channel   Let's use band-limited interpolation to reconstruct one decimated bandpass filter channel $y_k(sR)$, with band center at frequency $k\Omega$ for some band k. In order to have a picture to look at, let's plot some hypothetical output of $y_k(sR)$. This signal ordinarily is complex, but to keep it simple, let's just plot its magnitude $|y_k(sR)|$. Suppose the decimated sample magnitudes look like those in figure 10.34.

The steps are as follows:

1. Expand the channel data by padding it with zero-valued samples according to the rule:

$$y_k(n) = \begin{cases} y_k(n/R), & n = 0, \pm R, \pm 2R, \ldots, \\ 0 & \text{otherwise.} \end{cases} \tag{10.21}$$

Using the example with $R = 4$ produces figure 10.35. This expanded signal is an impulse train function at the original sampling rate. Recall from section 4.8 that the spectrum of such an impulse train signal is not band-limited (it is infinitely periodic in frequency).

Figure 10.34
Example decimated STFT channel output.

Figure 10.35
Example expanded STFT channel output.

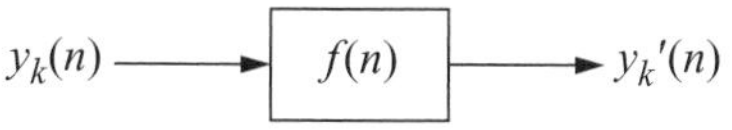

$$y_k(n) \longrightarrow \boxed{f(n)} \longrightarrow y_k'(n)$$

Figure 10.36
Anti-aliasing filter.

Figure 10.37
Reconstructed STFT filter channel.

2. If we limit the bandwidth of the impulse train function to the Nyquist frequency of the original sampling rate R, we will recover the original signal.

3. We do so by applying a lowpass anti-aliasing filter $f(n)$ (figure 10.36), which is designed to reject components above the Nyquist frequency, as described in the previous section. The result is shown in figure 10.37. Here we see again why anti-aliasing filters are also called reconstruction filters. The reconstructed output $y_k'(n)$ now characterizes the output of bandpass filter k at the original sampling rate.

The sum $y(n)$ of all reconstructed bandpass filter outputs $y_k(n)$ will equal the original input signal $x(n)$, assuming appropriate choices are made for N, $h(n)$, R, and $f(n)$, the reconstruction filter.

Equation (10.22) combines the decimation and interpolation steps.

$$\tilde{x}(n) = \sum_{s=-\infty}^{\infty} \left(\frac{1}{N} \sum_{k=-\infty}^{N-1} X_k(sR)e^{i2\pi kn/N} \right) f(n-sR), \qquad \textit{STFT Resynthesis} \quad (10.22)$$

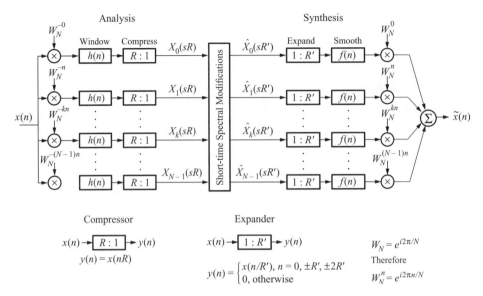

Figure 10.38
STFT analysis/synthesis diagram.

where R is the decimation factor, f is the reconstruction filter, and $\tilde{x}(n)$ is the reconstructed input signal (Crochiere 1980). For an in-depth discussion of multi-rate filter bank systems, the reader is referred to Vaidyanathan (1993).

10.2.8   Modifying STFT Resynthesis

Altogether, the analysis/synthesis STFT system can be characterized as shown in figure 10.38 (Portnoff 1978; Crochiere 1980). Separating the time compression factor R from the time expansion factor R' opens up the possibility of foreshortening or lengthening the resynthesis of $x(n)$ in time without altering its frequency content. Alternatively, we can change the frequency without changing the duration. Or, we can mix and match some of each. These details are taken up in the discussion of the phase vocoder (see section 10.3).

But even without understanding the details of what processing is required in the "short-time spectral modifications" box in figure 10.38, we can consider the effects of varying the values of R and R'. If $R = R'$, the input and output are identical: $x(n) = \tilde{x}(n)$.

If $R'/R > 1$, the output is dilated in time. If $\tilde{x}(n)$ is played back at the original sampling rate, the sound will appear to have been slowed down by a factor of R'/R in that the resulting spectrum will evolve more slowly. If instead, the sampling rate is sped up by a factor of R'/R so that the playback duration of $\tilde{x}(n)$ is the same as $x(n)$, the pitch of $\tilde{x}(n)$ will appear to have been shifted up by a factor of R'/R without affecting the duration.

If $R'/R < 1$, the output is contracted in time. If $\tilde{x}(n)$ is played back at the original sampling rate, the sound will appear to have been sped up by a factor of R/R' in that the resulting spectrum will evolve more quickly. If instead, the sample rating is slowed down by a factor of R/R' so that the playback duration of $\tilde{x}(n)$ is the same as $x(n)$, the pitch will appear to have been shifted down by R/R' without affecting the duration.

If we alter the compression/expansion ratios and the playback sampling rate dynamically, the frequency shifting and time dilation/contraction can change over time to achieve glissando and portamento effects.[6] However, we are limited to rational ratios of R/R' because both R and R' must be integers. Also, values of R and R' are limited by the constraints of the quantum sampling rate.

Unfortunately, there's one additional detail that this slightly simplistic summary omits: phase. Consider the case R'/R, where the spectrum evolves more slowly. Suppose we are time-dilating a sine tone pitched at A440. We must somehow invent extra periods (or fractional parts of periods) of the signal at this frequency in order to extend its duration in time without changing its frequency. If we wish to compress time, we must discard periods (or fractional parts of periods) from the signal. The methodology for this is taken up in the discussion of the phase vocoder (section 10.3).

Applications of the STFT   Apart from providing independent control over spectral and temporal aspects of sound, the STFT has a number of other important applications.

Noise Reduction   Suppose we analyze a harmonic tone, such as an oboe's, contaminated by a broadband noise in the background, such as tape hiss. Adopting the filterbank interpretation of the STFT, some of the filter bands will capture the analyzed harmonics plus noise while the rest of the bands that do not contain oboe harmonics capture only noise. If we suppress the outputs of the filter bands containing only noise, then the only noise remaining is that which lies within the same filter bands as the harmonics of the signal. Assuming that the intensity of each harmonic is sufficiently greater than the noise, and if the bandwidth of each filter lies within a critical band, then we can rely upon the ear to mask the remaining noise because it all lies within a critical band of each harmonic. The method is first to take the STFT of just the noise signal (for example, from the "silence" right before or after the performance) to obtain a noise threshold for each band. During analysis, if the intensity in a band is less than its threshold, the output of the band is set to zero; otherwise it is passed through. This technique depends upon the shape of the noise spectrum remaining relatively constant.

Data Reduction   If we know that a channel does not encode information that the ear can hear, we don't have to transmit or store it. Thus, we can use the rules of psychoacoustic masking to determine what spectral content in a signal can be perceived, and then only store or transmit the relevant frequency bands. This is one of the principles underlying psychoacoustic encodings such as MP3. See section 10.5.

Formant Regions and Frequency Shifting   In concluding this section, let me mention the singer and songwriter Ross Bagdasarian, the inventor and performer of the music for "Alvin and the Chipmunks." He developed a primitive but effective time scale modification system in the 1950s. Although his methods were definitely low-tech, there's still something to learn from them. To produce the chipmunks' voices, he recorded himself on a tape recorder singing *very slowly.*

When played back at normal speed, his voice sounded like a chipmunk's for the following interesting reason: our ears cue to the physical size of a singer by the formants the vocal tract superimposes on the glottal impulse function of the larynx. Our brains equate low formants with a big person and high formants with a little person. Interestingly, the effect still works, even for absurdly shifted formants, yielding the effect of a singing chipmunk.

One of the consequences of performing linear frequency transposition is that, like the voice of Alvin the Chipmunk, all components (and therefore all formants) are shifted linearly. But our ears infer the size of the sound source from the formants, and the formants are also shifted in frequency. So, to change pitch appreciably but preserve the perceived size of the sound source, the formant regions of the original signal must be determined and reapplied to the frequency-shifted signal by bandpass filtering. Linear predictive coding can be used to achieve this (see section 9.5.2).

10.3 Phase Vocoder

In the middle of figure 10.38, the box marked "short-time spectral modifications" is where we could put any additional spectrum-mangling methods we might imagine. But it's difficult to operate on the complex spectra $X_k(sR)$. It would be more convenient if these sequences of complex phase values could be converted into time functions of amplitude and frequency. This would provide an intuitive model of the signal, allowing us easily to perform whatever spectral surgery we could dream up. For example, such functions could then be used to control a bank of oscillators to synthesize sounds.

The principal embodiment of these ideas is the *phase vocoder,* a technique that specifies signals in terms of their short-time amplitude and phase spectra. Developed originally for voice telecommunications by Flanagan and Golden (1966), it has been adapted and used in a wide variety of musical applications.[7]

Picture just one information quantum in the STFT matrix shown in section 10.2. Let $z = X_k(n) = a + ib$ for some particular k and n. So z represents the complex-valued energy in the input signal localized in time by n and in frequency by k. Then $\mathrm{Re}\{z\} = a$, the real part of z, is the amount of cosine phase energy, and the imaginary part $\mathrm{Im}\{z\} = b$ is the amount of sine phase energy.

Through the Pythagorean theorem, we can interpret z in polar form as a magnitude $|z|$ at a particular phase angle $\angle z$.

If we hold k fixed, then $|X_k(n)|$ is a list of vector lengths in one frequency band at each time n, and $\angle X_k(n)$ is a list of vector directions. The vector direction is determined by how the cosine (a) and sine (ib) contributions happen to line up in each time/frequency quantum. We can use the magnitude information to control the amplitude envelopes of a set of oscillators through time and the phase information to fine-tune their frequencies.

For convenience in the following discussion, I assume that each frequency band $X_k(n)$ contains at most one frequency component. (This needn't be the case in practice.)

10.3.1 Calculating Amplitude

As we have seen, each STFT bandpass filter output can be represented as a complex spiral in time. The amplitude of each component is encoded in the time-varying magnitude of the spiral, which we can extract using the Pythagorean theorem:

$$|X_k(n)| \;=\; \sqrt{\mathrm{Re}\{X_k(n)\}^2 + \mathrm{Im}\{X_k(n)\}^2}\,. \qquad \textit{Time-Varying Amplitude} \quad (10.23)$$

The sequence $|X_k(n)|$ with k held constant is a sequence of amplitude estimates local to n.

10.3.2 Calculating Frequency

Since the frequency values in $X_k(n)$ are all demodulated to the baseband, the frequency of each component must be assembled by summing a fine and a coarse frequency parameter for each frequency band. The coarse parameter is simply the fixed band center frequency $f_k = k\Omega$ that modulates frequencies in the band back to the correct range.

The fine frequency parameter is based on the time-varying phase angles of $X_k(n)$ and is used to provide the exact frequency and phase within the band. Recall figure 10.21a, which shows how the STFT encodes frequencies above and below the band center frequency. If the analysis frequency equals the band center, then $X_k(n)$ is a constant function (figure 10.19a). We can decode the rotational position of the spiral at each time n by using the arctangent function, as follows. With $a_k(n) = \mathrm{Re}\{X_k(n)\}$, and $b_k(n) = \mathrm{Im}\{X_k(n)\}$, the instantaneous phase angle at time n is

$$\phi_k^{\mathrm{pv}}(n) \;=\; \angle X_k(n) \;=\; \tan^{-1}\frac{b_k(n)}{a_k(n)}\,. \qquad\qquad (10.24)$$

The "pv" in (10.24) reminds us that the arctangent function decodes angular displacement only in its *principal value* range: $\pm\pi$.

The frequency of the spiral corresponds to how rapidly its phase angle changes with time. By the definition of the derivative (section 6.1), we know that frequency is the derivative of instantaneous phase. Ideally, we'd compute the fine frequency parameter directly from the time derivative of the phase angle:

$$\dot{\phi}_k(n) \;=\; \frac{d}{dt}\phi_k(n) \;=\; \frac{a_k(n)\dot{b}_k(n) - b_k(n)\dot{a}_k(n)}{a_k^2(n) + b_k^2(n)}\,.$$

However, this approach requires a great deal of extra computation and introduces difficulties in case the signal or its first derivative has discontinuities (Moorer 1978; Gordon and Strawn 1985).

First Backward Difference Method   Dolson (1983) suggests using the first backward difference between successive phase values:

$$\Delta\phi_k(n) \;=\; \phi_k^{\mathrm{pv}}(n) - \phi_k^{\mathrm{pv}}(n-1). \qquad \textit{First-Order Phase Difference} \quad (10.25)$$

If we are dealing with a sequence decimated by D, then the Dth-order phase difference is

$$\Delta\phi_k(n) = \phi_k^{\text{pv}}(n) - \phi_k^{\text{pv}}(n-D), \tag{10.26}$$

where D represents the phase difference in radians per D samples. Thus $\Delta\phi_k(n)$ measures the frequency of the spiral encoded in each frequency band by measuring the angular displacement between successive sample points.

Unwrapping the Phase There's a small difficulty with using $\Delta\phi_k(n)$ to measure frequency: as noted, the arctangent function that defines it (equation 10.24) is limited to values in its principal value range, between $-\pi$ and π. But, as is evident from figure 10.21a, the phase angle for the spiral in a frequency band can increase or decrease without bound. The arctangent function will incorrectly wrap all phase angles that exceed the interval $\pm\pi$ back into the principal value range. This in turn will lead to errors in the first backward difference measurement for $\Delta\phi_k(n)$.

For example, suppose that we have successive phase angles x, y, z as marked in the circle in figure 10.39, separated by angular displacement σ. The grouping of these sample points indicates a counterclockwise motion $x \to y \to z$ along arcs A and B.

The frequency estimate, based on the radian velocity of arc A is

$$\Delta\phi_k(y) = \angle X_k(y) - \angle X_k(x) = \sigma.$$

But the frequency measurement based on the radian velocity of arc B will be incorrect because it crosses over the arctangent principal value barrier. The radian velocity of arc B is wrapped to $\Delta\phi_k(z) = \angle X_k(z) - \angle X_k(y) = \sigma - 2\pi$. The frequency estimate is misestimated by 2π (shown by arc C).

We can compensate for phase wrapping at the principal value boundary by adding or subtracting factors of 2π as necessary as the phase angle overflows or underflows the principal value

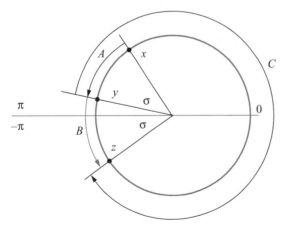

Figure 10.39
Unwrapping a phase.

range. We can relate the time-unwrapped phase of $X_k(n)$ to its principal value phase $\phi_k^{pv}(n)$ as follows:

$$\phi_k^{pv}(n) = ((\phi_k(n)))_{2\pi}, \tag{10.27}$$

where $((\dots))_{2\pi}$ means reducing modulo 2π. Equation (10.27) can be expressed in terms of an integer function $\iota_k(n)$ that denotes the integer number of jumps of 2π radians by which $\Delta\phi_k(n)$ must be adjusted. If we define $\iota_k(n)$ as

$$\iota_k(n) = \begin{cases} 1, & \Delta\phi_k(n) < -\pi, \\ 0 & \text{otherwise}, \\ -1, & \Delta\phi_k(n) > \pi, \end{cases}$$

then we can finally write an expression for the time-varying frequency:

$$\Theta_k(n) = f_k + \Delta\phi_k(n) + 2\pi\iota_k(n), \tag{10.28}$$

where f_k is the coarse frequency of the band center, $\Delta\phi_k(n)$ is the fine frequency, and $2\pi\iota_k(n)$ unwraps the phase at the boundaries of the arctangent principal value range.

To convert $\Theta_k(n)$ into frequency in Hertz, if we are operating on the decimated sequence $X_k(sR)$, we must scale all values by $1/R$ to compensate for decimation. Finally, we factor in the sampling rate: $f_s/(2\pi)$.

An Example Phase Vocoder Analysis   Figure 10.40 shows two phase vocoder channels derived from channels 15 and 28 of the "uh-oh" speech signal shown in figure 10.18. Figures 10.40a and 10.40d show the local amplitude in the channels, 10.40b and 10.40e show the local frequency (in radians, normalized to the center frequency of each band), and 10.40c and 10.40f show the complex spirals from which these analyses are made. The magnitude of the spiral encodes the amplitude curve, and the instantaneous frequency encodes the phase rate of change within its frequency band.

Notice that the two channels only have significant energy during the latter part (the "oh") of the utterance. The "uh" part at the beginning is encoded in other channels because it is pitched higher. Since there is no energy in these channels toward the beginning, the local frequency analysis is basically tracking noise during this time, and as a consequence we observe wild gyrations of the frequency estimates during the first half of figures 10.40b and 10.40e. Once there is significant energy in the signal, the frequency estimates settle down. We can clearly see that the "oh" part of the phrase glides down in frequency. During resynthesis, the gyrations of the frequency functions won't be audible because their corresponding amplitudes are so close to zero.

Driving an Oscillator Bank with the STFT   We can use the amplitude and frequency parameters to drive a bank of N oscillators:

$$y_k(n) = A_k(n)\sin[2\pi f_k nT + \phi_k(n)], \qquad \textit{Oscillator Bank} \tag{10.29}$$

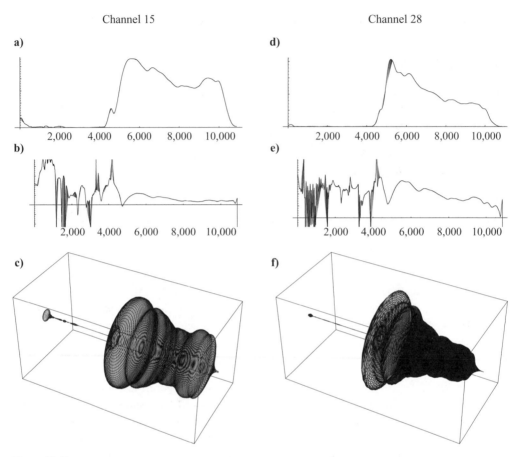

Figure 10.40
Phase vocoder channel outputs.

where $y_k(n)$ corresponds to the output of the kth oscillator, T is the sample period, and the sum of all oscillator outputs $y_k(n)$ is the final synthesized output:

$$y(n) = \sum_{k=0}^{N} y_k(n).$$

To wire this oscillator bank up to the analysis output of the STFT $X_k(n)$, we set

$$A_k(n) = |X_k(n)|$$

so that the local magnitude of STFT band k controls the amplitude of the kth oscillator through time. We set

$$f_k = \frac{k}{N}$$

so that the frequency of the kth oscillator corresponds to the frequency band center. We set

$$\phi_k(n) = \Delta\phi_k(n) + 2\pi\iota_k(n)$$

so that the local phase of STFT band k controls the phase offset of the kth oscillator through time to track its minute frequency fluctuations.

It is mathematically important to be able to resynthesize the input exactly so that the technique can be demonstrated to be a true transform. But artistically the identity transform is useless: what's the point of doing all this if all we get back is the same signal we put in? Interesting artistic applications consist of modifications to the sound that otherwise would be impossible. Consider these examples:

- By scaling the amplitude functions $A_k(n)$ either collectively or individually, we can arbitrarily transform the timbre of the sound.

- By scaling the frequency functions $\Theta_k(n)$ collectively, we can arbitrarily modify the pitch.

- By scaling the frequency functions $\Theta_k(n)$ individually, we can transpose the harmonics arbitrarily to create inharmonic spectra from harmonic spectra, and vice versa, among other possibilities.

- By controlling the rate at which we update the oscillators from these functions, we can arbitrarily dilate or contract time.

Timbre Interpolation, Timbre Morphing   Many exotic modifications beyond just time dilation and frequency shifting are available when the data are in the form of amplitude and frequency functions.

Interpolating between Instrumental Timbres   Suppose we have two instrument tones that we have analyzed with the phase vocoder into two sets of amplitude and frequency functions, for example, a bassoon tone $B\{A_k(n), \Theta_k(n)\}$ and a saxophone tone $S\{A_k(n), \Theta_k(n)\}$. We can create hybrid sounds based on these timbres by driving the oscillator bank in equation (10.29) by a weighted interpolation of these functions, where a weighting factor f determines the contribution of each timbre to the result. To linearly interpolate between two values, we can use the formula

$$\Upsilon(f, a, b) = f \cdot (a - b) + b, \qquad\qquad\qquad \textit{Linear Interpolation} \quad (10.30)$$

where f is a weighting factor that must be in the range $0 \leq f < 1$. As we adjust parameter f from 0 to 1, we want the timbre generated by the oscillator bank to go from pure bassoon to pure saxophone, and intermediate values should produce a hybrid timbre. Suppose that $f = 0.5$. Then if we drive the oscillator bank with

$$x_k(n) = \Upsilon(0.5, B\{A_k(n)\}, S\{A_k(n)\}) \sin[\Upsilon(0.5, B\{\Theta_k(n)\}, S\{\Theta_k(n)\})],$$

the resulting hybrid tone will sound like a cross between the two timbres.

Timbral interpolation is a big technological breakthrough. If we make an analogy between the discrete instrumental timbres of the orchestra and the integers, then timbral interpolation is the musical equivalent to the discovery of real numbers.

Also, gradually changing *f* through time constitutes a kind of timbral morphing, similar to figurative morphing in the visual arts. The degree of timbral plasticity that this approach offers is a new capability that is available to musicians for the first time in musical history.

Optimizing Intelligibility of Speech   We can use timbral dilation to slow the rate of speech, for instance, to study the accent pattern of a foreign tongue at a slower speed. But if we naively slow a speaker's voice down uniformly, he sounds drunk. The reason is that a drunk person tends to slow both vowels and consonants uniformly, whereas a nondrunk person speaks more slowly by lengthening just the vowels.

Since consonants are spectrally shaped noise signals and vowels have a harmonic spectrum, we might emulate the way the sober person slows his speech by evaluating the ratio of noise to harmonic content in the STFT. While this ratio is small (indicating mostly harmonic energy), we assume we are processing a vowel and dilate the signal; while the ratio is large (indicating a consonant), we assume we are processing a consonant and do not dilate.

Alternatively, we might wish to contract the speed of speech to increase the auditory information rate. A blind person could use such a system to increase the span of information she can digest through listening. Again, we must be careful with consonants, which should remain the same length. Basing the rate of contraction on consonant/vowel discrimination would help in this case as well. Practical time-warping systems using the STFT are now in widespread use in broadcasting, where it is frequently necessary to fit a recording of arbitrary length into a fixed time slot.

Implementing the STFT Using the FFT   Portnoff (1976) developed a way to use the efficient fast Fourier transform (FFT) to expedite STFT analysis. Because of how the FFT is implemented, the number of frequency bands N must be a highly composite number.[8] But the main hindrance is that equation (10.4) is not in the form of a discrete Fourier transform because of the product term $x(r)h(n-r)$, and it therefore cannot be directly computed with the FFT. Also, the limits on the summation in equation (10.4) are infinite, but in practice we are limited by the length of the window function $h(n)$. Portnoff shows how the STFT can be expressed in DFT form by *time domain aliasing* $x(r)h(n-r)$.

Here is equation (10.4) again for reference:

$$X_k(n) = \sum_{r=-\infty}^{\infty} x(r)h(n-r)e^{-i2\pi kr/N}.$$

Substituting $s = r - n$,

$$X_k(n) = \sum_{s=-\infty}^{\infty} x(n+s)h(-s)e^{-i2\pi k(n+s)/N},$$

Factoring the exponent, and extracting from the summation,

$$X_k(n) = e^{-i2\pi kn/N} \sum_{s=-\infty}^{\infty} x(n+s)h(-s)e^{-i2\pi ks/N}$$

Then the time domain aliasing. Substitute $s = Nl + m$:

$$X_k(n) = e^{-2\pi ikn/N} \sum_{l=-\infty}^{\infty} \sum_{m}^{N-1} x(n + Nl + m)h(-Nl - m)e^{-2\pi ik(Nl+m)/N}.$$

We can simplify the probe phasor by noting that $e^{-2\pi ik(Nl+m)/N} = e^{-2\pi ikm/N}$. Interchanging the orders of summation and grouping,

$$X_k(n) = e^{-2\pi ikn/N} \sum_{m=0}^{N-1} \left(\sum_{l=-\infty}^{\infty} x(n + Nl + m)h(-Nl - m) \right) e^{-2\pi ikm/N}.$$

For convenience, encapsulate the parenthesized expression into a new function $u_m(n)$:

$$u_m(n) = \sum_{l=-\infty}^{\infty} x(n + Nl + m)h(-Nl - m).$$

Express the Fourier transform of $u_m(n)$ as

$$X_k(n) = e^{-2\pi ikn/N} \sum_{m=0}^{N-1} u_m(n)e^{-2\pi ikm/N}. \tag{10.31}$$

Since equation (10.31) is the DFT of the N-point sequence $u_m(n)$, we can use the FFT to compute it, once it is formed.

Not content with this degree of efficiency, Portnoff goes on to show how the phasor calculation preceding the summation in equation (10.31) can be finessed without having to perform an actual multiplication (which can be a time-consuming operation on some computers) by relying on the shift theorem of the Fourier transform (see appendix section A.9). The *shift theorem* states in general that a circular shift in one domain corresponds to multiplication by a complex exponential in the other domain. In this case, we can achieve the effect of the multiplication by $e^{-2\pi ikn/N}$ in equation (10.31) simply by rotating $u_m(n)$ by a suitable amount before we compute its FFT. Recall that the DFT treats the N input samples as one period of an infinitely periodic waveform (see section 3.5). We can look upon the input signal $u_m(n)$ as one period of an infinitely periodic waveform of length N:

$$u_m(n) = u_{((m))_N}(n).$$

Based on this observation, define a new function $x_m(n)$ that rotates the sequence $u_{((m))_N}(n)$ by n:

$$x_m(n) = u_{((m-n))_N}(n).$$

Substituting this back into equation (10.31),

$$X_k(n) = \sum_{m=0}^{N-1} x_m(n)e^{-2\pi ikm/N}.$$

Based on this analysis, Portnoff's STFT synthesis method starts by multiplying the input signal x and the window function h, where the window function is assumed to be an even multiple of N plus 1. The resulting weighted sequence is partitioned into segments of length N. These segments are then added point by point. This constitutes the time domain aliasing. Then the sum is circularly shifted in m by n samples. Finally, the FFT is taken of the rotated sum, n is incremented to the next analysis frame, and the process is repeated until the whole signal is analyzed. The result is an efficient STFT. Corresponding techniques outlined in Portnoff's article are used to perform the inverse STFT.

Several important improvements to Portnoff's scheme have been developed. In particular, Dantus (1980) shows how Portnoff's scheme can be adapted to provide time dilation and contraction on a continuous basis rather than being limited to rational fractions of the interpolation and decimation rates.

10.3.3   Practical Considerations

Because it is based on the Fourier transform, the phase vocoder performs best when the input signal is mostly sinusoidal with slowly time-varying frequencies and amplitudes.

Frequency Precision   To use phase vocoder techniques on musical signals, it is generally desirable to adjust the number of frequency bins N so that each bin contains at most one component of the signal being analyzed. The inverse phase vocoder will resynthesize correctly if more than one component appears in the same frequency band (figure 10.23). However, multiple components cannot be manipulated independently if they appear in the same band. Similarly, though the phase vocoder can analyze and resynthesize polyphonic music (music with multiple instruments sounding concurrently), the components of separate harmonics that land in the same frequency band will be lumped together and cannot be independently manipulated.

In order to achieve a frequency resolution of, say, 12 Hz for each frequency band, if the signal was recorded at the standard rate of 44.1 kHz, then the window size of the STFT must be at least

$$N = \frac{44,100/2}{12} = 1,837.5 \text{ bands per frame.}$$

To use the FFT, this must be rounded up to the next highest power of 2 (in this case, 2048).

Temporal Content   There is a limit to how fine the frequency resolution can be made. As the analysis window size N grows, the Fourier transform averages frequency information over an increasingly large temporal span of the signal. If set too large, the window blurs together any fine temporal information that may be contained in the signal. Sharp transients are blunted if the analysis window is too wide. Thus in practice, a trade-off between temporal and frequency precision must be made, based on the spectral contents of the signal and the sampling rate.

Frequency Content   The Fourier transform is linear with frequency whereas the ear has logarithmic frequency sensitivity (see volume 1, section 6.9.6). From the ear's perspective, the Fourier transform overspecifies high frequencies and underspecifies low frequencies. With 12 Hz per bin as a case in point, the frequency margin of ±12 Hz around 1 kHz covers only a small fraction of a semitone; but at 100 Hz the frequency margin of ±12 Hz covers about a semitone. In order to achieve sufficient frequency precision, sounds with lower frequencies require longer analysis windows than higher-frequency sounds.

Phase Coherency When the phase vocoder is used to perform extreme time dilation, the results can have a kind of gurgling reverberant quality, as though the sound had been recorded underwater. This can occur because the lowpass filter $h(n)$ is not ideal, so there is some leakage of spectral components into adjacent frequency bands, where they set up beats against each other (see volume 1, section 6.7). To mitigate this problem, Puckette (1995) suggested a scheme that "locks" or entrains the phases of adjacent spectral bands containing significant energy. Laroche and Dolson (1999) suggested an improvement on this scheme that looks for "regions of influence" in a spectrum. A component establishes a region of influence if it has significantly greater energy than the surrounding frequency bands. The boundary of a region of influence can be determined to be either the midpoint between two peaks or the lowest point in the valley between them. In any event, the phases of the neighboring bins are entrained to the phases of the peak bins within each region of influence to eliminate the reverberant quality of time-dilated signals.

10.3.4 Signal Modeling and Synthesis

Classifying components within a spectrum as to their significance to the ear also allows us more easily to manipulate the spectral model of a sound for artistic purposes.

Sinusoidal Modeling Synthesis Quasi-stationary sounds can be modeled by extracting only the sinusoidal components of the signal. McAulay and Quatieri (1984) developed a speech analysis/synthesis system that searches the magnitude spectrum of the STFT frames for peaks that indicate sinusoidal components. For these peaks only, the sinusoidal parameters of amplitude, frequency and phase are extracted. The peaks are linked across successive frames, forming tracks that represent the trajectories of the sinusoidal components.

For a set of amplitude functions $A_j(n)$ and instantaneous phases $\Phi_j(n)$ (which encode frequency directly as phase), the model is

$$x(n) = \sum_{j=0}^{N} A_j(n)\cos[\Phi_j(n)]. \qquad\qquad \textit{M-Q System} \quad (10.32)$$

To resynthesize the sound, the magnitude and instantaneous phase functions are smoothed by interpolation to reduce discontinuities at frame boundaries.

By design, this system is restricted to analyzing only quasi-periodic signals that must vary slowly with time. Such a system can be used to identify tones against a background of noise, to aid in the automatic recognition and analysis of music, for example.

Spectral Modeling Synthesis   Serra (1989) extended the M-Q system to incorporate the noise component in the analysis signal. He considered sound to be made up of a deterministic component (the sinusoidal components) and a residual nondeterministic or stochastic component containing all the nonsinusoidal noise in the signal. Thus, his model is

$$x(n) = \sum_{j=0}^{N} A_j(n)\cos[\Phi_j(n)] + \varepsilon(n), \qquad \textit{Spectral Modeling Synthesis} \quad (10.33)$$

where $A_j(n)$ and $\Phi_j(n)$ are the instantaneous amplitude and phase as before, and $\varepsilon(n)$ is a function of the nonsinusoidal portion of the analyzed signal.

To extract the noise signal $\varepsilon(n)$, he first performed M-Q analysis and synthesis of the input signal. He then subtracted the M-Q synthesis signal from the original analysis signal. The remainder is by definition the residual nonsinusoidal remnant containing transients and noise.

When $\varepsilon(n)$ has been extracted, it can either be used directly or it can itself be modeled in various ways, including by fitting the spectrum of the noise signal to the shape of a filter driven by synthetic noise. By so doing, we no longer need to preserve the noise signal itself, and the analysis reduces the noise signal to a set of filter coefficients describing how the noise spectrum changes through time. Thus, in the end, we have a compact and plastic representation for analyzing and resynthesizing arbitrary signals. The sound is modeled entirely by functions of amplitude, phase, and noise spectrum. Time/pitch transformations are easy to perform. Models of different sounds can be morphed into one another by choosing between or interpolating between sets of these functions.

Since it is based on the M-Q system, it has the same restrictions and is limited to slowly time-varying quasi-stationary signals, at least for the deterministic portion of the sound.

Transient Modeling   Verma and Meng (2000) realized an improvement on spectral modeling synthesis by separating significant temporal information such as transients from the stationary background noise contained in the residual signal $\varepsilon(n)$. The analysis proceeds as before, with M-Q analysis/resynthesis, then subtracts the resynthesis from the original to generate the residual noise signal. Then the significant time-varying elements in the noise signal are identified, parameterized, and subtracted, creating a second residual signal that contains only the nontransient noise in the signal. Finally, the parameters of the nontransient noise signal are determined.

Transient detection exploits time/frequency domain symmetry: an impulse in the time domain produces a periodic function in the spectral domain, whereas a periodic function in the time domain produces an impulsive function in the frequency domain. This creates a spectral signature that can be used to discriminate between the impulsive transient signals and the background noise in $\varepsilon(n)$.

10.4   Improving on the Fourier Transform

At the beginning of this chapter, I mentioned that there were two important realizations by Gabor. The first led to the development of the STFT, and the second is that $1 \le \Delta t \Delta f$, implying that only the area of the information quantum is fixed, not its shape. In section 10.3.3 I mentioned that the

Fourier transform overspecifies high frequencies and underspecifies low ones because it is a linear time frequency representation (TFR), whereas the ear has logarithmic TFR. By adjusting the shape of the information quantum, we can use Gabor's second insight to adapt the Fourier transform to better serve the ear in this regard.

10.4.1 Wavelet Transform

We can use Gabor's quantum of sound to probe for the frequency content in a signal at any particular time in a way that always minimizes Heisenbergian uncertainty. Given the two independent parameters, α, the width of the Gaussian envelope, and f, the frequency of the phasor (equation (10.3)), we can optimize α and f so that when we are observing transient high frequencies, we set α to a large value (which narrows the Gaussian envelope in time and widens it in frequency), and when we are observing low frequencies, we set α to a small value (which widens the Gaussian envelope in time and narrows it in frequency). This way, we can adaptively shape the geometry of the Heisenberg box however we want. Within this window, we can set frequency f to select the frequency of interest. This allows us figuratively to see both the spectral forest and the temporal trees simultaneously.

These *multirate analysis* (MRA) techniques provide an adjustable time/frequency trade-off (Mallat 1989; 1998). Like a variable-resolution sonic microscope, MRA techniques allow us optimally to observe time and frequency aspects of a signal simply by adjusting the proportions of the Heisenberg box, depending upon what we want to analyze.

This flexibility allows us to represent musical signals in a very compact way. For example, if some of the time/frequency content of a signal falls below a threshold that we choose, we can discard the contents. This would be like eliminating frequency bins with low energy levels from signal spectra analyzed with the Fourier transform, but here we are able to discard time *and* frequency information, providing much greater data reduction.

Resurrecting Brahms   Variations on the same technique can also be used to remove noise from musical signals. A particularly vivid example was provided by Berger, Coifman, and Goldberg (1994). A wax cylinder recording was made in 1889 of Johannes Brahms playing the piano. A 78 rpm phonorecord of the original wax cylinder still exists. But the sound quality is horribly distorted, to the extent that it is difficult to tell even that the recording is music. Using wavelet transform techniques, they managed to tease out enough information to allow them to identify the piece Brahms played (his *First Hungarian Dance*) and even to transcribe his performance (which turned out to be an improvised variation on his published score).

Wavelet Functions   These days, Gabor's elementary signal is considered to be a kind of wavelet function where *wavelet* literally means a small wave. In general, wavelet functions are of finite duration (in mathematical jargon: they have compact support) and are oscillatory. For the same reason that many different families of windowing functions have been developed over time, many families of wavelet functions have been developed for particular applications, with names such as Daubechies, Coiflet, Haar, and Symmlet. There is a continuous wavelet transform (CWT) and a discrete wavelet transform (DWT), just as there are the continuous Fourier transform and discrete

Fourier transform. There is even a fast wavelet transform (FWT), which is faster than the fast Fourier transform (FFT). Whereas the computational complexity of the FFT is of $O(n \log_2(n))$, the FWT is only $O(n)$.

10.4.2   Analysis with Wavelets

The key to wavelets is to remember how the Fourier transform extracts frequencies (see section 3.1.7). Frequency extraction is based on the observation that when like-signed numbers are multiplied, the product is always positive. To detect a particular frequency, we multiply the input signal and a probe signal point by point, creating a product function; then all the points in the resulting product function are summed. The magnitude of the sum will be large in proportion to the correlation between the analyzed signal and the probe signal. The Fourier transform is simply a function showing how the magnitude of this sum varies with frequency. The STFT shows that the same frequency extraction technique can be applied even if the probe signal is not of infinite extent, so we can observe spectral dynamics through time.

In fact, it turns out that the probe signal need not even be a sinusoid. For example, suppose the size of an arbitrarily shaped probe signal just happens to be highly correlated with a particular time/frequency section of the input signal. Then the sum of the product of this probe signal and the input signal will still be a large value. Since such an arbitrary probe signal is local to the region of the signal to which it is applied (in the sense that it decays to zero when sufficiently far from its center), a narrow probe signal can be optimized to detect temporal features, and a wide probe signal can be used to detect spectral features. Figure 10.41 shows a plot of the wavelet spectral density for the speech signal "uh-oh" that we analyzed previously using the STFT (see figures 10.8 and 10.18).

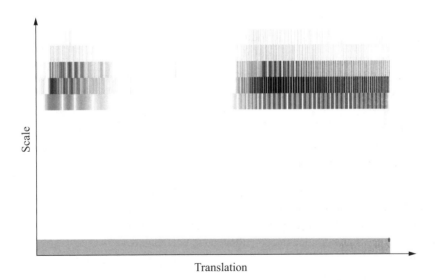

Figure 10.41
Wavelet analysis of "uh-oh".

10.4.3 MRA Transform Techniques

The *wavelet transform* is a powerful MRA technique that automatically provides appropriate time frequency resolution for all components in the signal. Although it appears to be very different from the Fourier transform, it is in fact based upon the same fundamental principle of frequency detection.

We can abstract the essential elements of the Fourier transform into a vocabulary that will allow us to compare it with MRA techniques.

- The Fourier transform *expands* the input signal onto *basis functions.*

This means that the Fourier transform breaks down the input signal $s(t)$ into a set of harmonic functions $S(f)$. It does so by multiplying the signal with sinusoidal basis functions at all frequencies under test.

- The basis functions can be constructed by *scaling* and *translating* a *prototypical function.*

With the Fourier transform, the probe phasor basis functions are all instances of a prototypical function: the complex exponential $e^{i\theta}$. This prototypical function can be adjusted in frequency (scaled) and shifted in time (translated) to analyze different aspects of the signal.

In contrast, the wavelet transform expands the input signal onto basis functions that are not sinusoidal. Whereas the Fourier transform probe signal is always a phasor, the basis functions for a wavelet transform can be constructed from one of several different prototypical functions, depending upon the application. The prototypical function is sometimes called the *mother wavelet,* meaning that all other basis functions are simply translations and scalings of this function. Does this sound familiar? The "mother idea" of the mother wavelet is Gabor's elementary signal (see section 10.1, especially figure 10.7). The Gabor wavelet is also called the Morlet wavelet after Jean Morlet. Different families of wavelets have been developed for various applications. The Gabor wavelet is useful for continuous and periodic signals. Because the wavelet is finite, we can use it to select a particular temporal region of the signal through translation. Because it is oscillatory, we can use it to select a particular spectral region of the signal through scaling. Thus translations and scalings of the mother wavelet serve to isolate temporal and spectral regions of the signal being examined.

Figure 10.42 shows examples of mother wavelets developed by Daubechies (1988). Their self-similar construction make them useful for application to fractal data. The Daubechies wavelet is used in the JPEG 2000 compression standard, a lossy method of data compression.[9]

The following example uses the Daubechies wavelet to remove broadband noise from an underlying periodic signal. Figure 10.43a shows the original signal, a sawtooth wave with three harmonics with a Hann window. Figure 10.43b shows its wavelet spectral density.

Figure 10.44a shows noise added to the sawtooth wave. Figure 10.44b is its wavelet spectral density. Even in the presence of noise equal to the amplitude of the signal, the wavelet spectral density clearly shows the presence of the underlying periodic signal.

Figure 10.45a is a thresholded version of figure 10.44b. Thresholding removes energy below a certain level. Figure 10.45b shows the inverse wavelet transform of the thresholded wavelet spectrum.

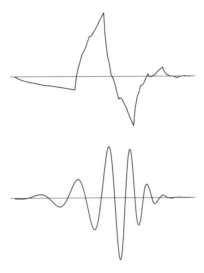

Figure 10.42
Daubechies mother wavelets.

a)   b)

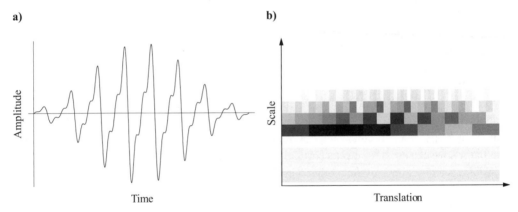

Figure 10.43
Windowed sawtooth wave and its wavelet spectral density.

a)

b)

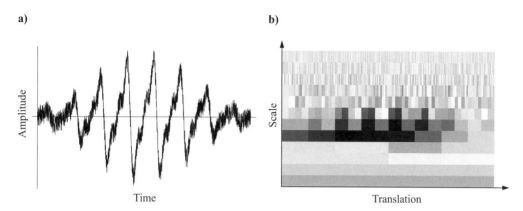

Figure 10.44
Noise added to sawtooth, and its wavelet spectral density.

a)

b)

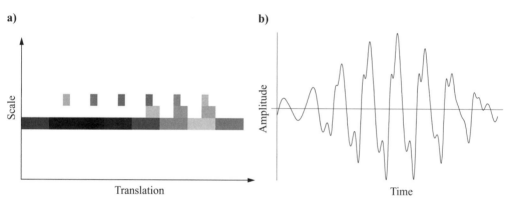

Figure 10.45
Thresholded spectral density and inverse wavelet transform.

a)   b)

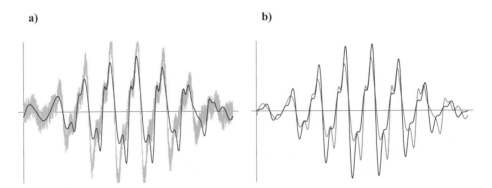

Figure 10.46
Comparison of reconstructed signals to originals.

Figure 10.46a shows a comparison of the reconstructed signal and the noisy original. Figure 10.46b compares the reconstructed signal with the original. We observe some inevitable phase distortion due to interactions between the superimposed noise signal and the thresholding function, but on the whole the reconstruction has captured many features of the original.

The wavelet transform is a generalization of the Fourier transform because it allows us to shape the Heisenberg box as needed to examine both fine temporal detail and fine frequency detail. This generality is not without cost, however. In comparison to the "point and shoot" nature of the Fourier transform, the wavelet transform is more like a photographer's studio complete with darkroom and an array of cameras and lenses for every occasion. Generality comes at the cost of increased complexity of operation.

10.5 Psychoacoustic Audio Encoding

As was discussed in section 1.11.1, the data rate requirements for uncompressed high-quality digital audio are technologically daunting. Only a little over 5 seconds of uncompressed 16-bit stereo audio can be encoded per megabyte of storage or transmission bandwidth. These requirements are a consequence of classical sampling theory discussed in chapter 1.

After the introduction of the compact disc in the early 1980s, the need soon arose for CD-quality audio at significantly lower data rates. From communications theory we know that most musical signals are highly periodic and so are highly redundant. From psychoacoustics we know that components that are masked are irrelevant to the ear. From signal processing, we know that frequency components of most music vary relatively slowly with time. These insights have been exploited by frequency domain coding techniques to dramatically reduce data rate requirements for high-quality audio storage and transmission.

In order for these techniques to become widely available, standard methods for audio encoding and decoding and a common interchange format had to be developed. The International Organization for

Standardization (ISO) and the International Electrotechnical Commission (IEC) instituted the Moving Picture Experts Group (MPEG) in the late 1980s to develop standards for video and audio to meet this need. MPEG-1 was the first international standard for compressed high-quality digital audio (ISO/IEC 11172). The ubiquitous MP3 audio file format is part of the MPEG-1 standard (MPEG-1 Layer III). The combination of MP3, the Internet, personal computers, and portable players has radically altered the distribution model of music, with dramatic impact on the music industry and our personal lives.

The MPEG coding technique exploits irrelevancy, redundancy, and frequency domain representation of signals to obtain significant reductions in audio data rate while preserving high quality. The key insight is that one can adjust the data rate used to encode particular frequency components based on their relevance to the ear. If a frequency component is masked by other components, we can decrease its allocation of coding bandwidth by reducing the number of bits used to quantize its value. Though this increases the quantization error for such components, careful attention to masking thresholds allows us to prevent the quantization noise introduced from being audible. Coding bandwidth that is saved this way can be assigned to more prominent frequency components as needed, preserving overall quality, or if not needed, then the overall data rate can be reduced.

Frequency domain coding begins by passing the signal through a bank of bandpass filters so that the signal is parsed into N different frequency bands. It is convenient to recall the filterbank view of the STFT for this step. MPEG encoding exploits the sampling rate reduction of the individual filter outputs by decimation. The signal output by each filter band is then quantized with an appropriate number of bits, allocating the fewest bits to those bands that are least audible because of masking. This way, the bit allocation of components matches their relevance to the ear. Because of the use of psychoacoustics for this step, these techniques are sometimes called psychoacoustic coders. These techniques are lossy because the variable quantization discards a significant amount of (presumed unhearable) information about the input signal during encoding. The primary difficulty with psychoacoustical codecs is that the masking curves they rely on are statistical in nature, and what worked for one listener or type of source material may not work for another. Variations in listening conditions also can affect perception of coding quality, as can variations in the implementation of the codec.

After the irrelevancies are stripped out, MP3 removes redundancies in the signal by entropy coding. This step examines the regularity of the codes used to represent the frequency components, and makes common codes shorter so that the average bit rate can be further reduced. The more often a certain code appears, the more bandwidth can be saved by substituting a shorter equivalent code for it. By minimizing redundancy this way, entropy coding can substantially reduce the overall data rate without any further loss of information. Huffman coding is a method to create substitution codes based on the probability of occurrence of each code word. It produces a variable length code where the most frequent symbols are encoded with the shortest substitutions. This method is most successful where the various input symbols are not all equally probable. A Huffman code is created by recursively assigning code bits to distinguish between the lowest probability symbols until all symbols are accounted for. The decoder must be made aware of the code substitutions made by the encoder. It can either recreate the allocation given the probabilities of the code words received, or the table of allocations can be provided to the decoder. See Bosi and Goldberg (2003).

Additional compactness can be achieved (at the cost of additional computation) through vector quantization where groups of sequential symbols are concatenated together into a new set of longer symbols before Huffman encoding is performed. This allows the Huffman coding to exploit correlations between adjacent symbols. Because the underlying audio is a continuous signal that often shows strong local correlation, this can result in substantial additional data rate savings. Entropy coding eliminates redundancy in the signal by deriving a new representation that requires fewer bits for the same information.

The result of the encoding process is a time-ordered sequence of frames each of which characterizes the spectrum of the input signal at a particular moment with all possible irrelevancy and redundancy removed. This signal is then sent to a decoder where the coded signal is dequantized and the individual frequency bands are combined to recreate the original signal. Since the signal was decimated when it was encoded, the decoder must use interpolation to restore the original sampling rate. By using bandlimited interpolation, the decoder also removes the aliased components introduced by downsampling.

The subband encoding process for MP3 is similar to the processing shown in figure 10.38. The input signal is bandpass filtered into subsampled filter output channels using a polyphase filter bank according to a method developed by Rothweiler (1983). His method allows for near-perfect reconstruction of the decoded signal. The term *polyphase* describes an efficient way of combining the bandpass filtering directly with subsampling in one filter structure. The polyphase structure is essentially the same as that shown in figure 10.38 where a single lowpass filter with cutoff $f_s/(4N)$ is modulated by N phasors to become N bandpass filters with bandwidth $f_s/(2N)$ each.

Subband coding for MP3 is defined as

$$S_k = \sum_{i=0}^{511} x_i C_i \cos \theta_{k,i} \tag{10.34}$$

where S_k is a block of 32 spectral estimates derived from 512 audio samples x_i in a FIFO buffer in *time-reversed order*. C_i is the prototype lowpass filter impulse response defined in the MP3 specification, and $\cos \theta_{k,i}$ is a cosine wave with frequency proportional to k and phase i. The reason for time reversing the input audio samples is so that equation (10.34) convolves x_i with the impulse response C_i, thereby performing filtering. The term $\theta_{k,i}$ is defined by the MPEG audio standard in paragraph C.1.3 as

$$\theta_{k,i} = (2k+1)(i-16)\frac{\pi}{64} \tag{10.35}$$

where cosine frequencies are indexed by $k = 0, 1, ..., 31$ and cosine phase is indexed by $i = 0, 1, ..., 63$.

Equation (10.34) is the brute force approach to subband coding. A great deal of attention has been paid to optimizing it to reduce the computational cost of subband coding. While the MPEG standard stipulates data formats, it does not specify a particular algorithm to use. However, it does provide a medium-complexity optimized example implementation based on the approach of Rothweiler. Note that in equation (10.34) only the term $\theta_{k,i}$ depends upon k, which determines the order of the individual subband filters. Extracting this term from the summation allows us to factor the operations in a way that takes advantage of symmetries of the cosine function, thereby reducing computation and memory requirements. The cosine is periodic in 2π radians. According to equation (10.35), the lowest frequency, when $k = 0$, is a cosine function containing four periods of a cosine wave of 128 points each in the space of the 512-sample input buffer. For every k, four cosine terms with identical value appear in 512 points: $\cos\theta_{k,0}$, $\cos\theta_{k,i+127}$, $\cos\theta_{k,i+255}$, and $\cos\theta_{k,i+381}$. Because successive integer values for k in equation (10.35) generate odd multiples of the fundamental analysis frequency, these groups of four points will always be equal for k in the range of 0 to 31. We can permute the order of operations as follows to economize processing. We gather together and sum those four product terms Z_i that are multiplied by a common cosine value, then multiply each sum of products by its common cosine term as follows:

$$S_k = \cos\theta_{k,i} \sum_{n=0}^{3} Z_{i+128n},$$

for $i = 0, 1, ..., 127$, and $k = 0, 1, ..., 31$. This optimized approach requires 1/4 of the multiply-accumulate operations and 1/4 of the memory of the brute force approach.

Other even more efficient optimized algorithms have been developed, including pseudo quadrature mirror filtering (PQMF) (Nussbaumer and Vetterli 1984) that incorporated the fast Fourier transform, and a fast subband filtering technique by Konstantinides (1994) that used a fast discrete cosine transform developed by Lee (1984). Krasner (1980) suggested that the number of channels N should be chosen to meet or exceed the number of critical bands (see volume 1 section 6.9). MPEG Level I and Level II perform QMF filtering with the number of filters $N = 32$. MPEG Level III (MP3) cascades a 32-band QMF filter with an 18-band modified discrete cosine transform (MDCT) for a total of $N = 32 \cdot 18 = 576$ channels. In terms of figure 10.38, the result of subband encoding is $X_N(sR)$, the decimated bandpass filter channels. The box in that figure labeled "Short-time Spectral Modifications" is where the psychoacoustic encoding, variable bit rate quantization, and Huffman encoding steps are performed. The decoding steps are the inverse of the encoding steps, with an interpolator restoring the original sampling rate and performing sample reconstruction. Bosi and Goldberg (2003) provide a good description of the processing steps.

The MPEG-2 standard (ISO/IEC 13818), finalized in 1994, defined a multichannel extension to MPEG-1 audio, and defined additional lower sampling rates. This was due in part to the introduction of multichannel audio systems such as Dolby AC-3 (known now as Dolby Digital) and the use of lower data rates for Internet audio. Dolby AC-3, like MPEG, defines a perceptual coding system for audio. Unlike MPEG, AC-3 was designed from the beginning as a multichannel system specialized for projection of surround digital audio in theaters. AC-3 is used in such applications as North American HDTV and DVD. Other perceptual audio codecs include ATRAC developed by Sony Corporation, MPEG-4 AAC used by Apple Computer's iPod portable music player and on-line services, Windows Media Audio (WMA) from Microsoft Corporation, and others.

One important difficulty with all of the coding schemes discussed in this section is more legal than technical. The commercial interests that developed these encoders hold basic patents on the underlying technology. This has led to the development of alternatives such as Ogg Vorbis which is an open-source free audio codec available on the Internet. Though Ogg Vorbis encoding is of very high quality and is quite popular in the open-source community, MP3 format is still entrenched as the dominant format worldwide for better or worse, and the legal position of Ogg Vorbis—or indeed any similar system—may not be clear until the expiration of the fundamental patents.

Psychoacoustic masking thresholds only scratch the surface of what can potentially be exploited to yield improved audio compression. One improvement is to incorporate source models into the coder (see section 9.8) because source models can usually encode audio much more efficiently than receiver models such as MP3. The MPEG-4 specification (ISO/IEC 14496) includes representation of source modeled audio data such as musical scores and synthesis patches, and can generate synthesized sound. MPEG-4 incorporates a synthesis language called Structured Audio Orchestra Language (SAOL) that resembles the Music N synthesis languages created by Max Mathews (1969) or the CSound synthesis language (Boulanger 1999): a score and an orchestra of instruments can be included in the MPEG data to synthesize sounds. Like the synthesis patches developed in chapter 9, instruments are dynamically networked collections of signal processing primitive operations. The signal processing required to synthesize the sounds may be implemented in hardware or software. Instruments can generate synthesized sounds from formulas (e.g., FM, physical-modeling, granular synthesis, subtractive synthesis, FOF, and so on) or can manipulate pre-stored sounds, such as wave tables. Instruments can even be controlled via MIDI, if desired.

The disadvantage of source models is that they must explicitly embody information that receiver models can take for granted, such as the acoustics of a performance space or a model of instrumental performance. But the combination of source models and receiver models in a unified framework allows one to use the strengths of each approach where appropriate.

Another potential improvement that may be available in codecs of the future is to incorporate deeper insights from psychoacoustics. For example, in volume 1 section 6.2, I mentioned the auditory scene analysis work of Albert Bregman. Our hearing can perform sound source separation which allows us, for example, to differentiate the various objects in our auditory environment, to discriminate between separate instruments performing polyphonic music, or to distinguish among the conversations at a noisy cocktail party. How our hearing performs this feat is not well understood, but were it to be determined, its exploitation could provide a considerably greater degree of audio compression. This would entail a combination of source modeling and receiver modeling. The sound field to be encoded would first be parsed into its individual salient acoustical objects each of which would be modeled individually, achieving improved realism as well as economy of representation.

Summary

The Fourier transform window size trades off temporal resolution for frequency resolution, and vice versa. While the Fourier transform provides temporal information, it is buried in the phase spectrum. We can improve on the Fourier transform to provide musical analysis tools that have both "time language" and "frequency language."

We begin with a two-dimensional representation of sound called an information diagram with time and frequency on the x-axis and y-axis, respectively. Harmonic oscillation is represented in an information diagram as a single vertical line; an impulse is represented as a single horizontal line. Nonideal signals with some effective frequency bandwidth and some effective duration occupy a region in the information diagram called a Heisenberg box. We can represent sounds as contiguous areas in time and frequency this way if we require input signals to be analytic so that only positive frequencies remain. Then the center of the Heisenberg box represents a single mean frequency and a mean time.

The Heisenberg box suggests graphically the physical limitations we experience with all frequency-based phenomena such as sound: we can't know an instantaneous frequency at an instantaneous time. Of course, we can know a local frequency at a local time, but we must choose to have better temporal resolution and poorer frequency resolution, or vice versa. This implies that only the area of the Heisenberg box is fixed, not its shape. The information diagram of the Fourier transform contains one complex datum (or two real data) per Heisenberg box, and so these boxes represent quanta of information.

A Gaussian window is used to represent the extent of the Heisenberg box because it possesses a small product of effective duration and width. An elementary signal is a phasor with a Gaussian envelope. Complex sounds can be thought of as a matrix or a tiling of Heisenberg boxes.

The short-time Fourier transform (STFT) is to the Fourier transform what a movie is to a still picture. We must extend the Fourier transform in order to localize a particular analysis frame in time as well as frequency. We obtain a matrix of Heisenberg boxes where the columns are successive Fourier transforms and the rows are particular frequency bands of those transforms. If we inspect the columns, we are taking the windowed Fourier transform view. If we inspect the rows, we are taking the filterbank view. The filterbank and windowed Fourier interpretations are mathematically equivalent.

If we have used an appropriate windowing function during analysis, we can recover the exact original sequence by applying the inverse short-time Fourier transform. For the inverse to exactly reconstruct the input signal, certain (rather easy) requirements are placed on the windowing filter.

The STFT provides a very general representation of the input sequence, allowing us to model any signal's dynamic spectral characteristics. But it causes an explosion of data. Also, there's a great deal of redundancy in the calculations. We can dramatically improve this by running each frequency channel of the STFT at its own sampling rate. If the application allows, we can also skip the analysis window by multiple samples. And, of course, we can speed things up using the FFT. We reconstruct the missing data by band-limited interpolation. Interpolating the STFT bandpass filter outputs restores the original signal.

We can also modify the STFT analysis before resynthesizing it to perform spectral modifications on the input. For example, we can convert these sequences of complex phase values into time functions of amplitude and frequency. This technique is called the phase vocoder. We can easily dilate or contract the spectral evolution of a signal without modifying the pitch, or change the pitch without modifying the spectral evolution this way. There are some practical limitations on the use of this technique, depending upon the application.

Because of psychoacoustic masking effects, we may not perceive all the spectral information a sound provides. If the energy in a frequency band lies below a masking threshold, we can remove the masked components from the sound description, thereby providing a psychoacoustic encoding of the audio signal. This is used extensively for MPEG and MP3 encoding, among other techniques.

Classifying components within a spectrum as to their significance to the ear also allows us more easily to manipulate the spectral model of a sound for artistic purposes. For example, M-Q analysis/synthesis models quasi-stationary sounds by extracting only the sinusoidal components of the signal. This can be used to identify tones against a background of noise. Subtracting the M-Q synthesis signal from the original analysis signal, we obtain a residual nonsinusoidal remnant containing transients and noise. The residual can be used directly, or it can be abstracted into a shaped-noise signal. The result is a compact and plastic representation for analyzing and resynthesizing arbitrary signals. We can further separate significant temporal information such as transients from the stationary background noise contained in the residual signal. Significant time-varying elements in the noise signal are identified, parameterized, and subtracted, creating a second residual signal that contains only the nontransient noise in the signal. Finally, the parameters of the nontransient noise signal are determined.

Gabor's original insight into concurrent time/frequency representation of signals has led to the development of an array of tools known as the wavelet transform, which provide direct control over the geometry of the Heisenberg box. Wavelet transform techniques are the logical extension of short-time Fourier techniques, insofar as they allow us to acquire both high-precision temporal and high-precision frequency content.

After the introduction of the compact disc in the early 1980s, the need arose for CD-quality audio at significantly lower data rates. From communications theory, we know that most musical signals are highly periodic, and so are highly redundant. From psychoacoustics we know that components that are masked are irrelevant to the ear. From signal processing, we know that frequency components of most music vary relatively slowly with time. One can adjust the data rate used to encode particular frequency components based on their relevance to the ear. One can further strip out redundancy using entropy coding. These insights were exploited by frequency domain coding techniques such as MPEG audio to dramatically reduce data rate requirements for high-quality audio storage and transmission. MP3 audio file format is specified in MPEG-1 Layer III. Many other frequency coding systems have been developed over time. Because they are psychoacoustically based, the performance of psychoacoustic coders is by definition subjective.

Suggested Reading

Specific Topics

Dolson, Mark. 1987. "The Phase Vocoder: A Tutorial." *CMJ: Computer Music Journal* 10(4): 14–27.

———. 1989. "Fourier-Transform-Based Timbral Manipulations." In *Current Directions in Computer Music Research,* ed. Max V. Mathews and John R. Pierce, 105–112. Cambridge, Mass.: MIT Press.

Gordon, John W., and John Strawn. 1985. "An Introduction to the Phase Vocoder." In *Digital Audio Signal Processing: An Anthology,* ed. John Strawn, 221–270. Middleton, Wisc.: A-R Editions.

Jaffe, David. 1987. "Spectrum Analysis Tutorial." *CMJ: Computer Music Journal* 11(2/3): 9–24.

Laroche, Jean. 1998. "Time and Pitch Scale Modification of Audio Signals." In *Applications of Digital Signal Processing to Audio and Acoustics,* ed. Mark Kahrs and Karlheinz Brandenburg. Norwell, Mass.: Kluwer.

Smith, Steven. 1997. *The Scientist and Engineer's Guide to Digital Signal Processing.* San Diego: California Technical Publishing.

Vaidyanathan, P. P. 1993. *Multirate systems and filter banks.* Englewood Cliffs: Prentice Hall.

Fundamentals

Dudley, Homer. 1939. "The Vocoder." *Bell Laboratories Record* 17: 122–126. Also in *IEEE Transactions on Acoustics, Speech, and Signal Processing* ASSP-29 (3): 347–351, 1981.

Moorer, James A. 1978. "The Use of the Phase Vocoder in Computer Music Applications." *Journal of the Audio Engineering Society* 26 (January/February): 42–45.

Portnoff, Michael R. 1976. "Implementation of the Digital Phase Vocoder Using the Fast Fourier Transform." *IEEE Transactions on Acoustics, Speech, and Signal Processing* ASSP-24 (3): 243–248.

———. 1978. "Time-Scale Modification of Speech Based on Short-Time Fourier Analysis." *IEEE Transactions on Acoustics, Speech, and Signal Processing* ASSP-29 (3): 374–390, 1981.

———. 1980. "Time-Frequency Representation of Digital Signals and Systems Based on Short-Time Fourier Analysis." *IEEE Transactions on Acoustics, Speech, and Signal Processing* ASSP-28 (1): 55–69.

———. 1981. "Short-Time Fourier Analysis of Sampled Speech." *IEEE Transactions on Acoustics, Speech, and Signal Processing* ASSP-29 (3): 364–373.

Press, William H., Brian P. Flannery, Saul A. Teukolsky, and William T. Vetterling. 1988. "Fast Fourier Transform." In *Numerical Recipes in C: The Art of Scientific Computing,* 496–536. Cambridge: Cambridge University Press.

MPEG Tutorials

Bosi, Marina, and Richard E. Goldberg. 2003. *Introduction to Digital Audio Coding and Standards.* Dordrecht, The Netherlands: Kluwer Academic Publishers.

Brandenberg, Karlheinz. 1999. "MP3 and AAC Explained." Presented at the AES 17th International Conference, Florence, Italy, Sept. 2–5. Available at http://www.cselt.it/mpeg/tutorials.htm.

Painter, Ted, and Andreas Spanias. 2000. "Perceptual Coding of Digital Audio." *Proc. IEEE* 88(4).

Pan, Davis. 1995. "A Tutorial on MPEG/Audio Compression." *IEEE Multimedia Magazine* 2(2): 60–74.

Epilogue

Our culture is what we have learned to our advantage about the ways of the world. There are as many methods of cultural transmission as there are kinds of knowledge.

It might be as simple as telling tales at night around a camp fire. Polynesian sailors navigated among the thousands of South Pacific islands by reckoning with symbolic maps they made from slats of bamboo. The Australian aborigines combine place and time, and link themselves to their ancestors in the dreamtime via songlines. The druids at Stonehenge encoded the grinding of the equinoctal plane against the ecliptic plane by a regular arrangement of giant stones imported from a far country. The Elysian mysteries of the ancient Greeks were imparted to novices by looking at a reflection in a wine glass. American Indians send their young out on vision quests. This is how the Promethian flame of culture is passed on among the generations. Cultural lore is not fixed but changes with every generation as new truths are discovered and old truths are forgotten. Each age tells what it knows to the next based on its memory of the old stories, tempered in the light of its unique experiences.

Cultural transmission is necessary because of the chaotic nature of our home in the universe. Everything here tends to disorder, including our hard-won distillations of truth. Only the tough kernels survive in the form of stories, songs, and equations.

What a priviledge it is to be able to stand up for a brief season and see the stars, or to be enchanted by the beauty of mathematics and music! How fortunate we are when life grants us the opportunity to let its beauty fill our souls, and to glimpse eternity without and within. We respond by pouring out our hearts in art, science, and literature. This book is one such outpouring.

But like our lives, our creations are temporary, though they may survive us for a while. What each age leaves behind is subjected to the vagaries of entropy. After a period of time, we begin to lose the glue that binds the creations to the contour of the lives that created them, and they come to resemble nothing so much as artifacts scattered around the fireplace of an ancient campground whose inhabitants are lost in the mists of time. Even as you read this sentence, the dust of forgetfulness is settling on the one before. This is the price of living in a temporal world from which we derive so much pleasure.

Gripped by the specter of our own temporality, is it any wonder that we dedicate ourselves to telling the stories, tending the flame, or singing the songs? Playing music and performing mathematics are ways we demonstrate our allegiance to the truths we have discovered, truths that we know transcend the temporal world we call home so briefly.

The stories we have of the mathematics of music from the Pythagoreans to the present resemble the artifacts of that ancient campground fireplace. We stumble upon their broken and fragmented remains and must use the glue of our own corporeal experience to make them whole again. Our own first-person experience guides us to an understanding of the insights and motivations of those who went before us.

The reward we obtain for our effort is that we can then use their eyes to see ourselves more clearly against the backdrop of nature, and to rediscover our one common humanity. Thus we recreate the myth of music in our time, the story of mathematics in our age. This book is one such telling of the great story of the mathematics of music and the music of mathematics.

Appendix

Music is the pleasure the human soul experiences from counting without being aware that it is counting.
—Gottfried Wilhelm von Leibniz

A.1 About Algebra

Algebra is the branch of mathematics that studies general properties of numbers. The name comes from the book title *kitab al-jabr w'al-muquabala,* loosely translated "book of restoration and reduction" from a text by Persian mathematician al-Khwarizmi (c. 780–c. 850), whose writings introduced the Indian number system into the West (Nelson 2003).

Algebra is about solving equations and, more deeply, about the conditions under which equations can be solved. Table 2.1 showed a sequence of simple equations, each of which required for its solution the introduction of a more encompassing numerical system, from whole numbers to signed integers to rational numbers to real numbers to irrational numbers and finally to complex numbers. Each new number system had to be developed because its predecessors were not *algebraically closed,* that is, for each number system there were always solutions that it failed to contain. The natural question when one sees a progression like this is to ask, Will it ever end? Who is to say that there aren't solutions to equations that even complex numbers can't represent? This is the question posed by the *fundamental theorem of algebra.*

It was the mighty mathematician Carl Friedrich Gauss (1777–1855) who answered this question in his Ph.D. thesis of 1799, and the answer he gave was good news. In fact, he supplied four different proofs over the course of his life. He established that every equation having complex coefficients and of degree greater than or equal to 1 has at least one complex root. Since every equation with complex coefficients has at least one complex solution, that means *the field of complex numbers is algebraically closed.* The consequence of this great discovery is that the search for a completely self-contained number system comes to an end with complex numbers.

A.1.1 Exponents

If p and q are any real numbers, a and b are positive real numbers, m and n are positive integers:

$$a^p a^q = a^{p+q}, \qquad \frac{a^p}{a^q} = a^{p-q}, \qquad (a^p)^q = a^{pq}, \qquad (a^m)^{1/n} = a^{m/n},$$

$$a^{-p} = \frac{1}{a^p}, \qquad \left(\frac{a}{b}\right)^{1/n} = \frac{a^{1/n}}{b^{1/n}}, \qquad a^0 = 1 \quad \text{if } a \neq 0, \qquad (ab)^p = a^p b^p.$$

A.1.2 Logarithms

If $a^p = x$, where a is not 0 or 1, then p is called the logarithm of x to the base a. If x, y, and a are real numbers, then by this definition and the rules given above for exponents, we can write

$$\log_a xy = \log_a x + \log_a y, \qquad \log_a \frac{x}{y} = \log_a x - \log_a y, \qquad \log_a x^y = y \log_a x.$$

The irrational number e is called the natural base of the logarithms, and $\log_e x$ is also written as $\ln x$. When written without specifying a base, log implies base e, that is, the natural logarithm, $\ln x = \log x = \log_e x$. The value of e is irrational, and its first few digits are $2.718281828\ldots$, but see section 2.4.2 for a precise way to calculate it to an arbitrary precision.

To change the base of a logarithm, use the formula $\log_a x = \log_b x / \log_b a$.

A.2 About Trigonometry

Trigonometry is the study of *trigons,* otherwise known as triangles, especially right triangles that are inscribed within a circle (figure A.1).

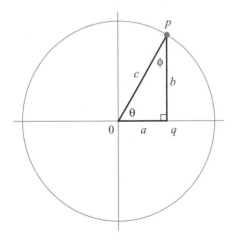

Figure A.1
Right triangle inscribed in a circle.

The ratio of the diameter to the circumference of a circle is the irrational number $\pi = 3.14 \ldots$. Because the radius is half the length of the diameter, the circumference is 2π times the radius.

Angles are commonly measured in degrees, and the angle corresponding to a complete rotation is 360°. There are 2π radians or 360° in a circle (see volume 1, section 5.2.2). An angle can be measured either clockwise or counterclockwise from a starting point. Conventionally, positive angles are measured counterclockwise from the positive horizontal (x) axis of the circle, and negative angles are measured clockwise. Thus, for example, if we assume a circle is on a wall in front of us (or more probably, on a blackboard in a geometry classroom), an angle of 0° conventionally points to the right along the positive horizontal axis, 90° points straight up, –90° points straight down, and 270° = –90°.

A.2.1 Sine Relation

Suppose we constructed a triangle like the one in figure A.1 so that the length of c (which is both the hypotenuse of the triangle and the radius of the circle) is fixed, but sides a and b are "elastic," so that they can grow and shrink. Also suppose that point p is able to move around the circle and that point q is constrained to follow it such that the angle $0qp$ is always a right angle. Last, the inner apex of the triangle is always at 0, the center of the circle. These rules basically mean that we are limited to right triangles inscribed in a circle with the triangle's base resting on the horizontal axis. As the angle θ increases, and point p moves counterclockwise around the circle, the triangle changes shape in a characteristic way (figure A.2). If we study the way in which the ratio b/c changes as the angle θ changes, we observe that this relation corresponds to sinusoidal motion. That is, the radius c, its angle to the horizontal axis θ, and the ratio b/c are connected to each other by the *sine relation:*

$$\sin \theta = \frac{b}{c}.$$ *Sine Relation* (A.1)

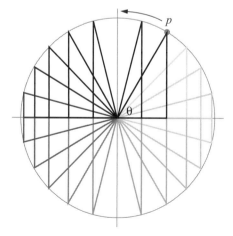

Figure A.2
Family of triangles inscribed in a circle.

Consider for example when $\theta = 0$. Then c lies along the positive horizontal axis, $a = c$ in length, and $b = 0$, since the "triangle" has no height. Hence $\sin 0 = 0/c = 0$. When $\theta = 90°$, c lies along the positive vertical axis, $b = c$, and $a = 0$. Hence $\sin 90° = 1/1 = 1$. See volume 1, section 5.4, for more details.

A.2.2   Cosine Relation

Consider the relation between sides a and c in figure A.1. The angle θ and the ratio a/c are connected to each other by the *cosine relation:*

$$\cos \theta = \frac{a}{c},$$ *Cosine Relation* (A.2)

which is similar to the sine relation except that its values are shifted by $90°$, that is, $\cos \theta = \sin(\theta + 90°)$. This makes sense, because the cosine involves the ratio a/c instead of b/c, and side a is orthogonal (that is, at a $90°$ angle) to side b; hence it precedes the sine wave by $90°$.

A.2.3   Tangent Relation

Consider the ratio b/a, the ratio of the two "elastic" sides of the triangle in figure A.1. The angle θ and the ratio of b/a are connected to each other by the *tangent relation:*

$$\tan \theta = \frac{b}{a}.$$ *Tangent Relation* (A.3)

When $\theta = 45°$, the triangle is an isosceles right triangle and $a = b$, so $\tan 45° = 1$. When $\theta = 0°$, $\tan \theta = 0/a = 0$. But when $\theta = 90°$, $\tan \theta = b/0 = \infty$.

A.2.4   Relating Tangent, Sine, and Cosine

We can define these relations given the preceding definitions, as follows:

$$\tan \theta = \frac{b}{a}, \qquad \sin \theta = \frac{b}{c}, \qquad \cos \theta = \frac{a}{c}, \qquad \frac{b}{a} = \frac{b}{c} \div \frac{a}{c}.$$

$$\therefore \tan \theta = \frac{\sin \theta}{\cos \theta}.$$ *Tangent, Sine, and Cosine Relation* (A.4)

A.2.5   Reciprocal Trigonometric Functions

We form the reciprocals for sine, cosine, and tangent by reversing the order of their ratios. Each of these reciprocals has its own name, although in my experience, they are seldom used.

$$\cot \theta = \frac{1}{\tan \theta} = \frac{a}{b},$$ *Cotangent* (A.5)

$$\sec \theta = \frac{1}{\cos \theta} = \frac{c}{a},$$ *Secant* (A.6)

$$\csc \theta = \frac{1}{\sin \theta} = \frac{c}{b}.$$ *Cosecant* (A.7)

A.2.6   Inverse Trigonometric Relations

The trigonometric functions determine the angle of the hypotenuse θ from the sides a, b, and c. But what if we know the angle θ and want to use it to find the proportions of the triangle?

The inverse trigonometric functions determine the ratio of the sides from the angle of the hypotenuse against the positive horizontal axis. For instance, the inverse of the sine function is the arcsine function, also written arcsin x, asin x, or $\sin^{-1} x$, where x is a ratio of two sides. The cosine and tangent functions are similarly named, for example, $\arctan x =$ atan $x = \tan^{-1} x$.

But how do we define these functions? At first you might think, just inscribe a triangle in a circle with angle θ, measure its sides, then find their ratio. But because we are measuring angles on a circle, there are actually many angles—infinitely many at multiples of $360°$—that correspond to any particular proportion of sides. For example, if $x = b/a = 1/1$, then clearly the triangle is an isosceles right triangle and atan $x = 45°$, but it is also true that atan $x = 45° \pm (k \cdot 360°)$, where k is an integer. So the inverse trigonometric functions are ambiguous.

But, in general, all we usually want is the angle when $k = 0$. So we define a range of *principal values* that covers just these angles. The principal values of the arctangent, arccosine, and arcsine are as follows:

$-90° < \tan^{-1} x < 90°$ arctangent

$0 \leq \cos^{-1} x \leq 180°$ arccosine

$-90° \leq \sin^{-1} x \leq 90°$ arcsine

A.3   Series and Summations

A *series* is any summation of a repeating pattern of terms. An example of an *arithmetic series* is $2 + 4 + 6 + 8 + \cdots$. Each subsequent term is computed by adding or subtracting a constant amount to the immediately preceding term.

A simple *geometric series* might be $2 + 4 + 8 + 16 + \cdots$. Each subsequent term is computed by multiplying or dividing the immediately preceding term by a constant amount.

Mathematicians have developed a useful shorthand for representing series, called *sigma notation*. Here is an example. The equation

$$s = \sum_{n=1}^{4} 5 \cdot n$$

is shorthand for the equivalent expression: $s = (5 \cdot 1) + (5 \cdot 2) + (5 \cdot 3) + (5 \cdot 4)$.

We can use it to form the sum of arithmetic and geometric series. The symbol Σ, the Greek character sigma, is used by mathematicians to represent the sum of a sequence of terms. The expression to the right of the sigma $(5 \cdot n)$ is the *summand*. The numbers below and above the sigma are the *limits of summation,* and the variable n is the *index*.

This example,

$$s(t) = \sum_{n=1}^{4} 5nt, \tag{A.8}$$

can be written equivalently as

$$s(t) = (5 \cdot 1t) + (5 \cdot 2t) + (5 \cdot 3t) + (5 \cdot 4t). \tag{A.9}$$

Equation (A.9) is called the *expansion* of equation (A.8). Every point t of the function s is described by the entire summation.

The examples above are finite series because the sequences of terms are finite. In the case of an infinite sequence of terms,

$$x_1, x_2, \ldots, x_n, \ldots$$

the correspondig infinite series is

$$x_1 + x_2 + \cdots + x_n + \cdots = \sum_{n=1}^{\infty} x_n.$$

The nth term x_n of a series is the *general term*. An infinite series is *convergent* if its value tends toward a finite sum, otherwise it is *divergent*.

A.4 Trigonometric Identities

This section provides proofs of the trigonometric identities used in this book.

Recall from Euler's formula that

$$e^{iz} = \cos z + i \sin z \qquad\qquad \textit{Euler's Formula, Positive Frequencies} \tag{A.10}$$

and that

$$e^{-iz} = \cos z - i \sin z \qquad\qquad \textit{Euler's Formula, Negative Frequencies} \tag{A.11}$$

(see section 2.5). By equations (2.52) and (2.54) we know that

$$\cos \theta = \frac{e^{i\theta} + e^{-i\theta}}{2} \qquad\qquad \textit{Euler's Cosine Identity} \tag{A.12}$$

and

$$\sin \theta = \frac{e^{i\theta} - e^{-i\theta}}{2i}.$$ *Euler's Sine Identity* (A.13)

We can derive a number of useful formulations often used in signal processing from these equations as follows.

A.4.1 Product of Cosines

We can find the product of $\cos \theta$ and $\cos \phi$ as follows:

$$\cos \theta \cos \phi = \frac{e^{i\theta} + e^{-i\theta}}{2} \cdot \frac{e^{i\phi} + e^{-i\phi}}{2}$$

$$= \frac{e^{i\theta}e^{i\phi} + e^{-i\theta}e^{-i\phi} + e^{i\theta}e^{-i\phi} + e^{-i\theta}e^{i\phi}}{4}$$

$$= \frac{e^{i\theta}e^{i\phi} + e^{-i\theta}e^{-i\phi}}{4} + \frac{e^{i\theta}e^{-i\phi} + e^{-i\theta}e^{i\phi}}{4}$$

$$= \frac{e^{i(\theta+\phi)} + e^{-i(\theta+\phi)}}{4} + \frac{e^{i(\theta-\phi)} + e^{-i(\theta-\phi)}}{4}.$$

Applying Euler's cosine identity equation (A.12) to both terms produces

$$\cos \theta \cos \phi = \frac{\cos(\theta + \phi) + \cos(\theta - \phi)}{2}.$$ *Product of Two Cosines* (A.14)

A.4.2 Product of Sines

Similarly, the product of two sines can be determined as follows:

$$\sin \theta \sin \phi = \frac{e^{i\theta} - e^{-i\theta}}{2i} \cdot \frac{e^{i\phi} - e^{-i\phi}}{2i}$$

$$= \frac{-e^{i\theta}e^{-i\phi} - e^{-i\theta}e^{i\phi} + e^{i\theta}e^{i\phi} + e^{-i\theta}e^{-i\phi}}{4i^2}$$

$$= \frac{e^{i\theta}e^{-i\phi} + e^{-i\theta}e^{i\phi} - e^{i\theta}e^{i\phi} - e^{-i\theta}e^{-i\phi}}{4}$$

$$= \frac{e^{i\theta}e^{-i\phi} + e^{-i\theta}e^{i\phi}}{4} - \frac{e^{i\theta}e^{i\phi} + e^{-i\theta}e^{-i\phi}}{4}$$

$$= \frac{e^{i\theta-\phi} + e^{-i\theta-\phi}}{4} - \frac{e^{i\theta+\phi} + e^{-i\theta+\phi}}{4}.$$

Applying Euler's cosine identity equation (A.12) to these two terms produces:

$$\sin\theta\sin\phi = \frac{\cos(\theta-\phi)-\cos(\theta+\phi)}{2}. \qquad \textit{Product of Two Sines} \quad (A.15)$$

A.4.3   Product of Sine and Cosine

We can find the product of $\sin\theta$ and $\cos\phi$ as follows:

$$\sin\theta\cos\phi = \frac{e^{i\theta}-e^{-i\theta}}{2i}\frac{e^{i\phi}+e^{-i\phi}}{2}$$

$$= \frac{e^{\theta i}e^{\phi i}-e^{-\theta i}e^{-\phi i}+e^{\theta i}e^{-\phi i}-e^{-\theta i}e^{\phi i}}{4i}$$

$$= \frac{e^{\theta i}e^{\phi i}-e^{-\theta i}e^{-\phi i}}{4i}+\frac{e^{\theta i}e^{-\phi i}-e^{-\theta i}e^{\phi i}}{4i}$$

$$= \frac{e^{i(\theta+\phi)}-e^{-i(\theta+\phi)}}{4i}+\frac{e^{i(\theta-\phi)}-e^{-i(\theta-\phi)}}{4i}.$$

Applying Euler's sine identity equation (A.13) to both terms produces

$$\sin\theta\cos\phi = \frac{\sin(\theta+\phi)+\sin(\theta-\phi)}{2}. \qquad \textit{Product of Sine and Cosine} \quad (A.16)$$

A.4.4   Equal Angles

When $\theta = \phi$, the preceding formulas reduce to the following.

When $\theta = \phi$, $\cos\theta\cos\phi = \cos^2\theta$, and by equation (A.14),

$$\cos^2\theta = \frac{\cos(\theta+\theta)+\cos(\theta-\theta)}{2} = \frac{\cos(2\theta)+1}{2}. \qquad \textit{Cosine Squared} \quad (A.17)$$

When $\theta = \phi$, $\sin\theta\sin\phi = \sin^2\theta$, and by equation (A.15),

$$\sin^2\theta = \frac{\cos(\theta-\theta)-\cos(\theta+\theta)}{2} = \frac{1-\cos 2\theta}{2}. \qquad \textit{Sine Squared} \quad (A.18)$$

When $\theta = \phi$, by equation (A.16),

$$\sin\theta\cos\theta = \frac{\sin(\theta+\theta)+\sin(\theta-\theta)}{2} = \frac{\sin 2\theta}{2}. \qquad \begin{array}{l}\textit{Sine and Cosine Product} \\ \textit{with Equal Angles}\end{array} \quad (A.19)$$

A.4.5   Sine and Cosine of a Sum

We want to find $\sin(\theta + \phi)$ and $\cos(\theta + \phi)$. Here is a method to find them both at once (Sawyer 1943). Start with equation (A.10), $e^{iz} = \cos z + i \sin z$. Now let $z = \theta + \phi$ so that

$$\cos(\theta + \phi) + i\sin(\theta + \phi) = e^{i\theta + i\phi}$$

$$= e^{i\theta}e^{i\phi}$$

$$= (\cos\theta + i\sin\theta)(\cos\phi + i\sin\phi)$$

$$= (\cos\theta\cos\phi - \sin\theta\sin\phi) + i(\sin\theta\cos\phi + \cos\theta\sin\phi).$$

Now, if two complex numbers $a + bi$ and $c + di$ are equal, then $a = c$ and $b = d$ (that is, the real and imaginary parts must be equal separately). Separating the real and imaginary quantities on each side of the equals sign into two separate equations,

$$\cos(\theta + \phi) = \cos\theta\cos\phi - \sin\theta\sin\phi,$$

$$i\sin(\theta + \phi) = i(\sin\theta\cos\phi + \cos\theta\sin\phi).$$

Notice that i appears on both sides of the second equation, allowing us to drop it (so long as we keep this equation separate from the other), finally yielding

$$\cos(\theta + \phi) = \cos\theta\cos\phi - \sin\theta\sin\phi$$

$$\sin(\theta + \phi) = \sin\theta\cos\phi + \cos\theta\sin\phi.$$

Sine and Cosine of a Sum   (A.20)

A.4.6   Sine and Cosine of a Difference

Equation (A.20) can be used to determine the sine and cosine of a difference. First, substitute $-\phi$ for ϕ, and then recall from section 2.6.12 that $\cos -x = \cos x$ and $\sin -x = -\sin x$.

$$\cos(\theta - \phi) = (\cos\theta)(\cos-\phi) - (\sin\theta)(\sin-\phi)$$

$$= (\cos\theta)(\cos\phi) - (\sin\theta)(-\sin\phi) \qquad\qquad\text{(A.21)}$$

$$= (\cos\theta)(\cos\phi) + (\sin\theta)(\sin\phi).$$

$$\sin(\theta - \phi) = (\sin\theta)(\cos-\phi) + (\cos\theta)(\sin-\phi)$$

$$= (\sin\theta)(\cos\phi) - (\cos\theta)(\sin\phi). \qquad\qquad\text{(A.22)}$$

A.4.7   Sine Squared of the Sum of Two Signals

$$\sin^2(at + bt) = \sin^2 at + 2(\sin at)(\sin bt) + \sin^2 bt$$

$$= \frac{1}{2}(1 - \cos 2at) + [\cos(a - b)t - \cos(a + b)t] + \frac{1}{2}(1 - \cos 2bt). \qquad\text{(A.23)}$$

A.4.8   Sine Squared Plus Cosine Squared

Look again at Figure A.1, which has a radius c equal to 1. By the definitions for sine and cosine given previously, we know that the length of sides a and b are determined by the cosine and sine, respectively, of the angle formed by radius and the positive x-axis, that is,

$a = \cos \theta,$

$b = \sin \theta.$

By the Pythagorean theorem we know that $c^2 = a^2 + b^2$, and because $c = 1$ for the unit circle,

$$\cos^2 \theta + \sin^2 \theta = 1. \tag{A.24}$$

Also, note that $b = \sin \theta = \cos \phi$. Therefore,

$$\cos^2 \theta + \cos^2 \phi = 1. \tag{A.25}$$

Generalizing, for a circle of radius r, we have

$$r^2 \cos^2 \theta + r^2 \sin^2 \theta = r^2. \tag{A.26}$$

A.5   Modulo Arithmetic and Congruence

If it's 1:30, and your friend says she'll meet you in 45 minutes, what time will it be when she joins you? If you answered 2:15, then you used modulo arithmetic to obtain the answer. Since there are 60 minutes in the hour, time-based calculations must keep the number of minutes in that range.

We could formalize the example this way. Using "minute arithmetic," we could write $75 \equiv ((15))_{60}$ or $75 \equiv 15 \bmod 60$. This says 75 is *congruent* to 15, modulo 60.

In general, if the difference between two integers r and b can be divided without remainder by another number m, then r and b are *congruent* modulo m. It's written as follows:

$$r \equiv ((b))_m \quad \text{if} \quad \frac{b-r}{m} \text{ is an integer.} \tag{A.27}$$

In the example, the quotient of $(75 - 15)/60$ is an integer, so 75 and 15 are congruent modulo 60.

A common use of modulo arithmetic is to obtain the remainder of integer division. The value b is called the base, and r is called the *remainder*. The FORTRAN programming language provides a way to obtain the remainder of two numbers with the function mod(b, m), whereas the C and C++ programming languages define it with a binary infix operator %. MUSIMAT, the programming language invented for this book (see volume 1, section 9.3), defines the remaindering operation as follows:

```
Integer Mod(Integer b, Integer m) {
    While (b >= m) {b = b - m; }
```

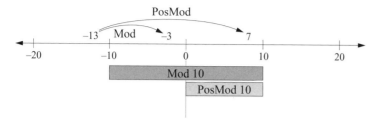

Figure A.3
Signed and unsigned modulus operation.

```
    While (b <= -m) {b = b + m;}
    Return(b);
}
```

Note that Mod() can operate on and return negative values. For example, $((-1))_{10} = -1$ and $((11))_{-10} = 1$. In general, the return value will be $-m < n < m$ for $((n))_m$. There are times when it would be convenient to force the remainder r to be a positive modulus number even if the base b is negative. For example, in MUSIMAT, the index of a List must be a positive integer. So we invent a version of Mod() that returns only the positive wing of modulo values, as follows:

```
Integer PosMod(Integer j, Integer k) {
    While (j >= k) {j = j - k;}
    While (j < 0) {j = j + k;}
    Return(j);
}
```

For example, Print(Mod(-13, 10)); prints –3, whereas Print(PosMod(-13, 10)); prints 7. Figure A.3 provides a visualization.

We see that both Mod() and PosMod() preserve the position within the modulus interval, but PosMod() also requires the value to be positive. If we have a List of ten elements numbered 0 to 9, we can provide a base of any positive or negative value b to PosMod(b, 10), and it will coerce the remainder to lie within the valid range of the List.

A.6   Finite Difference Approximations

Refer to volume 1, figure 4.2. The three slopes shown in that figure are approximations to the slope of the tangent at point B. That is, AC is the central difference, AB the backward difference and BC the forward difference. We can write the central difference approximation to the tangent slope at u_t as

$$\frac{\partial u}{\partial x} = \frac{\Delta u}{\Delta t} = \frac{u_{i+1} - u_{i-1}}{t_{i+1} - t_{i-1}}.$$ *Central Difference Approximation* (A.28)

A function $f(x)$ whose value and derivatives are known at some point x can be estimated at a nearby point $x + \Delta x$ using the Taylor series:

$$f(x + \Delta x) = f(x) + \Delta x f'(x) + \frac{\Delta x^2}{2!}f''(x) + \cdots.$$

Similarly, the value of the function at $x - \Delta t$ is given by

$$f(x - \Delta x) = f(x) - \Delta x f'(x) + \frac{\Delta x^2}{2!}f''(x) - \cdots.$$

We can estimate the first derivative of f by subtracting these equations and averaging:

$$f'(x) = \frac{f(x + \Delta x) - f(x - \Delta x)}{2\,\Delta x}.$$

Adding the equations together, we get an estimate of the second derivative:

$$f''(x) = \frac{f(x + \Delta x) + f(x - \Delta x) - 2f(x)}{\Delta x^2}.$$ (A.29)

Applying equation (A.26) to the wave equation, its Taylor series expansion at the point $t + \Delta t$ is

$$u_{t+\Delta t} = u_t + \Delta t \frac{\partial u_t}{\partial t} + \frac{1}{2!}\Delta t^2 \frac{\partial^2 u_t}{\partial t^2} + \frac{1}{3!}\Delta t^3 \frac{\partial^2 u_t}{\partial t^3} + \frac{1}{4!}\Delta t^4 \frac{\partial^2 u_t}{\partial t^4} + \cdots,$$

and the Taylor series expansion at the point $t - \Delta t$ is

$$u_{t-\Delta t} = u_t - \Delta t \frac{\partial u_t}{\partial t} + \frac{1}{2!}\Delta t^2 \frac{\partial^2 u_t}{\partial t^2} - \frac{1}{3!}\Delta t^3 \frac{\partial^2 u_t}{\partial t^3} + \frac{1}{4!}\Delta t^4 \frac{\partial^2 u_t}{\partial t^4} - \cdots.$$

Now if we sum the series, the terms with alternating signs drop out, and we are left with

$$u_{t+\Delta t} + u_{t-\Delta t} = 2u_t + \Delta t^2 \frac{\partial^2 u_t}{\partial t^2} + \frac{1}{12}\Delta t^4 \frac{\partial^2 u_t}{\partial t^4} + \cdots.$$

Rearranging, we get

$$\frac{\partial^2 u_t}{\partial t^2} = \frac{u_{t+\Delta t} - 2u_t + u_{t-\Delta t}}{\Delta t^2} + \varepsilon(\Delta t^2),$$

where the function

$$\varepsilon(\Delta t^2) = -\frac{1}{12}\Delta t^2 \frac{\partial^2 u_t}{\partial t^4} - \cdots$$

represents the part of the Taylor series we will ignore for the purposes of our estimate, thus making it a truncation error. By dropping this term, we end up with the second-order central difference *approximation* of the second derivative,

$$\frac{\partial^2 u_t}{\partial t^2} \approx \frac{u_{t+\Delta t} - 2u_t + u_{t-\Delta t}}{\Delta t^2}.$$ *Second-Order Central Difference Approximation of the Second Derivative*   (A.30)

A.7   Walsh-Hadamard Transform

The *Walsh transform* is a binary analog of the Fourier transform that uses rectangular waves as the basis functions. That is, instead of using a probe phasor such as $e^{ik\omega n/N}$, the Walsh transform uses rectangular waves. In particular, if a sampled function $f(n)$ is defined at points $n = 0, 1, \ldots, N-1$ such that $N = 2^q$, then the Walsh transform $W(\mu)$ is given by

$$W(\mu) = \frac{1}{N}\sum_{n=0}^{N-1} f(n) \prod_{k=0}^{q-1} (-1)^{b_k(n) b_{q-1-k}(u)},$$ *Walsh Transform* (A.31)

where $\mu = 0, 1, \ldots, N-1$. The function defined by the terms $b_k(x)$ is called a *Walsh-Hadamard function* and consists of square waves of varying frequencies, duty cycles, and phases. The terms $b_k(x)$ correspond to the kth magnitude in the binary representation of x. For example, because the binary representation of 5 is 101, then $b_0(5) = 1$, $b_1(5) = 0$, and $b_2(5) = 1$.

The inverse Walsh transform is given by

$$f(n) = \sum_{\mu=0}^{N-1} W(\mu) \prod_{k=0}^{q-1} (-1)^{b_k(n) b_{q-1-k}(u)},$$ *Inverse Walsh Transform* (A.32)

where $n = 0, 1, \ldots, N-1$.

The Hadamard transform involves a permutation of the order of Walsh basis functions but is basically the same idea. Although they are not identical, it is common to refer to the Walsh-Hadamard transform, even if that is not strictly correct.

$$H(\mu) = \frac{1}{N}\sum_{n=0}^{N-1} f(n)(-1)^{\sum_{k=0}^{M-1} b_k(n) b_k(\mu)}.$$ *Hadamard (Ordered) Transform* (A.33)

$$f(n) = \sum_{\mu=0}^{N-1} H(\mu)(-1)^{\sum_{k=0}^{M-1} b_k(n)b_k(\mu)} \quad . \qquad \textit{Inverse Hadamard (Ordered) Transform} \quad (A.34)$$

The attraction of Walsh-Hadamard is computational efficiency. The fast Fourier transform (FFT) reduces redundant calculations from the discrete Fourier transform (DFT); Walsh-Hadamard improves it further by removing all trigonometric calculations. The binary values of the basis functions are ideally suited to implementation on digital computers. The component basis functions for a transform must only be orthogonal, and there are many classes of such functions, including rectangular waves.

A.8   Sampling, Reconstruction, and the Sinc Function

Sampling can be seen as multiplying the continuous input signal being sampled with an impulse train function (see section 4.8). In the frequency domain, the spectrum of the input function is convolved with the spectrum of the impulse train function, which is another impulse train of infinite extent in the frequency domain with reciprocal spacing. Thus, the input spectrum is replicated infinitely around each spectral pulse.

If we wish to reconstruct the original continuous input signal, we must remove these spectral copies. To do so, we can multiply the spectrum by a function that is 1.0 over the principal spectrum and zero elsewhere. We can define the rectangular function as

$$\Pi_a(x) = \begin{cases} 1, & |x| \le a/2, \\ 0, & |x| > a/2, \end{cases} \qquad \textit{Rectangular Function} \quad (A.35)$$

where a is the extent of the rectangle. If we set $a = f_s$, the rectangle covers only the frequencies within the Nyquist frequency limit.

We know that multiplying the rectangle and the sampled spectrum corresponds to convolving the sampled signal with the inverse Fourier transform of the rectangle. We can derive the time domain equivalent of the frequency domain rectangular function as follows.

Denoting the Fourier transform as $\mathcal{F}\{\ \}$, we can write

$$\mathcal{F}\{\Pi_a(x)\}(\omega) = \int_{-\infty}^{\infty} \Pi_a(x)e^{i2\pi\omega x}dx.$$

Since all values outside the rectangle are 0, drop $\Pi_a(x)$ and evaluate the integral over its extent a:

$$\mathcal{F}\{\Pi_a(x)\}(\omega) = \int_{-a/2}^{a/2} e^{i2\pi\omega x}dx$$

Take the derivative:

$$\mathcal{F}\{\Pi_a(x)\}(\omega) = \frac{1}{i2\pi\omega}(e^{i\pi a\omega} - e^{-i\pi a\omega})$$

By Euler's formula,

$$\mathcal{F}\{\Pi_a(x)\}(\omega) = a\frac{\sin \pi a\omega}{\pi a\omega}$$

Finally, by definition,

$$\Pi_a(x) \Leftrightarrow a \cdot \text{sinc}\, a\omega.$$

Since $\text{sinc}\, 0 = \sin 0/0$, and dividing by zero is forbidden by most computers, a more practical definition of the sinc function is

$$\text{sinc}\, x = \begin{cases} (\sin \pi x)/x, & x \neq 0, \\ 1, & x = 0. \end{cases} \qquad \textit{Sinc Function, Computer-Friendly} \quad (A.36)$$

Assuming the function was sampled according to the sampling theorem, such that $f_s \geq 2f_{max}$, where f_{max} is the highest frequency in the function being sampled, then the sampled function can be perfectly reconstructed by convolving it with a sinc function.

Notice that the rectangular function in the spectral domain is a perfect "box-car" lowpass filter because it effectively eliminates any frequencies outside of its extent a. Thus the operation of convolving a sequence with the sinc function corresponds to lowpass-filtering it.

A.8.1 Why Signals with $1/f$ Spectra Are Fractal

One of the fundamental results of convolution theory states that the Fourier transform of the convolution of two signals g and h is just the product of the individual Fourier transforms: $g * h \Leftrightarrow G(f)H(f)$, where

$$g * h \equiv \int_{-\infty}^{\infty} g(\tau)h(t - \tau)\, d\tau.$$

The *correlation* of two functions is correspondingly $\text{corr}(g, h) = G(f)H^*(f)$, where

$$\text{corr}(g, h) = \int_{-\infty}^{\infty} g(\tau + h)h(\tau)d\tau$$

and $H^*(f)$ means the conjugate of $H(f)$. Correlation is a function of the lag t. If g is correlated with itself, then the function is called *autocorrelation*, and its transform pair, according to the Weiner-Khinchin theorem, is

$$\text{corr}(g, g) = |G(f)|^2.$$

The quantity $|G(f)|^2$, called the power spectrum, characterizes the energy in the signal at each frequency. Thus,

$$\langle f(t)f(t+\tau)\rangle = \frac{1}{2\pi}\int_{-\infty}^{\infty} S_f(\omega)e^{i\omega t}\,d\omega. \tag{A.37}$$

The power spectrum of a signal equals the Fourier transform of its autocorrelation.

This interesting result means that a relation between different regions of a power spectrum implies a relation between regions of the signal's autocorrelation.

A.8.2 Scale Invariance and 1/*f* Similarity

Let the spectrum $S(f) = 1/f$. Then its corresponding time domain function $s(t) = F^{-1}(S(f)) = F^{-1}(1/f)$. But now scale temporal resolution by α so that

$$s(\alpha t) = F^{-1}\left(\frac{1}{\alpha}S\left(\frac{f}{\alpha}\right)\right)$$

$$= F^{-1}\left(\frac{1}{\alpha}\cdot\frac{\alpha}{f}\right)$$

$$= F^{-1}\left(\frac{1}{f}\right),$$

which means $s(t) = s(\alpha t)$. In other words, for a signal with a power spectrum having a $1/f$ spectral tendency, the autocorrelation function is independent of time scale; that is, it shows self-similarity and is fractal. The statement $s(t) = s(\alpha t)$ means only that the signals are statistically equivalent because they are related only through their power spectrums.

A.9 Fourier Shift Theorem

Modulating a signal with frequency a shifts the resulting spectrum by a.

$$\mathcal{F}\{f(t)e^{-iat}\} = F(\omega-a), \qquad\qquad \textit{Fourier Frequency Shift Theorem} \tag{A.38}$$

where a is a constant and t is time.

Shifting a signal in time by duration a modulates the resulting spectrum by a.

$$\mathcal{F}\{f(t-a)\} = F(\omega)e^{-iat}, \qquad\qquad \textit{Fourier Time Shift Theorem} \tag{A.39}$$

where a is a constant and t is time.

The following derivation uses the discrete Fourier transform.

Take the Fourier transform of a time-shifted signal:

$$\mathcal{F}\{x(n-a)\} = \sum_{n=0}^{N-1} x(n-a)e^{-i2\pi nk/N}.$$

Let $m = n - a$, then shift the Fourier transform rather than the input signal:

$$\mathcal{F}\{x(n-a)\} = \sum_{m=-a}^{N-1-a} x(m)e^{-i2\pi(m+a)k/N}.$$

Split the summation in the exponent into two terms:

$$\mathcal{F}\{x(n-a)\} = \sum_{m=-a}^{N-1-a} x(m)e^{-i2\pi mk/N}e^{-i2\pi ka/N}.$$

$e^{-i2\pi ka/N}$ does not depend upon the index m, so extract it from the summation:

$$\mathcal{F}\{x(n-a)\} = e^{-i2\pi ka/N} \sum_{m=0}^{N-1} x(m)e^{-i2\pi mk/N}$$

$$= e^{-i2\pi ka/N} X(k).$$

A time shift of a samples in the signal $x(n)$ corresponds to a phase shift of $-2\pi ka/N$ in the phase spectrum of $X(k)$. This phase shift is linear in frequency. The magnitude spectrum is the same as the unshifted signal.

A.10 Spectral Effects of Ring Modulation

The following demonstration of the spectral effects of ring modulation (see section 9.3.2) uses the Fourier shift theorem. Suppose we have a musical signal $x(t)$ and modulate it with a cosine wave. Since by Euler's formula,

$$\cos(\omega_c t) = \frac{e^{i\omega_c t} + e^{-i\omega_c t}}{2},$$

we can write the product of the two signals as

$$f(t) = \frac{x(t)(e^{i\omega_c t} + e^{-i\omega_c t})}{2}.$$

The Fourier transform of $f(t)$ can be written as

$$\mathcal{F}\{f(t)\} = \mathcal{F}\left\{x(t)\frac{e^{i\omega_c t} + e^{-i\omega_c t}}{2}\right\}$$

$$= \mathcal{F}\left\{x(t)\frac{e^{i\omega_c t}}{2}\right\} + \mathcal{F}\left\{x(t)\frac{e^{-i\omega_c t}}{2}\right\}.$$

By the Fourier frequency shift theorem, equation (A.35), where $X(\omega) = \mathcal{F}\{x(t)\}$,

$$\mathcal{F}\{f(t)\} = \frac{1}{2}[X(\omega + \omega_c) + X(\omega - \omega_c)].$$

A.11   Derivation of the Reflection Coefficient

Equation (8.42) for v_0^- and equation (8.43) for v_1^- show the effect of the characteristic impedances on either side of a boundary on the velocity of the waves reflecting from the boundary. Let us begin by finding v_0^-. Recall from equation (8.41) that

$$v = 2\frac{Z_0 v_0^+ + Z_1 v_1^+}{Z_0 + Z_1}$$

and because $v = v_0^+ + v_0^-$, clearly, $v_0^- = v - v_0^+$.

$$v_0^- = 2\frac{Z_0 v_0^+ + Z_1 v_1^+}{Z_0 + Z_1} - v_0^+ \qquad\qquad \text{Substitute for } v.$$

$$= 2\frac{Z_0 v_0^+ + Z_1 v_1^+}{Z_0 + Z_1} - v_0^+\frac{Z_0 + Z_1}{Z_0 + Z_1} \qquad\qquad \text{Find common denominator.}$$

$$= \frac{2Z_0 v_0^+ + 2Z_1 v_1^+ - Z_0 v_0^+ - Z_1 v_0^+}{Z_0 + Z_1} \qquad\qquad \text{Expand terms.}$$

$$= \frac{2Z_1 v_1^+}{Z_0 + Z_1} + \frac{2Z_0 v_0^+ - Z_0 v_0^+ - Z_1 v_0^+}{Z_0 + Z_1} \qquad\qquad \text{Group like terms.}$$

$$= \frac{2Z_1 v_1^+}{Z_0 + Z_1} + \frac{Z_0 - Z_1}{Z_0 + Z_1}v_0^+ \qquad\qquad \text{Simplify and factor.}$$

$$= \frac{2Z_1}{Z_0 + Z_1}v_1^+ - \frac{Z_1 - Z_0}{Z_0 + Z_1}v_0^+. \qquad\qquad \text{See equation (8.42).}$$

Likewise, because $v = v_1^+ + v_1^-$, and, $v_1^- = v - v_1^+$, then

$$v_1^- = 2\frac{Z_0 v_0^+ + Z_1 v_1^+}{Z_0 + Z_1} - v_1^+ \qquad \text{Substitute for } v.$$

$$= 2\frac{Z_0 v_0^+ + Z_1 v_1^+}{Z_0 + Z_1} - v_1^+ \frac{Z_0 + Z_1}{Z_0 + Z_1} \qquad \text{Find common denominator.}$$

$$= \frac{2Z_0 v_0^+ + 2Z_1 v_1^+ - Z_0 v_1^+ - Z_1 v_1^+}{Z_0 + Z_1} \qquad \text{Expand terms.}$$

$$= \frac{2Z_0 v_0^+}{Z_0 + Z_1} + \frac{2Z_1 v_1^+ - Z_0 v_1^+ - Z_1 v_1^+}{Z_0 + Z_1} \qquad \text{Group like terms.}$$

$$= \frac{2Z_0}{Z_0 + Z_1} v_0^+ + \frac{Z_1 - Z_0}{Z_0 + Z_1} v_1^+ \qquad \text{Simplify and factor. See equation (8.43).}$$

Finally, if the common term appearing in both equations (8.42) and (8.43) is defined as the reflection coefficient

$$R = \frac{Z_1 - Z_0}{Z_0 + Z_1},$$

then equations (8.42) and (8.43) can be further simplified to just $v_0^- = (1 + R)v_1^+ - Rv_0^+$ and $v_1^- = (1 - R)v_0^+ + Rv_1^+$, as given in equation (8.45). To see this, let's start with the definition of v_0^- and go backwards to (8.42), as follows.

$$v_0^- = (1 + R)v_1^+ - Rv_0^+$$

$$= v_1^+ + \frac{Z_1 - Z_0}{Z_0 + Z_1} v_1^+ - Rv_0^+ \qquad \text{Substitute.}$$

$$= \frac{Z_0 + Z_1}{Z_0 + Z_1} v_1^+ + \frac{Z_1 - Z_0}{Z_0 + Z_1} v_1^+ - Rv_0^+ \qquad \text{Find common denominator.}$$

$$= \frac{Z_0 v_1^+ + Z_1 v_1^+ + Z_1 v_1^+ - Z_0 v_1^+}{Z_0 + Z_1} - Rv_0^+ \qquad \text{Multiply out.}$$

$$= \frac{2Z_1}{Z_0 + Z_1} v_1^+ - \frac{Z_1 - Z_0}{Z_0 + Z_1} v_0^+ \qquad \text{Group terms. Compare to equation (8.42).}$$

The term v_1^- has a similar derivation.

Notes

Foreword

1. Max V. Mathews, "The Digital Computer as a Musical Instrument," *Science* 142 (3592): 553–557, 1963.
2. The Samson Box was built by Systems Concepts with a grant to CCRMA from the National Endowment for the Arts.

Chapter 1

Epigraph. Titus Maccius Plautus (254–184 B.C.E.), Roman comic playwright.

1. Not all real-time systems are hard-deadline. Paying utility bills is an example of a soft-deadline real-time system: utilities usually allow a grace period before cutting off services if the due date is not met.

2. Harry Nyquist (1889–1976), mathematician, physicist, electrical and communications engineer. In addition to Nyquist, E. T. Whittaker, Claude Shannon, and Vladimir A. Kotelnikov all made substantial contributions to sampling theory.

3. The operator $\lfloor \ \rfloor$ returns the largest integer less than the value enclosed. For example, if $x = 5.3$, then $\lfloor x \rfloor = 5$. The $\lfloor \ \rfloor$ operator is called the floor function.

4. *Centroid:* the point in a geometrical figure whose coordinates are the arithmetic mean of the coordinates making up the figure (Nelson 2003).

5. MIDI, which stands for Musical Instrument Digital Interface, is a standard way of communicating musical performance gesture information. A MIDI sequencer is a hardware device or software program running on a personal computer that can receive MIDI messages from a MIDI-equipped instrument such as a keyboard. When the performer presses keys, switches under the keyboard trigger MIDI messages that are transmitted to the sequencer. The sequencer records each MIDI message together with the time it was received. The performance can be recreated by sending the recorded sequence of messages at the appropriate times to a MIDI-equipped music synthesizer. Sometimes these musical components—keyboard, sequencer, and synthesizer—all reside in the same hardware.

6. For an interesting perspective on the significance of the microstructure of musical performance, see Clynes (1983).

7. Hum arises when a power supply, which converts alternating current from the electrical mains into direct current used by an audio device, leaves a residual of the alternating current in its direct-current output. Hum may also contain strong harmonics. This residual is added to the signal being amplified, resulting in a loss of quality.

8. I've heard it said that a design goal for the compact disc was to be able to fit Beethoven's Ninth Symphony on one disc.

9. These conflicting usages of *megabyte* are still widespread and unreconciled, so far as I know. As with other terminological conflicts, the best approach is to cite the standard when referencing it. So I write "SI megabyte" to mean 10^6 and just plain "megabyte" to mean 2^{20}.

10. Actually, a great deal more data is packed on an audio CD, including error-correcting codes and control codes. The interested reader is referred to Pohlmann (1992).

11. Now anyone who has a fairly recent-vintage computer can do computer music. The laptop computer I'm writing this book on is many times more powerful than the expensive computer system at the Stanford Artificial Intelligence Laboratory where I began my studies in the 1970s. That computer consisted of dozens of refrigerator-sized boxes. It used so much power that it required its own electrical substation. Huge air conditioners were needed for cooling. Disk drives were the size of small woodsheds. It was all terribly expensive and required a cadre of programmers to keep it working. The cost and consequent scarcity of computers severely limited the numbers of people who could participate in the field.

12. My emphasis. Dawkins (1989, 192).

Chapter 2

Epigraphs. The "Scholia Enchiriadis," a ninth-century manuscript; quoted in James (1993, 83). Jacques Hadamard (1865–1963), French mathematician; the quotation is taken from Hadamard (1945; 1996 ed., 123).

1. This book indicates the imaginary number by the letter i, *as* physicists and mathematicians do. Electrical engineers prefer the letter j because they use i to represent electrical current.

2. I suppose the decision to use the y-axis for imaginary numbers was arbitrary, but it is customary. Some mathematicians refer to the complex plane as an Argand diagram.

3. The angle is also sometimes called the *argument* of z in mathematics literature.

4. If you want to see this derivation in action, it might be more intuitive to start from $(a + bi)(a - bi)$ and multiply it out.

5. Abraham de Moivre (1667–1754), French pioneer in the development of analytic trigonometry and probability.

6. Brook Taylor, (1685–1731), English mathematician who invented the method of expanding functions in terms of polynomials that we now call Taylor series.

7. Note that we must invoke infinity in order to establish the equalities in (2.22) and (2.23). Putting the concept of infinity to such practical uses is the most remarkable defining characteristic of our species, in my opinion. Even considered alongside such hallmarks as opposable thumbs and speech, it is the human capacity not just to imagine infinity but to use it for practical outcomes that most impresses me. Spinoza wrote, *Mens aeterna est quaternus res sub specie aeternitatis* (The mind is eternal insofar as it conceives things from the standpoint of eternity). Quoted in Monk (1990, 140–141).

8. Leonhard Euler (1707–1783). Among his numerous invaluable contributions to mathematics, he introduced e, i, and $f(x)$ for f as a function of x.

9. Imagine that the bicycle is upside-down and without a chain so that both its wheels are able to spin freely in either direction. Also ignore effects of friction, and assume that, once set in motion, the wheels continue spinning at the same rate.

10. Dividing by i rotates the expression 90° clockwise, as does multiplying the expression by $-i$ (see section 2.3.6).

11. It is curious how the term *analytic* came to mean "contains no negative frequencies" in signal processing. In the study of complex variables, if a real function has derivatives of all orders and has a Taylor series expansion at every point, it is called analytic. However, as Smith (2003) points out (in footnote on page 41), all bandlimited signals are analytic by this definition because they are all sums of finite-frequency sinusoids. Though it seems unlikely, perhaps the signal processing usage is unrelated to its usage in complex variables.

12. David Hilbert (1862–1943), German mathematician whose work lay at the intersection of geometry, logic, and number theory. In his famous 1900 speech before the Paris International Conference, he set out 23 outstanding problems in mathematics that he urged his colleagues to address. Known now as Hilbert's problems, some were solved quickly; others, such as identifying the axioms of physics, may never be completed.

Chapter 3

Epigraph. Hermann von Helmholtz (1821–1894), a German scientist whose contributions spanned physics, biology, and acoustics.

1. Joseph Fourier (1768–1830), French mathematician.

2. This code example follows the common programmer's convention of using uppercase I for the imaginary number i, and uppercase E for the base of the natural logarithms e.

3. Because of numerical precision limits of digital computers, small round-off errors sometimes creep into DFT/IDFT calculations. But if the input signal is real, any nonzero values in the imaginary part of the IDFT can be ignored.

4. Note that the Im$\{x\}$ function returns the imaginary component of a complex number as a real number. For example, Im$\{i\}$ is the real number 1.

5. Of course, the total energy of the spectrum goes down in proportion to the number of zero-valued samples we add because we're essentially adding silence to the signal. But we can compensate by scaling up the amplitude of the resulting spectrum accordingly.

6. Julius Ferdinand von Hann (1839–1921), an Austrian meteorologist, is seen as the father of modern meteorology. In signal processing, the Hann window is a window function named after him by R. B. Blackman and John Tukey in "Particular Pairs of Windows," in *The Measurement of Power Spectra, From the Point of View of Communications Engineering* (New York: Dover, 1959, 98–99). In this article, the use of the Hann window is called "hanning," e.g., "hanning" a signal means to apply the Hann window to it. This may have led to misconceptions about the name of the Hann window, leading many to call it the "Hanning window." Adapted from http://en.wikipedia.org/wiki/Julius_Hann.

7. In this context, the centered dot is used to stand for all possible function indexes simultaneously. If we think of the elements of discrete function x as a set of train cars stretching from minus to plus infinity, then saying $x(n)$ chooses one of them (and the choice depends upon the value of n). The centered dot notation chooses them all.

Chapter 4

Epigraph. Havelock Ellis (1859–1939), British doctor, psychologist, and social reformer. Quoted in Schaaf (1978).

1. Do not confuse ★ used for multiplication in MUSIMAT and other programming languages with * used for convolution.

2. Computer systems specifically for fast convolution have been developed. See, for example, Foster (1988).

3. If the impulses are of infinitesimal duration yet have an area of 1, they must compensate for their infinitesimal shortness by being infinitely high. Since this doesn't graph very well on a page, I make the impulses look a bit fatter and quite a bit shorter.

4. The interested reader is referred to Oppenheim and Schafer (1975).

5. This rule applies only for real functions. For complex functions, the right-hand side should be F(ω)G($-\omega$).

6. Reminder: *signal* stands for time domain function, and spectrum stands for frequency domain function.

Chapter 5

Epigraph. Alfred North Whitehead (1861–1947), British-American philosopher, physicist, and mathematician. The quotation is taken from Whitehead (1925).

1. You can either trust me on this one or consult your local supplier of mathematical truths under the subject heading "Fundamental Theorem of Algebra."

2. In practical applications, there is usually a floor below which the signal becomes insignificant. But mathematically, it is infinite so long as $b \neq 0$.

Chapter 6

Epigraph. Pythagoras (582–496 B.C.E.), Greek mathematician and philosopher.

1. Not all functions can be differentiated, but this requirement need not concern us here. A treatment of this requirement can be found in any good book on differential calculus.

2. Charles Seagrave, private communication. For a standardized definition of "jerk," see ISO 2041 (1990), Vibration and Shock—Vocabulary, 2.

3. In fact, pianos often display two decay slopes: an initial rapid decay and a longer sustained decay. This effect is a result of the piano's hammer striking multiple strings tuned in unison. All strings vibrate in phase immediately after the hammer blow, maximizing energy dissipation at first; but because of inevitable mistunings between the unison strings, their phases drift through time. When, for example, two strings are 180° out of phase, most energy is conserved between the two strings, and dissipation is reduced, thereby sustaining the tone.

4. *Gnarly* is a surfing term that characterizes the magnitude of an ocean wave that is about to eat you and your surfboard for lunch.

Chapter 7

Epigraph. Sir Isaac Newton (1642–1727). Quoted in Brewster (1855, vol. 2, ch. 27).

1. If you are wondering why c^2 is in the numerator instead of the denominator, note that in order to put equation (7.20) into canonical form, its terms are flipped around, and c^2 has actually been moved to the other side of the equals sign, hence is inverted.

Chapter 8

Epigraph. Aristotle (384–322 B.C.E.), Greek philosopher. Prob. xix. 29.

1. Georg Simon Ohm (1789–1854), mathematician, physicist, and educator. He is known best for "Ohm's law," based on his research linking the electrical current through a material to the potential difference applied across the material (resistance). He also proposed what became known as Ohm's law of acoustics, a psychoacoustic theory that the ear derives pitch by performing Fourier analysis on acoustical signals (see volume 1, section 6.4.1). This presents a potential confusion in terminology because in this chapter I adapt Ohm's law of electrical current to study acoustical current. Note that Ohm's law of acoustics is a theory of psychoacoustics, whereas the Ohm's law treated in this chapter is a theory of physics, unrelated to his other theory except by his name.

2. In practice, we must add to this mass some of the nearby air outside the tube, which is influenced by the movement of the air in the cylinder, called the end correction, discussed previously in the section on the Helmholtz resonator (volume 1, section 8.3.3). Also, we must ignore the effects of the inertia of the piston itself for this illustration, but for a realistic application we'd have to include its mass as well as other factors like friction.

3. In SI units, the impedance is 1 rayl if one unit of pressure produces one unit of velocity. However, "rayl" is also used in CGS units to represent impedance as 1 dyne-second per cubic centimeter. Thus, one CGS rayl equals 10 SI rayls. As a consequence of this confusion, one must always qualify whether the reference is to CGS rayls or SI rayls.

4. A more careful way to say the same thing is, The ratio of the chord length to the arc length for some small subtended angle tends to unity in the limit as the radius of the sphere goes to infinity, and the wave front is then plane across that chord.

5. An "experiment" of this kind was conducted by the legendary French horn performer Dennis Brain at one of the side-splittingly funny concerts staged in Britain in the 1950s by Gerard Hoffnung. At the Music Festival Concert in 1956, Brain played a Leopold Mozart horn concerto on rubber hosepipes. Highly recommended entertainment. Available at http://www.musicweb-international.com/classrev/2002/Nov02/HoffnungConcerts.htm.

Chapter 9

Epigraph. Octavius Magnus Aurelius Cassiodorus, Senator (ca. 485–ca. 575), Institutiones, II, iii, paragraph 21, in Strunk (1950, 88).

1. For audio reproduction through digital-to-analog conversion (DAC), the amplitude term would be set based on the numeric precision of the converter. For example, to generate a tone at maximum amplitude in a two's complement 16-bit DAC, set $A = 2^{16-1} - 1$ (see section 1.5.3).

2. In the fully general case, one should account for phase offset in equation (9.4). It is often omitted in practice on the belief that phase is not perceptually significant. See volume 1, section 6.4.1.

3. Phase offset is also neglected here. Even if phase does have an effect, the cost of carrying its calculation is generally deemed not to warrant the expense for engineering reasons. I think the jury is still out on the question of the importance of phase to perception.

4. For example, if there are 30 components in a piano tone lasting 1 second, then in general we will require 30 functions of time for harmonic amplitudes and 30 functions of time for harmonic frequencies, so the ratio of control functions to final waveform function is 60/1. At a sampling rate of 44.1 kHz, using 16-bit sample values, these functions would require about 5 MB per second of storage—a nontrivial amount of information for one note. A much better approach for realistic representation of complex tones is presented in chapter 10.

5. The convergence of a truncated Fourier series in the vicinity of Gibbs' horns can be improved by applying weighting functions to the coefficients, as described by Lipót Fejér. If N harmonics are to be synthesized in a truncated Fourier synthesis, then the amplitude of the kth harmonic should be scaled by $(N - k)/N$.

6. The terms *carrier* and *modulator* relate to the use of this technique in AM and FM radio transmission. In radio communications theory, the carrier oscillator produces a fixed frequency at an assigned location in the broadcast spectrum that "carries" the sound of the audio signal encoded by the modulator.

7. I was privileged to study and work directly with John Chowning during this time at the Center for Computer Research in Music and Acoustics as a graduate student in the Music Department. The laboratory facilities were housed at the Stanford Artificial Intelligence Laboratory in the Donald C. Power building.

8. For function composition to be valid, the domain of g must be equal to or a subset of the range of f.

9. Pafnuty Lvovich Chebyshev (1821–1894). He objected to being described as a "splendid Russian mathematician" (although he certainly was), preferring to think of himself as a "world-wide mathematician."

10. Intelligibility and sound quality are not the same. A recording of speech does not have to be of high quality to be intelligible. Beyond a certain point, quality of speech is important in telecommunications; after all, if the listener can't understand it, what's the use of transmitting speech? But quality generally takes a back seat to data reduction.

11. "At each point we must ask ourselves, 'Would I pay $5.95 for a record of this voice?'" Moorer (1979). The price cited for a record certainly dates this quotation.

12. This technique works best if the sound being filtered is spectrally bright. Otherwise, if there are no musical components to energize a formant, there will be no effect.

13. In the physical sciences, entropy is a measure of the ways that the energy of a system can be distributed among the motions of its particles. In information theory, it is a measure of the number of states required to represent the information in a system. The concepts are related but not entirely analogous. See "Information Theory and the Mathematics of Expectation" in volume 1, section 9.15.

14. Federal Standard 1015. The U.S. Department of Defense Federal-Standard-1015/NATO-STANAG-4198-based 2400 bit per second linear prediction coder (LPC-10) was republished as a Federal Information Processing Standards Publication 137 (FIPS Pub 137). It is described in Thomas E. Tremain, "The Government Standard Linear Predictive Coding Algorithm: LPC-10," *Speech Technology Magazine,* April 1982, 40–49. There is also a section about FS-1015 in Panos E. Papamichalis, *Practical Approaches to Speech Coding* (Upper Saddle River, N.J.: Prentice-Hall, 1987).

15. Federal Standard 1016. The U.S. Department of Defense Federal-Standard-1016- based 4800 bps code excited linear prediction voice coder version 3.2 (CELP 3.2) is described in Joseph P. Campbell, Jr., Thomas E. Tremain, and Vanoy C. Welch, "The Federal Standard 1016 4800 bps CELP Voice Coder," in *Digital Signal Processing* (New York: Academic Press, 1991), vol. 1, no. 3, 145–155.

16. Also see articles in the section titled "Time-Frequency Representations of Musical Signals" in De Poli, Piccialli, and Roads (1991).

17. The Phonogène, developed by Jacques Poullin and Pierre Schaeffer, is sometimes related to Gabor's device in the literature, but it was quite a different invention. The Phonogène could transpose the pitch of a tape loop to any of the 12 degrees of the chromatic scale from a keyboard. The keyboard selected one of 12 capstans of different diameters. A two-speed motor provided octave transposition. Unlike Gabor's device, the Phonogène did not provide independent control over tempo and frequency.

18. The term *flanging* came from analog tape recording technology. The same material is played simultaneously on two tape recorders. Since no two tape recorders have exactly the same speed, one drifts with respect to the other, creating a delay. To get the delay to change more rapidly, the operator slows one of the recorders by dragging a thumb on the *flange* of the tape reel, which is the outermost part of the reel. As the relative speed changes, one hears a characteristic hollow swish sound of the comb filtering.

19. That's why, standing outside the hall of a loud rock concert, you hear mostly bass: high frequencies are absorbed by the air and walls; low frequencies are admitted by the walls into the surrounding environment.

20. A cross-over network sends highpass filtered audio to the tweeter, and lowpass filtered audio to the woofer, of a high-fidelity loudspeaker system.

Chapter 10

Epigraph. Quoted from Gabor (1947).

1. Calling the Heisenberg box a "box" is slightly misleading; the region of information is just an area, not a shape. Heisenberg's uncertainty principle can be related directly to waveforms and signals. In quantum mechanics, particles exhibit wavelike behavior and can be interpreted as probability wave packets. The position of a particle is related to the point of maximum amplitude of the wave packet. Similarly, the velocity of a packet is related to its energy, which is directly proportional to its frequency of vibration. This is analogous to the relationship between velocity and position of the particle. In chapter 3, we saw that frequency cannot be measured at a single moment in time because frequency is the rate of repetition, and measurement must occur over a finite (non-zero) period of time sufficient to characterize the rate of repetition. This implies that, for signals, precision in time must be traded off for precision in frequency, and vice versa. Correspondingly, for particles, precision in location must be traded off for precision in velocity. To accurately determine particle velocity requires a longer observation window, but this increases the uncertainty of the particle's exact position. See Dirac (1958).

2. This is a consequence of the fact that we are dealing with sampled signals; continuous signals are not periodic in frequency (see section 4.8).

3. The requirements of the sampling theorem are the same regardless of the domain in which we are sampling. Any frequencies in a signal being sampled that exceed the Nyquist rate will be *aliased in frequency,* thereby appearing in the wrong place in the spectrum. The same rule applies to undersampling in the spectral domain. For example, suppose we have a continuous spectrum $X(f)$ that corresponds to a continuous time domain signal $x(t)$. Suppose we purposefully undersample $X(f)$, resulting in a sampled spectrum $X'(k)$. Certainly, the time domain signal $x'(t)$ that we reconstruct from $X'(k)$ will not be identical to $x(t)$ because $x'(t)$ is based on the undersampled spectrum $X'(k)$. The distortions in $x'(t)$ are *aliased in time.* The moral is that the sampling theorem applies regardless of the domain being sampled.

4. Note, that's not "analysis frames per second" or "samples per second"; it's *quanta*—or Heisenberg boxes—per second.

5. Originally, *decimate* referred to the killing of every tenth person, a punishment used in the Roman army for mutinous legions.

6. Glissando is a continuous gliding pitch. Portamento is a gliding pitch quantized to the tones of a scale. A violinist performs glissando by sliding a finger up the string. A guitarist performs chromatic portamento by sliding a finger over the frets; a pianist performs diatonic portamento by sliding a finger across the white keys.

7. The original channel vocoder was developed by Homer Dudley in the 1930s. He named the vocoder as a contraction of VOice enCODER.

8. A number is *highly composite* if its prime factors are small. For the FFT, the number of frequency bands is usually an integral power of 2.

9. See document ISO/IEC 15444-1:2000, "Information Technology—JPEG 2000 Image Coding System—Part 1: Core Coding System."

Appendix

Epigraph. Gottfried Wilhelm Leibniz (1646–1716). Quoted in Rose (1988).

Glossary

Absolute value (complex)   Distance from a point to the origin of the complex plane.

Absolute value (real)   Distance from a point to the origin along the real axis.

Acceleration   (1) First derivative of velocity with respect to time. (2) Time rate of change of velocity, or time rate of change of the time rate of change of displacement.

Accuracy   *See* Precision/accuracy.

Acoustical capacitance   Ratio of change in volume to change in pressure.

Acoustical capacitor   Air mass that stores energy in the relative pressure of the air's particles.

Acoustical current   Rate at which a volume of air flows. Velocity or rate of flow of a volume of air.

Acoustical inductor   Air mass that stores energy when accelerated.

Acoustical ohm   A measure of the impedance of air: 1 pascal second per cubic meter.

Acoustical power   Product of pressure and current. Square of current times impedance. Square of pressure divided by impedance.

Admittance   Inverse of impedance.

Algebra   Branch of mathematics that studies general properties of numbers and generalizations of these properties.

Algebraically closed   Property of a number system containing all required solutions.

Aliasing   Arises when frequencies outside a range are indistinguishable from frequencies within a range because of the way sampling affects observation.

Alternating current   Flow that reverses direction.

Amplitude modulation   Instantaneous amplitude variation of a signal.

Analog-to-digital conversion   A way of converting a continuously varying signal into a corresponding sequence of discrete sample quanta.

Analytic signal   Complex signal that has no negative frequencies.

Angular velocity   Rate at which a phasor spins.

Ansatz   A German word, used to describe a solution to a problem that is guessed.

Anti-aliasing filter   Lowpass filter designed to remove frequencies outside the Nyquist frequency limit.

Bandwidth   Distance in frequency between two points surrounding an amplitude peak that have half as much power (–3 dB) as the peak.

Bandwidth of a filter   Breadth of the band of frequencies affected by the filter.

Baseband   All frequencies in the range of plus and minus one half of the sampling rate.

Canonical   Describing a rule or equation that summarizes or idealizes other rules or equations in a fundamental way.

Capacitive reactance   A counterforce proportional to displacement.

Causal   Describing a system that references only current and past (not future) input and past (not current or future) output.

Certainty   Proportional to precision and sampling rate.

Characteristic impedance   (1) A measure of the extent to which a medium impedes the progress of a signal moving through it. (2) Product of a medium's density and the speed of sound.

Charge   Current flow times duration.

Circular motion   Root motion governing simple harmonic motion and sinusoidal motion.

Clipping   Nonlinear distortion that arises when the input signal exceeds the upper limit of the system's dynamic range.

Codec   Device that encodes and decodes a signal.

Complex number   Sum of a real and an imaginary number. The imaginary part of the sum is distinguished by i.

Complex plane   Plane spanned by the real number line and the imaginary number line.

Composition   Art of controlled expectation.

Conjugate   Negation of the imaginary part of a complex number.

Conjugate symmetry   Arises when a positive frequency phasor and a negative frequency phasor are tied to the same angular displacement.

Conservative forces   Forces that store energy via elasticity and inertia.

Continuous   Describing a system that retains its identity no matter how it is divided. Length, area, and volume are continuous by this definition.

Convolution theorem   Convolution in the time domain corresponds to multiplication in the frequency domain, and vice versa.

Cosine   Sum of two phasors in conjugate symmetry.

Critical damping   *See* Damping.

DAC   *See* Digital-to-analog conversion.

Damping   Dissipative counterforce proportional to the velocity of a mass. Ratio of dissipation to other forces. In underdamping, dissipation is small compared to other forces; in critical damping, dissipation matches other forces; in overdamping, dissipation exceeds other forces.

Danielson-Lanczos lemma   A way to divide the discrete Fourier transform recursively into ever smaller DFTs. The first step toward obtaining the fast Fourier transform.

Data reduction   A method of identifying and removing redundant or undetectable content.

Decimate   To remove elements from a sequence in a regular pattern, such as to remove every tenth element.

Demodulation   Multiplying phasors to lower their frequency.

de Moivre's theorem   Raising a complex number with unity magnitude to a power n is equivalent to multiplying the angle of the complex number by n.

Derivative   (1) Rate of change of a function with respect to its independent variable. (2) Slope of a line tangent to the function at the point of interest.

Differential equations   Expressions relating a primary quantity such as displacement to derived quantities such as velocity and acceleration.

Digital-to-analog conversion   A means to convert a discrete set of sample quanta into a corresponding continuously variable function. If we think of a sample sequence as a "dehydrated" version of the original analog signal, we use a DAC to "rehydrate" it to recover the original analog signal.

Direct current   Flow that does not reverse direction.

Discretization   Fixing a point on a continuous function so that it becomes individually discernible.

Duty cycle   Ratio of a rectangular function's nonzero interval to the duration of its period.

Eigenvalues   Frequencies of the normal mode vibrations of a system.

Entropy   Number of discrete states required to characterize a system.

Field equation   Formula governing the local motion at each point in a vibrating body.

Filtering   Frequency-selective amplification.

Fourier analysis   A way to measure the strengths of the individual components of a harmonic signal.

Fourier synthesis   A way to create a waveform from a specification of the strengths of its various harmonics.

Frequency resolution of the fast Fourier Transform   Ratio of the length of the Fourier transform to the sample rate. This determines the distinguishability of frequencies.

Frequency response   Ratio of a filter's output amplitude to its input amplitude.

Gibbs' horns Slight overshoot and ringing that occurs at the ends of the vertical excursion of some Fourier series waveforms.

Harmonic A signal whose frequency is an integer multiple of a fundamental frequency.

Harmonicity ratio In FM, the ratio of modulating frequency to carrier frequency.

Heisenberg box Hearable region in the information diagram with some effective duration and effective bandwidth.

Helmholtz motion Characteristic stick/slip sawtooth wave motion of the bowed violin string.

Hermitian Defining a spectrum whose real part is an even function and whose imaginary part is an odd function.

Homogeneous Describing a polynomial all of whose terms are of the same degree. If a homogeneous polynomial is set equal to zero, that is said to be a homogeneous equation.

Impedance (1) How hard one must push for the momentum obtained. (2) Ratio of pressure to current.

Impulse function Function equal to 1 when indexed at 0, and equal to 0 elsewhere. An impulsive force is a large force acting over a brief moment, such as a door slamming.

Impulse response Sum of all delayed and scaled reflections in a reverberant system.

Impulse train Periodic impulse function with period equal to the distance between impulses.

Inductance Property of systems that store energy by acceleration.

Inductive reactance Counterforce that is proportional to acceleration of a mass.

Inertance Inertia of an air mass that opposes change in volume velocity.

Information A quantity that relates to the number of facts necessary to convey a measurement.

Integrator Device that adds the past value to the current input value and holds.

Linear Describing a system in which any weighted combination of inputs results in a proportional combination of outputs.

Magnitude Length of a vector without regard to its direction.

Magnitude (complex) Square root of the sum of the squares of the real and imaginary components.

Magnitude spectrum (1) Square root of the sum of the squares of all cosinusoidal and sinusoidal components of a signal. (2) Magnitude of a complex spectrum or complex frequency response.

Modulation Multiplying phasors to raise their frequency.

Music "The discipline which treats of numbers in their relation to those things which are found in sound."—Cassiodorus.

Natural numbers Whole numbers greater than zero.

Node A point along a modal vibration where contrary vibrating forces cancel.

Noise An unwanted signal added to one being measured.

Nonconservative (dissipative) forces Forces that transfer and transform energy as heat and sound.

Nyquist frequency Highest frequency representable without aliasing.

Ohm's law Flow of current related directly to pressure and inversely to resistance. Not to be confused with Ohm's law of acoustics.

Order of a differential equation This is determined by the highest derivative used in the equation.

Order of a filter This is determined by the maximum of the number of input terms or output terms of its defining equation.

Overdamping *See* Damping.

Phasor A representation of a complex number in terms of the complex exponential.

Physical model Study of the interactions of vibrating systems that are the basis of natural sounds, including energy propagation and resonance.

Pole Location of a positive infinity above the Z plane.

Power spectrum (1) Fourier transform of a signal's autocorrelation. (2) Same as magnitude spectrum but without the square root operation. (3) Square of the magnitude spectrum.

Precision/accuracy Precision is the amount of information we have about a measurement. Acccuracy relates to whether the measurement is true.

Pressure Force per unit of area.

Probe phasor Frequency-detecting phasor used by the Fourier transform to detect the presence of energy at specific frequencies.

Propagation   Sound travel (spreading) in a medium where there is no impediment. *See also* Transmission.

Q   Ratio of the energy stored in a resonator because of conservative forces (reactance) to the energy dissipated because of nonconservative forces (resistance) per radian of oscillation.

Quadrature   (1) A 90° phase relation between two periodic quantities varying with the same period. (2) Quality of signals that are lock-stepped together at a 90° phase difference, such as sine and cosine projection of circular motion.

Quantization   Reducing the true magnitude of a measured object to the available precision and discarding the difference between the true value and its quantized approximation.

Quantization error   Difference between the input value of a quantizer and the corresponding output value of the dequantizer.

Rate of change   Slope that is tangent to a function at the point of interest.

Reactance   Opposition to applied force.

Real time   The period during which a task can be performed in a timely manner with respect to the larger dynamical system of which it is a part.

Recursive   Describing a system in which past outputs are used by the same system directly or indirectly as inputs.

Resistance   A measure of how pressure affects velocity.

Resonance   Tendency of a system to vibrate more strongly at a particular frequency.

Resonance bandwidth   Distance in frequency between two points that have half as much power (–3 dB) as the peak.

Resonant frequency   Frequency where the current flow is at a maximum, and the amplitude is highest.

Sampling   (1) Selection process that arises when an observer momentarily regards an aspect of an object as a consequence of a triggering event. (2) Measuring a system discontinuously, "every so often." (3) Multiplying the continuous input signal with an impulse train function. (4) Periodic observation and recording of the instantaneous values of a continuous function.

Series   Sum of the terms of a sequence.

Shift theorem   A circular shift in one domain corresponds to multiplication by a complex exponential in the other domain.

Sign bit   Most significant bit of a two's complement number.

Signal-to-noise ratio   Amount of noise in a system related to the amplitude of its largest useful signal.

Simple harmonic motion   Projection of circular motion onto one-dimensional displacement.

Sinusoidal motion   Projection of simple harmonic motion through time.

Spectrum   (1) Function of the strengths of the various frequency components contained in a signal, ordered by frequency. (2) Specification (in the form of a function or a list) of the strengths of the harmonics of a waveform. (3) Set of all eigenvalues of a vibration.

Spectrum of a constant   Magnitude of a constant at 0 Hz.

Stiffness   Restoring force that is proportional to the amount of bend in a string or membrane.

Superposition   Addition of signals point by point.

Transfer function   (1) Function that shows how much of the input is transferred to the output, depending on frequency. (2) Product of the frequency response and the phase response.

Transmission   Sound travel when constrained to one dimension, such as by a tube. *See also* Propagation.

Underdamping   *See* Damping.

Unit circle   Circle on the complex plane that is a unit distance from the complex origin.

Velocity   (1) First derivative of displacement with respect to time. (2) Time rate of change of displacement.

Vibrato   (1) A periodic pitch modulation around a target pitch. (2) Correlated perturbation of harmonic frequency.

Wave equation   Describes, for some point on an object, how its displacement from equilibrium changes from moment to moment, based on the forces in its immediate neighborhood.

Waveguide   Bidirectional transmission line that can carry signals from a transmitter to a receiver, and vice versa.

z plane   Complex plane.

Zero   Location of a zero of transmission on the Z plane.

References

Arfib, Daniel. 1979. "Digital Waveshaping Synthesis of Complex Spectra by Means of Multiplication of Nonlinear Distorted Sine Waves." *Journal of the Audio Engineering Society* 27 (10): 757–768.

Benade, Arthur H. 1976. *Fundamentals of Musical Acoustics.* New York: Oxford University Press.

Beranek, L. L. 1986. *Acoustics.* Rev. ed. Melville, N.Y.: American Institute of Physics.

Berger, J., R. Coifman, and M. Goldberg. 1994. "A Method of Removing Noise from Old Recordings." *Proceedings of the International Computer Music Conference,* Aarhus, Denmark.

Boulanger, R. 1999. *The C Sound Book.* Cambridge, Mass.: The MIT Press.

Bracewell, R. 1999. *The Fourier Transform and Its Applications.* 3d ed. New York: McGraw-Hill.

Brewster, David. 1855. *Memoirs of the Life, Writings, and Discoveries of Sir Isaac Newton.* London: Hamilton, Adams. London: Johnson Reprint Corp., 1965.

Chowning, John. 1973. "The Synthesis of Complex Audio Spectra by Means of Frequency Modulation." *Journal of the Audio Engineering Society* 21 (7): 526–534. Also in *Foundations of Computer Music,* ed. Curtis Roads and John Strawn. Cambridge, Mass.: MIT Press, 1985.

———. 1989. "Frequency Modulation Synthesis of the Singing Voice." In *Current Directions in Computer Music Research,* ed. Max V. Mathews and John R. Pierce. Cambridge, Mass.: MIT Press.

Clynes, Manfred. 1983. "Expressive Microstructure in Music, Linked to Living Qualities," in *Studies of Music Performance,* ed. J. Sundberg. No. 39, 76–181. Stockholm: Royal Swedish Academy of Music.

Cooley, J. W., and J. W. Tukey. 1965. "An Algorithm for the Machine Calculation of Complex Fourier Series." *Mathematics of Computation* 19: 297–301.

Crochiere, R. E. 1980. "A Weighted Overlap-Add Method of Short-Time Fourier Analysis/Synthesis." *IEEE Transactions on Acoustics, Speech, and Signal Processing* ASSP-28: 99–102.

Danielson, G. C., and C. Lanczos. 1942. "Some Improvements in Practical Fourier Analysis and Their Application to X-ray Scattering from Liquids." *Journal of the Franklin Institute* 233: 365–380, 435–452.

Dantus, Samuel H. 1980. "Non-uniform Time-Scale Modifications of Speech." M.S. and E.E. thesis, Department of Electrical Engineering and Computer Science, MIT, Cambridge, Mass.

Dashow, James. 1980. "Spectra as Chords." *CMJ:* Computer Music Journal 4 (1): 43–52.

Daubechies, I. 1988. "Orthonormal Bases of Compactly Supported Wavelets." *Communications on Pure and Applied Mathematics* 41: 909–996.

Dawkins, Richard. 1989. *The Selfish Gene.* 2d. ed. Oxford: Oxford University Press.

De Poli, Giovanni, Aldo Piccialli, and Curtis Roads, eds. 1991. *Representations of Musical Signals.* Cambridge, Mass.: MIT Press.

Dirac, P. A. M. 1958. *The Principles of Quantum Mechanics.* New York: Oxford University Press.

DMCA (Digital Millennium Copyright Act). 1998. http://www.copyright.gov/laws/.

Dolson, Mark. 1983. "A Tracking Phase Vocoder and Its Use in the Analysis of Ensemble Sounds." Ph.D. dissertation, California Institute of Technology, Pasadena.

Flanagan, J. L., and R. M. Golden. 1966. "Phase Vocoder." *Bell System Technical Journal, November,* p. 1493.

Foster, S. H. 1988. *Convolvotron User's Manual.* Groveland, Calif.: Crystal River Engineering.

Gabor, Dennis. 1946. "Theory of Communication." *Journal of the Institute of Electrical Engineers* 93: 429–457.

———. 1947. "Acoustical Quanta and the Theory of Hearing." *Nature* 159 (4044): 591–594.

Gerzon, M. A. 1976. "Unitary (Energy-Preserving) Multichannel Networks with Feedback." *Electronics Letters* 12 (11): 278–279.

Gordon, John W., and John Strawn. 1985. "An Introduction to the Phase Vocoder." In *Digital Audio Signal Processing: An Anthology,* ed. John Strawn. Middleton, Wisc.: A-R Editions. See especially appendix A.

Hadamard, Jacques. 1945. *Essay on the Psychology of Invention in the Mathematical Field.* Princeton, N.J.: Princeton University Press. Republished as *The Psychology of Invention in the Mathematical Field.* New York: Dover, 1954; and as *The Mathematician's Mind: The Psychology of Invention in the Mathematical Field.* Princeton, N.J.: Princeton University Press, 1996.

Helmholtz, Hermann. 1863. *On the Sensations of Tone.* 2d English ed., 1885. Trans. Alexander J. Ellis based on the 4th German ed., 1877. New York: Dover, 1954.

Hiller, Lejaren, and Pierre Ruiz. 1971. "Synthesizing Musical Sounds by Solving the Wave Equation for Vibrating Objects: Parts I and II." *Journal of the Audio Engineering Society* 19 (6): 462–470, 19 (7): 542–551.

ISO/IEC 11172. 1993. *Coding of Moving Pictures and Associated Audio for Digital Storage Media at up to about 1.5 Mbit/s.*

ISO/IEC 13818. 1994–1997. *Generic Coding of Moving Pictures and Associated Audio.*

Jaffe, David, and Julius Smith. 1983. "Extensions of the Karplus-Strong Plucked-String Algorithm." *CMJ: Computer Music Journal* 7 (2): 76–87. Also in *The Music Machine,* ed. Curtis Roads. Cambridge, Mass.: MIT Press, 1989.

James, Jamie. 1993. *The Music of the Spheres: Music, Science, and the Natural Order of the Universe.* New York: Grove Press.

Jot, Jean-Marc. 1997. "Efficient Models for Reverberation and Distance Rendering in Computer Music and Virtual Audio Reality." *Proceedings of the International Computer Music Conference,* Thessaloniki, Greece, September.

Karplus, Kevin, and A. Strong. 1983. "Digital Synthesis of Plucked-String and Drum Timbres." *CMJ: Computer Music Journal* 7 (2): 43–55. Also in *The Music Machine,* ed. Curtis Roads. Cambridge, Mass.: MIT Press, 1989.

Konstantinides, Konstantinos. 1994. "Fast Subband Filtering in MPEG Audio Coding." *IEEE Signal Processing Letters,* 1(2).

Krasner, Michael A. 1980. "The Critical Band Coder—Digital Encoding of Speech Signals Based on the Perceptual Requirements of the Auditory System." *Proc. IEEE ICASSP80,* vol. 1, 327–311. Piscataway, N.J.: IEEE Press.

Kreyszig, Erwin. 1968. *Advanced Engineering Mathematics.* New York: Wiley.

Laroche, Jean, and Mark Dolson. 1999. "New Phase-Vocoder Techniques for Pitching, Harmonizing, and Other Exotic Effects." *Proceedings of the 1999 IEEE Workshop on Applications of Signal Processing to Audio and Acoustics,* New Paltz, N.Y., October.

LeBrun, Mark. 1979. "Digital Waveshaping Synthesis." *Journal of the Audio Engineering Society* 27 (4): 250–266.

Lee, B. G. 1984. "A New Algorithm to Compute the Discrete Cosine Transform." *IEEE Trans. Acoust. Speech Signal Processing ASSP* 32(6): 1243–1245.

Makhoul, J. 1975. "Linear Prediction: A Tutorial Review." *Proceedings of the IEEE* 63: 561–580.

Mallat, S. G. 1989. "A Theory for Multiscale Signal Decomposition: The Wavelet Representation." *IEEE Transactions on Pattern and Machine Intelligence* 11 (7): 674–693.

———. 1998. *Wavelet: A Wavelet Tour of Signal Processing.* New York: Academic Press.

Markel, J. D., and A. H. Gray. 1976. *Linear Prediction of Speech.* New York: Springer-Verlag.

Mathews, Max V. 1969. *The Technology of Computer Music.* Cambridge, Mass.: MIT Press.

McAulay, R. J., and T. F. Quatieri. 1984. "Magnitude-Only Reconstruction Using a Sinusoidal Speech Model." *Proceedings of the IEEE International Conference on Acoustics, Speech, and Signal Processing,* San Diego.

McNabb, Michael. 1981. "Dreamsong: The Composition." *CMJ: Computer Music Journal* 5 (4): 36–53.

Monk, Ray. 1990. *Ludwig Wittgenstein: The Duty of Genius.* New York: Macmillan.

Moore, Gordon E. 1965. "Cramming More Components onto Integrated Circuits." *Electronics* 38 (April 19).

Moore, F. Richard. 1990. *Elements of Computer Music*. Upper Saddle River, N.J.: Prentice-Hall.

Moorer, James A. 1978. "The Use of the Phase Vocoder in Computer Music Applications." *Journal of the Audio Engineering Society* 26 (January/February): 42–45.

———. 1979a. "About This Reverberation Business." *CMJ: Computer Music Journal* 3 (2): 13–28. Also in *Foundations of Computer Music*, ed. Curtis Roads and John Strawn. Cambridge, Mass.: MIT Press, 1985.

———. 1979b. "The Use of Linear Prediction of Speech in Computer Music Applications." *Journal of the Audio Engineering Society* 27 (3): 134–140.

Morse, P. M. 1960. *Vibration and Sound*. New York: McGraw-Hill Book Co.

Nelson, David, ed. 2003. *The Penguin Dictionary of Mathematics*. 3d ed. London: Penguin Books.

Nussbaumer, H. J., and M. Vetterli. 1984. "Computationally Efficient QMF Filter Banks." *Proc. International Conference IEEE ASSP*. Piscataway, N.J.: IEEE Press.

Nyquist, Henry. 1928. "Certain Topics in Telegraph Transmission Theory." *Transactions of the American Institute of Electrical Engineers* 47 (April): 617–644.

Olson, Harry F. 1952. *Music, Physics, and Engineering*. New York: Dover. Rev. ed. 1967.

Oppenheim, A. V., and R. W. Schafer. 1975. *Digital Signal Processing*. Upper Saddle River, N.J.: Prentice-Hall.

Pohlmann, Ken. 1992. *The Compact Disc Handbook*. 2d ed. Middleton, Wisc.: A-R Editions.

Portnoff, Michael R. 1976. "Implementation of the Digital Phase Vocoder Using the Fast Fourier Transform." *IEEE Transactions on Acoustics, Speech, and Signal Processing* ASSP-24 (3): 243–248.

———. 1978. "Time-Scale Modification of Speech Based on Short-Time Fourier Analysis." Sc.D. dissertation, Department of Electrical Engineering and Computer Science, MIT, Cambridge, Mass. Also in *IEEE Transactions on Acoustics, Speech, and Signal Processing* ASSP-29 (3): 374–390, 1981.

Press, William H., Brian P. Flannery, Saul A. Teukolsky, and William T. Vetterling. 1988. *Numerical Recipes in C: The Art of Scientific Computing*. Cambridge: Cambridge University Press.

Puckette, M. 1995. "Phase-Locked Vocoder." *Proceedings of the IEEE Conference on Applications of Signal Processing to Audio and Acoustics,* Mohonk, N.Y.

Rabiner, L. R., and R. W. Schafer. 1978. *Digital Processing of Speech Signals*. Upper Saddle River, N.J.: Prentice-Hall.

Ramsay, Douglas C. 1996. *Principles of Engineering Instrumentation*. London: Arnold.

Risset, Jean-Claude. 1969. "An Introductory Catalog of Computer-Synthesized Sounds." Internal memorandum. Murray Hill, N.J.: Bell Telephone Laboratories.

Roads, Curtis. 1988. "Introduction to Granular Synthesis." *CMJ: Computer Music Journal* 12 (2): 11–13.

———. 2001. *Microsound*. Cambridge, Mass.: MIT Press.

Rodet, Xavier. 1984. "Time-Domain Formant-Wave Function Synthesis." *CMJ: Computer Music Journal* 8 (3): 9–14.

Rodet Xavier, Yves Potard, and Jean-Baptiste Barrière. 1984. "The CHANT Project: From Synthesis of the Singing Voice to Synthesis in General." *CMJ: Computer Music Journal* 8 (3): 15–31. See also http://musicweb.koncon.nl/ircam/en/artificial/chant.html#.

Rose, N., ed. 1988. *Mathematical Maxims and Minims*. Raleigh, N.C.: Rome Press.

Rothweiler, Joseph H. 1983. "Polyphase Quadrature Filters—A New Subband Coding Technique." *Proc. International Conference IEEE ASSP* vol. 27, 1280–1283. Piscataway, N.J.: IEEE Press.

Ruiz, Pierre M. 1969. "A Technique for Simulating the Vibrations of Strings with a Digital Computer." Master's thesis, Department of Music, University of Illinois, Urbana-Champaign.

Sawyer, W. W. 1943. *Mathematician's Delight*. London: Penguin Books. Republished 1991.

Schaaf, William L. 1978. *A Bibliography of Recreational Mathematics*. Vol. 1. Washington, D.C.: National Council of Teachers of Mathematics.

Schafer, R. W., and L. R. Rabiner. 1973. "A Digital Signal Processing Approach to Interpolation." *Proceedings of the IEEE* 61 (6): 692–702.

Schelleng, John C. 1974. "The Physics of the Bowed String." *Scientific American* 230 (January): 87–95.

Schottstaedt, Bill. 1977. "The Simulation of Natural Instrument Tones Using Frequency Modulation with a Complex Modulating Wave." *CMJ: Computer Music Journal* 1 (4): 46–50. Also in *Foundations of Computer Music,* ed. Curtis Roads and John Strawn. Cambridge, Mass.: MIT Press, 1985.

———. 2003. *An Introduction to FM.* http://ccrma-www.stanford.edu/software/snd/snd/fm.html.

Schroeder, Manfred R. 1970. "Digital Simulation of Sound Transmission in Reverberant Spaces: Part 1." *Journal of the Audio Engineering Society* 47 (2): 424–431.

Serra, Xavier. 1989. "A System for Sound Analysis/Transformation/Synthesis Based on a Deterministic Plus Stochastic Decomposition." Ph.D. dissertation, Center for Computer Research in Music and Acoustics, Stanford University.

Smith, Julius O. III. 1985. "An Introduction to Digital Filter Theory." In *Digital Audio Signal Processing: An Anthology,* ed. John Strawn. Middleton, Wisc.: A-R Editions.

———. 1987a. *Efficient Simulation of the Reed-Bore and Bow-String Mechanisms.* Technical Report STAN-M-35. Department of Music, Stanford University. http://ccrma.stanford.edu /STANM/stanm/node3.html.

———. 1987b. *Music Applications of Digital Waveguides.* Technical Report STAN-M-39. Department of Music, Stanford University. http://ccrma.stanford.edu/STANM/stanm/node3.html.

———. 1996. "Discrete-Time Modeling of Acoustic Systems with Applications to Sound Synthesis of Musical Instruments." http://www-ccrma.stanford.edu/~jos/ccrmapubs.html.

———. 2003. *Mathematics of the Discrete Fourier Transform (DFT).* W3K Publishing. http://www-ccrma.stanford.edu/~jos/ccrmapubs.html.

———. 2004. "Digital Waveguide Modeling of Musical Instruments." Draft. Center for Computer Research in Music and Acoustics (CCRMA), Stanford University. http://ccrma.stanford.edu/~jos/pasp/.

———. 2005. "Finite Difference Methods in Musical Instrument Modeling." In preparation. Center for Computer Research in Music and Acoustics (CCRMA), Stanford University. http://www-ccrma.stanford.edu/~jos/ccrmapubs.html.

Strunk, Oliver, ed. 1950. *Source Readings in Music History.* New York: W. W. Norton. 6th rev. ed. 1998.

Sundberg, Johan. 1991. "Synthesized Singing." In *Representations of Musical Signals,* ed. Giovanni De Poli, Aldo Piccialli, and Curtis Roads. Cambridge, Mass.: MIT Press.

Suzuki, H. 1987. "Model Analysis of a Hammer-String Interaction." *Journal of the Acoustical Society of America* 82 (4): 1145–1151.

Thompson, D'Arcy Wentworth. 1917. *On Growth and Form.* Cambridge: Cambridge University Press.

Truax, B. 1977. "Organizational Techniques for C:M Ratios in Frequency Modulation." *CMJ: Computer Music Journal* 1 (4): 39–45. Also in *Foundations of Computer Music,* ed. Curtis Roads and John Strawn. Cambridge, Mass.: MIT Press, 1985.

———. 1988. "Real-Time Granular Synthesis with a Digital Signal Processing Computer." *CMJ: Computer Music Journal* 12 (2): 14–26.

Vaidyanathan, P. P. 1993. *Multirate Systems and Filter Banks.* Englewood Cliffs: Prentice Hall.

Verma, T. S., and T. H. Y. Meng. 2000. "Extending Spectral Modeling Synthesis with Transient Modeling Synthesis." *CMJ: Computer Music Journal* 24 (4): 47–59.

Whitehead, Alfred N. 1925. *Science and the Modern World.* London: Macmillan. Republished New York: Simon and Schuster, 1967.

Xenakis, Iannis. 1971. *Formalized Music.* Bloomington: Indiana University Press.

Index of Equations and Mathematical Formulas

Subject Index

μ-Law codec, 41

AAC, 506
absolute value, 57, 142, 214
absolute value operator, 142
absorber, perfect sound, 351
absorption, 178, 430–431, 433
absorption loss, 444
AC. *See* alternating current
AC-3, 505
acausal, 219
accelerando, 32
acceleration, 264, 267–268,
   275–276, 302, 307–316, 333–346
accumulator, 244
accuracy, 29
acoustical current, 326
acoustical insulation, 350
acoustical ohm, 328–329, 344
acoustical power, 343
acoustical resistor, 328
acoustical systems, 325
acoustics, 325
   concert hall, 425
   musical instrument, 326
acquisition time, 6
ADC. *See* analog-to-digital
   conversion, analog-to-digital
   converter
admittance, 344, 433
AGC. *See* automatic gain control
A-Law codec, 41
algebra, 50, 280, 286
algebraically closed, 513
aliasing, 11–13, 15–16, 18–19, 22,
   42, 186, 377, 476–477
   consequences of, 16
   frequency domain, 477
   time domain, 477, 492, 494
   time-domain, 493
aliveness, 400
all poles model, 414–415
allpass, 261, 381
   filter, 199, 259, 261

alternating current, 184,
   334–338, 340
Alvin the Chipmunk, 486
AM, 388
ambience, 45
amperes, 340
amplification, 200
amplitude control, 367, 400
amplitude envelope, 386
amplitude modulation, 386–389,
   472, 479
amplitude spectrum, 122
analog systems, 34
analog tape recorder, 195–199
analog waveform, 20–21
analog-to-digital conversion, 9, 11,
   20, 45–46, 427
analog-to-digital converter, 11, 16,
   18, 24, 34, 40, 44, 46, 427
analogy, 3
analysis
   bins, 131, 154
   black-box, 204
   of non-harmonic signals, 142
   spectrum, 104
   white-box, 204
analysis frame, 416–417, 459,
   479, 494
analysis frame rate, 473, 477
analysis spectrum, 191
analysis window, 117, 131,
   139–144, 460–461, 473, 479,
   494–495
analysis/synthesis, 411, 484
analysis-based synthesis, 422
analytic, 534
analytic signal, 95–98, 100
anechoic chamber, 178, 350
angle, 57, 61, 63, 68, 255, 515
   of a complex number, 212
   of projection, 73
   of the transfer function,
      212, 240
angular coordinate, 71

angular velocity, 12, 19, 87, 282,
   287, 289–290, 293, 295
ansatz, 277
aperture time, 6
arccosine, 517
arcsine, 517
arctangent, 100, 121–122, 487, 517
   principal value range, 487–489
area
   Heisenberg box, 456
   of a tube, 357
argument
   as an angle, 59, 61
   of a function, 68, 134, 137,
      225–226, 405
arithmetic, 363, 522
artificial reverberation, 168
astronomy, 363
asynchronous, 28
ATRAC, 506
attack and decay, 367
attack time, 401
attenuation, 38, 42–43, 190, 200,
   258, 427–428, 446
audio limiter, 150
audio recording, 9, 19, 39, 136
auditory cues, 179
auditory information rate, 492
auditory system, 179
autocorrelation, 192–193, 527
automatic gain control,
   217–218
auxiliary equation, 280

Bagdasarian, Ross, 485
balancing the exponents, 213
band center, 421, 461, 464, 469–472,
   482, 487–491
band-limited, 19–20, 22, 480
bandwidth, 131, 154, 257,
   477–478, 483
   computational, 40
   of control parameters, 449
   data, 40–41